AF581744

Selected Papers from the 5th International Electronic Conference on Sensors and Applications

Selected Papers from the 5th International Electronic Conference on Sensors and Applications

Special Issue Editors

Stefano Mariani

Francesco Ciucci

Dirk Lehmhus

Thomas B. Messervey

Alberto Vallan

Stefan Bosse

Francisco Falcone

MDPI • Basel • Beijing • Wuhan • Barcelona • Belgrade • Manchester • Tokyo • Cluj • Tianjin

Special Issue Editors
Stefano Mariani
Politecnico di Milano
Italy

Thomas B. Messervey
Research to Market Solution s.r.l.
Italy

Francisco Falcone
Public University of Navarre
Spain

Francesco Ciucci
The Hong Kong University of Science and Technology
China

Alberto Vallan
Politecnico di Torino
Italy

Dirk Lehmhus
University of Bremen
Germany

Stefan Bosse
University of Bremen
Germany

Editorial Office
MDPI
St. Alban-Anlage 66
4052 Basel, Switzerland

This is a reprint of articles from the Special Issue published online in the open access journal *Sensors* (ISSN 1424-8220) (available at: https://www.mdpi.com/journal/sensors/special_issues/5th_ecsa).

For citation purposes, cite each article independently as indicated on the article page online and as indicated below:

LastName, A.A.; LastName, B.B.; LastName, C.C. Article Title. *Journal Name* **Year**, *Article Number*, Page Range.

ISBN 978-3-03928-849-6 (Hbk)
ISBN 978-3-03928-850-2 (PDF)

Contents

About the Special Issue Editors . . . . **vii**

Preface to "Selected Papers from the 5th International Electronic Conference on Sensors and Applications" . . . . **xi**

Carlo Camerlingo, Mikhail Lisitskiy, Maria Lepore, Marianna Portaccio, Daniela Montorio,Salvatore Del Prete and Gilda Cennamo
Characterization of Human Tear Fluid by Means of Surface-Enhanced Raman Spectroscopy
Reprinted from: *Sensors* **2019**, *19*, 1177, doi:10.3390/s19051177 . . . . **1**

Ghazal Rouhafzay and Ana-Maria Cretu
An Application of Deep Learning to Tactile Data for Object Recognition under Visual Guidance
Reprinted from: *Sensors* **2019**, *19*, 1534, doi:10.3390/s19071534 5 . . . . **11**

Jorge Otero, Ivan Felis, Miguel Ardid and Alicia Herrero
Acoustic Localization of Bragg Peak Proton Beams for Hadrontherapy Monitoring Reprinted from: *Sensors* **2019**, *19*, 1971, doi:10.3390/s190919715 . . . . **25**

Tiago M. Fernández-Caramés, Oscar Blanco-Novoa, Iván Froiz-Míguez and Paula Fraga-Lamas
Towards an Autonomous Industry 4.0 Warehouse:A UAV and Blockchain-Based System for Inventory and Traceability Applications in Big Data-Driven Supply Chain Management
Reprinted from: *Sensors* **2019**, *19*, 2394, doi:10.3390/s19102394 . . . . **39**

Rosa Ma Alsina-Pagès, Ferran Orga, Francesc Alías and Joan Claudi Socoró
A WASN-Based Suburban Dataset for Anomalous Noise Event Detection on Dynamic Road-Traffic Noise Mapping
Reprinted from: *Sensors* **2019**, *19*, 2480, doi:10.3390/s19112480 . . . . **71**

Tri Dev Acharya, Anoj Subedi and Dong Ha Lee
Evaluation of Machine Learning Algorithms for Surface Water Extraction in a Landsat 8 Scene of Nepal Reprinted from: *Sensors* **2019**, *19*, 2769, doi:10.3390/s19122769 . . . . **93**

Pau Bergadà and Rosa Ma Alsina-Pagès
An Approach to Frequency Selectivity in an Urban Environment by Means of Multi-Path Acoustic Channel Analysis
Reprinted from: *Sensors* **2019**, *19*, 2793, doi:10.3390/s19122793 . . . . **109**

Aldo Ghisi and Stefano Mariani
Effect of Imperfections Due to Material Heterogeneity on the Offset of Polysilicon MEMS Structures
Reprinted from: *Sensors* **2019**, *19*, 3256, doi:10.3390/s19153256 . . . . **133**

Paula Fraga-Lamas, Mikel Celaya-Echarri, Peio Lopez-Iturri, Luis Castedo, Leyre Azpilicueta, Erik Aguirre, Manuel Suárez-Albela, Francisco Falcone and Tiago M. Fernández-Caramés
Design and Experimental Validation of aLoRaWAN Fog Computing Based Architecture forIoT Enabled Smart Campus Applications
Reprinted from: *Sensors* **2019**, *19*, 3287, doi:10.3390/s19153287 . . . . **153**

Tiago M. Fernández-Caramés, Iván Froiz-Míguez, Oscar Blanco-Novoa and Paula Fraga-Lamas
Enabling the Internet of Mobile Crowdsourcing Health Things: A Mobile Fog Computing, Blockchain and IoT Based Continuous Glucose Monitoring System for Diabetes Mellitus Research and Care
Reprinted from: *Sensors* **2019**, *19*, 3319, doi:10.3390/s19153319 . . . . **183**

Mikel Celaya-Echarri, Leyre Azpilicueta, Peio López-Iturri, Erik Aguirre and Francisco Falcone
Performance Evaluation and Interference Characterization of Wireless Sensor Networks for Complex High-Node Density Scenarios
Reprinted from: *Sensors* **2019**, *19*, 3516, doi:10.3390/s19163516 5 . . . . **207**

Martin A. Aulestia Viera, Paulo R. Aguiar, Pedro Oliveira Junior, Felipe A. Alexandre, Wenderson N. Lopes, Eduardo C. Bianchi, Rosemar Batista da Silva, Doriana D'addona and Andre Andreoli
A Time–Frequency Acoustic Emission-Based Technique to Assess Workpiece Surface Quality in Ceramic Grinding with PZT Transducer
Reprinted from: *Sensors* **2019**, *19*, 3913, doi:10.3390/s19183913 . . . . **231**

Felipe A. Alexandre, Paulo R. Aguiar, Reinaldo Götz, Martin Antonio Aulestia Viera, Thiago Glissoi Lopes and Eduardo Carlos Bianchi
A Novel Ultrasound Technique Based on Piezoelectric Diaphragms Applied to Material Removal Monitoring in the Grinding Process
Reprinted from: *Sensors* **2019**, *19*, 3932, doi:10.3390/s19183932 . . . . **251**

Jiří Přibil, Anna Přibilová and Ivan Frollo
Analysis of the Influence of Different Settings of Scan Sequence Parameters on Vibration and Noise Generated in the Open-Air MRI Scanning Area
Reprinted from: *Sensors* **2019**, *19*, 4198, doi:10.3390/s19194198 . . . . **275**

Stefan Bosse and Uwe Engel
Real-Time Human-In-The-Loop Simulation with Mobile Agents, Chat Bots, and Crowd Sensing for Smart Cities
Reprinted from: *Sensors* **2019**, *19*, 4356, doi:10.3390/s192043565 . . . . **289**

Imanol Picallo, Hicham Klaina, Peio Lopez-Iturri, Erik Aguirre, Mikel Celaya-Echarri, Leyre Azpilicueta, Alejandro Eguizábal, Francisco Falcone and Ana Alejos
A Radio Channel Model for D2D Communications Blocked by Single Trees in Forest Environments
Reprinted from: *Sensors* **2019**, *19*, 4606, doi:10.3390/s19214606 . . . . **327**

Ivan Felis Enguix, Marta Sánchez Egea, Antonio Guerrero González and David Arenas Serrano
Underwater Acoustic Impulsive Noise Monitoring in Port Facilities: Case Study of the Port of Cartagena
Reprinted from: *Sensors* **2019**, *19*, 4672, doi:10.3390/s19214672 . . . . **347**

Rodolfo Rocha Vieira Leocádio, Alan Kardek Rêgo Segundo and Cibelle Ferreira Louzada
A Sensor for Spirometric Feedback in Ventilation Maneuvers during Cardiopulmonary Resuscitation Training
Reprinted from: *Sensors* **2019**, *19*, 5095, doi:10.3390/s19235095 . . . . **361**

About the Special Issue Editors

Stefano Mariani received his M.S. degree (cum laude) in Civil Engineering in 1995 and Ph.D. degree in Structural Engineering in 1999 from the Polytechnic University of Milan. He is currently Associate Professor at the Department of Civil and Environmental Engineering of the Polytechnic University of Milan. He was a research scholar at the Danish Technical University in 1997, Adjunct Professor at Penn State University in 2007, and Visiting Professor at the Polytechnic Institute of New York University in 2009. He serves on the Editorial Boards of *Algorithms, International Journal on Advances in Systems and Measurements, Inventions, Machines, Micro and Nanosystems, Micromachines,* and *Sensors*. He has been recipient of the Associazione Carlo Maddalena Prize for graduate students (1996), and of the Fondazione Confalonieri Prize for Ph.D. students (2000). His main research interests are in the reliability of MEMS that are subject to shocks and drops; structural health monitoring of composite structures through MEMS sensors; numerical simulations of ductile fracture in metals and of quasi-brittle fracture in heterogeneous and functionally graded materials; extended finite element methods; the calibration of constitutive models via extended and sigma-point Kalman filters; and multiscale solution methods for dynamic delamination in layered composites.

Francesco Ciucci is Associate Professor (with tenure) at the Hong Kong University of Science and Technology (HKUST), where he also serves as Associate Director of the Building Energy Research Center (Guangzhou HKUST, China). Prof. Ciucci graduated cum laude from Politecnico di Milano, Italy, and Ecole Centrale de Paris, France, with degrees in Aerospace Engineering and Engineering Science, respectively. He pursued his Ph.D. in Engineering and Applied Sciences at the California Institute of Technology, USA, where he was supported by the prestigious Rotary Ambassadorial Scholarship and Bechtel Fellowship. Prof. Ciucci conducted postdoctoral work at the Heidelberg Graduate School of Mathematical and Computational Methods for the Sciences of the University of Heidelberg, Germany. There, he was supported by a Heidelberg Graduate School Fellowship and obtained the prestigious Marie Curie Reintegration Grant. Prof. Ciucci has managed numerous research projects and has overseen 8 Ph.D. and 7 research MSc students who are now serving the engineering communities worldwide. Among them, 2 are teaching at research universities. Prof. Ciucci's current research centers around solid-state energy technologies, including materials used in high-temperature fuel cells and solid-state batteries. Prof. Ciucci has raised more than HKD 28 million (euro 3.2 million) in funding in the past 8 years and currently manages a group consisting of 13 postgraduate students and 4 postdoctoral scholars. His articles have been featured in many high-impact journals, including *Nature Chemistry, Advanced Energy Materials, and Chemical Reviews.*

Dirk Lehmhus joined the Fraunhofer Institute for Manufacturing Technology and Advanced Materials (IFAM) in 1998 following his diploma in Mechanical Engineering from the University of Hanover. Having received his Ph.D. from the University of Bremen's Faculty of Production Engineering for studies on optimizing manufacturing and predominantly mechanical properties of aluminum foams produced according to the powder compact melting or Fraunhofer process, he took over the role of Managing Director of the University of Bremen's ISIS Sensorial Materials Scientific Centre, dedicated to the coordination of research activities in the area of material-integrated intelligent systems. In 2015, he was granted a scholarship for a short term lectureship at New York University by the German Academic Exchange Service, DAAD. In January 2018, he re-joined Fraunhofer IFAM as Senior Scientist in the area of casting technology. Dr. Lehmhus serves on the Editorial Boards of Metals and Materials. His main research interests are in material-integrated intelligent systems, structural health monitoring, additive manufacturing, powder metallurgy, and metal casting technology.

Thomas B. Messervey has over 25 years of engineering, military, and business experience, which includes military service in the US Army Corps of Engineers, industrial experience with the Italian Engineering Company, D'Appolonia, teaching excellence at the United States Military Academy at West Point, and coaching services as the EU Facilitator for the Intelligent Manufacturing Systems program. He is CEO and co-founder of R2M Solution "Research to Market", which is an innovation and consulting company that acts as an accelerator for technology transfer that is mostly active in the fields of sustainability, construction, energy, the smart grid, and robotics. He also currently serves on the executive committee of Green Building Council Italia (GBC Italia). He has served as a reviewer and expert for the European Commission on various topics related to innovation and impact, and has served as a member and reviewer for various associations including the International Association of Bridge Maintenance and Safety (IABMAS), the International Association of Life Cycle Civil Engineering (IALCCE), the *International Journal of Safety and Security Engineering, Structure and Infrastructure Engineering*, and MDPI *Sensors*. He is passionate about leadership, mentorship, and the implementation of new value propositions that make growth and sustainability possible.

Alberto Vallan received his M.S. degree in Electronic Engineering from the Politecnico di Torino, Italy, in 1996 and Ph.D. degree in Electronic Instrumentation from the University of Brescia, Italy, in 2000. He is currently Associate Professor in Electrical and Electronic Measurements with the Department of Electronics and Telecommunications of the Politecnico di Torino. From 2000 to present, he has been a lecturer in courses concerning Electronic Measurements and Sensors. His research interests are focused on the development and characterization of fiber optical sensors and measuring instruments for biomedical and industrial applications. He is a Senior Member of the IEEE/Instrumentation & Measurement Society.

Stefan Bosse studied physics at the University of Bremen. He was awarded his Ph.D./doctoral degree (Dr. rer. nat.) in Physics in the year 2002 at the University of Bremen, where he also conducted his postdoctorate (Habilitation) and Venia Legendi in Computer Science in the year 2016. Since 2016, he has been teaching and researching as a Privatdozent at the University of Bremen, Department of Computer Science, and since 2018, he has been Interim Professor at the University of Koblenz-Landau, Faculty Computer Science, Institute of Software Technologies. His main research area is in distributed artificial intelligence in general and, in particular, information processing in massive parallel and distributed systems using agent-based approaches combined with machine learning and agent-based simulation. A broad range of fields of application and domains are addressed: Material Science, Materials Informatics, Smart Materials, IoT, Production Engineering, Social Science, Crowd Sensing, and Geo Science. He has conducted projects in the internationally recognized ISIS Scientific Centre for Intelligent Sensorial Materials, pushing interdisciplinary research and closing the gap between technology and computer science. He is currently Principle Investigator and Researcher at the DFG founded interegio and interdisciplinary Research Unit FOR3022 (Ultrasonic Monitoring of Fibre Metal Laminates Using Integrated Sensors), Subproject 4 (Automated Model-free Damage Diagnostic).

Francisco Falcone received his Telecommunication Engineering Degree in 1999 and his Ph.D. in Communication Engineering in 2005, both from the Universidad Pública de Navarra (UPNA) in Spain. From February 1999 to April 2000, he was Microwave Commissioning Engineer at Siemens-Italtel, deploying microwave access systems. From May 2000 to December 2008, he was Radio Access Engineer at Telefónica Móviles, performing radio network planning and optimization tasks in mobile network deployment. As co-founding member, he served as Director of Tafco Metawireless, a spin-off company from UPNA from January until May 2009. In parallel, he was Assistant Lecturer in the Electrical and Electronic Engineering Dept., UPNA, from February 2003 to May 2009. In June 2009, he became Associate Professor in the EE Dept., being Dept. Head from January 2012 until July 2018. From January 2018 to May 2018, he was Visiting Professor at the Kuwait College of Science and Technology in Kuwait. He is also affiliated with the Institute for Smart Cities (ISC) at UPNA, which hosts around 140 researchers, and is currently Head of the ICT section. His research interests are related to computational electromagnetics applied to analysis of complex electromagnetic scenarios, with a focus on the analysis, design, and implementation of heterogeneous wireless networks to enable context aware environments. He has participated in over 70 public and private funded projects. He has over 500 contributions in indexed international journals, book chapters and conference contributions. He is Senior Member IEEE. He has been awarded the CST 2003 and CST 2005 Best Paper Award, Ph.D. Award from the Colegio Oficial de Ingenieros de Telecomunicación (COIT) in2006, Doctoral Award UPNA, 2010, 1st Juan Gomez Peñalver Research Award from the Royal Academy of Engineering of Spain in 2010, XII Talgo Innovation Award 2012, IEEE 2014 Best Paper Award, 2014, ECSA-3 Best Paper award, 2016 and ECSA-4 Best Paper Award, 2017.

Preface to "Selected Papers from the 5th International Electronic Conference on Sensors and Applications"

This issue of Sensors is a collection of selected papers from the proceedings of the 5th International Electronic Conference on Sensors and Applications (ECSA-5), held online on 15–30 November 2018, through the sciforum.net platform developed by MDPI. The annual ECSA conference was initiated in 2014 as an online only conference to allow unrestricted participation from all over the world, forgoing concerns regarding travel and related expenditures. This conference is of particular interest since the research concerning sensors has been rapidly expanding, and a platform for rapid and direct exchanges about the latest research findings can provide a further boost in the development of novel ideas.In total, 18 technical contributions were published after standard peer reviewing. Collectively, they present recent advances in the areas of chemical sensors, biosensors, physical sensors, sensor networks, and cutting-edge topics like smart cities, smart sensing systems and structures, structural health monitoring (SHM), and wearable sensors.

Stefano Mariani, Francesco Ciucci, Dirk Lehmhus, Thomas B. Messervey, Alberto Vallan, Stefan Bosse and Francisco Falcone

Special Issue Editors

Article

Characterization of Human Tear Fluid by Means of Surface-Enhanced Raman Spectroscopy †

Carlo Camerlingo [1,*], Mikhail Lisitskiy [1], Maria Lepore [2], Marianna Portaccio [2], Daniela Montorio [3], Salvatore Del Prete [4] and Gilda Cennamo [5]

1 Consiglio Nazionale delle Ricerche, SPIN-CNR, Via Campi Flegrei 34, 80078 Pozzuoli, Italy; carlo.camerlingo@spin.cnr.it
2 Dipartimento di Medicina Sperimentale, Università della Campania "L. Vanvitelli", 80138 Naples, Italy; maria.lepore@unicampania.it (M.L.); marianna.portaccio@unicampania.it (M.P.)
3 Dipt. di Neuroscienze e Scienze Riproduttive ed Odontostomatologiche, Universitá di Napoli 'Federico II', 80121 Naples, Italy; da.montorio@gmail.com
4 CISME—Centro Interdipartimentale di Microscopia Elettronica, Universitá di Napoli 'Federico II', 80100 Naples, Italy; saldelp@gmail.com
5 Dipt. di Sanitá Pubblica, Universitá di Napoli 'Federico II', 80131 Naples, Italy; xgilda@hotmail.com
* Correspondence: carlo.camerlingo@spin.cnr.it; Tel.: +39-081-8675044
† This paper is an extended version of our paper published in Camerlingo, C.; Lisitskiy, M.; Lepore, M.; Portaccio, M.; Montorio, D.; Prete, S.D.; Cennamo, G. A Preliminary Investigation on Human Tears by Means of Surface Enhanced Raman Spectroscopy. *Proceedings* **2019**, *4*, 18.

Received: 21 January 2019; Accepted: 4 March 2019; Published: 7 March 2019

Abstract: Tears are exceptionally rich sources of information on the health status of the eyes, as well as of whole body functionality, due to the presence of a large variety of salts and organic components whose concentration can be altered by pathologies, eye diseases and/or inflammatory processes. Surface enhanced Raman spectroscopy (SERS) provides a unique method for analyzing low concentrations of organic fluids such as tears. In this work, a home-made colloid of gold nanoparticles has been used for preparing glass substrates able to efficiently induce an SERS effect in fluid samples excited by a He–Ne laser (λ = 633 nm). The method has been preliminary tested on Rhodamine 6G aqueous solutions at different concentrations, proving the possibility to sense substance concentrations as low as few µM, i.e., of the order of the main tear organic components. A clear SERS response has been obtained for human tear samples, allowing an interesting insight into tear composition. In particular, aspartic acid and glutamic acid have been shown to be possible markers for two important human tear components, i.e., lactoferrin and lysozyme.

Keywords: SERS; tear; biomedical sensors; lactoferrin; lysozyme

1. Introduction

Human tears represent an exceptionally rich source of information regarding the health status of eyes and, more generally, of whole body functionality. This is mainly due to the presence in tears of a large variety of salts and organic components (including proteins, lipids, metabolites, nucleic acids, and electrolytes) whose concentrations can be altered by pathologies, eye diseases and/or inflammatory processes [1,2]. An increasing attention is presently given to the analysis of this human body fluid. The small amounts of substance considered and the typical low concentration of organic compounds hamper the access to direct analysis by biochemical methods even if the improvement of technology is overcoming some of the main obstacles and the route to a wider and reliable use of tear diagnostics is in progress. In this framework, micro-Raman spectroscopy can have an important contribution. Vibrational spectroscopies have been shown to be extremely useful to analyze biological samples providing valuable insight into the nature of samples and biofluids [3–6] and into the determination of the chemical structure of specific molecules [7]. Raman spectroscopy has been also used for investigating tears [8–10]. More recently, the use of Surface Enhanced Raman Spectroscopy (SERS) has been proposed with interesting perspectives in terms of sensitivity and specificity of the signal response [11–13]. In the work of P. Hu et al. [11] SERS has been implemented by using silver nanoparticles and applied to the analysis of dried human tears. The spectroscopy of dried fluids, the so-called drop-coating deposition Raman spectroscopy (DCDRS), has many advantages with respect to liquid sample analysis and has been applied fruitfully to tear investigation [8,9]. SERS mechanisms add new potentialities to DCDRS. In the case of tears, SERS response has some differences with respect to DCDRS signal and the observed modes are mainly assignable to the proteins [11]. In particular, P. Hu et al. [11] evidenced a correlation between the intensity of SERS modes at 757 and 1557 cm^{-1} spectral positions and the lysozyme content, in agreement with a previous work of J. Filik and N. Stones [14]. Gold coated nanostructured substrates have been used by S. Choi et al. to perform SERS investigation of dried tears aimed to characterize the spectral changes occurring in adenoviral conjunctivitis-infected subjects [12]. In that work, the principal component analysis (PCA) allowed a global analysis of the spectrum configuration that was able to distinguish between normal and infected tear fluids confirming that SERS measurements can offer valuable diagnostic tools. For this reason, we investigate the possibility of performing SERS measurements on human tears using a cheap and easy method to prepare home-made substrates, differently from reported works that used more sophisticated and expensive substrates [11,12] not suitable for large screening purposes. In this present work, human tear samples have been examinated by SERS using home-made SERS subtrates based on gold nanoparticles using a procedure simpler and faster than the one proposed in [11], which requires the mixing of each tear sample with colloids. In order to test the efficiency and sensibility of the method, preliminary SERS measurements have been performed on Rhodamine 6G aqueous solutions at different concentrations.

2. Materials and Methods

2.1. Tears Collection

Tear specimens have been collected from healthy volunteers. Informed consent has been preliminarly obtained from them. Smooth edge sterile capillary glass tubes (ACCU-FILL 90 MICROPET by Becton, Dickinson and Co, Clay Adams CA USA) with internal diameter = 0.2 mm, outside diameter = 1 mm and length = 12.7 mm have been used. During the collection, the volunteers' lower eyelids were gently pulled down and the tip of the open capillary tube placed in contact with the tear meniscus without irritating the conjunctiva. A minimum of 1.0 μL tear was collected. The samples were stored at 4 °C and tested by SERS within 7 days. In this work, eight healthy patients aged between 33 and 85 years have been considered. The list of patients, their birth year and sex is reported in Table 1.

Table 1. List of patients.

id. Number	Birth Year	Sex
A	1985	f
B	1980	f
C	1972	f
D	1956	f
E	1952	m
F	1948	m
G	1943	f
H	1933	m

2.2. SERS Implementation

SERS has been induced by using home-made Gold nanoparticles (GNP) onto conventional microscope slides. A colloid of GNPs has been obtained by a conventional citrate reduction method [15,16]. A 0.01% $HAuCl_4$ solution was reduced by 1% sodium citrate with vigorous stirring at near boiling temperature. The amount of sodium citrate was defined in order to obtain GNPs with controlled size with particle diameter in the range of 20–50 nm. Preliminary tests have shown that the GNPs with a diameter equal to ~27 nm give optimum SERS features [16].

Particle size, preparation stability and Raman signal of the resulting preparations have been investigated by means of absorption spectroscopy, Transmission Electron Microscop (TEM), and Dynamic Light Scattering (DLS). In particular, absorption spectroscopy was used to get information on the plasmon resonance peak and to study the stability of the preparations [17,18]. TEM and DLS were employed to estimate the size of the GNPs. Absorption spectra of the prepared GNPs were recorded by UV–VIS spectroscopy by a two-beams spectrophotometer (LS25, Perkin Elmer, Waltham, MA, USA).

Hydrogen tetrachloroaurate ($HAuCl_4$), and trisodium citrate were purchased from Sigma-Aldrich (Sigma-Aldrich Co., St. Louis, MO, USA). All chemicals were used as received. Rhodamine 6G were purchased from Sigma-Aldrich (St. Louis, MO, USA).

2.3. SERS Measurements

A small amount of samples to be analyzed was dropped on the dried residual of the GNP colloid, and quickly measured. SERS measurements were performed by using a Jobin–Yvon system from Horiba Scientific ISA (Osaka, Japan), equipped with a TriAx 180 monochromator, a liquid nitrogen cooled charge-coupled detector and a 1800 grooves/mm grating (final spectral resolution: 4 cm^{-1}). The spectra were recorded in air at room temperature using a 17 mW He–Ne laser source (wavelength 632.8 nm). Accumulation times in the 30–180 s range were used. The laser beam was focused to a 2-µm spot size on the sample through an Olympus microscope equipped with 100× optical objective. Raman spectroscopy measurements were also done for comparison, by using conventional microscope slides as substrates. Acquisition times and modalities were similar to those used for SERS measurements.

2.4. Data Analysis

Before being analyzed and compared, the spectra were numerically elaborated in order to eliminate background signal, limit the signal noise and normalize data. For this purpose, we used mathematical algorithms based on "wavelet" functions. Widely used in the signal analysis, "wavelets" provide an efficient basis for the hierarchical spectrum representation through the so called Discrete Wavelet transform (DWT) [19]. Biorthogonal wavelets based on the B-spline function were used in the framework of software package 'wavelet toolbox' of MATLAB program (by MathWorks Inc.). Low and high scale components of the signal DWT were not be included in the final reconstructed signal, taking away the background and part of the uncorrelated noise, respectively. Finally, the amplitude of the spectral data was normalized by using vector normalization, i.e., by scaling the spectral data set suitably to obtain the standard deviation with respect to the average value equal to 1. The spectra were analyzed in terms of convoluted lorentzian-shaped vibrational modes by performing a best-fit procedure based on the Levenberg–Marquardt nonlinear least-square methods.

3. Results

Some preliminary SERS measurements on Rhodamine 6G diluted in deionized water at different concentrations were performed in order to quantify the effect of the Raman response enhancement. SERS spectra were collected on the Rhodamine–water solutions in similar conditions, by using an acquisition time of 30 s. The obtained spectra are reported in Figure 1a, arbitrarily shifted along the y-axis in order to improve the readability. They refer to Rhodamine 6G solutions with concentrations of

$c = 0.0025, 0.005, 0.01, 0.1, 0.5, 1$ and 5 mM, respectively. The intensities of the spectra referring to lower concentrations (red lines) are reported as being magnified ten times with respect to the remaining spectra (blue lines). The spectra observed are consistent with data reported in literature [20] and prove the high sensitivity of the method. The dependence of SERS signal intensity on the Rhodamine 6G concentration is shown in Figure 1b, where the intensity values of the spectral mode at 1504 cm^{-1} are reported as a function of the solution concentration c. The data have been obtained by fitting the Raman spectra of Figure 1 by Lorentzian functions. For low-concentration values the SERS signal intensity increases approximately linearly with concentration, as visible in Figure 1b where the linear dependence is represented by a dotted line. When concentration is larger than 10 µM SERS signal is lower than the value expected in the case of a linear dependence on c, indicating a saturation trend probably related to a filling effect of GNP surfaces involved in the SERS mechanism. A SERS efficiency of about four times 10^3 has been estimated by comparing the SERS data of $c = 5$ µM solution with the signal obtained by conventional Raman spectroscopy on the $c = 5$ mM solution.

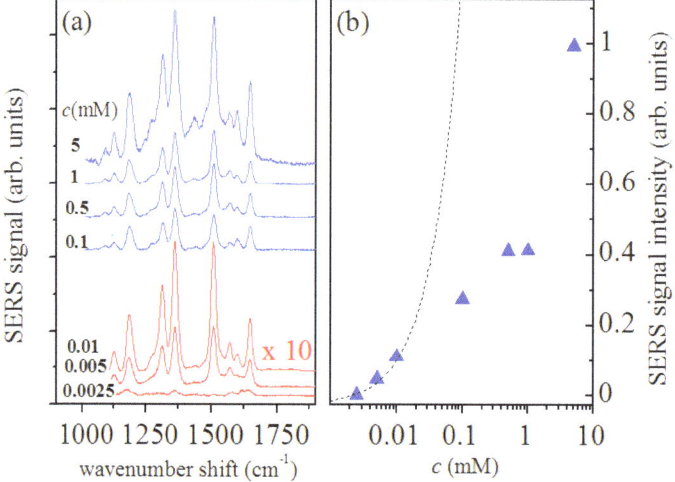

Figure 1. (**a**) SERS of Rhodamine aqueous solution at different concentration in the range of 0.0025–5 mM. The intensity of spectra reported in red are amplified by a factor 10. The spectra are reported arbitrarly shifted along the y-axis. (**b**) dependence of SERS signal intensity (SERS mode at 1504 cm^{-1}) on the Rhodamine 6G concentration (log scale). The linear dependence is represented by the dotted line.

SERS signals from tears were collected in conditions similar to those used for the preliminary tests on Rhodamine 6G. In this case, the acquisition time was typically 180 s. Three or more acquisitions were performed for each sample, using different points of the sample area. The relative standard deviation of the SERS signals obtained from each sample was typically lower than 5%. The spectrum reported in Figure 2a is obtained by averaging SERS responses of tear samples obtained from the eight considered healthy patients. In order to quantify similarities or possible differences among the samples and the repeatability of the measurement, the standard deviation with respect to the average values was calculated for each point of the average SERS signal, and the variation range of SERS response, due to patients' individual peculiarities, is reported in Figure 2b. A mean standard deviation $\sigma = 6.3 \pm 3.3\%$ was calculated with respect to averaged signals (bottom spectrum of Figure 2). The signal dispersion is generally lower than 10% of the SERS signal, even if a larger signal deviation range occurs in some points of the spectra, at about 1050, 1336, 1523 and 1624 cm^{-1}. Amide I and Amide III Raman bands are clearly seen in the reported spectrum at the wavenumber Raman-shifts of $\sim$1600 cm^{-1} and $\sim$1250 cm^{-1}, respectively. In Figure 3a, the result of the spectrum deconvolution in terms of Lorentzian components (reported as blue lines) obtained by a numerical fitting procedure is reported. The main component modes are determined and listed in Table 2, with a temptative assignment of the vibrational sources, based on SERS data of proteins available in the literature [12,21,22]. The SERS signal of Figure 2a is compared with the conventional Raman spectrum measured on a human tear dried drop (Figure 2b). Acquisition times and modalities were similar to those used for SERS measurements, but a significant signal intensity increase and improvement in the spectral resolution are noticed in the SERS signal when the two spectra are compared. It is worth to note that the spectroscopy measurements here reported employed lower laser power and/or photon energy (i.e., a lower Raman scattering efficiency) with respect to those used in the works of Filik [14] and Hu [11] on dried tear drops. This evidences that the observed spectra are due to SERS mechanisms instead of a mere optimization of the Raman spectroscopy collection, as in the case of DCDRS experiments.

Table 2. Principal contributions in SERS spectrum of human tears and their assignment according to Refs. [12,21,22].

nr	Center (cm^{-1})	Assignment	Component
1	484	Ring def.	amino acids
2	645	COO^- wag.	amino acids
3	730	COO^- def.	amino acids
4	855	C-C str.	amino acids
5	944	C-C str.	amino acids
6	1000	symm. ring CC str.	phenylalanine
7	1121	NH_3^+ def.	amino acids
8	1165	N-H wag.	amino acids
9	1243	Amide III β-sheet, CH_2 wag.	protein
10	1291	Amide III α-helix CH_2 wag.	protein
11	1342	C-H def.	Aspartic acid, amino acids
12	1435	CO^- symm. str.	Glutamic acid, amino acids
13	1533	C-C str.	amino acids
14	1567	NH_2 sciss.	amino acids
15	1624	Indole N-H, C=O str.	amino acids
16	1661	Amide I α-helix	protein

(def.: deformation; wag.: wagging; str.: stretching; sciss.:scissoring).

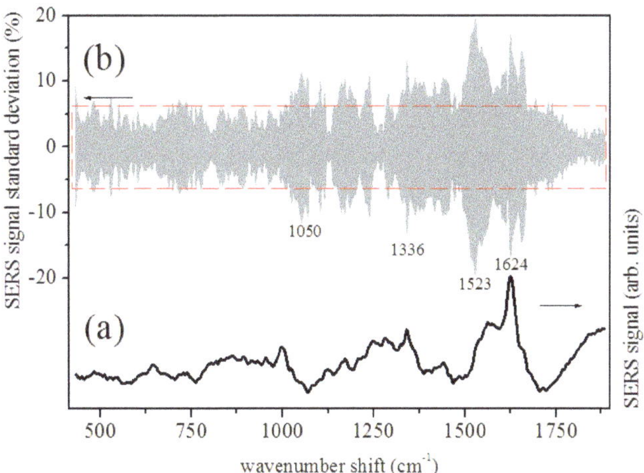

Figure 2. (**a**) Human tear SERS signal. (**b**) SERS signal standard deviation with respect to the average signal (bottom spectrum) of tears. The red box indicated the mean value of the signal standard deviation (6.3 ± 3.3%).

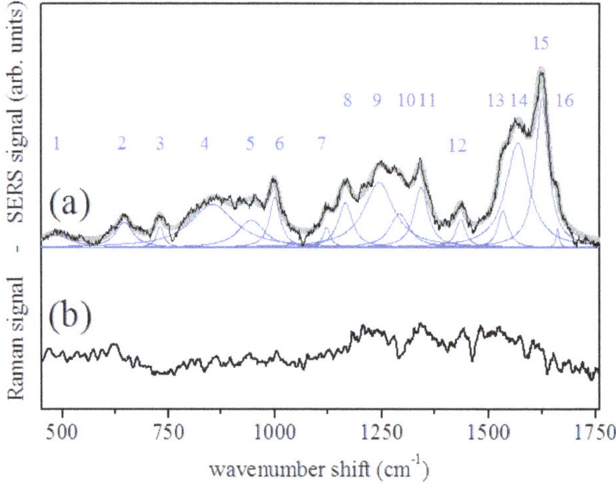

Figure 3. (**a**) SERS spectrum of human tears. The experimental data were fitted by a convolution of Lorentzian functions (numbered blue peaks) representing the main Raman modes occurring in the sample (see Table 2 for a temptative assignment of the Raman modes. (**b**) Conventional Raman spectrum of dried human tears (the signal in (**a**,**b**) are arbitrary scaled).

4. Discussion

The modes featured by the spectrum of tears reported in Figure 3 and listed in Table 2 are mainly concerning proteins and can be temptatively assigned to amino acids [22]. The main components occurring in tears are immunoglobulins (IgA), lactoferrin, lysozyme, lipocalin and albumine [10,23]. The molecular weights and the typical concentrations in human tears of these substances are listed in Table 3. The concentrations are of the order of few mg/mL [10], thus, compatible with SERS sensitivity estimated by Rhodamine 6G measurements, even if each substance could have a different response efficiency regard to SERS mechanisms.

Table 3. Main components of human tears.

Component	Mol. Weight (KDa)	c (mg/mL) [10]	c (μM)	Description
lactoferrin	80	1.8–2.7	23–34	iron binding glycoprotein
lysozyme	14	1.6–2.5	111–172	single chain polypeptide
lipocaline	20	1.2–2.95	62–145	low mol. weight protein
immunoglobulins (IgA)	162	0.2–0.3	1.5–1.9	glycoproteins (antibodies)
albumine	66	1.3	20	single peptide chain

Among the tear components, lactoferrin (LF) and lysozyme (LZ) have an important role for eye functionality providing defence mechanisms against infective agents [1]. A significant decrease of their levels has been reported in patients suffering from inflammatory Dry Eye Disease (DED) [24–26]. LF is an iron-binding protein present in almost all body biofluids. It has a proven anti-bacterial and anti-inflammatory ability [27]. The main components of the LF are Glutamic and Aspartic acids, Leucine, Arginine, Lysine, Valine and Phenylalanine [28]. LZ is an antimicrobial enzyme constituted by a single chain polypeptide. Its main components are Aspartic acid, Alanine, Glycine, Arginine, Serine, Leucin and threonine [29]. A direct correlation of SERS signal to the LF and LZ is not seen due to the complex and rich composition of tears. However, the peak assignments done in Table 2 for Glutamic acid (at 1243 cm^{-1} and 1434 cm^{-1}) and for Aspartic acid (1342 cm^{-1}) provide potential markers for LF and LZ, respectively. These modes are in agreement with Raman spectra reported for single-component solutions of LF and LZ [11,14] and with SERS response of LZ [30]. In particular, the SERS spectrum of LZ reported by Jun Hu et al. [30] exhibited a particularly intense peak at 1358 cm^{-1} stronger than the ones observed at 1280 and 1432 cm^{-1} . Furthermore, modes at 730 cm^{-1} and 1567 cm^{-1} should be related to LZ, in agreement with Ref. [14,30]. In a recent SERS study on human tears, W.S. Kim et al. found a correlation between SERS signal intensity ratio at 1342 cm^{-1} and 1242 cm^{-1} and the infection state of eye [13]. The authors noticed an increase of the I_{1342}/I_{1242} ratio value in patients affected by Adenovirus (I_{1342}/I_{1242} = 1.13) or Herpes Simplex (I_{1342}/I_{1242} = 3.73) diseases. These values are significatively higher than the I_{1342}/I_{1242} = 0.8 estimated for reference healthy patients. This feature could be explained in terms of a decrease of the LF level originated by the infection state. LZ level should not change significantly because, in agreement with oculist phenomenology, patients with blepharitis, conjunctivitis, and keratitis had normal mean LZ content of tears while patients with herpes simplex keratitis had low LZ values [31]. In our case, we estimated an average value of $I_{1342}/I_{1242} = 0.9 \pm 0.3$ for the considered height healthy patients.

5. Conclusions

The potentiality of SERS for characterizing tears has been investigated by using home-made fabricated Gold-nanoparticle-based substrates. The method has been previously tested and characterized on water diluted Rhodamine 6G samples. Human tear fluids from eight healthy patients have been considered. The SERS response results are mainly related to amino acids and provided a valuable source of information even if the interpretation is not immediate and more work should be done in order to have a more complete and exhaustive data comprehension. Nevertheless, assignments for two components of tears have been determined at the spectral positions of 1243 cm^{-1} and 1434 cm^{-1} (lactoferrin) and 1342 cm^{-1} (lysozyme). As widely reported in the literature and discussed above, the concentrations of both these components are affected by eye health state, and a change in SERS intensity is expected in the case of pathologies. A quantitative assessment of main tear components (in particular, Lysozyme and Lactoferrin) is undoubtedly a demanding issue and a challenge for the future progress of SERS as a diagnostic method. The promising results that are reported here allow us to estimate a future development of research activity towards the implementation of methods suitable to a widespread application of SERS methods in oculistic practice. The development of cheap and friendly methods for SERS implementation, as the one considered in this work, is a first step towards this aim. An interesting future perspective is constituted by the development of soft substrates for SERS implementation, properly designed by using paper or tissues embedded in metallic nanoparticles [13,32,33].

Author Contributions: Conceptualization, C.C., M.L. (Mikhail Lisitskiy), D.M. and G.C.; Formal analysis, C.C., M.L. (Mikhail Lisitskiy) and M.L. (Maria Lepore); Investigation, C.C., M.P., S.D.P. and G.C.; Methodology, M.L. (Maria Lepore) and M.P.; Resources, D.M., S.D.P. and G.C.; Writing—original draft, C.C. and M.L. (Maria Lepore); Writing—review & editing, M.L. (Mikhail Lisitskiy), M.P., D.M., S.D.P. and G.C.

Conflicts of Interest: The authors deckare no conflict of interest.

References

1. Hagan, S.; Martin, E.; Enríquez-de-Salamanca, A. Tear fluid biomarkers in ocular and systemic disease: Potential use for predictive, preventive and personalised medicine. *EPMA J.* **2016**, *7*, 15. [CrossRef] [PubMed]
2. Börger, M.; Funke, S.; Bähr, M.; Grus, F.; Lingor, P. Biomarker sources for Parkinson's disease: Time to shed tears? *Basal Ganglia* **2015**, *5*, 63–69. [CrossRef]
3. d'Apuzzo, F.; Perillo, L.; Delfino, I.; Portaccio, M.; Lepore, M.; Camerlingo, C. Monitoring early phases of orthodontic treatment by means of Raman spectroscopies. *J. Biomed. Opt.* **2017**, *22*, 115001. [CrossRef] [PubMed]

4. Camerlingo, C.; Zenone, F.; Perna, G.; Capozzi, V.; Cirillo, N.; Gaeta, G.M.; Lepore, M. An investigation on micro-Raman spectra and wavelet data analysis for Pemphigus Vulgaris follow-up monitoring. *Sensors* **2008**, *8*, 3656–3664. [CrossRef]
5. Rusciano, G.; Zito, G.; Pesce, G.; Del Prete, S.; Cennamo, G.; Sasso, A. Assessment of conjunctival microvilli abnormality by micro-Raman analysis. *J. Biophotonics* **2016**, *9*, 551–559. [CrossRef]
6. Gąsior-Głogowska, M.; Komorowska, M.; Hanuza, J.; Mączka, M.; Zając, A.; Ptak, M.; Będziński, R.; Kobielarz, M.; Maksymowicz, K.; Kuropka, P.; et al. FT-Raman spectroscopic study of human skin subjected to uniaxial stress. *J. Mech. Behav. Biomed. Mater.* **2013**, *18*, 240–252.
7. Reichebacher, M.; Popp, J. Challenges in molecular structure determination. In *Vibrational Spectroscopy XX*; Springer: Heidelberg, Germany, 2013; pp. 63–143.
8. Filik, J.; Stone, N. Analysis of human tear fluid by Raman spectroscopy. *Anal. Chim. Acta* **2008**, *616*, 177–184. [CrossRef] [PubMed]
9. Filik, J.; Stone, N. Investigation into the protein composition of human tear fluid using centrifugal filters and drop coating deposition Raman spectroscopy. *J. Raman Spectrosc.* **2009**, *40*, 218–224. [CrossRef]
10. Kuo, M.-T.; Lin, C.-C.; Liu, H.-Y.; Chang, H.-C. Tear Analytical Model Based on Raman microspectroscopy for investigation of infectious diseases of the ocular surface. *Investig. Ophthalmol. Vis. Sci.* **2011**, *52*, 4942–4950. [CrossRef]
11. Hu, P.; Zheng, X.S.; Zong, C.; Li, M.H.; Zhang, L.Y.; Li, W.; Ren, B. Drop-coating deposition and surface-enhanced Raman spectroscopies (DCDRS and SERS) provide complementary information of whole human tears. *J. Raman Spectrosc.* **2014**, *45*, 565–573. [CrossRef]
12. Choi, S.; Moon, S.W.; Shin, J.-H.; Park, H.-K.; Jin, K.-H. Label-free biochemical method for early detection of Adenoviral conjuctivitis using human tear biofluids. *Anal. Chem.* **2014**, *86*, 11093–11099. [CrossRef]
13. Kim, W.-S.; Shin, J.-H.; Park, H.-K.; Choi, S. A low-cost, monometallic, surface-enhanced Raman scattering-functionalized paper platform for spot-on bioassays. *Sens. Actuators B Chem.* **2016**, *222*, 1112–1118. [CrossRef]
14. Filik, J.; Stone, N. Drop coating deposition Raman spectroscopy of protein mixtures. *Analyst* **2007**, *132*, 544–550. [CrossRef] [PubMed]
15. Frens, G. Particle size and sol stability in metal colloids. *Colloid Polym. Sci.* **1972**, *250*, 736–741. [CrossRef]
16. Camerlingo, C.; Portaccio, M.; Tatè, R.; Lepore, M.; Delfino, I. Fructose and pectin detection in fruit-based food products by Surface-Enhanced Raman Spectroscopy. *Sensors* **2017**, *17*, 839. [CrossRef]
17. Haiss, W.; Thanh, N.T.K.; Aveyard, J.; Ferni, D.G. Determination of Size and Concentration of Gold nanoparticles from UV-Vis Spectra. *Anal. Chem.* **2007**, *79*, 4215–4221. [CrossRef] [PubMed]
18. Njoki, P.N.; Lim, I.S.; Mott, D.; Park, H.-Y.; Khan, B.; Mishra, S.; Sujakumar, R.; Luo, J.; Zhong, C.-J. Size correlation of optical and spectroscopic properties for gold nanoparticles. *J. Phys. Chem. C* **2007**, *111*, 14664–14669. [CrossRef]
19. Camerlingo, C.; Zenone, F.; Gaeta, G.M.; Riccio, R.; Lepore, M. Wavelet data processing of micro-Raman spectra of biological samples. *Meas. Sci. Technol.* **2006**, *17*, 298–303. [CrossRef]
20. Ameer, F.S.; Pittman, C.U., Jr.; Zhang, D. Quantification of Resonance Raman Enhancement Factors for Rhodamine 6G (R6G) in Water and on Gold and Silver Nanoparticles: Implications for Single-Molecule R6G SERS. *J. Phys. Chem. C* **2013**, *117*, 27096–27104. [CrossRef]
21. Madzharova, F.; Heiner, Z.; Gühlke, M.; Kneipp, J. Surface-Enhanced Hyper-Raman Spectra of Adenine, Guanine, Cytosine, Thymine, and Uracil. *J. Phys. Chem.* **2016**, *120*, 15415–15423. [CrossRef]
22. Stewart, S.; Fredericks, P.M. Surface-enhanced Raman spectroscopy of amino acids adsorbed on an electrochemically prepared silver surface. *Spectrochim. Acta Part A* **1999**, *55*, 1641–1660. [CrossRef]
23. Fullard, R.J.; Sbyder, C. Protein levels in nonstimulated and stimulated tears of normal human subjects. *Investig. Ophthalmol. Vis. Sci.* **1990**, *31*, 1119–1126.
24. Mackie, I.A.; Seal, D.V. Confirmatory tests for the dry eye of Sjögren's syndrome. *Scand. J. Rheumatol. Suppl.* **1986**, *61*, 220–223. [PubMed]
25. Boersma, H.G.; van Bijsterveld, O.P. The lactoferrin test for the diagnosis of keratoconjunctivitis sicca in clinical practice. *Ann. Ophthalmol.* **1987**, *19*, 152–154. [PubMed]
26. Goren, M.B.; Goren, S.B. Diagnostic tests in patients with symptoms of keratoconjunctivitis sicca. *Ann. J. Ophthalmol.* **1988**, *106*, 570–574. [CrossRef]

27. González-Chávez, S.A.; Arévalo-Gallegos, S.; Rascón-Cruz, Q. Lactoferrin: Structure, function and applications. *Int. J. Antimicrob. Agents* **2009**, *33*, 301e1–301e8. [CrossRef]
28. Querinjean, P.; Masson, P.L.; Heremans, J.F. Molecular Weight, Single-chain structure and Amino acid composition of human Lactoferrin. *Eur. J. Biochem.* **1971**, *20*, 420–425. [CrossRef]
29. Thompson, A.R. Amino acid sequence in lysozyme. 1. Displacement chromatography of peptides from a partial hydrolysate on ion-exchange resins. *Biochem. J.* **1955**, *60*, 507–515. [CrossRef]
30. Hu, J.; Sheng, R.S.; Xu, Z.S.; Zheng, Y. Surface enhanced Raman spectroscopy of lysozyme. *Spectrochim. Acta* **1995**, *51A*, 1087–1096.
31. Saari, K.M.; Aine, E.; Posz, A.; Klockars, M. Lysozyme content of tears in normal subjects and in patients with external eye infections. *Graefes Arch. Clin. Exp. Ophthalmol.* **1983**, *221*, 86–88. [CrossRef]
32. Yokoyama, M.; Nishimura, T.; Yamada, K.; Jeong, H.; Kido, M.; Sakurai, Y.; Ohno, Y. Raman spectroscopy of tear fluid with paper substrates for point-of-care therapeutic drug monitoring. *J. Nurs. Sci. Eng.* **2015**, *2*, 25–31.
33. Yamada, K.; Endo, T.; Imai, H.; Kido, M.; Jeong, H.; Ohno, Y. Effectiveness of surface enhanced Raman spectroscopy of tear fluid with soft substrate for point-of-care therapeutic drug monitoring. In *Optical Diagnostics and Sensing XVI: Toward Point-of-Care Diagnostics*; Coté, G.L., Ed.; International Society for Optics and Photonics: Bellingham, DC, USA, 2017; Volume 9715, p. 97150E.

Article

An Application of Deep Learning to Tactile Data for Object Recognition under Visual Guidance †

Ghazal Rouhafzay * and Ana-Maria Cretu

Department of Systems and Computer Engineering, Carleton University, Ottawa, ON K1S 5B6, Canada; acretu@sce.carleton.ca

* Correspondence: ghazal.rouhafzay@carleton.ca

† This paper is an extended version our paper published in Rouhafzay, G.; Cretu, A.-M. Data-Driven A Visuo-Haptic Framework for Object Recognition Inspired by Human Tactile Perception. In Proceedings of the 5th International Electronic Conference on Sensors and Applications, Online, 15–30 November 2018; Sciforum Electronic Conference Series; Volume 4, doi:10.3390/ecsa-5-05754.

Received: 27 February 2019; Accepted: 25 March 2019; Published: 29 March 2019

Abstract: Drawing inspiration from haptic exploration of objects by humans, the current work proposes a novel framework for robotic tactile object recognition, where visual information in the form of a set of visually interesting points is employed to guide the process of tactile data acquisition. Neuroscience research confirms the integration of cutaneous data as a response to surface changes sensed by humans with data from joints, muscles, and bones (kinesthetic cues) for object recognition. On the other hand, psychological studies demonstrate that humans tend to follow object contours to perceive their global shape, which leads to object recognition. In compliance with these findings, a series of contours are determined around a set of 24 virtual objects from which bimodal tactile data (kinesthetic and cutaneous) are obtained sequentially and by adaptively changing the size of the sensor surface according to the object geometry for each object. A virtual Force Sensing Resistor array (FSR) is employed to capture cutaneous cues. Two different methods for sequential data classification are then implemented using Convolutional Neural Networks (CNN) and conventional classifiers, including support vector machines and k-nearest neighbors. In the case of conventional classifiers, we exploit contourlet transformation to extract features from tactile images. In the case of CNN, two networks are trained for cutaneous and kinesthetic data and a novel hybrid decision-making strategy is proposed for object recognition. The proposed framework is tested both for contours determined blindly (randomly determined contours of objects) and contours determined using a model of visual attention. Trained classifiers are tested on 4560 new sequential tactile data and the CNN trained over tactile data from object contours selected by the model of visual attention yields an accuracy of 98.97% which is the highest accuracy among other implemented approaches.

Keywords: Haptic exploration; visual attention; visuo-haptic interaction; tactile object recognition; Convolutional Neural Network

---

1. Introduction

As the second decade of the 21st century is ending, artificial intelligence and machine learning solutions are finding their place more and more in daily life. Besides the wide application of machine learning in data processing, planning, and decision-making, intelligent machines and robots are expected to perform all the physical tasks that humans do. These robots are required to reproduce human capabilities to be reliable substitutes for them. To achieve this, a huge research effort is devoted to both discover human brain functions and develop computational models to align technology on biological patterns. As such, the creation of artificial sense of touch is an important topic of interest. Many different tactile sensors are nowadays designed, produced, and available on the market [1].

However, data acquisition and processing techniques that make their use viable in practical applications are still required to evolve toward higher levels of efficiency.

In the case of human tactile perception, neuroscientists have revealed six different exploratory procedures that humans are make use of in order to perceive a stimulus by touch, among which "contour following" and "enclosure" are used for object recognition as they help exploring the shape of objects [2]. Klatsky et al. [3] mentioned the contribution of two sensory modalities in tactile perception, namely, cutaneous and kinesthetic cues. Cutaneous cues sensed by mechanoreceptors in the skin can obtain information about the texture, roughness, vibration, and temperature of a surface, while kinesthetic cues are provided by joints, bones, and muscles and supply information about the weight or the object's shape [3].

On the other hand, haptic exploration of objects is believed to be more reliable when visual data is available [4]. Moreover, Amedi et al. [5] refer to multimodal cells in human brain responding to both visual and tactile data, concluding the close contribution of human visual and tactile systems.

With inspiration from the visuo-haptic contribution in human sensorial loop and in accordance with the contour following strategy that humans use for object recognition, in this work we simulate the process of tactile data acquisition from a dataset of 3D models where the tactile data includes both of the two tactile sensory modalities employed by humans (cutaneous and kinesthetic). To align the process of tactile data acquisition with reality, an adaptive procedure is introduced to reduce the size of sensor while increasing the spatial resolution to probe the locations where a real sensor of a determined size will not be able to acquire data due to geometrical features. A similar probing strategy is also used in humans when using fingertips to touch finer details of objects and the palm for larger surfaces. The acquired sequential data from object contours are then used for the purpose of object recognition. A computational model of visual attention (i.e., a biologically inspired model selecting relevant areas in a visual scene for further exploration and analysis, as in human visual system) is advantageously employed to guide the process of contour following and results are compared to the case where visual information is not available (blind contour following in Section 5). The originality of the work with respect to the literature is found, more specifically, in the technique we propose to employ visual data for object recognition.

The main contributions of the work are as follows. (1) Simulation of tactile images (cutaneous cues) using a virtual tactile sensor relying on working principle of Force Sensing Resistor (FSR) arrays. (2) Adaptive simulation of tactile images according to geometry of objects. (3) Guidance of the process of contour following by engaging visual data from an enhanced model of visual attention. (4) Classification of sequential tactile data using different machine learning approaches including Convolutional Neural Networks (CNN), support vector machines (SVM), and k-Nearest Neighbors (kNN). (5) Fusion of cutaneous and kinesthetic data for making decision on object classes based on the probability values obtained using the CNNs.

The paper is structured as follows. Section 2 briefly discusses the current literature on the topic. A concise explanation of the framework we are proposing is provided in Section 3. The process of tactile data acquisition is detailed in Section 4. Section 5 provides more information about the implementation of contour following. The classification of the acquired tactile data is discussed in Section 6. Results are reported and discussed in Section 7, and Section 8 concludes the work.

2. State-Of-The-Art

Biologically inspired cognitive architectures are a challenging research area aiming to enhance machine intelligence solutions. Many researchers in the field of robotics target visuo-haptic interaction present in humans to design more intelligent robots with capability of sensing and exploring the environment in the way humans do. Despite the vast advancements in processing and learning visual data and the huge research interest in evolving the artificial sense of touch, the optimal integration of visual and haptic information is not yet achieved.

From the psychophysics and neuroscience side, many researchers are trying to explain how the tactile and visual information contribute in humans to interpret their environment. Klatzky et al. suggest that both vision and touch rely on shape information for object recognition [3]. Other researchers study different exploratory procedures that humans apply for tactile object recognition [2] and reveal the superiority of tactile perception in presence of vision [4]. Demarais et al. [6] studied the performance of visual, tactile, and bimodal exploration of objects for both learning and testing procedures for object identification.

From the cognitive computation and robotic side, several researchers are aiming to achieve an optimal integration of visual and tactile data. Magosso [7] trained a neural network reproducing a variety of visuo-haptic interactions, including the improvement of tactile spatial resolution using visual data, resolving conflict situations, and compensating poor unisensory information by cross-modal data. Gao et al. [8] trained a deep neural network by learning both visual and haptic features, confirming the idea that the integration of visual and haptic data outperforms the case where the two sensory data features are employed separately. Burka et al. [9] designed and constructed a multimodal data acquisition system emulating human vision and touch senses. Their sensor suite includes an RGB-D vision sensor, an ego motion estimator, and contact force and contact motion detectors. Kroemer et al. [10] trained a robot using both visual and tactile data to discriminate different surfaces by touch. Calandra et al. [11] trained a deep convolutional neural network to learn regrasping policies from visuo-tactile data. The network was then used to predict the probability of success when grasping, based on a set of grasping configurations. Van Hoof et al. [12] trained a robot by reinforcement learning using an autoencoder to perform tactile manipulations based on visual and tactile data, separately. Fukuda et al. [13] designed and produced a biocompatible tactile sensor used in laparoscopic surgeries, employing both visual and tactile feedback.

On the other hand, the technology, processing, and interpretation of tactile data itself attract huge research interest. The latest advancements in technology of tactile sensors are listed in the paper of Chi et al. [1]. Liu et al. [14] took advantage of joint sparse coding to classify tactile sequences from their dissimilarities as computed by dynamic time wrapping. A sequence of tactile data acquired as palpations on a set of seven objects using a five finger robotic hands is used in Gorges et al. [15]. Song et al. [16] designed and used a tactile sensor constructed from a thin polyvinylidene fluoride film to classify different textures. In another research [17], they took advantage of a similar sensor to evaluate fabric surfaces. They trained a support vector machine over data extracted by fast Fourier-transform followed by a principal component analysis to reduce the dimensionality.

In our previous work [18], we exploited a computational model of visual attention to guide the process of tactile probing by collecting imprints sequentially from a sequence of eye fixations. In this work, using inspiration from the haptic exploration of objects by humans for object recognition [3], we follow object contours to capture sequences of tactile data. The visual information in form of a set of visually interesting points, determined by the enhanced model of visual attention presented in our previous work [19], is advantageously employed to help selecting the contours which can enhance the recognition rate. The tactile exploration of objects is also brought closer to exploration strategies in humans by adaptively changing the size of tactile sensor according to geometrical features of the object.

3. Framework

The proposed tactile object recognition framework is summarized in Figure 1. Starting from 24 object models from 8 classes, we first constructed a dataset of sequential tactile data. Relying on the contour following exploratory procedure employed by humans to perceive the general shape of objects and for object identification, in this work the sequence of tactile data is generated by following a complete contour of each model where the contour following is implemented either blindly or guided by a computational model of visual attention. The main objective is to show that the recognition rate can be improved by engaging visual data. Two different scenarios were then implemented to classify sequential data. In the first scenario, two Convolutional Neural Networks were used to learn the features from sequences of tactile images (videos of tactile imprints) and sequences of normal vectors to object surface (cutaneous cues). In the second scenario, a series of features is extracted from a set of time series extracted from tactile videos using wavelet-decomposition and then a conventional learning algorithm, such as support vector machines and K-nearest neighbors are trained and tested for object classification. Each of the mentioned time series monitors the alternation in a specific tactile feature while the tactile sensor moves along the contour of objects. These features are themselves extracted using directional contourlet transformation [20]. The rest of the paper details the process of tactile data acquisition as well as the classification.

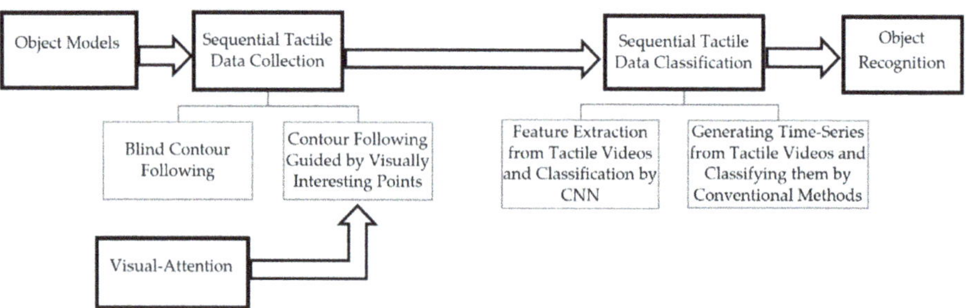

Figure 1. Framework for sequential tactile object recognition.

4. Tactile Data Acquisition

Psychological studies on human tactile perception reveal the contribution of several forms of tactile information to be interpreted to understand a stimulus. Cutaneous data captured by skin can provide information about the temperature, vibration, roughness, and local deformations on the surface. On the other hand, kinesthetic data from muscles, joints, and bones can help make an estimation of the approximate weight and global shape of an interacted object.

In order to recognize an object by touch, roughness, and discontinuities of object's surface (as cutaneous cues) and the finger movements to track the global shape (as kinesthetic cues) contribute together. On the other hand, contour following is the main exploratory procedure [2] that humans use to recognize an object. Accordingly, the following sections detail how, in this work, the cutaneous and kinesthetic cues are simulated and applied to reproduce the human sense of touch for robots.

4.1. Cutaneous Cues (Tactile Imprints)

In piezoresistive tactile sensors, which are widely accepted as a promising solution for tactile object recognition [21], the deformation in the sensor surface when subjected to an external force and in touch with an object can modify the resistance of a bridge circuit producing a proportional differential voltage. This differential voltage is then further processed to generate a tactile image. Inspired from working principle of piezoresistive sensors, we have developed a virtual tactile sensing module where the deformation measure in the surface of the tactile array is simulated as the distance between object surface and all the cells on a determined tangential plane to the object surface when the distance between the center of the plane and the object is zero. Figure 2 illustrates the simulated tactile sensor as a blue plane as well as an example of locally captured deformation profile (Figure 2b) and the produced tactile image (Figure 2c).

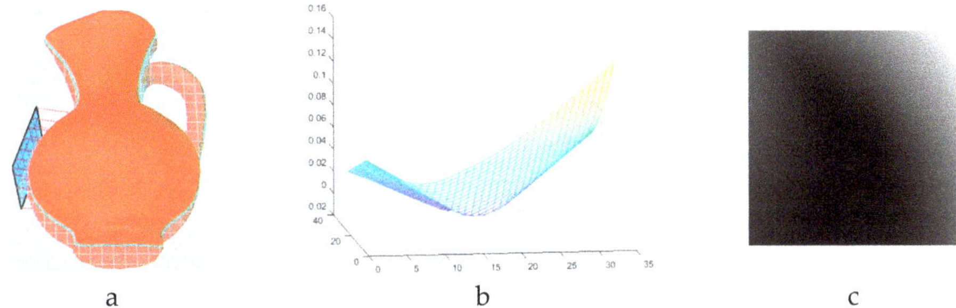

Figure 2. Adaptive modification of tactile sensor size.

4.2. Adaptive Probing

In this work, the locations on the surface of objects, from which tactile data are captured, are previously determined using object contours (blindly or based on a model of visual attention as it will be discussed in Section 5), and the center of the sensor is considered to be positioned at these locations. Consequently, in concave surfaces, such as the example in Figure 3, the sensor surface intersects the object resulting in negative distance values between the object and sensor. Since a real rigid backing FSR sensor cannot acquire tactile data from such probing cases, we follow the haptic exploration strategy by humans, where tactile information from larger surfaces is obtained by the palm and fingertips are used for finer details and concave surfaces. Accordingly, in this work, we have adaptively adjusted the sensor size to capture the local tactile data. A real counterpart robotic hand can be designed and produced using multiple FSR arrays with different sizes placed in palm and different phalanges. Alternatively, the Barrett robotic hand [22] can be purchased equipped with tactile sensing pads across fingers (smaller pad) and palm (larger pad), which is in accordance to the use of sensors we make in this work. In order to keep the size of the tactile image consistent during the experimentations (i.e., 32×32 in the current work), the distance between the sensing points is diminished; therefore, a higher local precision is achieved since the same number of sensing elements are assigned to touch a smaller surface of the object. As such, a tactile imprint of size 32×32 is obtained for finer details of objects as well.

Figure 3. Adaptive modification of tactile sensor size.

4.3. Kinesthetic Cues

As previously mentioned, kinesthetic cues can supply crucial information regarding the shape and size of the explored objects that are not perceivable by human skin. Drawing inspiration from kinesthetic cues contributing in human sense of touch, such as the angle between finger phalanges and the trajectory of finger motions when exploring an object, we have computed and used the normal vectors to the object surface in the process of object recognition. When probing an object with a real tactile sensor, the normal vectors to the surface are similarly computed and used to bring the sensor in contact with the object.

4.4. Sequential Tactile Data Collection

According to psychological research, when exploring an object with the hands in order to recognize it, humans tend to follow object contours to understand the global shape of the object leading to object recognition [2,3]. Relying on this biological fact, we move the tactile sensor along a complete contour of the object to simulate both cutaneous and kinesthetic cues. As a result, a video of tactile imprints for cutaneous cues is generated for each contour following, where the number of consecutive frames of the video is subsampled to 25 frames to reduce the high computational cost of data processing. Similarly, a trajectory of normal vectors and the 3D coordinate of probing locations are computed.

5. Contour Following

As previously mentioned, this work relies on the classification of sequential tactile data collected around contours of objects. Blind contour following and contour following guided by visually interesting points are the two strategies that are explored in this work to investigate the idea that the contour over which tactile probing takes place can play a decisive role in recognition rate.

Object contours are determined using 3D planes intersecting the object. Finding the equation of each plane, the set of points belonging both to the object and the plane, form a contour around the object. In order to find the equation of a plane in 3D space, three distinct noncolinear points are required. In this work, all the planes produced to determine probing paths are set to pass through the center of the models to avoid the selection of local contours around object extremities. As such, the center of each model is chosen as one of the three required points for formation of all planes. It is worth mentioning that such an implementation does not necessarily require visual data since supplementary tactile explorations such as the grasp stabilization method used in Regoli et al. [23] or a reinforcement learning as described in Pape et al. [24] can assist in determination of such contours. The acquisition of tactile information by exploration is both expensive in time and robot programming effort. Besides, it could lead to the acquisition of unnecessary data. All these, together with the possible advantage of visual cues in selection of more informative contours, incited us to consider the two data acquisition strategies as follows.

In the case of blind contour following, besides the central point of the model, the two other points are randomly selected from the vertices of the object model; in the case where contours are guided by the model of visual attention [19], the two other points are selected randomly from the set of visually interesting points.

The computational model of visual attention presented in [19] is adopted in this work to determine visually interesting points. The model uses the virtual camera of Matlab to collect a series of images from each object such that the complete coverage of the object surface is ensured. The obtained images are then decomposed into nine channels which are believed to contribute in guidance of attentions in humans, including color opponency, DKL color space, intensity, contrast, orientation, curvature, edges, entropy, and symmetry. The contribution weight of each channel is then learned based on a set of ground-truth points identified by a set of users. The extracted visual features are finally integrated according to the computed weights as described by equation 1, where $Smap$ is the computed saliency map, $w_{col}, \ldots\ w_{sym}$ are the contribution weight of each feature, respectively, and $C_{col}, \ldots\ C_{sym}$ are the feature maps illustrated in Figure 4.

$$Smap = \frac{w_{col} \cdot C_{col} + w_{con} \cdot C_{con} + w_{curv} \cdot C_{curv} + w_{DKL} \cdot C_{DKL} + w_{edg} \cdot C_{edg} + w_{ent} \cdot C_{ent} + w_{int} \cdot C_{int} + w_{ori} \cdot C_{ori} + w_{sym} \cdot C_{sym}}{\sum w_{Conspicuity\ Maps}} \quad (1)$$

Subsequently, the brightest regions on the resulted feature map (saliency map) after setting the intensity of image background to zero are identified using a nonmaximum suppression paradigm leading to determination of visually interesting points on images. A 2D to 3D projection algorithm recuperates the 3D coordinates of each salient point, which are used in the current study to guide the determination of object contours. A detailed description about the computation of each channel as well as further details on the computational model of visual attention are available in Rouhafzay et al. [19] for interested readers.

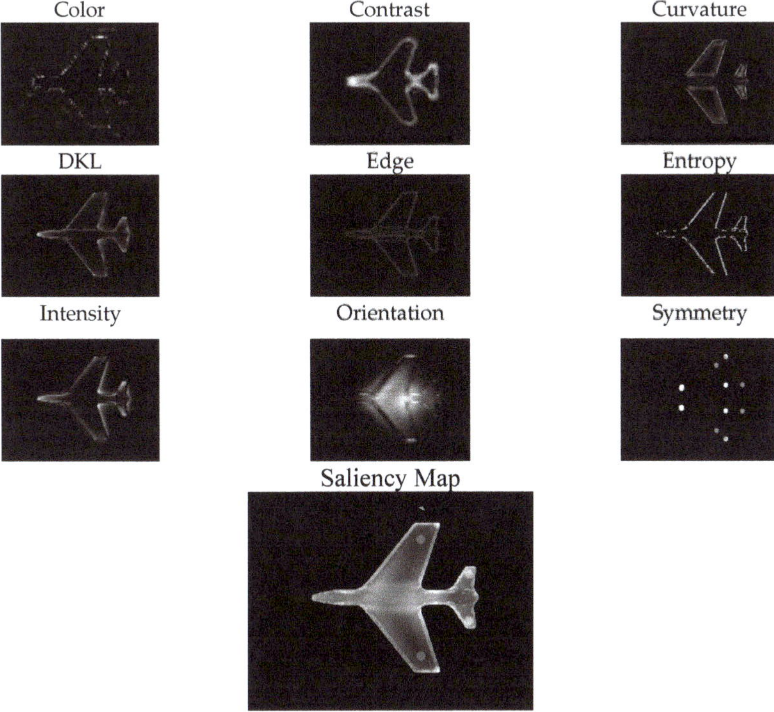

Figure 4. The nine channels contributing to the model of visual attention.

Figure 5a illustrates three examples of probing paths formed by random points (blind contour following), while examples of paths guided by visually interesting points are depicted in Figure 5b.

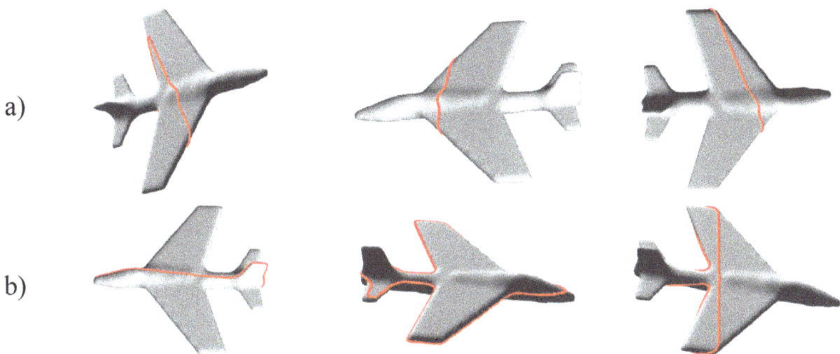

Figure 5. (**a**) Examples of blind contour following paths for model of Plane. (**b**) Examples of guided contour following paths by visually interesting points for model of Plane.

Since the number of vertices on a contour is very large and the collection of tactile data from all those vertices is neither efficient nor necessary, the obtained set of vertices is first subsampled to 25 points with equal distances between them and then cutaneous and kinesthetic data as explained in Section 4 are captured. Six of the twenty-five consecutive frames of the tactile video captured from the model of plane are depicted in Figure 6a, while Figure 6b illustrates an example of a set of normal vectors to the surface.

Figure 6. (**a**) Example of six consecutive frames of the tactile video captured from the model of Plane. (b) Example of sequence of normal vectors.

6. Sequential Tactile Data Classification

Once both cutaneous and kinesthetic cues are acquired for the objects, we implement two different approaches for object recognition by classifying the sequences of tactile data to determine if the acquired results for all techniques confirm the superiority of the model of visual attention. Convolutional neural networks allow us to feed the acquired tactile images directly; it automatically performs both feature extraction and classification of the tactile data. In order to use the two other classifiers, i.e., support vector machine and K-nearest neighbors, we need to extract ourselves the relevant features from tactile imprints and verify how these features are altered by moving the sensor around the object. This is the main distinction between the two approaches used in this paper and which are discussed in the next subsections.

6.1. Feature Extraction from Tactile Data and Classification by Convolutional Neural Networks

In the first approach we train two separate convolutional neural networks (CNN) to learn the features from the video of tactile images (cutaneous cues) and the sequence of normal vectors to the surface (kinesthetic cues). All tactile images captured by the virtual FSR sensor, are 32×32 grayscale images and 25 frames are considered for each exploration (contour).

The first CNN (dedicated to cutaneous cues) takes benefit from two convolution layers with 3D kernels followed by a batch normalization layer speeding up the learning process. No pooling layer is added as the size of tactile images is not so large to require down sampling. Two fully connected layers are then exploited to learn the relationship between the extracted features through the filters in the convolution layers. A SoftMax layer finally outputs the probability distribution values over predicted output classes. The network is trained for 50 epochs.

The second CNN (dedicated to cutaneous cues) has $25 \times 3 \times 1$ data in its input layer. Thus, it can be implemented with 2D kernels in convolution layers. We set up a similar architecture for the second CNN, i.e., two convolution layers with a batch normalization in between followed by two fully connected layers, and a SoftMax layer generating the probability distribution results predicted for the output layer.

Each of these CNNs are applied after training to tactile data captured over identical contours as a test sample and output the probabilities that the sample belongs to each of the eight object classes. The winning class has the highest probability among all. In order to integrate the results obtained by cutaneous and kinesthetic cues, for each probing sequence from the test data, we sum up the obtained probability values computed by the two CNNs for each class and choose the class with the highest probability as the winning class, as illustrated in Figure 7.

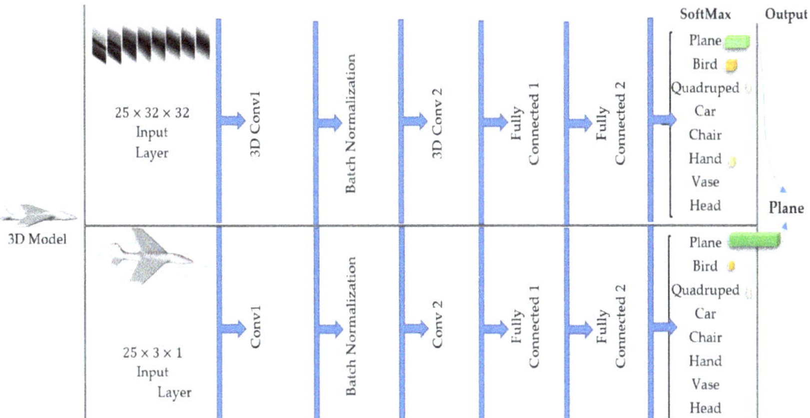

Figure 7. The two convolutional neural network (CNN) structures and the decision on output class.

6.2. Feature Extraction from Time Series by Wavelet Decomposition and Classification by SVM and KNN

In the second approach we simplify the tactile video classification such that a set of features from videos can be directly fed into conventional classifiers. In the previous approach, we took benefit from a CNN with a 3D kernel to learn features of tactile data. Here we employ a 16 directional contourlet transformation [20] for feature extraction from each tactile imprint. As such, a feature vector of size 16 is computed as the standard deviation of each directional subband. At this point, the normal vectors to probing locations (kinesthetic cues) are added to the obtained feature vector to create a 1 × 19 feature vector for each probing point from the contour. The variation of each of these 19 features by moving the tactile sensor along a contour creates 19 time-series of length 25. Consequently, we exploit a 3-level wavelet decomposition for each of the 19 time-series using the Daubechies 2 wavelet, to extract features characterizing how the 19 tactile features vary when the tactile sensor moves along the object contour. Then the root mean squared (rms) value, standard deviation, and skewness of the wavelet coefficients for each level, as well as those of the sequence itself, are concatenated to produce a final feature vector. To avoid the negative effect of high dimensional data on the performance of the classifiers, a feature selection method selecting the most relevant features based on their information gain is first employed to select the most informative features and then the size of the acquired feature vector is reduced to five using a self-organizing-map. Figure 8 summarizes the data processing strategy for object recognition using a conventional classifier.

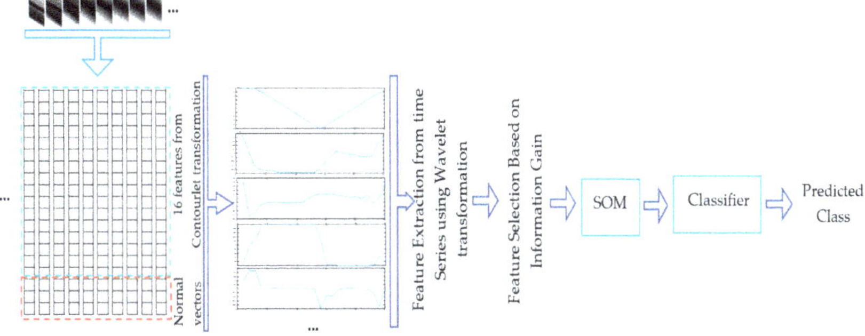

Figure 8. Process of object classification using conventional classifiers.

7. Experimental Results

Figure 9 illustrates the 3D objects used in this study. Twenty-four object models belonging to eight classes are selected from a popular dataset [25]. We have applied the proposed framework to acquire sequential tactile data by blind contour following as well as using contour following paths guided by visually interesting points and the obtained sequences are classified using the previously explained approaches.

Since CNN, like any other deep learning solution, requires a large data set for training; we have collected a total of 22,800 sequences, from which 20% (4560) is used for testing and the obtained accuracy is reported in Table 1. The process of tactile data acquisition is simulated using the MATLAB programming platform and its statistics and machine learning toolbox are used for training and testing Convolutional Neural Networks.

In the case of the approach in Section 6.2, the acquired data set is first standardized using z-score before being fed into SVM and kNN. After splitting the data into 80:20 samples for developing the classifier and testing it, the conventional classifiers are trained and validated using 5-fold cross-validation. The kNN classifier takes benefit from the Euclidean distance metric to determine nearest neighbors and assigns the label of winning vote among the 10 nearest ones to test data. The support vector machine employs an RBF kernel function with $\gamma = 1$, $\epsilon = 0.5$, and $C = 1$ hyperparameters.

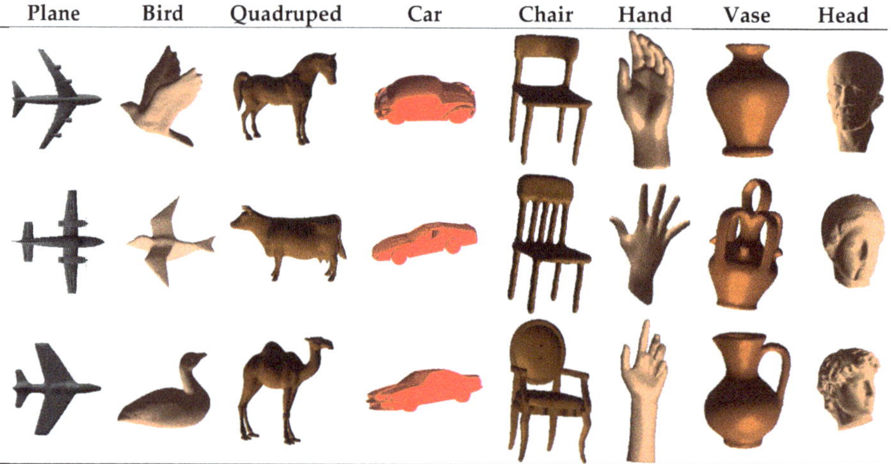

Figure 9. Objects used for experiments.

The accuracy values for the 20% of the data kept out for testing, which includes 4560 test sequences, are reported in Table 1 for the three classifiers. Confusion matrices are also provided in Figure 10, in which the eight object classes are numbered as class one to eight in the same order in which they are presented in Figure 8.

The accuracy values confirm that the use of visually interesting points to determine object contours has a positive impact on classification accuracy for all classifiers and in the case of Convolutional Neural Networks the accuracy is improved by 15.68%. Furthermore, CNNs show a good capability in extracting and learning features from tactile data. It is worth mentioning that the random guess in our experiments is $\frac{1}{8}$ or 12.5%, and the best performance achieved in this work is 98.97%, i.e., 86.47% above the random guess.

Table 1. Classification accuracy of test data.

	Contour Following Guided by Visually Interesting Points	Blind Contour Following
CNNs	98.97%	83.29%
kNN	86.07%	73.11%
SVM	88.44%	77.21%

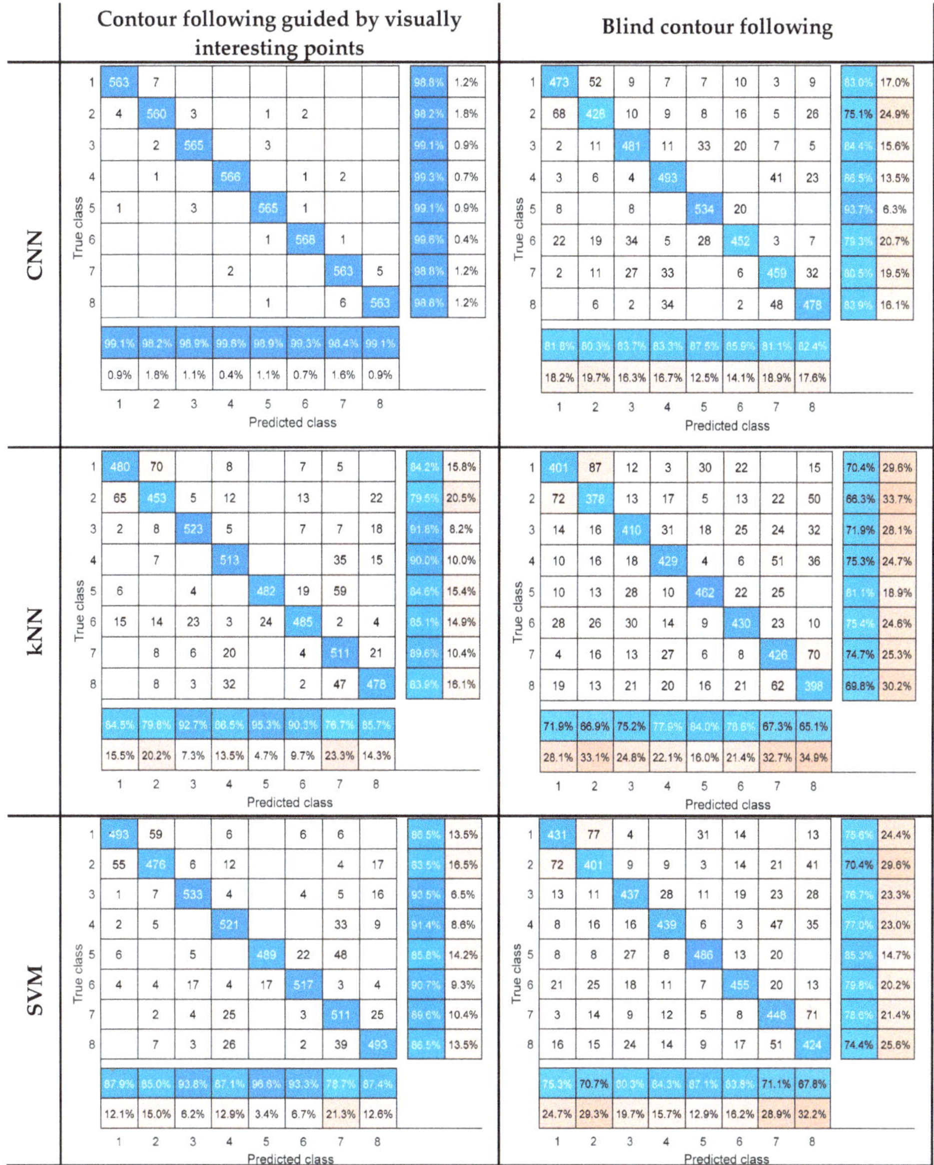

Figure 10. Confusion matrices.

According to the confusion matrices, the visual data helps making a cleaner discrimination among some object classes, so the confusion occurs only among the classes with more tactile similarities.

This can be explained by the fact that visually interesting points lead to the selection of contours which are more informative about the object characteristics. For example, a round or oval shape contour can be followed on almost all objects if we blindly follow a local path around object extremities, while visual data guides the process by selection of different contours simplifying the object recognition process.

8. Conclusions

In this work, we proposed a novel framework for robotic tactile object recognition with inspiration from human tactile object exploration. The two sensing modalities in human tactile perception including kinesthetic and cutaneous cues are simulated for a set of object models and an enhanced model of visual attention intervenes to guide the process of touching. The originality of the work described in this paper is that the visual data does not directly interfere in the process of object recognition but is employed to select the most informative contours of the object that enhance the accuracy of classifiers. Two different approaches using convolutional neural networks and two other conventional classifiers (kNN and SVM) are employed to classify the tactile data for object recognition. The obtained results are compared to the case where visual data are not used confirming the fact that visual attention can improve the process of tactile data acquisition. An accuracy of 98.97% is the highest performance achieved in this study using CNN. A future application of the current work is the integration of the proposed intelligent algorithm in the decision system of a robot that can make use of its vision to select contours and then use its hand equipped with FSR tactile sensors of different sizes in finger phalanges, fingertips, and palm to touch objects and recognize them. Since for real implementation of the proposed framework the occurrence of noise may affect the performance, we will take advantage of a denoising autoencoder assisting to reconstruct the corrupted inputs.

Author Contributions: G.R. and A.-M.C. conceived and designed the experiments; G.R. performed the experiments; G.R. and A.-M.C. analyzed the data and wrote the paper.

Funding: This work is supported in part by the Natural Sciences and Engineering Research Council of Canada (NSERC) Discovery Grant Program and by the Ontario Graduate Scholarship (OGS) program.

Conflicts of Interest: The authors declare no conflicts of interest.

References

1. Chi, C.; Sun, X.; Xue, N.; Li, T.; Liu, C. Recent Progress in Technologies for Tactile Sensors. *Sensors* **2018**, *18*, 948. [CrossRef] [PubMed]
2. Lederman, S.J.; Klatzky, R.L. Haptic perception: A tutorial. *Atten. Percept. Psychophys.* **2009**, *71*, 1439–1459. [CrossRef] [PubMed]
3. Klatzky, R.L.; Lederman, S.J.; Metzger, V.A. Identifying objects by touch: An "expert system. *Percept. Psychophys.* **1985**, *37*, 299–302. [CrossRef] [PubMed]
4. Klatzky, R.L.; Lederman, S.J.; Matula, D.E. Haptic Exploration in the Presence of Vision. *Hum. Percept. Perform.* **1993**, *19*, 726–743. [CrossRef]
5. Amedi, A.; Malach, R.; Hendler, T.; Peled, S.; Zohary, E. Visuo-haptic object-related activation in the ventral visual pathway. *Nat. Neurosci.* **2001**, *4*, 324–330. [CrossRef] [PubMed]
6. Desmarais, G.; Meade, M.; Wells, T.; Nadeau, M. Visuo-haptic integration in object identification using novel objects. *Atten. Percept. Psychophys.* **2017**, *79*, 2478–2498. [CrossRef] [PubMed]
7. Magosso, E. Integrating Information from Vision and Touch: A Neural Network Modeling Study. *IEEE Trans. Inf. Technol. Biomed.* **2010**, *14*, 598–612. [CrossRef] [PubMed]
8. Gao, Y.; Hendricks, L.; Kuchenbecker, K.J. Deep learning for tactile understanding from visual and haptic data. In Proceedings of the IEEE international conference robotics and automation (ICRA), Stockholm, Sweden, 16–21 May 2016.
9. Burka, A.; Hu, S.; Helgeson, S.; Krishnan, S.; Gao, Y.; Hendricks, L.A. Proton: A Visuo-Haptic Data Acquisition System for Robotic Learning of Surface Properties. In Proceedings of the IEEE International Conference on Multi-sensor Fusion and Integration for Intelligent Systems (MFI), Baden-Baden, Germany, 19–21 September 2016.

10. Kroemer, O.; Lampert, C.H.; Peters, J. Learning Dynamic Tactile Sensing with Robust Vision-Based Training. *IEEE Trans. Robot.* **2011**, *27*, 545–557. [CrossRef]
11. Calandra, R.; Owens, A.; Jayaraman, D.; Lin, J.; Yuan, W.; Malik, J. More Than a Feeling: Learning to Grasp and Regrasp Using Vision and Touch. *IEEE Robot. Autom. Lett.* **2018**, *3*, 3300–3307. [CrossRef]
12. Van Hoof, H.; Chen, N.; Karl, M.; Van der Smagt, P.; Peters, J. Stable reinforcement learning with autoencoders for tactile and visual data. In Proceedings of the IEEE/RSJ International Conference on Intelligent Robots and Systems, Daejeon, South Korea, 9–14 October 2016.
13. Fukuda, T.; Tanaka, Y.; Kappers, A.M.L.; Fujiwara, M.; Sano, A. Visual and tactile feedback for a direct-manipulating tactilesensor in laparoscopic palpation. *Int. J. Med Robot. Comput. Assist. Surg.* **2018**, *14*, e1879. [CrossRef] [PubMed]
14. Liu, H.; Guo, D.; Sun, F. Object Recognition Using Tactile Measurements: Kernel Sparse Coding Methods. *IEEE Trans. Instrum. Meas.* **2016**, *65*, 656–665. [CrossRef]
15. Gorges, N.; Navarro, S.E.; Goger, D.; Worn, H. Haptic Object Recognition using Passive Joints and Haptic Key Features. In Proceedings of the IEEE International Conference on Robotics and Automation, Anchorage, AK, USA, 3–7 May 2010.
16. Song, A.; Han, Y.; Hu, H.; Li, J. A Novel Texture Sensor for Fabric Texture Measurement and Classification. *IEEE Trans. Instrum. Meas.* **2014**, *63*, 1739–1747. [CrossRef]
17. Hu, H.; Han, Y.; Song, A.; Chen, S.; Wang, C.; Wang, Z. A Finger-Shaped Tactile Sensor for Fabric Surfaces Evaluation by 2-Dimensional Active Sliding Touch. *Sensors* **2014**, *14*, 4899–4913. [CrossRef] [PubMed]
18. Rouhafzay, G.; Cretu, A.-M. A Visuo-Haptic Framework for Object Recognition Inspired by Human Tactile Perception. *Proceedings* **2019**, *4*, 47.
19. Rouhafzay, G.; Cretu, A.-M. Perceptually Improved 3D Object Representation Based on Guided Adaptive Weighting of Feature Channels of a Visual-Attention Model. *3D Res.* **2018**, *9*, 29. [CrossRef]
20. Do, M.N.; Vetterli, M. The Contourlet Transform: An Efficient Directional Multiresolution Image Representation. *IEEE Trans. Image Process.* **2005**, *14*, 2091–2106. [CrossRef] [PubMed]
21. Pasca, C.; Payeur, P.; Petriu, E.M.; Cretu, A.-M. Intelligent Haptic Sensor System for Robotic Manipulation. In Proceedings of the Instrumentation and Measurement Technology Conference, Como, Italy, 18–20 May 2004.
22. Barrett TECH. Available online: http://www.barrett.com/features-and-benefits (accessed on 15 March 2019).
23. Regoli, N.; Jamali, N.; Metta, G.; Natale, L. Controlled Tactile Exploration and Haptic Object Recognition. In Proceedings of the 18th IEEE International Conference on Advanced Robotics, Hong Kong, China, 10–12 July 2017.
24. Pape, L.; Oddo, C.M.; Controzzi, M.; Cipriani, C.; Förster, A.; Carrozza, M.C.; Schmidhuber, J. Learning tactile skills through curious exploration. *Front. Neurorobotics* **2012**, *6*, 6. [CrossRef] [PubMed]
25. Dutagaci, H.; Cheung, C.P.; Godil, A. A Benchmark for Automatic Best View Selection of 3D Objects. National Institute of Standards and Technology (NIST). Available online: https://www.itl.nist.gov/iad/vug/sharp/benchmark/bestview/data/EXP_MODELS.zip (accessed on 12 November 2018).

Article

Acoustic Localization of Bragg Peak Proton Beams for Hadrontherapy Monitoring †

Jorge Otero [1,*], Ivan Felis [2], Miguel Ardid [1] and Alicia Herrero [3]

1 Institut d'Investigació per a la Gestió Integrada de les Zones Costaneres (IGIC), Universitat Politècnica de València (UPV), Gandia, 46730 València, Spain; mardid@fis.upv.es
2 Centro Tecnológico Naval y del Mar (CTN), Fuente Álamo, 30320 Murcia, Spain; ivanfelis@ctnaval.com
3 Institut de Matemàtica Multidisciplinar, Universitat Politècnica de València (UPV), 46022 València, Spain; aherrero@mat.upv.es
* Correspondence: jorotve@upv.es; Tel.: +34-963-877-000 (ext. 43681)
† This is an extension version of conference paper: Otero, J.; Ardid, M.; Herrero, A; Felis, I. Acoustic Location of Bragg Peak for Hadrontherapy Monitoring. In Proceedings of the 5th International Electronic Conference on Sensors and Applications, 15–30 November 2018.

Received: 31 March 2019; Accepted: 23 April 2019; Published: 26 April 2019

Abstract: Hadrontherapy makes it possible to deliver high doses of energy to cancerous tumors by using the large energy deposition in the Bragg-peak. However, uncertainties in the patient positioning and/or in the anatomical parameters can cause distortions in the calculation of the dose distribution. In order to maximize the effectiveness of heavy particle treatments, an accurate monitoring system of the deposited dose depending on the energy, beam time, and spot size is necessary. The localized deposition of this energy leads to the generation of a thermoacoustic pulse that can be detected using acoustic technologies. This article presents different experimental and simulation studies of the acoustic localization of thermoacoustic pulses captured with a set of sensors around the sample. In addition, numerical simulations have been done where thermo-acoustic pulses are emitted for the specific case of a proton beam of 100 MeV.

Keywords: hadrontherapy; acoustic localization; Bragg peak; thermoacoustic; piezoelectric ceramic

1. Introduction

Localization of a source is a technique in which a source is located by detecting propagated signals received in several sensors and the analysis of them [1]. There are many localization techniques proposed for wireless sensor networks [2,3]. However, in this article, a three-dimensional localization to solve the estimation of an acoustic source in a homogeneous medium is introduced. The use of acoustic sensors to locate sound sources in such practical systems is of great interest but needs further development and improved performance systems. This research has significant potential for many applications in medicine, physics, engineering, and underwater acoustics. The method to locate the tumor tissue is based on a computed tomography scan to find the area that will then be radiated by heavy particles in the Bragg peak region [4]. However, uncertainties in the patient positioning and/or in the anatomical parameters can increase the uncertainty during the radiotherapy. In these cases, acoustic source localization in medical applications has gained a lot of interest in recent years, which ought to be the necessity for improving the monitoring of tumor tissue in hadrontherapy treatments. Linear sensors can be employed for acoustic source localization in a noise environment using a time delay estimation. The method presented in this paper is based on the TDOA (time difference of arrival) [5] technique that performs very well in the localization of an acoustic event in both two-dimensional and three-dimensional spaces decreasing the error while increasing the number of sensors. The acoustic signal is generated and detected by piezoelectric sensors in known positions and using a DAQ system to record the signal. Differences in the signal propagation path from the source lead to different phases in the detected signal. Therefore, cross-correlation analysis is used to estimate the delays of arrivals accurately [6], even in conditions with low signal-to-noise ratio.

The pressure source localization of the Bragg peak in hadrontherapy can also be used to identify the regions of local heat released due to energy deposition. In this paper, we focus on the objective of monitoring the position for hadrontherapy through the Bragg proton beam acoustic localization. This pressure is related to the beam energy, the temporal pulse width, the size of the beam, and the number of protons by pulse, so, to some extent, it might be used in the future for dose sensitivity as well, but this aspect is out of the scope of the paper. For this reason, as a first approach, the source assessed in this article presents a pressure above the threshold of detection [7] for beams of a few million protons per spill of energies from 20 up to 200 MeV. Both in simulation and experiment, homogenous and isotropic medium is used and the wall effect is neglected since the direct signal arrives earlier to the Omni-directional receiver than the reflected signal.

2. Overview of Approach

Techniques based on cross-correlation and generalized correlation (GCC) [8] have been employed to determine the time difference of arrival of the signals (TDOA) given its computational cost and accuracy of the results. To obtain a better estimate of the TDOA, it is convenient to filter the signal before its integration, as shown in Figure 1 [6].

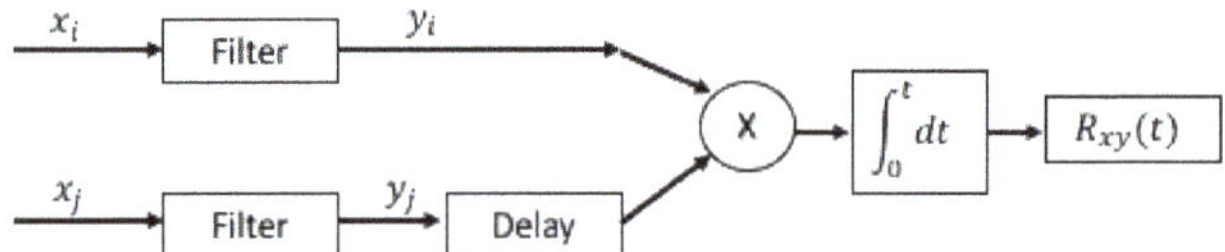

Figure 1. Scheme for obtaining the time of arrival (TOA).

The cross-correlation $R_{x_ix_j}$ between the signal x_i and x_j filtered by the filters H_i and H_j, is expressed as a function of the power spectral density $G_{x_ix_j}$, as shown below.

$$R^{GCC}_{x_ix_j}(t') = \int_{-\infty}^{+\infty} H_i(f)H_j^*(f)G_{x_ix_j}(f)e^{i2\pi ft'}df = \int_{-\infty}^{+\infty} \varphi^{GCC}(f)G_{x_ix_j}(f)e^{i2\pi ft'}df \quad (1)$$

where {*} indicates a conjugated complex and $\varphi^{GCC}(f)$ is a frequency-dependent weight function. Due to finite observations, we can only obtain an estimation of $G_{x_ix_j}(f)$ [9]. Therefore, to obtain the TDOA, the following expression will be used [6].

$$\hat{R}^{GCC1}_{x_ix_j}(t') = \int_{-\infty}^{+\infty} \varphi^{GCC}(f)\hat{G}_{x_ix_j}(f)e^{i2\pi ft'}df \quad (2)$$

where $\hat{G}_{x_ix_j}(f)$ is the obtained estimation of $G_{x_ix_j}(f)$. For each pair of sensors, the TDOA is taken as the time delay that maximizes the cross-correlation between the filtered signals of both sensors, that is: $\hat{\tau}^{GCC}_{ij} = \arg\left(\max_{t'}\left\{\hat{R}^{GCC1}_{x_ix_j}(t')\right\}\right)$.

A general model for three-dimensional (3-D) estimation of a source using M receivers is developed. To obtain the location of the source, we start by knowing the spatial position (x_i, y_i, z_i) of a certain number of sensors. Let $(x_s,\ y_s,\ z_s,)$, the position of the source to be located, the distance between the source and the i-th sensor will be:

$$d_i = \sqrt{(x_i - x_s)^2 + (y_i - x_s)^2 + (z_i - x_s)^2} \quad (3)$$

The range difference in distance between the i-th receiver and the first receiver, d_{i1}, is given by:

$$d_{i1} = c \cdot \tau_{i1} = d_i - d_1 = \sqrt{(x_i - x_s)^2 + (y_i - y_s)^2 + (z_i - z_s)^2} - \sqrt{(x_1 - x_s)^2 + (y_1 - y_s)^2 + (z_1 - z_s)^2} \quad (4)$$

where c is the sound velocity in the medium, d_1 is the distance between the first receiver and the source, and τ_{i1} is the estimated TDOA between the first receiver and the i-th receiver [10].

Equation (4) considered for all the sensors form a nonlinear equation system whose solution can be found by several ways. After studying different resolution methods, it was decided to use the Newton-Raphson method since it offers very good results and computation time.

Newton-Raphson Method

To get the position of the thermoacoustic source inside the medium, we have solved the nonlinear system using the Newton-Raphson method [11] by means of partial derivatives. Consider a system of m equations and n unknowns.

$$f_m(x_1, x_2, x_3, \ldots, x_n) = 0 \quad (5)$$

This system can be written in vector form as $f(x) = 0$, where f is a vector of m dimensions and x is a vector of n dimensions. To solve this system of equations, we have to find a vector x such that the function $f(x)$ equals the null vector. If we call η to the solution of the system and x_r to an approximation of it, we can develop f in Taylor series around this approximation as:

$$f(x) = f(x_r) + \nabla f(x_r) \cdot (x - x_r) + \cdots \quad (6)$$

Since $f(\eta) = 0$ then we get, as an approximation, the following.

$$0 \approx f(x_r) + \nabla f(x_r) \cdot (\eta - x_r) \quad (7)$$

Now, we can define the vector x_{r+1} as this approximation, which is closer to the root than x_r. We can continue with the iterative method to obtain approximations closer and closer to the solution. To write the iterative method, the term $\nabla f(x_r)$ is replaced by the Jacobian of the function f, that is:

$$J(x_r) = \begin{bmatrix} \frac{\partial f_1}{\partial x_1} & \frac{\partial f_1}{\partial x_2} & \cdots & \frac{\partial f_1}{\partial x_n} \\ \frac{\partial f_2}{\partial x_1} & \frac{\partial f_2}{\partial x_2} & \cdots & \frac{\partial f_2}{\partial x_n} \\ \vdots & \vdots & \vdots & \vdots \\ \frac{\partial f_m}{\partial x_1} & \frac{\partial f_m}{\partial x_2} & \cdots & \frac{\partial f_m}{\partial x_n} \end{bmatrix} \quad (8)$$

which is a $n \times m$ matrix. Then, it is possible to obtain a new value of x_{r+1} by solving the following relationship.

$$J(x_r)(x_{r+1} - x_r) = f(x_r) \quad (9)$$

Then, iteratively, we can approximate more and more the x_{r+1} to η until a solution error $|x_r - x_{r+1}|$ previously fixed is reached.

This method has been compared with other algorithms for solving systems of nonlinear equations and has shown some advantages over the rest like good accuracy and low computing cost. However, if the initial solution value of the system differs greatly from the real solution, then the method does not converge conveniently. Figure 2 shows the convergence of the location algorithm as a function of the distances between the initial point of the method, the result of the reconstruction of the position and the real position for a total of 10,000 simulations.

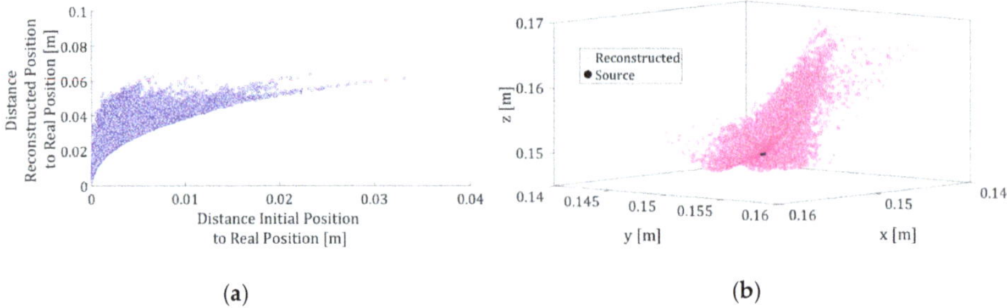

Figure 2. (**a**) The distance between the initial position for the algorithm and the real position of the source is shown on the abscissa axis, while the axis of the ordinates shows the distance between the signal reconstructed by the algorithm and the real position of the source; (**b**) The reconstructed positions for each of the 10,000 simulations are shown together with the real position (black).

3. Thermoacoustic Simulation

3.1. Bragg Peak

The main physical advantage of heavy particles as compared to photons is their characteristic depth-dose profile, the known Bragg curve in honor of Sir William Henry Bragg who investigated the energy deposition of alpha particles, which form a Radium source in the air at the beginning of the last century [12]. While the photon dose decreases exponentially with penetration depth according to the absorption law for electromagnetic radiation, the depth-dose of heavy charged particles exhibits a flat plateau region with a low range of the particles [13]. This paper uses an analytical approach presented by T. Bortfiel in 1996 [4], and the numerical representation valid for protons with energies between 10 MeV and 200 MeV. Thus, the energy of single protons along a Z-axis in a homogeneous medium (water) is considered. The total energy released in the medium per unit mass in the Z-axis is shown below.

$$T(z) = -\frac{1}{\rho}\left(\phi(z)\frac{dE(z)}{dz} + \frac{d\phi(z)}{dz}E(z)\right) \tag{10}$$

where $\phi(z)$ is the proton flow, that is, the number of protons per cm^2, $E(z)$ is the energy deposited on the Z-axis, and ρ represents the mass density of the medium. The method makes use of a midpoint where a certain fraction γ of the energy released in nuclear interactions is absorbed locally while the rest is ignored. Then, the total absorbed energy $\hat{D}(z)$, will be given by:

$$\hat{D}(z) = -\frac{1}{\rho}\left(\phi(z)\frac{dE(z)}{dz} + \gamma\frac{d\phi(z)}{dz}E(z)\right) \tag{11}$$

A relationship between the initial energy $E(z=0) = E_0$ and the range $z = R_0$ in the medium can be approximated as $R_0 = \alpha E_0^p$ for $p = 1.5$. This relation is valid for protons with energy close to 250 MeV. The factor α is proportional to the square root of the effective atomic mass of the medium [4]. Using the inverse for $R_o \leq 0.5$ cm and assuming E_o to be given in units of MeV, the best fit parameters for $E_o(R_o)$ are $p = 1.77$, $\alpha = 2.2 \times 10^{-3}$ for the proton in water [4]. The remaining energy $E(z)$ at an arbitrary depth $z \leq R_o$ fails to travel the distance $R_o - z$ according to the range-energy relationship shown below.

$$E(z) = \frac{1}{\alpha^{1/p}}(R_0 - z)^{1/p} \tag{12}$$

For energies above 20 MeV, there is non-negligible probability that protons may be lost from the beam due to nuclear interactions. This non-elasticity was studied and tabulated by Janni [14] as a function of the residual range $(R_0 - z)$. The proton flow $\phi(z)$ can be written by using the equation below.

$$\phi(z) = \phi_0 \frac{\beta}{1 + \beta R_0} \tag{13}$$

where ϕ_0 is the primary fluence and the slope parameter β was determined to be $\beta = 0.012\ \text{cm}^{-1}$ [4]. Thus, the distribution of the deposition of energy along the depth range can be expressed as the equation below.

$$D(z) = \Phi_0 \frac{e^{\zeta^2/4} \sigma^{\frac{1}{p}} \Gamma\left(\frac{1}{p}\right)}{\sqrt{2\pi} \rho p \alpha^{\frac{1}{p}} (1 + \beta R_0)} \times \left[\frac{1}{\sigma} L_{-1/p}(-\zeta) + \left(\frac{\beta}{p} + \gamma\beta + \frac{\varepsilon}{R_0} \right) L_{-1/p-1}(-\zeta) \right] \tag{14}$$

where Γ represents the gamma function, $\zeta = (R_0 - z)/\sigma$, with a σ value of $0.012R_0^{0.935}$, and ε represents a relatively small fraction of the fluence Φ_0 in the peak. Figure 3 shows the distribution of the dose as a function of the range for a different proton energy.

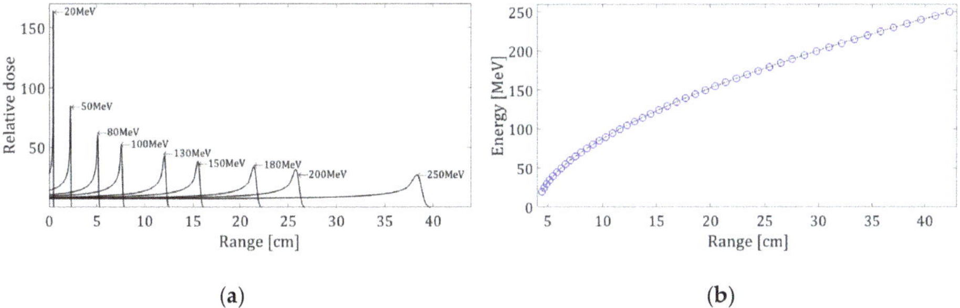

(a) (b)

Figure 3. Bragg peak for different energies. (**a**) The deposition of the dose varies according to the energy of the proton. The maximum of the Bragg peak varies according to the energy; (**b**) The relationship Range–Energy for protons in water is shown.

3.2. Thermoacoustic Model

In the thermoacoustic case, an excited point source emits a pressure wave proportional to the first time derivate of the excitation pulse [15]. Hence, a Gaussian excitation pulse leads to a bipolar acoustic emission consisting of a positive compression, which results in an increase in pressure. This is followed by a negative rarefaction, which is a decrease of pressure. The positive and negative pressure peaks are not only due to the heating and cooling of the medium, but the variation of the heating rate also plays a role. The medium expands or contracts according to its coefficient of thermal volumetric expansion α'. As a result, a pressure wave is observed. The pressure wave from an energy deposition in a region can be understood as the sum of the individual responses that would be observed from decomposing the spatial deposition into point sources. The resulting pressure signal depends on the time derivative of the excitation pulse. The amplitude of the wave depends on the energy deposited, the number of protons per pulse of the beam, and the temporal shape of the excitation pulse. A dose of 1 Gy generates a ~ 240 μK temperature increase in water [15]. Ignoring heat diffusion and cinematic viscosity, the wave equation that describes the pressure p at a time t and position $\vec{r}$, is shown below [16–19].

$$\vec{\nabla}^2 p(\vec{r}, t) - \frac{1}{c_s^2} \cdot \frac{\partial^2 p(\vec{r}, t)}{\partial t^2} = -\frac{\alpha'}{C_p} \cdot \frac{\partial^2 \epsilon(\vec{r}, t)}{\partial t^2} \tag{15}$$

where c_s (ms^{-1}) represents the speed of sound in the middle, C_p ($J\,kg^{-1}K^{-1}$) is the specific heat capacity, and $\epsilon(\vec{r},t)$ ($J\,s^{-1}m^{-3}$) is the energy density deposited in the medium. Equation (15) can be solved using the Kirchhoff integral as shown below.

$$p(\vec{r},t) = \frac{1}{4\pi}\frac{\alpha'}{C_p}\int_V \frac{dV'}{\left|\vec{r}-\vec{r}'\right|}\cdot\frac{\partial^2}{\partial t^2}\epsilon\left(\vec{r}',t-\frac{\left|\vec{r}-\vec{r}'\right|}{c_s}\right) \quad (16)$$

where $p(\vec{r},t)$ denotes the hydrodynamic pressure at a given place and time. The values for the thermoacoustic model were an energy of 100 MeV, a temporal profile of 1 μs, 3.4×10^6 protons per pulse, the beam with a size of 1 mm, and a sensor located 40 mm from the Bragg peak. The characteristics of the simulation are given by simulation results from different studies, as well as their application in clinical cases [7,15,20–25]. The values for this case are shown in Figure 4. As a result, the pressure obtained at the reception point will be the signal that will be emitted by the piezoelectric [26] transducer to simulate a bipolar source that will be located by the sensor array.

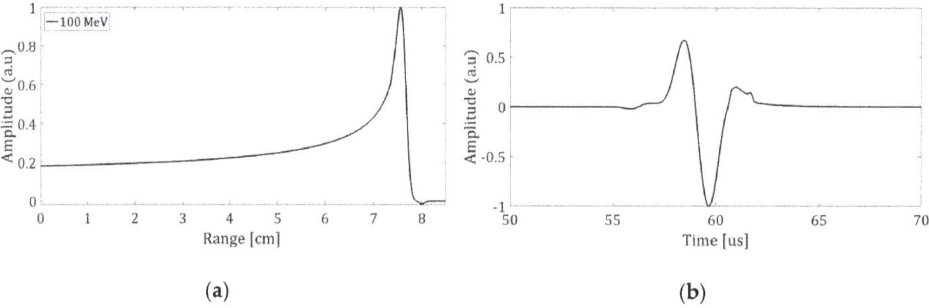

Figure 4. (**a**) Bragg curves with an initial energy of 100 MeV protons in water. The line represents the dose contribution from the fraction of protons that have nuclear interactions; (**b**) Pressure for a sensor located 4 cm from the Bragg peak on the axis of symmetry of the emission.

4. Experimental Setup

The experimental data measurements were made in the laboratories of the physics department at the Universitat Politècnica de València (Spain). There is a water tank with a volume of 0.64 m^3 with a programmable 3D axis system MOCO PI MICOS arm that was programmed to move the Reson TC4014 receiver hydrophone in the tank. The hydrophone has a receiving sensitivity of -186 ± 3 dB @1 V/μPa and a frequency response from 15 kHz to 480 kHz. The emitter hydrophone is a Reson TC4038 with a transmitting response of 110 dB @1 μPa/V @ 1m and a frequency response from 50 kHz to 800 kHz. Figure 5 shows the experimental setup with the transmitter and receiver inside the tank. A National Instruments data acquisition system was used with PXI type cards to generate the signal used as input of the linear E&I A150 amplifier that feeds the transmitter. Both the receiving and the feeding signal were captured. The latter was captured with an ×100 probe to avoid overloads in the system. All signals were stored at 10 Ms/s with a duration of 500 μs.

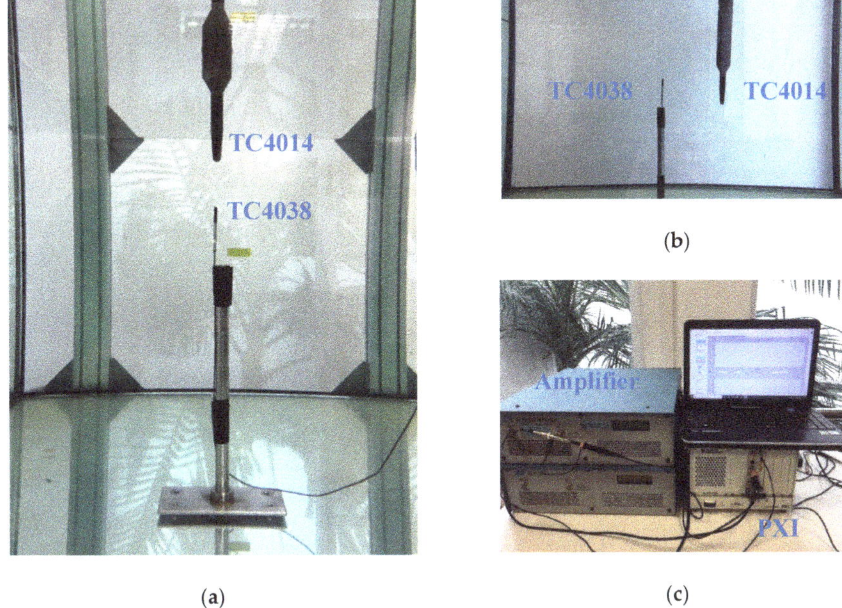

(a) (b) (c)

Figure 5. (**a**) The emitter and receiver are located as close as possible to each other to calibrate the motors; (**b**) The first position for measurements of sound speed and location; (**c**) System of generation and capture of signals.

Two experiments have been performed. First, a calibration is done to reduce the uncertainty due to the time of arrival from which the speed of the sound has been measured. For this, 12 different reception positions were assessed in a straight line on the emitter axis. By having the 12 measurements along the line, a time-distance linear adjustment is made whose slope value corresponds to the speed of sound. In addition, the calibration allows a time correction that results in a decrease of the arrival time of the signal, according to the linear adjustment obtained in the fit.

In the second part of the experiment, 12 reception points have been set that will correspond to the 12 sensor positions of the array. The mechanical axis moves the hydrophone to each of the points and then the signal is emitted and recorded 10 times. Figure 5b shows one of the measuring points where the Reson TC4014 sensor is fixed to the mechanical arm.

5. Studies and Results

5.1. Numerical Simulation

To evaluate the localization method described, the reconstruction of the location of a Gaussian pulse source of 50 µs is simulated from the reception of four sensors located on the lateral surface of different coordinates. Figure 6 shows the position of the sensors and the source in the space for the simulated model [6]. In this simulation, the sources are always into the volume covered by the coordinates of the sensors. Despite it being possible to locate the source with three sensors, the use of at least four sensors is incorporated to both improve the reliability and quality of the results.

Table 1. Positions of the sensors and the source in the simulations.

Axis	Sensors				Source (mm)		
	1	2	3	4	1	2	3
X	H/2	0.0	H/2	H	100	100	80
Y	0.0	H/2	H	H/2	100	180	100
Z	3H/4	H/2	H/2	H/4	100	150	180

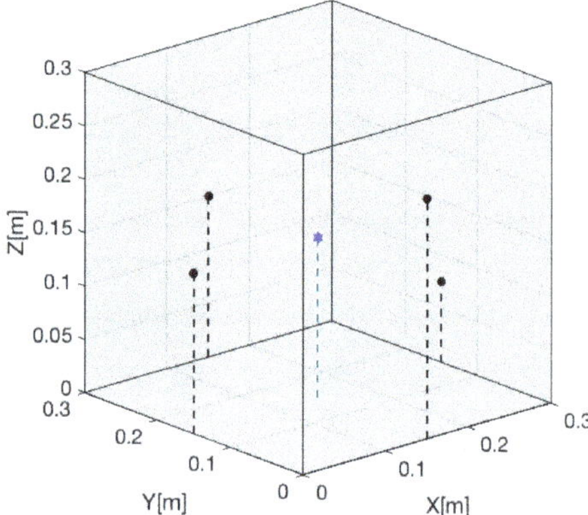

Figure 6. Volume proposed to evaluate the localization algorithm. In this case, four sensors (black points) have been situated on the sides of the cube. Inside, three events will be simulated in different positions. The positions of the sensors and sources are shown in Table 1. This figure shows the point source (blue point).

To evaluate the algorithm, the volume of the cube has been modified between 27.0×10^{-3} m^3 and 512.0×10^{-3} m^3. Positions of the sensors are shown in Table 1, where H represents the length of the cube, whose values were 200, 300, 400, 500, and 600 mm [6].

Table 2 shows the deviation of the position of the simulated source with respect to the real position of the source. The results can be expressed as a function of the distance between the reconstructed position and the real position of the source. Figure 7 shows the results, where the abscissa axis shows the volume in m^3, while the ordinates axis shows the distance in mm between the prediction of the algorithm and the real position. In addition, a fitting line to the results is shown.

Table 2. Real and estimated positions of three different sources depending on the volume case.

Coord.	Real Position (mm)	Estimated Position (mm) for the Volume (m^3)				
		27×10^{-3}	64×10^{-3}	125×10^{-3}	216×10^{-3}	343×10^{-3}
X	100	99.89 ± 0.07	100.03 ± 0.02	100.79 ± 0.55	101.97 ± 1.72	97.84 ± 1.50
Y	100	99.94 ± 0.04	100.02 ± 0.01	100.63 ± 0.45	102.26 ± 1.91	97.19 ± 2.02
Z	100	99.75 ± 0.17	99.85 ± 0.10	100.99 ± 0.70	103.90 ± 2.06	95.83 ± 2.91
X	100	100.63 ± 0.45	99.64 ± 0.03	100.19 ± 0.01	100.41 ± 0.01	98.94 ± 0.40
Y	180	179.45 ± 0.38	178.55 ± 0.10	180.68 ± 0.01	180.39 ± 0.01	179.44 ± 0.42
Z	150	150.52 ± 0.37	147.92 ± 1.55	151.68 ± 0.12	150.84 ± 0.01	148.31 ± 1.12
X	80	79.80 ± 0.13	80.87 ± 0.60	78.65 ± 1.02	78.43 ± 1.11	78.52 ± 1.01
Y	100	100.04 ± 0.03	101.04 ± 0.71	98.67 ± 0.92	96.33 ± 2.61	98.43 ± 1.10
Z	100	99.32 ± 0.47	101.81 ± 1.30	98.20 ± 1.32	97.05 ± 2.08	98.07 ± 1.42

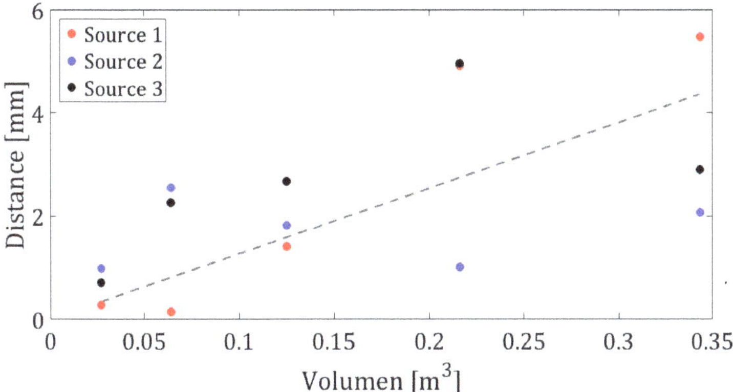

Figure 7. The positions of sources 1, 2, and 3 are shown in red, blue, and black, respectively. The dotted line is a fitting line to the results.

These reconstructed positions do not exceed 5 mm of the real position for all the studied cases. In addition, it is important to calculate the algorithm for a future application in real time. For this reason, Figure 8 shows the calculation times for 4, 6, 8, 10, and 12 sensors for a source position, depending on changes in volume.

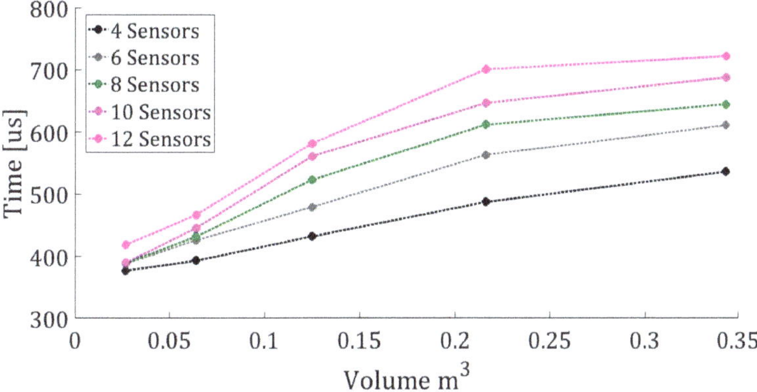

Figure 8. Calculation time with an Intel i5 processor for different sensors and volume.

Once satisfactory results of the localization algorithm have been obtained by simulating different known source points in known sensor positions, the localization method has been evaluated experimentally in the next section.

5.2. Experimental Localization with Thermoacoustic Signals

To validate the localization method, 12 reception positions (sensors) have been configured for a single source position. The signal emitted is the pulse obtained by the simulation of the thermoacoustic model, given by Equation (15). Table 3 shows the positions of the measuring points, which have all been referenced to the lower corner of the tank.

Table 3. Position of the source and sensor reception points.

Axis	Sources Position	Sensor Positions [cm]											
		1	2	3	4	5	6	7	8	9	10	11	12
X	54.0	70.5	70.5	56.5	42.5	42.5	42.5	56.5	70.5	70.5	70.5	42.5	42.5
Y	53.0	53.0	40.5	40.5	40.5	53.0	65.0	65.0	65.0	65.0	40.5	40.5	65.0
Z	38.0	31.0	31.0	31.0	31.0	31.0	31.0	31.0	31.0	43.0	43.0	43.0	43.0

Two inputs channels were used. Channel 1 recorded the emitted signal passed to the amplification system while channel 2 recorded the received signal of the emitter Reson TC4014. Both signals were used afterward for correlation. The receiver is fixed to the MOCO programmed axis that moves the hydrophone to each of the measurement positions from position 1 to position 12 in the direction shown in Figure 9. By doing this, we have the same information as when we using a 12-element sensor array.

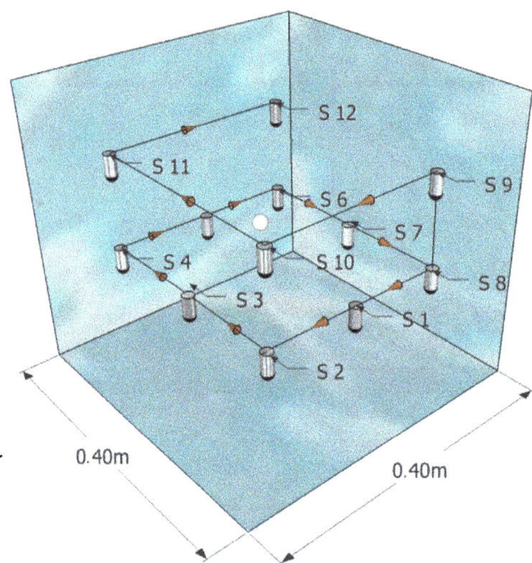

Figure 9. This figure shows the configuration of the receiver positions. The source is represented by a sphere (white) that is inside the volume generated by the sensors (cylinders of grey color with black tip) whose route is shown with the black lines and the conical red marks. In this figure, a smaller volume is represented inside the tank for its correct visualization.

The analysis of the measurements has taken into account the calibration of the measurement of the speed of sound 1492 m/s. This correction allows for a better detection precision in the TOA. In addition, the speed of the localization algorithm and the accuracy of the results will be evaluated, according to the number of reception points. Figure 10 shows an example of the signal emitted and the signal received as well as the correlation between them in order to obtain the TOA as the time of maximum amplitude of the correlation [9].

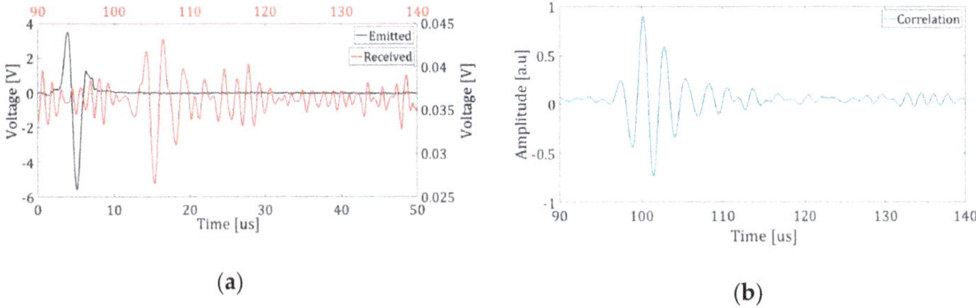

(a) (b)

Figure 10. The figure shows the process of detecting the signal from the correlation between the emitted and received signal. (**a**) The signal generated in the simulation, shown in black, is emitted by the transmitter. The captured signal for point 1 is shown in red; (**b**) The arrival time can be extracted from the maximum value of the correlation of signals.

Table 4 shows the estimated location of the source (mean and standard deviation) for different groups of sensors labelled, according to Figure 9 and Table 3. From Table 4, we observe that the results obtained with this configuration are satisfactory and below 1 mm. Furthermore, similar accurate results are also possible with fewer sensors such as four sensors surrounding the source and covering different directions isotropically. In cases that sensors are only in a specific region, larger offsets (several mm) may appear in specific coordinates.

Table 4. Reconstructed position for different groups of sensors.

Group	Number of Sensors	Sensor	Estimated Location [cm]		
			X	Y	Z
1	4	2, 4, 6, 8	53.08 ± 0.87	53.10± 0.31	30.25± 0.72
2	4	9, 10, 11, 12	53.17 ± 0.18	53.12 ± 0.25	27.31 ± 0.21
3	4	6, 8, 9, 12	54.45 ± 1.11	60.04 ± 4.61	37.79 ± 2.02
4	4	2, 4, 10, 11	53.11 ± 2.10	53.05 ± 0.50	38.28 ± 0.41
5	4	3, 4, 9, 11	53.98 ± 0.44	53.01 ± 0.44	37.98 ± 0.70
6	4	1, 4, 9, 12	53.99 ± 0.20	53.03 ± 0.70	37.97 ± 1.21
7	6	2, 5, 7, 9, 11, 12	53.99 ± 0.10	53.01 ± 0.26	38.00 ± 0.20
8	8	1, 3, 4, 6, 7, 8, 10, 11	54.01 ± 0.24	53.01 ± 0.12	38.01 ± 0.35
9	10	1, 3, 4, 5, 7, 8, 9, 10, 11, 12	54.00 ± 0.17	53.00 ± 0.14	38.01 ± 0.24
10	12	All of them	54.01 ± 0.17	53.00 ± 0.12	37.99 ± 0.38

6. Conclusions

In this paper, the proton range of the analytical medium has been used to calculate the distribution of the dose in space. Specifically, we have addressed values of time between 1 μs and 10 μs, energies between 20 MeV and 200 MeV, and protons per pulse from 3.1×10^8 to 8×10^8. In the same way, an analytical model to calculate the distribution of the initial pressure at a single point has been used with a computing program to find the numerical solution of the general wave equation and calculate the acoustic waves resulting from the initial pressure. Some simulation parameters include pulse widths, beam energy, and spatial and temporal configuration. Thus, it is possible to determine the parameters and frequency spectrum to select the frequency responses of the transmitter and the receiver for the experiment.

This study has described and tested a procedure for monitoring the location of a hadrontherapy acoustic source based on the detection of the signal through piezoelectric sensors and on a model for calculating the position of the energy deposition. The localization algorithm has been applied for different configurations of sensors. Thus, the results show a significant improvement when a greater number of sensors is used. For a minimal set of four sensors, the results are better if the sensors cover different directions of the space. The accuracy of the results improves as the number of sensors increases, as shown in Table 4. Although the calculation time increases with the number of sensors, the difference is not significant for any of the proposed cases. Thus, the results indicate that it would be possible to monitor in real time the hadrontherapy treatment acoustically. At first look, the case studied may be considered too simplistic since the human body is neither homogenous nor isotropic and sensors are not omnidirectional. Therefore, for a practical case in hadrontherapy, all these aspects should be considered when taking care of the geometry and tissues involved, the real response of sensors, etc. However, for most of the cases, the results in small differences in the technique proposed, and the situation considered in the paper is good, and most probably the best, for a first general assessment of the technique proposed.

Author Contributions: Conceptualization, J.E.O. and I.F. Investigation, J.E.O., M.A., and I.F. Methodology, I.F. and A.H. Project administration, M.A. Supervision, M.A., I.F., and A.H. Writing—review & editing, J.E.O., M.A., I.F., and A.H.

Funding: This research was funded by the Spanish Agencia Estatal de Investigación, grant number FPA2015-65150-C3-2-P (MINECO/FEDER).

Conflicts of Interest: The authors declare no conflict of interest.

References

1. Kundu, T. Acoustic source localization. *Ultrasonics* **2014**, *54*, 25–38. [CrossRef] [PubMed]
2. Moore, D.; Leonard, J.; Daniela, R.; Teller, S. Robust distributed network localization with noisy range measurements. In Proceedings of the 2nd International Conference on Enbedded Networked Sensor Systems SenSy 04, Baltimore, MD, USA, 3–5 November 2004.
3. Niculescu, D.S. Ad Hoc Positioning System (APS. In Proceedings of the Global Telecommunications Conference, San Francisco, CA, USA, 3–6 August 2001.
4. Bortfeld, T. An analytical approximation of the Bragg curve for therapeutic proton beams. *Med. Phys.* **1998**, *24*, 2024–2033. [CrossRef] [PubMed]
5. Gustafsson, F.; Gunnarsson, F. Positioning using time-difference of arrival measurements. In Proceedings of the International Conference on Acoustic, Speech and Signal Processing, Hong Kong, China, 6–10 April 2003.
6. Otero, J.; Ardid, M.; Felis, I.; Herrero, A. Acoustic location of Bragg peak for Hadrontherapy Monitoring. In Proceedings of the 5th International Electronic Conference on Sensors and Applications, 15–30 November 2018.
7. Ahmad, M.; Xiang, L.; Yousefi, S.; Xing, L. Theoretical detection threshold of the proton-acoustic range verification technique. *Med. Phys.* **2015**, *42*, 5735–5744. [CrossRef] [PubMed]
8. Knapp, C.; Carter, G. The generalized correlation method for estimation of time delay. *IEEE Trans. Acoust. Speech Signal Process.* **1976**, *24*, 320–327. [CrossRef]
9. Adrián-Martínez, S.; Ardid, M.; Bou-Cabo, M.; Felis, I.; Llorens, C.; Martínez-Mora, J.A.; Saldaña, M. Acoustic signal detection through the cross-correlation method in experiments with different signal to noise ratio and reverberation conditions. *Ad-Hoc Netw. Wirel.* **2015**, *8629*, 66–79.
10. Felis, I.; Martínez-Mora, J.; Ardid, M. Acoustic Sensor Design for Dark Matter Bubble Chamber Detectors. *Sensors* **2016**, *16*, 860. [CrossRef] [PubMed]
11. Chapra, S.; Canale, R. Raíces de ecuaciones. In *Métodos Numéricos para Ingenieros*; Médico D.F, Edifi cio Punta Santa Fe: Ciudad de México, México, 2007; pp. 142–167.
12. Bragg, W.; Kleeman, R. On the α-particles of radium and their loss of range in passing through various atoms and molecules. *Lond. Edingurgh Dublin Philos. Mag. J. Sci.* **1905**, *10*, 318–340. [CrossRef]
13. Schardt, D. Hadronteraphy. In *Basic Concepts in Nuclear Physics: Theory, Experiments and Applications*; Suiza, Springer: Berlin, Germany, 2015; pp. 55–86.
14. Janni, J.F. Energy loss, range, path lenght, time-of-flight, straggling, multiple scattering, and nuclear interaction probability: In two parts. Part 1. For 63 compounds Part 2. For elements 1 < Z < 92. *At. Data Nucl. Data Tables* **1982**, *27*, 147–339.
15. Jones, K.; Seghal, C.; Avery, S. How proton pulse characteristics influence protoacoustic determination of proton-beam range: Silumation studies. *Phys. Med. Biol.* **2016**, *61*, 2213–2242. [CrossRef]

16. Lai, H.M.; Young, K. Theory of the pulsed optoacoustic technique. *J. Acoust. Soc. Am.* **1982**, *72*, 2000. [CrossRef]
17. Sigrist, M.W. laser generation of acoustic waves in liquids and gases. *J. Appl. Phys.* **1986**, *60*, R83. [CrossRef]
18. Tam, A.C. Applications of photoacoustic sensing techniques. *Rev. Mod. Phys.* **1986**, *58*, 381–431.
19. Xiang, L.; Carpenter, C.; Pratx, G.; Kuang, Y.; Xing, L. X-ray acoustic computed tomography with pulsed x-ray beam from a medical linear accelerator. *Med. Phys. Lett.* **2013**, *40*, 010701. [CrossRef] [PubMed]
20. Assmann, W.; Kellnberger, S.; Reinhardt, S.; Lehrack, S.; Edlich, A.; Thirolf, P.G.; Moser, M.; Dollinger, G.; Omar, M.; Ntziachristos, V.; et al. Ionoacoustic characterization of the proton Bragg peak with submillimeter accuracy. *Med. Phys.* **2015**, *42*, 567–574. [CrossRef] [PubMed]
21. Bonis, G.D. Acoustic signals from proton beam interaction in water—Comparing experimental data and Monte Carlo simulation. *Nucl. Instrum. Methods Phys. Res. Sect. A* **2009**, *604*, 199–202. [CrossRef]
22. Kraan, A.C.; Battistoni, G.; Belcari, N.; Camarlinghi, N.; Cirrone, G.A.; Cuttone, G.; Ferretti, S.; Ferrari, A.; Pirrone, G.; Romano, F.; et al. Proton range monitoring with in-beam PET: Monte Carlo activity predictions and comparision with cyclotron data. *Phys. Med.* **2014**, *30*, 559–569. [CrossRef] [PubMed]
23. Patch, S.; Hoff, D.; Webb, T.; Sobotka, G.; Zhao, T. Two-stage ionoacoustic range verification leveraging Monte Carlo and acoustic simulations to stably account for tissue inhomogeneity and accelerator–specific time structure—A simulation study. *Med. Phys.* **2017**, *45*, 783–793. [CrossRef] [PubMed]
24. Lehrack, S.; Assmann, W.; Bertrand, D.; Henrotin, S.; Herault, J.; Heymans, V.; Stappen, F.V.; Thirolf, P.G.; Vidal, M.; Van de Walle, J.; et al. Submillimeter ionoacoustic range determination for protons in water at a clinical synchrocyclotron. *Phys. Med. Biol.* **2017**, *62*, 20–30. [CrossRef] [PubMed]
25. Hickling, S.; Lei, H.; Hobson, M.; Léger, P.; Wang, X.; El Naqa, I. Experimental evaluation of x-ray acoustic computed tomography for radiotherapy dosimetry applications. *Med. Phys.* **2016**, *44*, 608–617. [CrossRef] [PubMed]
26. Ardid, M.; Felis, I.; Martínez-Mora, J.A.; Otero, J. Optimization of Dimensions of Cylindrical Piezoceramics as Radio-Clean Low Frequency Acoustic Sensors. *J. Sens.* **2017**, *2017*, 8179672. [CrossRef]

Article

Towards an Autonomous Industry 4.0 Warehouse: A UAV and Blockchain-Based System for Inventory and Traceability Applications in Big Data-Driven Supply Chain Management †

Tiago M. Fernández-Caramés *, Oscar Blanco-Novoa, Iván Froiz-Míguezand Paula Fraga-Lamas *

Department of Computer Engineering, Faculty of Computer Science, Universidade da Coruña, 15071 A Coruña, Spain; o.blanco@udc.es (O.B.-N.); ivan.froiz@udc.es (I.F.-M.)

* Correspondence: tiago.fernandez@udc.es (T.M.F.-C.); paula.fraga@udc.es (P.F.-L.); Tel.: +34-981-167-000 (ext. 6051) (P.F.-L.)

† This paper is an extended version of our conference paper: Fernández-Caramés, T.M.; Blanco-Novoa, Ó.; Suárez-Albela, M.; Fraga-Lamas, P. A UAV and Blockchain-based System for Industry 4.0 Inventory and Traceability Applications. In Proceedings of the 5th International Electronic Conference on Sensors and Applications, 15–30 November 2018; Available online: https://ecsa-5.sciforum.net/.

Received: 31 March 2019; Accepted: 22 May 2019; Published: 25 May 2019

Abstract: Industry 4.0 has paved the way for a world where smart factories will automate and upgrade many processes through the use of some of the latest emerging technologies. One of such technologies is Unmanned Aerial Vehicles (UAVs), which have evolved a great deal in the last years in terms of technology (e.g., control units, sensors, UAV frames) and have significantly reduced their cost. UAVs can help industry in automatable and tedious tasks, like the ones performed on a regular basis for determining the inventory and for preserving item traceability. In such tasks, especially when it comes from untrusted third parties, it is essential to determine whether the collected information is valid or true. Likewise, ensuring data trustworthiness is a key issue in order to leverage Big Data analytics to supply chain efficiency and effectiveness. In such a case, blockchain, another Industry 4.0 technology that has become very popular in other fields like finance, has the potential to provide a higher level of transparency, security, trust and efficiency in the supply chain and enable the use of smart contracts. Thus, in this paper, we present the design and evaluation of a UAV-based system aimed at automating inventory tasks and keeping the traceability of industrial items attached to Radio-Frequency IDentification (RFID) tags. To confront current shortcomings, such a system is developed under a versatile, modular and scalable architecture aimed to reinforce cyber security and decentralization while fostering external audits and big data analytics. Therefore, the system uses a blockchain and a distributed ledger to store certain inventory data collected by UAVs, validate them, ensure their trustworthiness and make them available to the interested parties. In order to show the performance of the proposed system, different tests were performed in a real industrial warehouse, concluding that the system is able to obtain the inventory data really fast in comparison to traditional manual tasks, while being also able to estimate the position of the items when hovering over them thanks to their tag's signal strength. In addition, the performance of the proposed blockchain-based architecture was evaluated in different scenarios.

Keywords: UAV; drones; Industry 4.0; traceability; blockchain; inventory; supply chain management; logistics; RFID; smart contracts; DLT; IPFS

---

1. Introduction

The concept Industry 4.0 fosters the evolution of traditional factories towards smart factories through the use of some of the latest advances in paradigms and technological enablers like big data [1], augmented reality [2–5], robotics [6], cyber–physical systems [7], fog computing [8,9] or the Industrial Internet-of-Things (IIoT) [10]. For instance, the application of advanced Big Data analytics in supply chain management can help to improve decision making for all activities across the supply chain. In particular, in inventory tasks [11], big data can help organizations to design modern inventory optimization systems, predict inventory needs, respond to changing customer demands, reduce inventory costs, obtain a holistic view of inventory levels, optimize the flow and storage of inventory, or even reduce safety stock.

Unmanned Aerial Vehicles (UAVs) are also considered a key technology for smart factories, since they allow carrying out repetitive and dangerous tasks without almost any human intervention or supervision. In the last years, UAVs proved to be really useful in fields like remote sensing (e.g., mining), real-time monitoring, disaster management, border and crowd surveillance, military applications, delivery of goods, precision agriculture, infrastructure inspection or media and entertainment, among others [12,13]. In many of such fields, UAVs perform tasks that constitute one of the foundations of Industry 4.0: to collect as much data as possible from multiple locations dynamically. In addition, UAVs not only collect data, but are also able to store, process and exchange information with suppliers or with devices deployed in factories.

Industry 4.0 technologies have to be integrated horizontally so that manufacturers and suppliers can cooperate. In order for a company to dynamically determine its need for supplies, it is necessary to keep track of their stock. For such a purpose, many companies carry out a periodic inventory and decide whether more supplies have to be purchased. Unfortunately, in many companies, such an inventory is performed manually, making it a really costly, time-consuming and tedious task. There exists software to automate stock control, but when it is operated by humans, the process is prone to accounting errors and it is usually not carried out in real time. Therefore, the ideal inventory should be performed automatically, in real-time and in an efficient, flexible and safe way.

UAVs have been applied to inventory and traceability tasks in the past. For instance, some of the latest commercial UAV-based systems [14–16] make use of a scanner installed on a UAV that performs a predefined flight in order to read barcodes. In the literature, there are more ambitious solutions like the one presented in [17], which describes an autonomous UAV that makes use of Radio-Frequency IDentification (RFID) and self-positioning/mapping techniques based on a 3D Light Detection and Ranging (LIDAR) device.

Blockchain and other Distributed Ledger Technologies (DLTs) are also predicted to be essential for many industries [18,19], since they allow for storing the collected data (or a proof of such data) so that they can be exchanged in a secure way among entities that do not trust each other. Although blockchain can be considered to be still under development in many aspects [20,21], some of its applications for fields where trust is a required (e.g., finance) have been already deployed [22]. Moreover, blockchain and other DLT enable the creation of smart contracts, which can be defined as self-sufficient decentralized codes that are executed autonomously when certain conditions of a business process are met. Thus, the code of a smart contract translates into legal terms the control over physical or digital objects through an executable program. For instance, a smart contract may be used as a sort of communication mechanism with a supplier when certain materials run low and one expects more incoming work that would require them.

Specifically, this article includes the following contributions:

- This is one of the first articles where design and testing of an RFID-based drone for performing inventory tasks are described. In fact, we have not found in the literature any other solution that makes use of a similar identification technology.

- Besides recent literature on blockchain-based autonomous business activity for UAVs [23], to our knowledge, this article is the first that proposes the theoretical design and practical implementation of a UAV-based, versatile, modular and scalable architecture aimed at fostering cyber security (specially, data integrity and redundancy) by including together the use of blockchain and a decentralized database solution. Specifically, the proposed system can use a blockchain to receive the inventory data collected by UAVs, validate them, ensure their trustworthiness and make them available to the interested parties. Moreover, the system is able to use smart contracts to automate certain processes without human intervention.
- We also evaluated the performance of the proposed decentralized database and the implemented smart contracts in diverse scenarios.

The rest of this paper is structured as follows. Section 2 reviews the previous work related to the use of UAVs in industrial applications, as well as the state-of-the-art of technologies for industrial inventory and traceability. Section 3 details the architecture of the proposed system, while Section 4 describes its implementation. Finally, Section 5 is devoted to the experiments and Section 6 to the conclusions.

2. Related Work

The solution proposed and evaluated in this article requires four essential elements to keep product traceability and obtain the inventory of a warehouse:

- A labelling technology: items need to be attached to tags or labels that are associated with a unique identifier and, in some cases, with additional information on the item.
- An identification technology: the labels or tags attached to the items have to be read remotely to automate the inventory processes.
- A UAV: although most labelling technologies make use of handheld readers, this article proposes the use of a UAV to automate and to accelerate data gathering.
- Supply management techniques: data gathering, processing and storing processes need to be efficient when handling a relevant number of information.

The most relevant aspects of these four elements are first studied and then analyzed in the following sub-sections in order to later determine the main components of the designed system.

2.1. Labelling and Identification Technologies

Currently, one of the most popular identification technologies is barcodes, which are essentially a visual representation of Global Trade Item Number (GTIN) codes [24]. However, barcode readers need Line-of-Sight (LoS) with the barcode label to read it correctly and it can only be read at relatively short distances (just a few tens of centimeters). Despite the mentioned limitations, barcodes improved inventory speed in industrial scenarios with respect to traditional manual identification procedures. In addition, barcode label manufacturing cost is really low, since they only require specific software and a printer.

In the middle of the 1990s, barcodes evolved towards bidimensional codes known as BiDimensional (BiDi) or Quick Response (QR) codes, which are able to store data (usually more than 1800 characters) and can be read with a simple smartphone camera, thus, reducing the overall cost of the inventory/traceability system. Nonetheless, QR codes can only be read up to a distance that depends on the size of the QR marker (the reading distance is often estimated to be less than ten times of the QR code diagonal).

Since both barcodes and QR codes are limited by their reading distance and their need for LoS between the reader and the code, they evolved towards more sophisticated labels (Figure 1 illustrates such an evolution). Labels based on RFID are considered the next step in the inventory and traceability system evolution [25]. Such a kind of label can be read at a distance that goes from several centimeters to several tens of meters [26]. In fact, reading distance is commonly related to the type of RFID tag: passive tags (i.e., tags that do not depend on batteries for carrying out RFID communications) reading distance usually does not exceed 20 m, while active tag communications (i.e., the ones that do use batteries for carrying out RFID communications) can easily reach 100 m in unobstructed environments. In addition, certain RFID tags can store information, which may be useful in certain inventory and traceability processes. It must be also noted that, in the last years, academia and industry have evolved RFID tags in order to add to them sensing capabilities, thus, creating tags that are able to measure temperature [27], acceleration [28] or light [29].

Figure 1. Technological evolution of identification tags.

The next step in the tag evolution are the so-called smart labels [30], which are noticeably more complex than RFID tags and allow for providing industry 4.0 functionality like event detection [31], operator interactivity, a display for visual feedback, one or more communication technologies, positioning services or embed IoT sensors that collect environmental data that are relevant for the state of a product. Omni-ID's View smart labels [32] are one of the few on the market and embed an e-ink display, flash memory storage, active and passive UHF RFID transceivers, and an infrared receiver for beacon-based positioning.

Besides the identification technologies related to the previously mentioned labelling solutions, there are others that have been suggested. For instance, Near Field Communication (NFC) [33] has been used in inventory applications, but it is only dedicated to scenarios where a short reading distance (often less than 30 cm) is needed.

There are other technologies, which, have been actually conceived as communications technologies, but can also provide identification mechanisms. For instance, Bluetooth Low Energy (BLE) is a generic Wireless Personal Area Network (WPAN) technology that usually does not reach more than 10 m when used through smartphones or up to 100 m in industrial applications [34]. There are diverse BLE profiles and specific devices called beacons that can be useful in certain inventory and traceability applications, as well as in indoor locations [35,36] and IoT applications [37,38], since they transmit a periodic signal to be located and identified.

The Media Access Control (MAC) address of a WiFi device (i.e., a device compatible with the IEEE 802.11 family of standards, including its vehicular version [39,40]) can also be used for identification purposes, as well as other technologies less frequently used in inventory tags, like devices based on ultrasounds, infrared communications, ZigBee [41], Low-Power Wide-Area Network (LoRA) [42], Dash7 [43], Ultra Wide Band (UWB), WirelessHART [44], RuBee (IEEE standard 1902.1), SigFox [45], ANT+ [46] or IEEE 802.11ah, among others.

As a summary, the main characteristics of the most relevant identification technologies for inventory and traceability applications are shown in Table 1, which specifies their frequency band, usual maximum range, data rate, power consumption, relevant features and some examples of applications that have made use of each technology.

Table 1. Main characteristics of the most relevant communications and identification technologies for inventory and traceability applications.

Technology	Frequency Band	Max. Range in Optimal Conditions	Data Rate	Power type	Main Features	Main Limitations for Inventory Applications	Popular Applications
ANT+	2.4 GHz	30 m	20 kbit/s	Ultra-low power	Up to 65,533 nodes	Lack of commercial inventory tags	Health, sport monitoring
Barcode/QR	–	<4 m	–	No power	Very low cost, visual decoding	Need for LOS	Asset tracking and marketing
Bluetooth 5 LE	2.4 GHz	<400 m	1360 kbit/s	Low power	Batteries only last days to weeks	Batteries need to be recharged, shared communications radio frequency	Beacons, wireless headsets
DASH7/ISO 18000-7	315–915 MHz	<10 km	27.8 kbit/s	Very low power, alkaline batteries last months to years	Long reading distance, multi-year battery	Batteries need to be recharged, shared communications radio frequency	Smart industry and military
HF RFID	3–30 MHz (13.56 MHz)	a few meters	<640 kbit/s	No power	NLOS, no need for batteries	Relatively short reading range	Smart Industry, payments, asset tracking
Infrared (IrDA)	300 GHz to 430 THz	a few meters	2.4 kbit/s–1 Gbit/s	Low power	Low-cost hardware, security, high speed	Need for LOS, batteries may drain fast when transmitting continuously	Remote control, data transfer
IQRF	433 MHz, 868 MHz or 916 MHz	hundreds of meters	19.2 kbit/s	Low power	Long communications range	Shared communications radio frequency	Internet of Things and M2M applications
LF RFID	30–300 KHz (125 KHz)	<10 cm	<640 kbit/s	No power	NLOS, low cost	Very short reading distance (in general, a few centimeters)	Smart Industry and security access
NB-IoT	LTE in-band, guard-band	<35 km	<250 kbit/s	Low power	Long reading range	Dependent on third-party infrastructure	IoT applications
NFC	13.56 MHz	<20 cm	424 kbit/s	No power	Low cost	Short reading distance	Ticketing and payments
RuBee	131 KHz	20 m	8 kbit/s	Very low power	Magnetic propagation, multi-year battery life	Only one known manufacturer	Applications with harsh electromagnetic propagation
LoRa/LoRaWAN	2.4 GHz	kilometers	0.25–50 kbit/s	Low power	Long range, long battery life	Very few commercial inventory tags, more expensive than other alternatives	Smart cities, M2M applications
SigFox	868–902 MHz	50 km	100 kbit/s	Low power	Long range, global cellular network	Dependent on third-party infrastructure	Internet of Things and M2M applications
UHF RFID	30 MHz–3 GHz	tens of meters	<640 kbit/s	Very low power or no power	NLOS, wide range of suppliers, low cost	Propagation problems with metal and liquids (specially with high transmission frequencies)	Smart Industry, asset tracking and toll payment
Ultrasounds	>20 kHz (2–10 MHz)	<10 m	250 kbit/s	Low power	Based on sound wave propagation	Relatively short reading range	Asset positioning and location
UWB/IEEE 802.15.3a	3.1 to 10.6 GHz	< 10 m	>110 Mbit/s	Low power (batteries last hours to days)	Accurate positioning (centimeter accuracy)	Expensive hardware, propagation problems in metallic environments	Real Time Location Systems (RTLS), short-distance streaming
Wi-Fi (IEEE 802.11b/g/n/ac)	2.4–5 GHz	<150 m	up to 433 Mbit/s (one stream)	High power (batteries may last hours)	High speed, ubiquity	Short battery life	Internet access, broadband
Wi-Fi HaLow/IEEE 802.11ah	868-915 MHz	<1 km	>100 Kbit/s per channel	Low power	Long communications range	Not compatible with previous Wi-Fi standards, shared communications radio frequency	IoT applications
WirelessHART	2.4 GHz	<10 m	250 kbit/s	Low power (Batteries last several years)	Compatibility with HART protocol, standardized as IEC 62591	Shared communications radio frequency, lack of commercial inventory tags	Wireless sensor network applications
ZigBee	868–915 MHz, 2.4 GHz	<100 m	20–250 kbit/s	Very low power (batteries last months to years)	Easy to scale, up to 65,536 nodes	Relatively expensive hardware, potential interference from devices in the same frequency band	Smart Home and industrial applications

2.2. UAVs for Inventory and Traceability Applications

UAVs have been suggested for providing services in a number of industrial applications, like for critical infrastructure inspections [47–51] or sensor monitoring [52–54]. It is important to note that industrial environments require certain UAV features that may differ from other applications, like obstacle collision avoidance [55], automatic cargo transport [56,57] or logistics optimization for swarms of UAVs [58].

UAVs have already been used for traceability and inventory management applications. For example, in [59], the authors use a QR code-based UAV to detect items. Although the system shows an impressive precision of 98.08%, it requires LoS with QR codes, which in practice limits its application to certain scenarios. Another interesting article is [60], where UWB devices are used to accurately position (i.e., with an indoor location accuracy of only 5 cm) a UAV that performs inventory management tasks.

RFID is the identification technology embedded in a UAV in [61]. In such a paper, the scenario is an open storage yard that is monitored by a DJI Phantom 2 drone that carries a passive UHF reader embedded by a PDA. Researchers from MIT have also conceived a similar theoretical system where multiple UAVs carry RFID readers to aid inventory automation in a warehouse [62]. Finally, it is worth mentioning the work in [63], which proposes the collaboration of an Unmanned Ground Vehicle (UGV) and a UAV to perform inventory tasks in a warehouse. Specifically, the UGV is used as a ground reference for the indoor flight of the UAV, which embeds a camera that recognizes augmented reality markers. Thus, the UGV carries the UAV to places where there are items to be inventoried and then the UAV flies vertically to scan them, sending the collected data to a ground station.

2.3. Big Data for Inventory and Supply Chain Management

Big Data analytics can also be applied to optimize knowledge extraction and decision-making in inventory and supply management applications. A comprehensive review on the application of big data to such fields can be found in [64], with the management of safety stocks being one of the key aspects [65]. Moreover, other authors have already analyzed how to face the current challenges and opportunities that the transformation of supply chain management suppose in order to be driven by big data techniques [66].

With respect to practical deployments, in [67], the authors propose a big data analytics framework that processes the information collected from an RFID-enabled shop floor. The authors define, at a logistics management level, different Key Performance Indicators (KPIs) in order to evaluate different manufacturing objects and the main findings are converted into managerial guidance for decision making.

Regarding the use of big data analytics for inventory applications, several advantages can be pointed out [11]: it can help to take informed decisions related to inventory performance, it may aid to obtain a holistic view at inventory levels across the supply chain and with external stakeholders, and it enables to predict inventory needs and changes in customer demand using statistical forecasting techniques, as well as to reduce inventory costs dramatically.

2.4. Analysis of the State of The Art

Table 2 shows a comparison of the most relevant characteristics of the previously cited UAV-based inventory systems together with the ones of the system proposed in this article. As it can be observed after reviewing the different aspects of the state-of-the-art, it is possible to highlight several important shortcomings that motivated this article.

Table 2. Comparison of the main features of the most relevant UAV-based inventory systems and the proposed system.

Reference	Type of Solution	Labelling and Identification Technology	UAV Characteristics	Designed Architecture and Communications	Main Inventory Function	Experiments and Key Performance Indicators (KPIs)	Advanced Supply Management Data Techniques	Blockchain or Any Other DLT
[14]	Commercial solution by Hardis Group	Barcodes	Autonomous quadcopter with a high-performance scanning system and an HD camera. Battery life around 20 min (50 min to charge it).	It incorporates indoor localization technology. Automatic flight area and plan, 360° anti-collision system.	Automate inventory-taking and inventory control in warehouses	No available KPI	Automatic acquisition of photo data. Cloud applications to manage mapping, data processing, reporting, and the fleet of drones. Compatible with all WMS and ERPS and managed by a tablet app.	No DLT
[15]	Commercial solution, Geodis and delta drone	Barcodes	Autonomous quadcopter equipped with four HD cameras	Indoor geolocation technology, it operates autonomously during the hours the site is closed.	Plug and play solution, this solution also adapts to all types of Warehouse Management Systems (WMS)	Reading rates close to 100%.	Enables the counting and reporting of data in real time, the processing of data, and its restitution in the warehouse's information system.	No DLT
[16]	Commercial solution, Dron Scan	Barcodes	UAV equipped with a camera and a mounted display	DroneScan base station communicates via a dedicated RF frequency (not WiFi or Bluetooth) and has a range of over 100 m.	A Windows touch screen tablet allows the operator to receive live feedback both on screen and from audible cues as the drone scans and records data	50 times faster than manual capturing	All aspects of the imported data are customizable by modifying scripts, the customisation changes the way the system works and how the scanned data are processed. DroneScan software uploads scanned data and drone position information to the cloud (Azure IoT), to the customer systems (web services, RFC's API's or BAPI's) and exports the data to Excel. The imported data are used to re-build a virtual map of the warehouse so that the location of the drone can be determined.	No DLT
[17]	Academic solution	RFID, multimodal tag detection.	Autonomous Micro Aerial Vehicles (MAVs), RFID reader and two high-resolution cameras	Fast fully autonomous navigation and control, including avoidance of static and dynamic obstacles in indoor and outdoor environments.	Robust self-localization solely based on an onboard LIDAR at high velocities (up to 7.8 m/s)	-	-	No DLT
[19]	Academic solution	-	-	-	Endogenous risk management mechanism to improve supply chain's operation efficiency	Theoretical pharmaceutical factory supply chain topology structure based on blockchain.	Confront supply chain endogenous risk avoiding the credit risk caused by the information asymmetry among the enterprises inside the supply chain, and the risk caused by incomplete information acquisition inside the supply chain.	Blockchain and smart contracts
[23]	Academic solution	-	-	-	Autonomous economic system with UAV.	Although field trials were conducted with drones, no KPIs are available.	Architectural solution for organizing a business activity protocol for multi-agent systems.	Communication system between agents (DAOs) in a P2P network using Ethereum and smart contracts.
[59]	Academic solution	QR	IR-based camera, no additional description	Computer vision techniques (region candidate detection, feature extraction, and SVM classification) for barcode detection and recognition in factory warehouses.	Drone-assisted inventory management with an efficient detection framework to determine the localizations of 2D barcodes to improve path planning and reduce power consumption	Experiment performance results of 2D barcode images. The proposed method demonstrates a precision of 98.08% and a recall of 98.27% within a fused feature ROC curve.	-	No DLT
[60]	Academic solution, open-source code of the UWB hardware and MAC protocol software.	QR	-	Plug-and-play capabilities and minimal pre-existing infrastructure by combining two wireless technologies: sub-GHz for IoT-standardized long-range wireless communication backbone and UWB for localization.	A MAC protocol for an UWB localization system using battery-powered or energy harvesting operated anchors.	Experimental validation for two real-life scenarios: autonomous drone navigation in a warehouse mock-up and tracking of runners in sport halls. Theoretical evaluation of the design choices on overall system performance in terms of update rate, energy consumption, maximum communication range, localization accuracy and scalability.	-	No DLT
[61]	Academic solution	RFID	Phantom 2 vision DJI (weight 1242 g, maximum speed 15 m/s and up to 700 m)	Drone with a Windows CE 5.0 portable PDA (AT-880) that acts as a UHF RFID reader moves around an open storage yard.	Inventory checking in an open stock yard	Prototype. No performance experiments.	A data collection program detects and saves the information of passive tags obtained by a portable PDA. After the flight, the gathered tag data is transferred to the inventory checking server and is compared with the inventory data stored in database and classified in to four inventory states: normality, location error, missing, unregistered.	No DLT

Table 2. *Cont.*

Reference	Type of Solution	Labelling and Identification Technology	UAV Characteristics	Designed Architecture and Communications	Main Inventory Function	Experiments and Key Performance Indicators (KPIs)	Advanced Supply Management Data Techniques	Blockchain or Any Other DLT
[62]	Academic solution	RFID (EPC)	Draganfly commercial radio-controlled helicopters 82 × 82 cm, average flight time of 12 min	RFID readers attached to the simulated UAVs are assumed to have a 100% read-guarantee when EPC tags are within the reading range of the RFID reader.	Read the EPCs in the warehouse within the 12-min duration	Preliminary simulation results, three-dimensional graphical simulator framework has been designed using Microsoft XNA framework to represent a real warehouse	Coordinated distribution of the UAVs. Although six independent UAVs were deployed, they collectively failed to complete the task of finding all EPCs	No DLT
[63]	Academic solution	Barcodes, AR markers	UAVs and UGVs with LIDARs	UAV and UGV work cooperatively using vision techniques. The UGV acts as a carrying platform and as an AR ground reference for the indoor flight of the UAV. While the UAV is used as the mobile scanner.	Novel indoor warehouse inventory scheme to improve automation as well as the diminution of time consumption and injuries risks.	Experimental setup is to validate the visual guidance of the UAV taking the UGV as a ground. UAV need to be equipped with sensors to avoid collision with the racks during the scanning process.	-	No DLT
[67]	Academic solution	RFID	-	-	Physical Internet-based intelligent manufacturing shop floors	Experiments on logistics rules for optimizing the delivery time	Big data analytics framework that processes the information collected from an RFID-enabled shop floor	No DLT
Proposed system	Academic solution	RFID	Indoor/outdoor hexacopter designed from scratch as a trade-off between cost, modularity, payload capacity and robustness.	Modular and scalable UAV-architecture using WiFi infrastructure and ability to run decentralized applications.	Enable inventory and traceability applications focused on a holistic view at inventory levels across the supply chain and with external stakeholders	Prototype and performance experiments: inventory time in the warehouse under different circumstances, signal strength monitoring and performance of the implemented architecture (decentralized database and blockchain response latency)	Data distribution and enhanced cyber security (information integrity, tamper-proof data, ensured reliability and availability), efficient data storage and data versioning.	Decentralized database (OrbitDB) over InterPlanetary File System (IPFS) in a P2P network using Ethereum and smart contracts to automate certain processes.

First of all, few papers detail practical UAV-based applications for inventory and traceability applications, specially with RFID as the identification technology embedded in the UAV. Specifically, most of them do not take into consideration in the UAV design a trade-off between cost, modularity, payload capacity and robustness. Second, the emergence of Big Data analytics impose additional issues to emerging solutions that are still open for research like:

- Data volume. The amount of heterogeneous data generated by Industry 4.0 technologies has increased exponentially, especially the information collected from Industrial Internet of Things (IIoT) devices and remote entities. These data have to be hosted, distributed and computed across a number of organizations ensuring a certain degree of operational efficiency.
- Speed. Decisions should be made as quickly as possible (ideally, in real time), so there is a need for speeding up the processes related to them: data production, data collection, reliability, data transfer speed, data storage efficiency, knowledge extraction and analysis, as well as decision-making.
- Verification and veracity. There is a myriad of factors that may derive into collecting bad data (e.g., noise, inaccurate readings), so they should be verified so that only valid or true data are further processed.
- Versioning. Massive datasets should be linked and the accidental disappearance of important data should be prevented.
- Accessibility. Stakeholders must be able to access data through resilient networks, which enable persistent availability with or without Internet connectivity.

Although different alternatives have been proposed for facing some of the issues mentioned above, DLTs seem to be the most promising solution in order to guarantee operational efficiency and data transparency, authenticity and security. However, there is a shortcoming related to the fact that, although there are examples in the literature of DLTs applied to other sectors, to the knowledge of the authors, there are no examples of DLTs or blockchain-based architectures for the upcoming big data-driven supply chain applications.

Finally, it must be noted that the use of current DLTs, and specifically blockchain, involve certain weaknesses related to the immature status of the technology, like the lack of scalability, high energy consumption, low performance, interoperability risks or privacy issues [18–22]. Moreover, blockchain technologies face design limitations in transaction capacity (i.e., throughput and latency), in validation protocols or in the implementation of smart contracts. Therefore, architecture design should consider these constraints together with the specific decentralization requirements. Furthermore, as of writing, there are several open issues that require more research. For instance, no references were found in the literature on the evaluation of the performance of blockchain-based application architectures.

As a consequence of the previous analysis, the design of the system proposed in this article has been devised taking the previous shortcomings into consideration.

3. Design of the System

Figure 2 depicts the proposed communications architecture. In such an architecture, a UAV carries a Single-Board Computer (SBC) and a tag reader. The tag reader is used for collecting data from wireless tags that are attached to the items to be inventoried or traced. Regarding the SBC, it collects tag data from the tag reader, processes them and sends them through a wireless communications interface to a ground station. Then, to provide enhanced cybersecurity, the ground station can make use of its internal software modules to send the collected information to two possible destinations: to a decentralized remote storage network or to a blockchain.

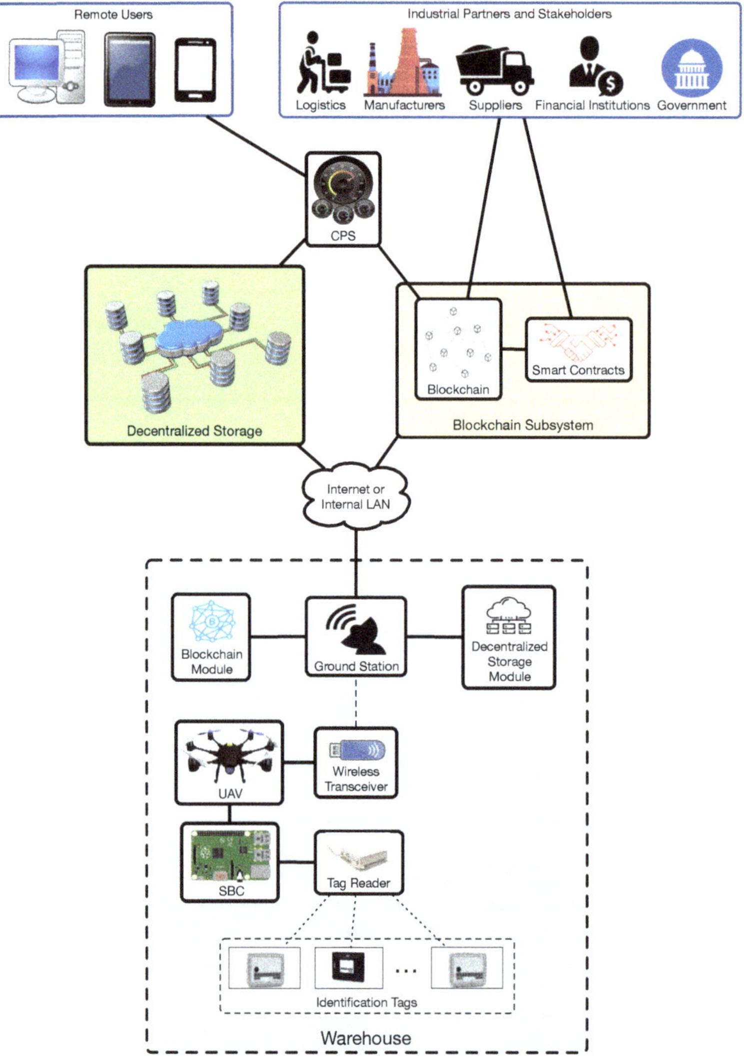

Figure 2. Proposed communications architecture.

In the case of sending the data to a blockchain, the ground station makes use of a software module that acts as a blockchain client. Therefore, the SBC is able to store in a secure way the collected data (or their hashes) into the remote blockchain, which also allows the proposed system to participate in smart contracts that enables the direct participation of suppliers, manufacturers, retailers or logistics operators [68]. It is important to note that very similar features may be provided by traditionally centralized solutions like databases, cloud-based services or complex-event processing software, but blockchain technologies include the following features that make them really attractive for industrial scenarios where third parties may be interested in carrying out audits or financial evaluations, or may need to verify the fulfillment of certain laws [21]:

- Transparency. Blockchain and smart contracts allow for providing access to inventory and traceability information to third-parties, which can monitor it and determine whether it has been tampered with.
- Application decentralization. Medium and large software deployments usually require to make use of centralized servers that are often expensive to deploy and maintain [69,70]. Blockchain is one of the technologies that enable application decentralization and, at the same time, avoids the involvement of middlemen that provide outsourced centralized solutions.
- Data authenticity. In many industries, it is essential to trust the inventory and traceability data received from suppliers, manufacturers or governments. Blockchain enables to implement mechanisms that provide accountability and, as a consequence, trustworthiness. In addition, it is worth noting that such data trustworthiness is essential for the effectiveness of the application of Big Data techniques.
- Data security. Blockchain allows for preserving the privacy and anonymity of the data exchanged with other entities, so that they remain private to non-authorized parties.
- Operational efficiency. Blockchain technologies are able to automate the verification of the attributes of a transaction in an inexpensive way.

With respect to the decentralized storage, it is included in the architecture in order to provide the following three main advantages that useful for enhancing the security of the inventory and traceability data:

- Redundancy. Decentralized storage systems, when properly synchronized, are able to create copies of the stored data automatically in multiple nodes, so the information is not available from a single source that may constitute a point of failure.
- Tolerance to cyber-attacks. Since the information can be available from multiple data storage nodes, if one or several of them are taken down by cyber-attacks (e.g., Denial of Service (DoS) or Distributed Denial of Service (DDoS) attacks), it may be possible to access it through the other nodes.
- Ability to run decentralized applications (dApps). Besides pure file storage, it is possible to develop and deploy dApps that provide the previously mentioned features (i.e., redundancy and increased security). In addition, it is possible to develop dApps that run on decentralized storage systems while cooperating with a blockchain [71].

Both the blockchain and the decentralized storage system essentially manage the same data types as in other traditional inventory solutions: sets of unique alphanumeric identifiers (UIDs). Thus, item UID association and information processing (e.g., to determine the number of items available of one specific type) need to be performed by an additional software layer that may be implemented through a Cyber-Physical System (CPS). Moreover, note that, although remote users may access the collected raw information directly from one of the nodes of the decentralized storage network or through a blockchain browser, it is usually more practical to make use of a user-friendly interface like the one provided by the CPS. Thus, a CPS collects and processes the data needed by remote users and presents them in a way that can be easily understood by remote users [72].

4. Implementation

4.1. UAV Implementation

UAVs vary widely in size, materials, components and configuration. In the design of the proposed UAV, the main objective was to develop a cost-effective, simple and modular initial prototype that can be easily adapted to different applications, scenarios and/or performance criteria.

Figure 3 depicts the main components of the designed UAV, while Table 3 shows a summary of their characteristics. A multipurpose UAV was designed with both indoor and outdoor flight capabilities in order to extend its use to different applications and maximize its exploitation opportunities. The chosen configuration offers a good trade-off between cost, payload capacity and reliability; an hexacopter configuration has the minimum number of motors in order to be able to recover and land in the case of a motor failure without damaging people or goods. More rotors would increase the reliability even further but would also increase the cost and its total weight.

Specifically, the UAV is mounted on an hexacopter frame of 550 mm of diameter mostly made out of carbon fiber except for the arms, which are made of plastic reinforced by carbon fiber rods in the interior. The flight controller is a PixHawk 2.4.8 that has been flashed with the well-known open-source firmware Ardupilot [73] The thrust to move the UAV is generated by six 920 Kv brushless motors controlled by six 30 A Electronic Speed Control (ESCs), which are powered by a four-cell 5 Ah Li-Po battery that also provides power to all the on-board electronics through a voltage conversion module. Besides the built-in sensors of the flight controller board, a UBLOX M8N GPS module was included to enable outdoor autonomous flights since in this industrial use case, there are storage areas outdoors in a nearby dock next to the warehouse where inventory can also be performed.

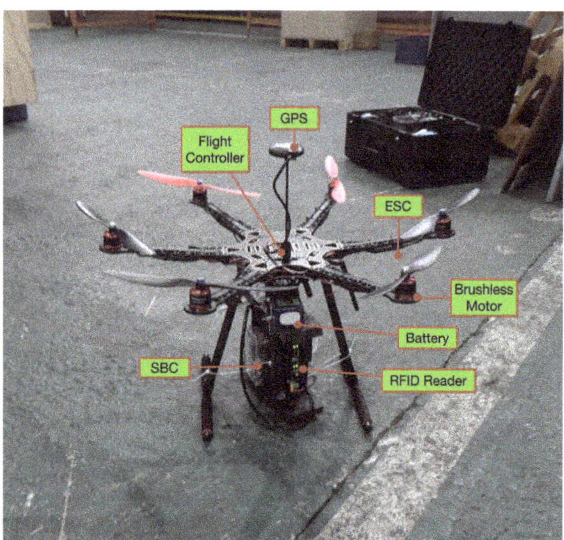

Figure 3. UAV used for the inventory and traceability system.

Table 3. Main features of the UAV components.

Components	Relevant Features
Flight controllers	Pixhawk 2.4.8 STM32F427 microcontroller STM32F103 coprocessor
Sensors	L3GD20 3-axis digital gyroscope LSM303D 3-axis accelerometer and magnetometer MPU6000 6-axis accelerometer and magnetometer MS5607 barometer GPS M8N
RFID reading system	NPR Active Track-2 OrangePI PC Plus (SBC)
Additional components	Frame with six arms (550 mm of wingspan) Brushless motors 920 Kv ESCs Simonk 30 A Propellers: 10 inch-diameter and 45 inch-pitch Battery: 5 Ah (capacity) and 45 c-rate (discharge rate)

4.2. Implemented Architecture

Figure 4 shows the actual implemented architecture of the proposed system. It can be observed that it is very similar to Figure 2, but, since this article is focused on assessing the performance of the UAV-based inventory system, it includes certain simplifications on aspects that are not evaluated (i.e., on the CPS).

As it can be observed in Figures 3 and 4, in order to perform the inventory, an RFID reader (NPR Active Track-2) is carried by the hexacopter. Specifically, active Ultra High Frequency (UHF) RFID has been selected as item identification technology due to its performance in environments with multiple metallic obstacles [26]. The RFID reader has been modified to reduce its weight by replacing its steel case with a lighter one made of foam, which protects the reader and reduces vibrations. In addition, it is worth pointing out that, as it can be observed in Figure 3, the RFID reader antennas were placed as far as possible from the hexacopter propellers, which can influence communications performance. Nonetheless, the location of the antennas may be further optimized.

Regarding the used RFID tags, they are active UHF tags from RF-Code [74] that can be read with the NPR Active Track 2 reader (e.g., Active Rugged Tag-175S). Such a reader is connected through an Ethernet cable to the SBC. The tag UIDs collected by the SBC are first stored locally in a JavaScript Object Notation (JSON) file and then sent through WiFi to the ground station server, which, for the experiments performed in this article, it was run in a laptop.

Figure 4 shows a high-level view of the implemented blockchain-based architecture. As it can be observed in the Figure, to persist the collected inventory data, the proposed system makes use of the decentralized database OrbitDB [75], which runs over InterPlanetary File System (IPFS) [76]. IPFS provides a high throughput content-addressed block storage model with content-addressed hyper links. This forms a generalized Merkle Directed Acyclic Graph (DAG), a data structure upon which one can build versioned file systems, blockchains, and even a Permanent Web. Specifically, the main features of IPFS are:

- It keeps every version of the files, making it simple to set up resilient networks.
- It removes duplication across the network, therefore saving in storage. Each network node stores only content it is interested in and some indexing information that helps to figure out who is storing what.
- Each file and all of the blocks within it are given a unique fingerprint (i.e., cryptographic hash).

- Every file can be found in an user-friendly way (e.g., by human-readable names) using a decentralized naming system called Inter-Planetary Name System (IPNS).
- It has no single point of failure and nodes do not need to trust each other.

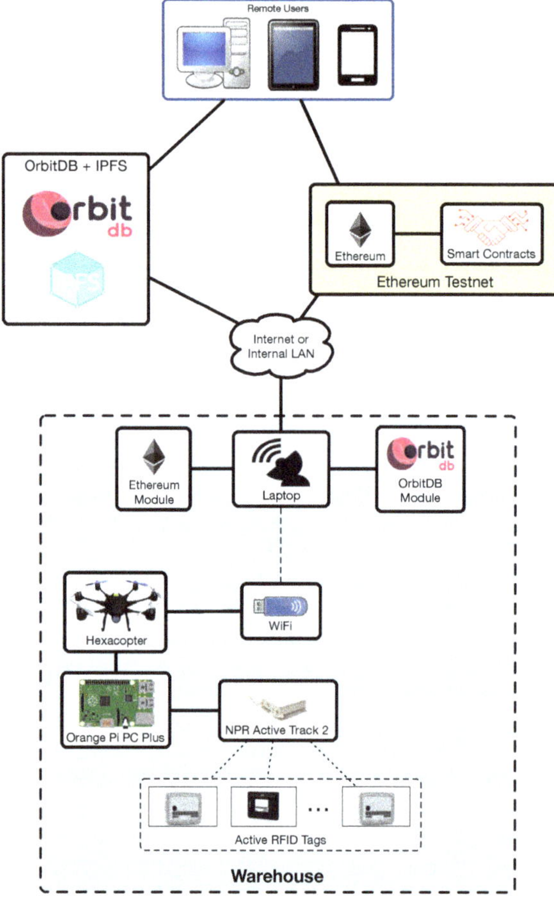

Figure 4. Implemented architecture.

OrbitDB is built on top of IPFS to create a serverless, distributed, peer-to-peer database in alpha-stage software. It uses IPFS as its data storage, as well as IPFS Pubsub to synchronize databases with peers automatically. In addition, OrbitDB provides various types of databases for different data models and use cases:

- log: an immutable (append-only) log with traceable history. Useful for "latest N" use cases or as a message queue.
- lfeed: a mutable log with traceable history. Entries can be added and removed.
- keyvalue: a key-value database.

- docs: a document database where JSON documents can be stored and indexed by a specified key. Useful for building search indices or version controlling documents and data.
- counter: Useful for counting events separate from log/feed data.

All the previously mentioned types of databases are implemented on top of ipfs-log, an immutable, operation-based Conflict-free Replicated Data Structure (CRDT) for distributed systems. If none of the OrbitDB database types match a dApp needs or if a specific functionality is needed, it allows for implementing a custom database easily.

With respect to the blockchain, two different Ethereum testnets (i.e., test networks that are isolated from the public Ethereum blockchain [77]) were used: Rinkeby [78] and Ropsten [79]. Rinkeby is a Proof-of-Authority (PoA) testnet created by the Ethereum team that makes use of the Clique PoA consensus protocol, where authorized signers are responsible for minting the blocks. In such a network blocks are created on average every 15 s and Ether cannot be mined (it is requested through a faucet [80]). In contrast, Ropsten is a Proof-of-Work (PoW) Ethereum testnet where Ether can be either mined or requested from a faucet [81]. Ropsten's blocks are usually minted in less than 30 s and, although the testnet reproduces with more fidelity than Rinkeby Ethereum's mainnet production environment, it is prone to Denial-of-Service (DoS) attacks (e.g., by increasing the block gas limit remarkably while sending large transactions through the network), which makes synchronization slow and makes clients consume a lot of disk space. Rinkeby and Ropsten testnets allow for executing smart contracts (compiled and deployed through Truffle [82]), which store in a string the JSON file with the inventory data and its hash, so the blockchain acts both as a immutable log and as a timestamping server.

In the implemented architecture, remote users are able to access the raw OrbitDB and Ethereum data, but, thanks to the proposed architecture, it is straightforward to develop a backend that collects data from OrbitDB and Ethereum, and presents them through a frontend.

4.3. Inventory Data Insertion and Reading Processes

Figure 5 illustrates all the components that are involved in the implementation of the part the architecture related to decentralized storage and the blockchain. Besides the previously mentioned components (i.e., OrbitDB, IPFS, the Ethereum testnets and Truffle), three another elements are used:

- Infura [83]: it provides an easy to use HTTP API for accessing Ethereum that can be even implemented by resource-constraint IoT devices.
- Node.js [84]: it is an open-source platform that allows for executing JavaScript code outside of a browser.
- Web3 Javascript API [85]: It is a collection of libraries that allow for interacting with Ethereum nodes. In the proposed architecture, the Web3 API is called from a Node.js instance that exchanges requests that are handled by Infura.

Figure 5 also shows the steps performed when inserting inventory data in OrbitDB and Ethereum. Before starting to store data, it is assumed that all the involved components are up and running, and that the different OrbitDB nodes synchronize their data periodically (this is illustrated by steps 0A and 0B, which indicate that every node is synchronized through a publish/subscribe scheme). Moreover, it is also assumed that the smart contracts haven been deployed through Truffle in one of the Ethereum's testnets. Then, when a ground station wants to upload new inventory data, it first appends them to OrbitDB, which returns a hash as an acknowledgement of the transaction. Such an OrbitDB hash, together with the inventory data and the Infura authentication token, are sent to Infura through a data insertion request. If the authentication token is valid, Infura would take the inventory data and the OrbitDB hash, and would execute the setData function of the smart contract, so that it updates the stored inventory data values and the OrbitDB hash. As a result of this operation, Ethereum returns a hash that can be used later to request the blockchain transaction information.

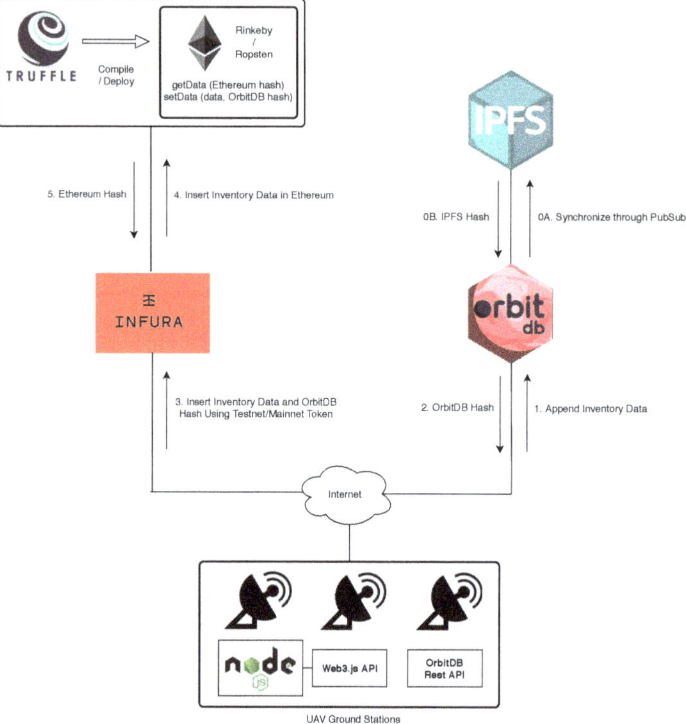

Figure 5. Inventory data insertion process and implemented architecture.

Figure 6 illustrates the steps required by an entity that wants to verify that the data stored in OrbitDB have not being altered after their insertion. In such a case, the entity only would need to first request the inventory data from OrbitDB and them perform through Infura a *GetData* request to the corresponding Ethereum's smart contract, which would return the inventory information stored on the blockchain and its OrbitDB hash. Then, the entity would only need to compare the inventory information collected from OrbitDB with the one from Ethereum, and determine whether it has been modified.

As it can be concluded from the proposed architecture and processes, they enable data trustworthiness through four different mechanisms:

- Information integrity can be verified by checking its hash. In addition, Ethereum and OrbitDB act as timestamping services, so it can be easily verified when the data were inserted.
- Since the data stored on the blockchain cannot be tampered without leaving a trace, the authenticity of the inventory information stored on OrbitDB can be easily checked.
- Every user transaction within OrbitDB is protected by asymmetric cryptography mechanisms that make use of a public and a private key.

- Similarly, the data that is exchanged with Ethereum are managed by Infura, which protects them through an API key and a secret key.

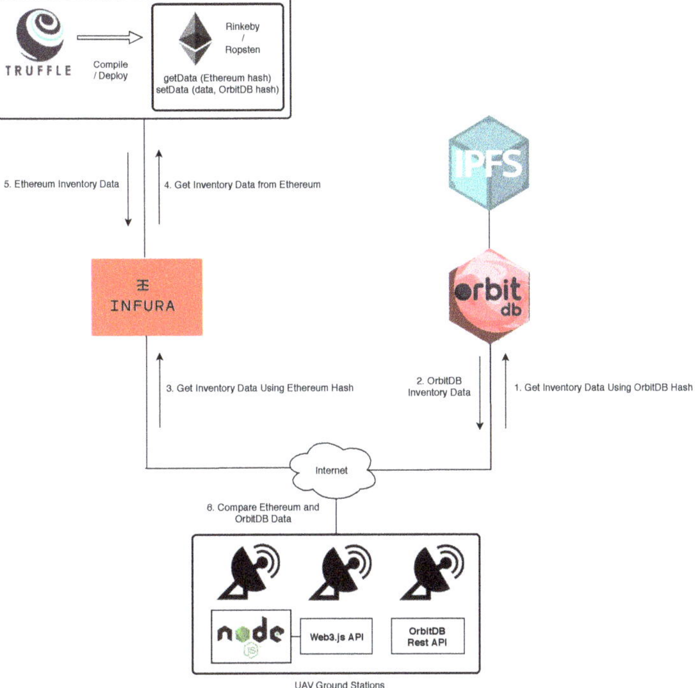

Figure 6. Inventory data verification process.

5. Experiments

5.1. Experimental Scenario

In order to evaluate the proposed system, it was tested in a big industrial warehouse that is shown in Figure 7. The warehouse is approximately 120 m long and 40 m wide, but, due to security reasons (the warehouse was in operation during the experiments), the tags were deployed in an isolated subarea of 50 m × 40 m.

As it can be observed in Figure 7, the warehouse stores items that are packaged into wooden, plastic or cardboard boxes. There are also items that are not packaged and, thus, only a plastic inventory label is attached to them. Due to such an item diversity, in order to provide a fair estimation of the performance of the proposed system, the deployed tags were attached to items made out of diverse materials, some of which are shown in Figure 8.

Figure 7. Warehouse where the experiments were performed.

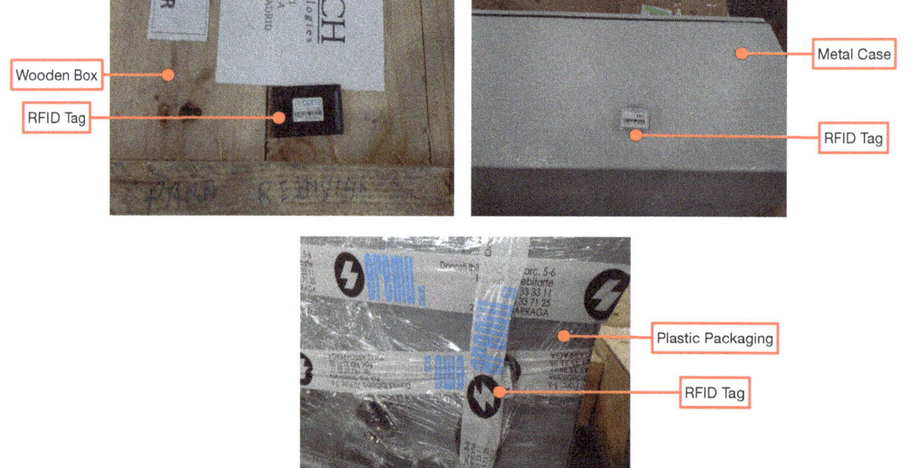

Figure 8. Warehouse material diversity.

A total of 13 different tags were attached to items scattered throughout the previously mentioned isolated area of the warehouse that was monitored with the drone. Figure 9 shows the location of the ground station during the tests, while Figure 10 illustrates one of the moments during the experiments. As it can be observed in Figure 10 and in the video of the Supplementary Materials, during the tests the drone was operated in manual mode in order to avoid possible physical security problems. Such an operation is expected to be automated thanks to the use of different on-board sensors and the placement of multiple image-based or RFID markers in the environment. In the same way, while during the tests the operator moved the drone to follow a circular path over the monitored area, in the future such a path will be prefixed by software through indoor waypoints.

Figure 9. Ground station setup.

Figure 10. One of the instants during the inventory tests.

5.2. Inventory Time

The firsts tests were carried out in order to determine how fast the inventory data could be collected. Figure 11 shows the percentage of read tags through time for four different tests. During such tests the tags remained at the same location and were attached to the same items. Considering the nature of the proposed experiments and the RFID reader reading range, the pilot experience and skills can be considered negligible in the results obtained.

An example of the collected data is shown in Table 4, which is related to Test 2 of Figure 11. Specifically, the table indicates the number and percentage of read inventory tags, and the time stamp collected by the drone every time a new tag is detected (its ID is indicated in the last column).

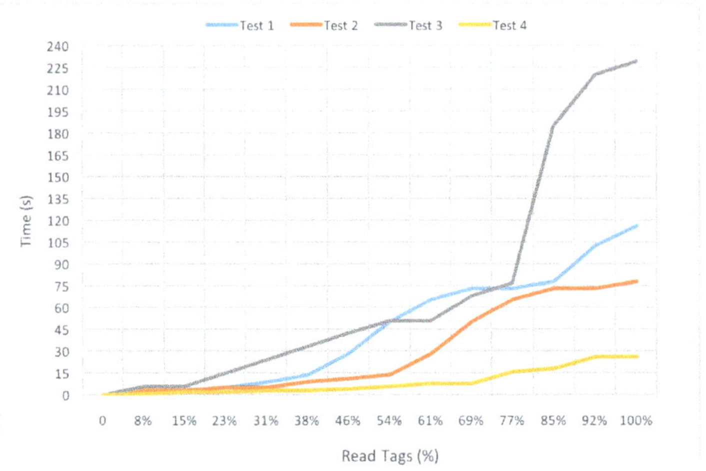

Figure 11. Percentage of read tags during four inventory flights.

Table 4. Example of the collected inventory data.

# Read Tags	% Read Tags	Timestamp (HH:MM:SS,ms)	New Read Tag ID
0	0	18:14:43,087 (Take-off time)	
1	7.692307692	18:14:46,058	LOCATE00380349
2	15.38461538	18:14:46,090	RFCBDG00011185
3	23.07692308	18:14:48,558	LOCATE00380364
4	30.76923077	18:14:48,589	RFCBDG00011185
5	38.46153846	18:14:52,748	LOCATE00380372
6	46.15384615	18:14:54,349	LOCATE00380349
7	53.84615385	18:14:57,129	RFCBDG00011188
8	61.53846154	18:15:11,403	LOCATE00380330
9	69.23076923	18:15:33,008	LOCATE00365573
10	76.92307692	18:15:49,288	LOCATE00375358
11	84.61538462	18:15:56,454	LOCATE00380359
12	92.30769231	18:15:56,456	LOCATE00380357
13	100	18:16:01,630	LOCATE00375356

During Tests 1 and 2, the drone departed from the same spot and followed a similar path. This can be observed in Figure 11: the curves for both tests are very similar. Nonetheless, the inventory was gathered faster during Test 2: it only took approximately 78 s to collect the data, which is less than the 116 s needed during Test 1.

Although the tags were scattered randomly throughout the monitored area of the warehouse (with the only restriction of attaching them to items of the three most common materials), the departing point of the drone is essential for accelerating the inventory collection. This can be observed in the curves of Figure 11 related to Tests 3 and 4. In the case of Test 3 the departing point was located in the opposite side of the beginning of the monitored area, where several tags were placed. This derived into requiring a total of 229 s to collect all the UIDs of the inventoried items. However, in Test 4 the drone was roughly in the middle of the monitored area, thus accelerating remarkably their detection (it only took 26 s to complete the inventory).

Except for Test 3 (due to its specific location), in the other cases, mainly because of the reading range and the omni-directional antennas of the reader, during the first 11 s (as the drone rose from the ground), it was possible to read roughly 30% of the tags. Moreover, in all tests, 50% of the items where inventoried in less than a minute and all of them in less than four minutes. These results are really promising, since the time required by a human operator to collect the same information is far greater than when using the proposed UAV-based system (the operator needs to walk through the area, locate the items and identify them manually).

5.3. Signal Strength Monitoring

As the drone hovers above the warehouse, besides tag UIDs, it also collects the Signal Strength Indicator (SSI) of such tags, which can potentially be used for locating industrial items and for creating signal strength maps [86].

An example of the SSI fluctuation perceived by the drone for one of the RFID tags during an inventory flight is shown in Figure 12. It can be observed that while the drone is hovering around the warehouse, in locations far from the tag, the collected SSIs fluctuate between −50 and −62 dBm. However, once the drone is close to the tag, SSI levels go up to around −40 dBm and, as the drone moves away from the tag, the collected SSI values go down again to be between −50 and −62 dBm. Therefore, if the location of the drone can be obtained through an indoor positioning system, due to the existent correlation, tag locations may be estimated at the same time as the inventory data are collected. However, it must be noted that, in order to design an indoor positioning system for industrial environments, additional factors should be considered, like the reflections, diffraction and refraction caused by surrounding materials, the presence of hostile electromagnetic sources, certain features of the scenario (e.g., presence of metals, water, exposure to liquids, acids, salinity, fuel or other corrosive substances, tolerance to high temperatures) or the actual reading distance, among others [26].

Finally, it is worth pointing out that just after flying over the tag there were several seconds (roughly between seconds 120 and 160) during which the tag SSIs were not received by the UAV, what was probably caused by a signal blockage related to the presence of large metallic item in the scenario. This must be taken into account in future developments in order to determine the optimal tag and item positions to maximize SSI reception.

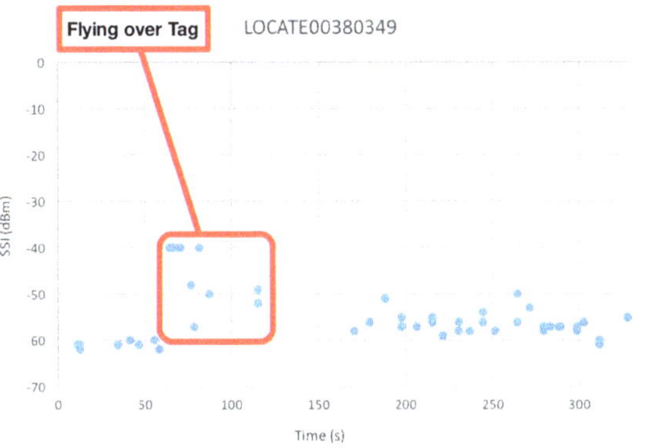

Figure 12. SSI evolution of a tag during an inventory flight.

5.4. Performance of the Implemented Architecture

5.4.1. Performance of the Decentralized Database

The performance of OrbitDB was measured in terms of response latency (i.e., how fast the inventory data were inserted into OrbitDB after being sent by a UAV ground station). Three scenarios (named A, B and C) were simulated in order to evaluate the effect of payload size and network delay, so small, medium and large payloads were sent to OrbitDB when it was running either in the same Intranet (i.e., the same private local network) as the ground station, and when it was running in a remote cloud on the Internet (i.e., for simulating the connection with other stakeholders such as suppliers or external audits). As a reference, it can be indicated that the minimum/average/maximum round-trip times to the machine that ran OrbitDB in the Intranet were 0.935/1.034/1.695 ms, while the same values for the connection between the ground station and the remote cloud were 37.948/39.362/51.172 ms. Scenarios A, B and C take place inside an Intranet. Scenario A corresponds exactly to the environment tested in Section 5.2: the inventory data were only 13 Tag IDs, which derived into a very small payload (around 1 KB). Scenarios B and C simulate inventory data of 5000 (around 30 KB) and 10,000 (around 67 KB) items, respectively. Regarding Scenarios D, E and F, they also simulate 13, 5000 and 10,000 items, respectively, but the inventory data is sent from the local network to a remote OrbitDB instance that runs on a cloud on the Internet.

These scenarios resulted into the six use cases characterized in Table 5, which shows, for each scenario, the number of inventoried tags, as well as the mean and variance of the obtained OrbitdB response latency. In addition, Figures 13 and 14 show the response latency for the six evaluated scenarios and for 2000 insertion requests per scenario when measuring the time elapsed since each request is issued, until the OrbitDB hash is obtained (i.e., from step 1 to step 2 of Figure 5).

Table 5. Mean and variance of the use cases considered in the OrbitDB performance test.

# Tag IDs/Network	Intranet	Internet
13	Scenario A: μ: 0.1576 s; σ^2: 0.0039 s	Scenario D: μ: 0.3648 s; σ^2: 0.0114 s
5000	Scenario B: μ: 0.1669 s; σ^2: 0.0032 s	Scenario E: μ: 0.5063 s; σ^2: 0.0265 s
10,000	Scenario C: μ: 0.1892 s; σ^2:0.0049 s	Scenario F: μ: 0.5553 s; σ^2: 0.0093 s

The results show that, in terms of response delay, OrbitDB insertions are really fast: in the worst tested case (Scenario F), the average inventory data insertion time requires roughly 0.55 s. In addition, it can be pointed out that, as expected, the larger the payload, the slower the insertion. However, the difference in response delay between the insertion of 13 and 10,000 tag IDs is negligible when OrbitDB runs on the Intranet (only 31.6 ms), while, in the same case but when OrbitDB was running of the remote cloud, the difference rose up to 190.5 ms. Moreover, at the view of the low variance values, it can be observed that response delay is very stable in both networks, although, obviously, it is clearly more stable for Scenarios A, B and C (i.e., on the Intranet).

Regarding Figures 13 and 14, they seem to show a clear average response time, although such a time oscillates depending on different factors, like the network load (i.e., the network is actually shared with other users). Despite such oscillations, it is possible to obtain the Probability Density Function (PDF) of the different response delays. Figure 15 shows the PDFs for the three Intranet scenarios, which follow a Generalized Extreme Value (GEV) distribution with different values for the parameters of location (μ), scale (σ) and shape (k) [87]. In the case of the Internet scenarios, Figure 16 shows that they seem to follow a multimodal distribution (e.g., bimodal), although in the Figure it was fitted to a kernel distribution [88]. Therefore, the mentioned PDFs may be used in the future to model the behavior of the different scenarios and then generate artificial samples from the fitted distributions to perform Monte Carlo simulations to test the theoretical load that can be supported by the proposed system.

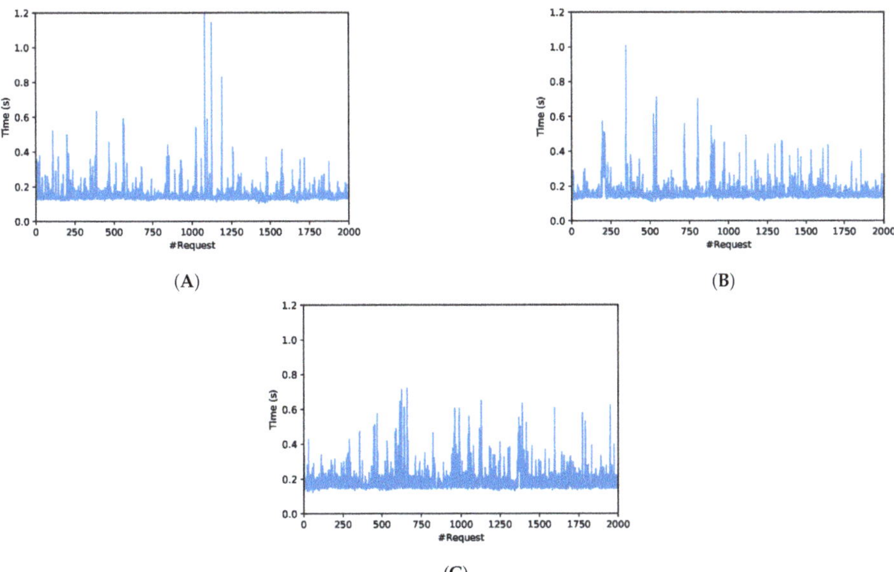

Figure 13. Response time in OrbitDB for (**A**) Scenario A, (**B**) Scenario B and (**C**) Scenario C.

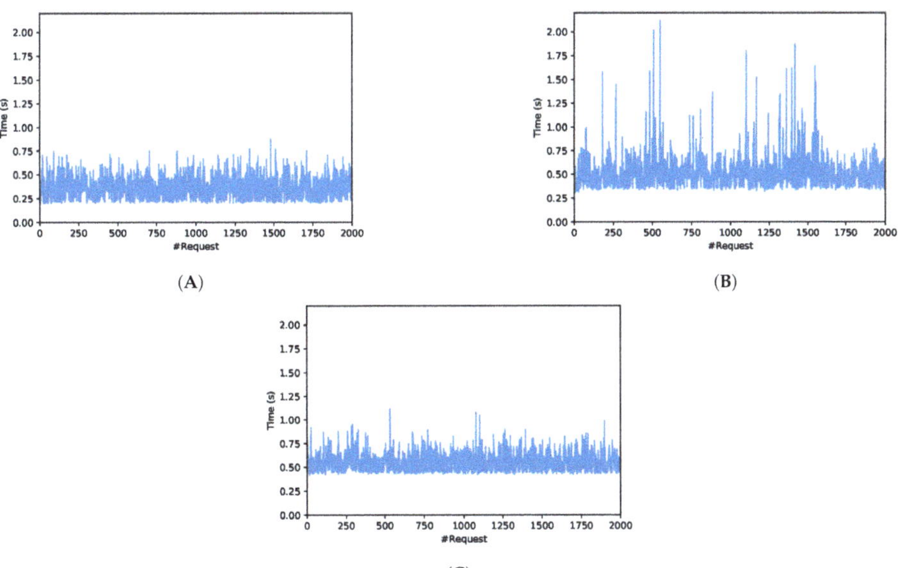

Figure 14. Response time in OrbitDB for (**A**) Scenario D, (**B**) Scenario E and (**C**) Scenario F.

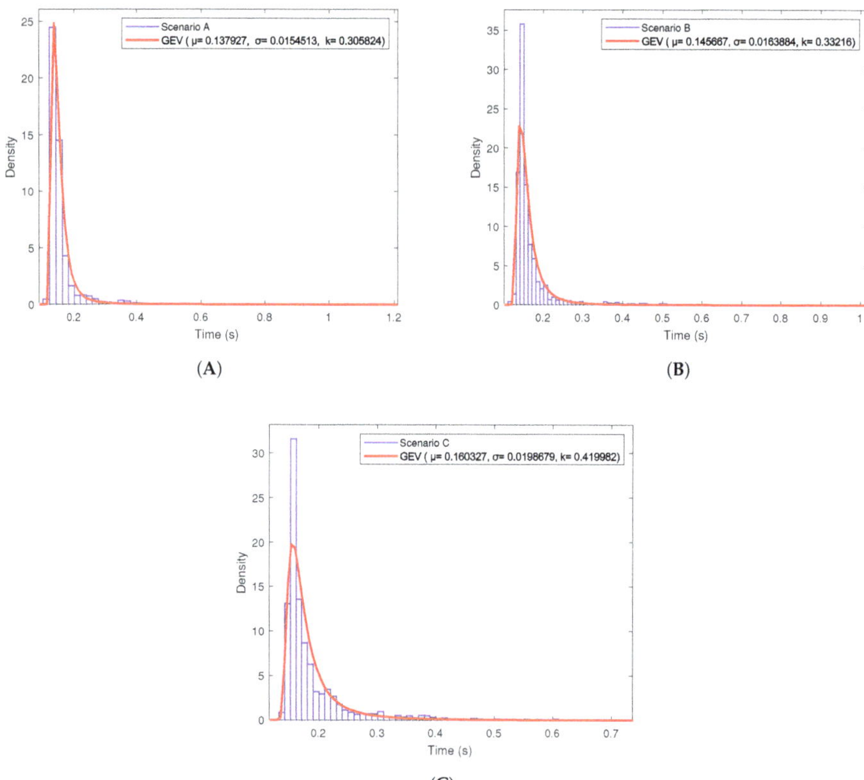

(A) (B)

(C)

Figure 15. Probability Density Function (pdf) in OrbitDB for (**A**) Scenario A, (**B**) Scenario B and (**C**) Scenario C.

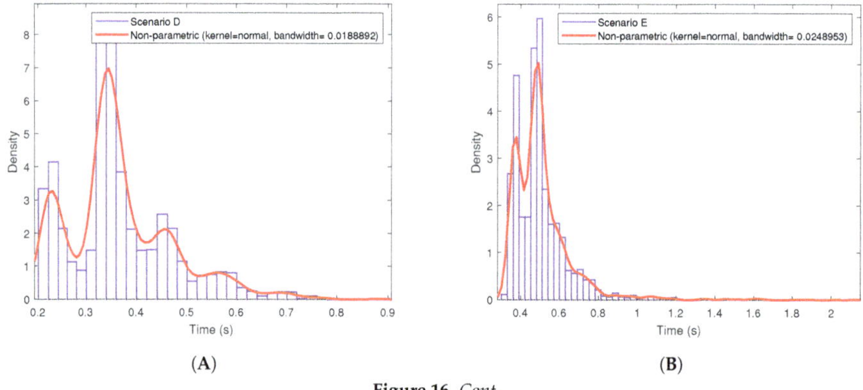

(A) (B)

Figure 16. *Cont.*

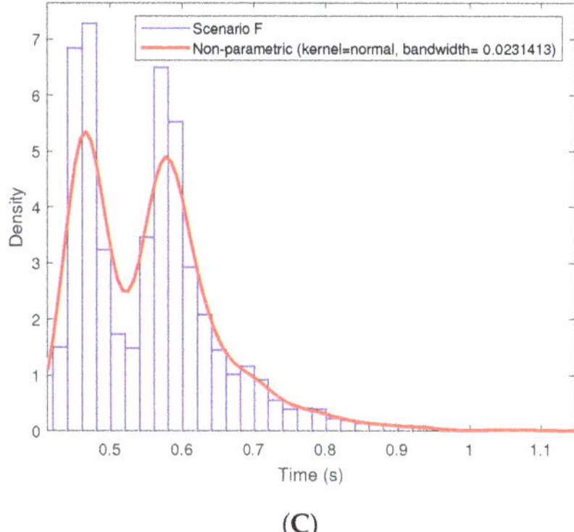

(C)

Figure 16. Probability Density Function (pdf) in OrbitDB for (**A**) Scenario D, (**B**) Scenario E and (**C**) Scenario F.

5.4.2. Performance of The Blockchain

It is also interesting to measure the blockchain response latency in order to detect a possible bottleneck of the architecture. While Rinkeby is a PoA testnet whose time to mint a block is set to an average of 15 s, the Ropsten testnet is based on PoW and thus the same time may vary noticeably.

Figure 17 shows the response latency of almost 100 transactions (9.4 KB) in the Ropsten testnet during a smart contract update of the proposed system. Specifically, the elapsed time is measured since the transaction is included in a block until it is validated (from step 3 to step 5 in Figure 5). As it can be observed in Figure 17, the response time varies significantly from less to 5 s up to more than 70 s, being the delays significantly higher than in the OrbitDB performance tests.

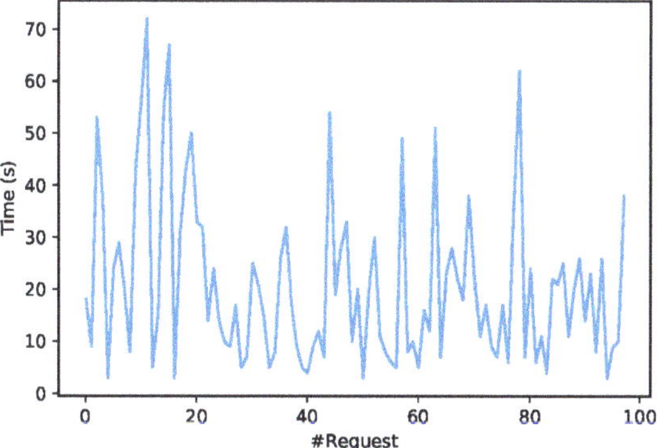

Figure 17. Ropsten testnet time response.

Figure 18 shows a caption of Etherscan where the specific details of the transactions performed on the smart contract used for the tests performed in this Section (0x3A99AFac4A32b29C17 Aeb7d0B4E3C4F28EA200c7). The details of the transaction include the account balance, the performed transactions, the addresses of the involved parties and the number of miners. Additional details are available on https://ropsten.etherscan.io/address/0x3a99afac4a32b29c17aeb7d0b4e3c4f28ea200c7.

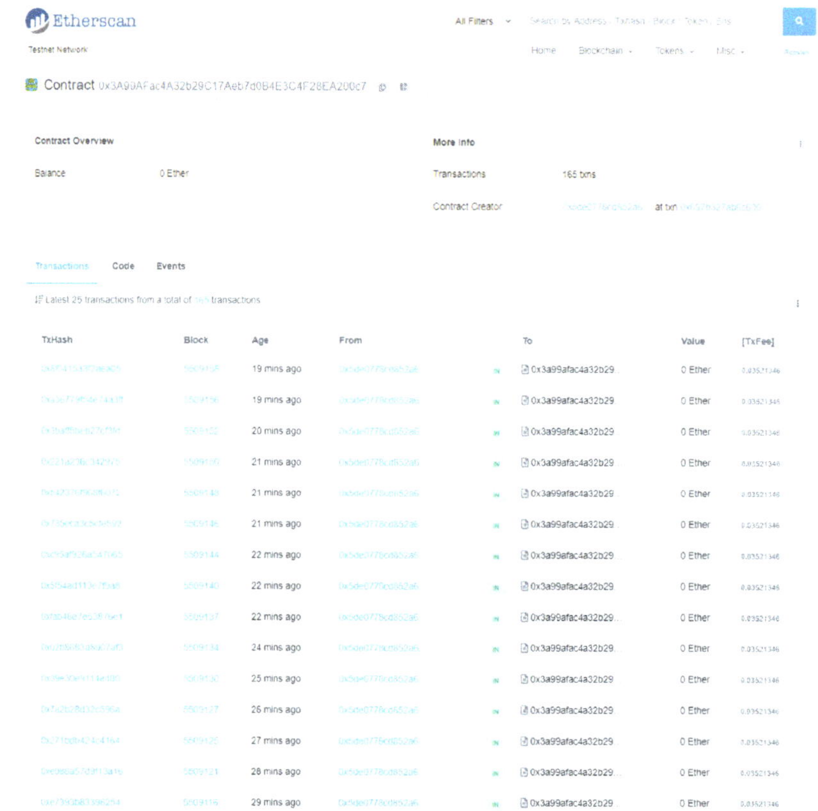

Figure 18. Caption of Ropsten Testnet transactions.

6. Conclusions

This article presented the design, implementation and evaluation of an UAV and blockchain-based system for Industry 4.0 inventory and traceability applications. After reviewing the most relevant initiatives related to industrial UAV applications and identification technologies, the architecture and components of the proposed UAV and RFID-based system were detailed. Such a system is able to collect and process inventory data in real-time and send them to a blockchain and to a decentralized storage network for providing enhanced cyber security, redundancy and the ability to run decentralized applications. Moreover, the system was able to use smart contract to automate certain processes without human intervention. The proposed system was tested in a real warehouse and the obtained results show that it is able to collect inventory data remarkably faster than a human operator and that it is possible to locate items in the warehouse by using their SSI. Furthermore, the performance of the proposed blockchain-based architecture was evaluated in diverse scenarios.

Supplementary Materials: The following video is available online at http://www.mdpi.com/1424-8220/19/10/2394/s1, Video S1: Blockchain-based UAV performing a warehouse inventory.

Author Contributions: T.M.F.-C. and P.F.-L. conceived and designed the system; O.B.-N. built the drone, O.B.-N., T.M.F.-C., I.F.-M. and P.F.-L. performed the experiments; P.F.-L. and T.M.F.-C. wrote the paper.

Funding: This work has been funded by the Xunta de Galicia (ED431C 2016-045, ED431G/01), Centro singular de investigación de Galicia accreditation 2016-2019 through the CONCORDANCE (Collision avOidaNCe fOR UAVs using Deep leArNing teChniques and optimized sEnsor design) project, the Agencia Estatal de Investigación of Spain (TEC2016-75067-C4-1-R) and ERDF funds of the EU (AEI/FEDER, UE).

Conflicts of Interest: The authors declare no conflict of interest.

References

1. Qi, Q.; Tao, F. Digital Twin and Big Data Towards Smart Manufacturing and Industry 4.0: 360 Degree Comparison. *IEEE Access* **2018**, *6*, 3585–3593. [CrossRef]
2. Blanco-Novoa, Ó.; Fernández-Caramés, T.M.; Fraga-Lamas, P.; Vilar-Montesinos, M.A. A Practical Evaluation of Commercial Industrial Augmented Reality Systems in an Industry 4.0 Shipyard. *IEEE Access* **2018**, *6*, 8201–8218. [CrossRef]
3. Wohlgemuth, W.; Triebfürst, G. ARVIKA: Augmented Reality for development, production and service. In Proceedings of the DARE 2000, Elsinore, Denmark, 12–14 April 2000.
4. Friedrich, W. ARVIKA-Augmented Reality for Development, Production and Service. In Proceedings of the International Symposium on Mixed and Augmented Reality, Darmstadt, Germany, 30 September–1 October 2002; pp. 3–4.
5. Fraga-Lamas, P.; Fernández-Caramés, T.M.; Blanco-Novoa, Ó.; Vilar-Montesinos, M.A. A Review on Industrial Augmented Reality Systems for the Industry 4.0 Shipyard. *IEEE Access* **2018**, *6*, 13358–13375. [CrossRef]
6. Robla-Gómez, S.; Becerra, V.M.; Llata, J.R.; González-Sarabia, E.; Torre-Ferrero, C.; Pérez-Oria, J. Working Together: A Review on Safe Human-Robot Collaboration in Industrial Environments. *IEEE Access* **2017**, *5*, 26754–26773. [CrossRef]
7. Fernández-Caramés, T.M.; Fraga-Lamas, P.; Suárez-Albela, M.; Díaz-Bouza, M.A. A Fog Computing Based Cyber-Physical System for the Automation of Pipe-Related Tasks in the Industry 4.0 Shipyard. *Sensors* **2018**, *18*, 1961. [CrossRef]
8. Bonomi, F.; Milito, R.; Zhu, J.; Addepalli, S. Fog Computing and its Role in the Internet of Things. In Proceedings of the First Edition of the MCC Workshop on Mobile Cloud Computing, Helsinki, Finland, 13–17 August 2012; pp. 13–16.
9. Fernández-Caramés, T.M.; Fraga-Lamas, P.; Suárez-Albela, M.; Vilar-Montesinos, M. A Fog Computing and Cloudlet Based Augmented Reality System for the Industry 4.0 Shipyard. *Sensors* **2018**, *18*, 1798. [CrossRef]
10. Xu, L.D.; He, W.; Li, S. Internet of Things in Industries: A Survey. *IEEE Trans. Ind. Inform.* **2014**, *10*, 2233–2243. [CrossRef]
11. Wang, G.; Gunasekaran, A.; Ngai, E.W.; Papadopoulos, T. Big data analytics in logistics and supply chain management: Certain investigations for research and applications. *Int. J. Prod. Econ.* **2016**, *176*, 98–110. [CrossRef]
12. Shakhatreh, H.; Sawalmeh, A.; Al-Fuqaha, A.I.; Dou, Z.; Almaita, E.; Khalil, I.M.; Othman, N.S.; Khreishah, A.; Guizani, M. Unmanned Aerial Vehicles: A Survey on Civil Applications and Key Research Challenges. *arXiv* **2018**, arXiv:1805.00881.
13. Hassanalian, M.; Abdelkefi, A. Classifications, applications, and design challenges of drones: A review. *Prog. Aerosp. Sci.* **2017**, *91*, 99–131. [CrossRef]
14. Hardis Group, EyeSee Official Webpage. Available online: http://www.eyesee-drone.com (accessed on 31 March 2019).
15. Geodis and Delta Drone Official Communication. Available online: www.goo.gl/gzeYV7 (accessed on 31 March 2019).

16. DroneScan Official Webpage. Available online: www.dronescan.co (accessed on 31 March 2019).
17. Beul, M.; Droeschel, D.; Nieuwenhuisen, M.; Quenzel, J.; Houben, S.; Behnke, S. Fast Autonomous Flight in Warehouses for Inventory Applications. *IEEE Robot. Autom. Lett.* **2018**, *3*, 3121–3128. [CrossRef]
18. Fraga-Lamas, P.; Fernández-Caramés, T.M. A Review on Blockchain Technologies for an Advanced and Cyber-Resilient Automotive Industry. *IEEE Access* **2019**, *7*, 17578–17598. [CrossRef]
19. Fu, Y.; Zhu, J. Big Production Enterprise Supply Chain Endogenous Risk Management Based on Blockchain. *IEEE Access* **2019**, *7*, 15310–15319. [CrossRef]
20. Fernández-Caramés, T.M.; Fraga-Lamas, P. A Review on the Use of Blockchain for the Internet of Things. *IEEE Access* **2018**, *6*, 32979–33001. [CrossRef]
21. Fernández-Caramés, T. M.; Fraga-Lamas, P. A Review on the Application of Blockchain for the Next Generation of Cybersecure Industry 4.0 Smart Factories. *IEEE Access* **2019**, *7*, 45201–45218. [CrossRef]
22. Yu, B.; Wright, J.; Nepal, S.; Zhu, L.; Liu, J.; Ranjan, R. IoTChain: Establishing Trust in the Internet of Things Ecosystem Using Blockchain. *IEEE Cloud Comput.* **2018**, *5*, 12–23. [CrossRef]
23. Kapitonov, A.; Lonshakov, S.; Krupenkin, A.; Berman, I. Blockchain-based protocol of autonomous business activity for multi-agent systems consisting of UAVs. In Proceedings of the 2017 Workshop on Research, Education and Development of Unmanned Aerial Systems (RED-UAS), Linkoping, Sweden, 3–5 October 2017; pp. 84–89.
24. Global Trade Item Number Official Web Page at GS1 Website. Available online: https://www.gs1.org/gtin (accessed on 31 March 2019).
25. Finkenzeller, K. *RFID Handbook: Fundamentals and Applications in Contactless Smart Cards and Identification*, 3rd ed.; John Wiley & Sons: Hoboken, NJ, USA, 2003.
26. Fraga-Lamas, P.; Noceda-Davila, D.; Fernández-Caramés, T.M.; Díaz-Bouza, M.; Vilar-Montesinos, M. Smart Pipe System for a Shipyard 4.0. *Sensors* **2016**, *16*, 2186. [CrossRef] [PubMed]
27. Ganesan, R.; Krumm, J.; Pankalla, S.; Ludwig, K.; Glesner, M. Design of an organic electronic label on a flexible substrate for temperature sensing. In Proceedings of the ESSCIRC, Bucharest, Romania, 16–20 September 2013.
28. Todd, B.; Phillips, M.; Schultz, S.M.; Hawkins, A.R.; Jensen, B.D. RFID threshold accelerometer. In Proceedings of the IEEE AUTOTESTCON, Salt Lake City, UT, USA, 8–11 September 2008.
29. Amin, E.M.; Bhattacharyya, R.; Sarma, S.; Karmakar, N.C. Chipless RFID tag for light sensing. In Proceedings of the IEEE Antennas and Propagation Society International Symposium, Memphis, TN, USA, 6–11 July 2014.
30. Fernández-Caramés, T.M.; Fraga-Lamas, P. A Review on Human-Centered IoT-Connected Smart Labels for the Industry 4.0. *IEEE Access* **2018**, *6*, 25939–25957. [CrossRef]
31. Fraga-Lamas, P.; Fernández-Caramés, T.M.; Noceda-Davila, D.; Díaz-Bouza, M.; Vilar-Montesinos, M.; Pena-Agras, J.D.; Castedo, L. Enabling automatic event detection for the pipe workshop of the shipyard 4.0. In Proceedings of the 2017 56th FITCE Congress, Madrid, Spain, 14–16 September 2017; pp. 20–27.
32. ProVIEW Smart Label Official Web Page. Available online: https://www.omni-id.com/rfid-in-manufacturing/ (accessed on 31 March 2019).
33. Coksun, V.; Ok, K.; Ozdenizci, B. *Near Field Communications: From Theory to Practice*, 1st ed.; John Wiley & Sons: Hoboken, NJ, USA, 2012.
34. Held, I.; Chen, A. Channel Estimation and Equalization Algorithms for Long Range Bluetooth Signal Reception. In Proceedings of the IEEE Vehicular Technology Conference, Taipei, Taiwan, 16–19 May 2010.
35. Fernández-Caramés, T.M., Rodas, J., Escudero, C.J., Iglesia, D.I. Bluetooth Sensor Network Positioning System with Dynamic Calibration. In Proceedings of the International Symposium on Wireless Communications Systems, Trondheim, Norway, 17–19 October 2007.
36. Rodas, J.; Fernández-Caramés, T.M.; Escudero, C.J.; Iglesia, D.I. Multiple Antennas Bluetooth System for RSSI Stabilization. In Proceedings of the International Symposium on Wireless Communications Systems, Trondheim, Norway, 17–19 October 2007.
37. Fraga-Lamas, P.; Fernández-Caramés, T.M.; Castedo, L. Towards the Internet of Smart Trains: A Review on Industrial IoT-Connected Railways. *Sensors* **2017**, *17*, 1457. [CrossRef]

38. Hernández-Rojas, D.L.; Fernández-Caramés, T.M.; Fraga-Lamas, P.; Escudero, C.J. Design and Practical Evaluation of a Family of Lightweight Protocols for Heterogeneous Sensing through BLE Beacons in IoT Telemetry Applications. *Sensors* **2017**, *18*, 57. [CrossRef]
39. Fernández-Caramés, T.M.; González-López, M.; Castedo, L. FPGA-based vehicular channel emulator for real-time performance evaluation of IEEE 802.11 p transceivers. *EURASIP J. Wirel. Commun. Netw.* **2010**, *2010*, 607467. [CrossRef]
40. Fernández-Caramés, T.M.; González-López, M.; Castedo, L. FPGA-based vehicular channel emulator for evaluation of IEEE 802.11p transceivers. In Proceedings of the International Conference on Intelligent Transport Systems Telecommunications, Lille, France, 20–22 October 2009.
41. ZigBee Alliance. Available online: http://www.zigbee.org (accessed on 31 March 2019).
42. Khutsoane, O.; Isong, B.; Abu-Mahfouz, A.M. IoT devices and applications based on LoRa/LoRaWAN. In Proceedings of the Annual Conference of the IEEE Industrial Electronics Society, Beijing, China, 29 October–1 November 2017.
43. Weyn, M.; Ergeerts, G.; Berkvens, R.; Wojciechowski, B.; Tabakov, Y. DASH7 alliance protocol 1.0: Low-power, mid-range sensor and actuator communication. In Proceedings of the IEEE Conference on Standards for Communications and Networking (CSCN), Tokyo, Japan, 28–30 October 2015.
44. Kim, A.N.; Hekland, F.; Petersen, S.; Doyle, P. When HART goes wireless: Understanding and implementing the WirelessHART standard. In Proceedings of the IEEE International Conference on Emerging Technologies and Factory Automation, Hamburg, Germany, 15–18 September 2008.
45. SigFox Official Web Page. Available online: https://www.sigfox.com (accessed on 31 March 2019).
46. ANT Wireless Official Web Page. Available online: https://www.thisisant.com (accessed on 31 March 2019).
47. Lu, J.; Xu, X.; Li, X.; Li, L.; Chang, C.-C.; Feng, X.; Zhang, S. Detection of Bird's Nest in High Power Lines in the Vicinity of Remote Campus Based on Combination Features and Cascade Classifier. *IEEE Access* **2018**, *6*, 39063–39071. [CrossRef]
48. Zhou, Z.; Zhang, C.; Xu, C.; Xiong, F.; Zhang, Y.; Umer, T. Energy-Efficient Industrial Internet of UAVs for Power Line Inspection in Smart Grid. *IEEE Trans. Ind. Inform.* **2018**, *14*, 2705–2714. [CrossRef]
49. Lim, G.J.; Kim, S.; Cho, J.; Gong, Y.; Khodaei, A. Multi-UAV Pre-Positioning and Routing for Power Network Damage Assessment. *IEEE Trans. Smart Grid* **2018**, *9*, 3643–3651. [CrossRef]
50. Wang, L.; Zhang, Z. Automatic Detection of Wind Turbine Blade Surface Cracks Based on UAV-Taken Images. *IEEE Trans. Ind. Electron.* **2017**, *64*, 7293–7303. [CrossRef]
51. Peng, K.; Liu, W.; Sun, Q.; Ma, X.; Hu, M.; Wang, D.; Liu, J. Wide-Area Vehicle-Drone Cooperative Sensing: Opportunities and Approaches. *IEEE Access* **2018**, *7*, 1818–1828. [CrossRef]
52. Rossi, M.; Brunelli, D. Autonomous Gas Detection and Mapping with Unmanned Aerial Vehicles. *IEEE Trans. Instrum. Meas.* **2016**, *65*, 765–775. [CrossRef]
53. Scilimati, V.; Petitti, A.; Boccadoro, P.; Colella, R.; Di Paola, D.; Milella, A.; Grieco, L.A. Industrial Internet of things at work: A case study analysis in robotic-aided environmental monitoring. *IET Wirel. Sens. Syst.* **2017**, *7*, 155–162. [CrossRef]
54. Misra, P.; Kumar, A.A.; Mohapatra, P.; Balamuralidhar, P. Aerial Drones with Location-Sensitive Ears. *IEEE Commun. Mag.* **2018**, *56*, 154–160. [CrossRef]
55. Li, H.; Savkin, A.V. Wireless Sensor Network Based Navigation of Micro Flying Robots in the Industrial Internet of Things. *IEEE Trans. Ind. Inform.* **2018**, *14*, 3524–3533. [CrossRef]
56. Olivares, V.; Córdova, F. Evaluation by computer simulation of the operation of a fleet of drones for transporting materials in a manufacturing plant of plastic products. In Proceedings of the 2015 CHILEAN Conference on Electrical, Electronics Engineering, Information and Communication Technologies (CHILECON), Santiago, Chile, 28–30 October 2015; pp. 847–853.
57. Zhao, S.; Hu, Z.; Yin, M.; Ang, K.Z.Y.; Liu, P.; Wang, F.; Dong, X.; Lin, F.; Chen, B.M.; Lee, T.H. A Robust Real-Time Vision System for Autonomous Cargo Transfer by an Unmanned Helicopter. *IEEE Trans. Ind. Electron.* **2015**, *62*, 1210–1219. [CrossRef]
58. Kuru, K.; Ansell, D.; Khan, W.; Yetgin, H. Analysis and Optimization of Unmanned Aerial Vehicle Swarms in Logistics: An Intelligent Delivery Platform. *IEEE Access* **2019**, *7*, 15804–15831. [CrossRef]

59. Cho, H.; Kim, D.; Park, J.; Roh, K.; Hwang, W. 2D Barcode Detection using Images for Drone-assisted Inventory Management. In Proceedings of the 15th International Conference on Ubiquitous Robots (UR), Honolulu, HI, USA, 26–30 June 2018.
60. Macoir, N.; Bauwens, J.; Jooris, B.; Van Herbruggen, B.; Rossey, J.; Hoebeke, J.; De Poorter, E. UWB Localization with Battery-Powered Wireless Backbone for Drone-Based Inventory Management. *Sensors* **2019**, *19*, 467. [CrossRef] [PubMed]
61. Bae, S.M.; Han, K.H.; Cha, C.N.; Lee, H.Y. Development of Inventory Checking System Based on UAV and RFID in Open Storage Yard. In Proceedings of the International Conference on Information Science and Security (ICISS), Pattaya, Thailand, 19–22 December 2016.
62. Ong, J.H.; Sanchez, A.; Williams, J. Multi-UAV System for Inventory Automation. In Proceedings of the 1st Annual RFID Eurasia, Istanbul, Turkey, 5–6 September 2007.
63. Harik, E.H.C.; Guérin, F.; Guinand, F.; Brethé, J.; Pelvillain, H. Towards An Autonomous Warehouse Inventory Scheme. In Proceedings of the IEEE Symposium Series on Computational Intelligence (SSCI), Athens, Greece, 6–9 December 2016.
64. Tiwari, S.; Wee, H.M.; Daryanto, Y. Big data analytics in supply chain management between 2010 and 2016: Insights to industries. *Comput. Ind. Eng.* **2018**, *115*, 319–330. [CrossRef]
65. Rossmann, B.; Canzaniello, A.; von der Gracht, H.; Hartmann, E. The future and social impact of Big Data Analytics in Supply Chain Management: Results from a Delphi study. *Technol. Forecast. Soc. Chang.* **2018**, *130*, 135–149. [CrossRef]
66. Zhong, R.Y.; Newman, S.T.; Huang, G.Q.; Lan, S. Big Data for supply chain management in the service and manufacturing sectors: Challenges, opportunities, and future perspectives. *Comput. Ind. Eng.* **2016**, *101*, 572–591. [CrossRef]
67. Zhong, R.Y.; Xu, C.; Chen, C.; Huang, G.Q. Big Data Analytics for Physical Internet-based intelligent manufacturing shop floors. *Int. J. Prod. Res.* **2017**, *55*, 2610–2621. [CrossRef]
68. Christidis, K.; Devetsikiotis, M. Blockchains and smart contracts for the Internet of Things. *IEEE Access* **2016**, *4*, 2292–2303. [CrossRef]
69. Koomey, J.; Brill, K.; Turner, P.; Stanley, J.; Taylor, B. *A Simple Model for Determining True Total Cost of Ownership for Data Centers*; White Paper; Uptime Institute: Seattle, WA, USA, 2007.
70. Middleton, S.G.; Marden, M. *Deploying an Effective Server Life-Cycle Strategy Will Minimize Costs: Leasing Is a Valuable*; White Paper; IDC: Framingham, MA, USA, 2015.
71. Cai, W.; Wang, Z.; Ernst, J.B.; Hong, Z.; Feng, C., Leung, V.C.M. Decentralized Applications: The Blockchain-Empowered Software System. *IEEE Access* **2018**, *6*, 53019–53033. [CrossRef]
72. Fraga-Lamas, P. Enabling Technologies and Cyber-Physical Systems for Mission-Critical Scenarios. May 2017. Available online: http://hdl.handle.net/2183/19143 (accessed on 31 March 2019).
73. Ardupilot Official Web Page. Available online: http://ardupilot.org/ardupilot/ (accessed on 31 March 2019).
74. RF-Code Official Web Page. Available online: https://www.rfcode.com/ (accessed on 31 March 2019).
75. OrbitDB: Peer-To-Peer Database for the Decentralized Web. Github Repository. Available online: https://github.com/orbitdb/orbit-db (accessed on 31 March 2019).
76. IPFS Official Webpage. Available online: https://ipfs.io/ (accessed on 31 March 2019).
77. Ethereum Official Webpage. Available online: https://www.ethereum.org/ (accessed on 31 March 2019).
78. Rinkeby's Official Website. Available online: https://www.rinkeby.io/ (accessed on 30 April 2019).
79. Ropsten's Official GitHub Page. Available online: https://github.com/ethereum/ropsten (accessed on 30 April 2019).
80. Rinkeby's Ether Faucet. Available online: https://faucet.rinkeby.io/ (accessed on 30 April 2019).
81. Metamask Ether Faucet. Available online: https://faucet.metamask.io/ (accessed on 30 April 2019).
82. Truffle Official Website. Available online: https://truffleframework.com (accessed on 30 April 2019).
83. Infura's Official Website. Available online: https://infura.io/dashboard (accessed on 30 April 2019).
84. Node.js Official Web Site. Available online: https://nodejs.org/en/ (accessed on 30 April 2019).
85. Web3 Official GitHub Page. Available online: https://github.com/ethereum/web3.js/ (accessed on 30 April 2019).

86. Fraga-Lamas, P.; Fernández-Caramés, T.M.; Noceda-Davila, D.; Vilar-Montesinos, M. RSS Stabilization Techniques for a Real-Time Passive UHF RFID Pipe Monitoring System for Smart Shipyards. In Proceedings of the 2017 IEEE International Conference on RFID (IEEE RFID 2017), Phoenix, AR, USA, 9–11 May 2017.
87. Kotz, S.; Nadarajah, S. *Extreme Value Distributions: Theory and Applications*; Imperial College Press: London, UK, 2000.
88. Kernel Distribution. Matlab Official Webpage. Available online: https://es.mathworks.com/help/stats/kernel-distribution.html (accessed on 30 April 2019).

Article

A WASN-Based Suburban Dataset for Anomalous Noise Event Detection on Dynamic Road-Traffic Noise Mapping

Rosa Ma Alsina-Pagès *, Ferran Orga, Francesc Alías and Joan Claudi Socoró

GTM—Grup de recerca en Tecnologies Mèdia, La Salle—Universitat Ramon Llull., C/Quatre Camins, 30, 08022 Barcelona, Spain; ferran.orga@salle.url.edu (F.O.); francesc.alias@salle.url.edu (F.A.); joanclaudi.socoro@salle.url.edu (J.C.S.)

* Correspondence: rosamaria.alsina@salle.url.edu; Tel.: +34-932-902-455

Received: 29 March 2019; Accepted: 27 May 2019; Published: 30 May 2019

Abstract: Traffic noise is presently considered one of the main pollutants in urban and suburban areas. Several recent technological advances have allowed a step forward in the dynamic computation of road-traffic noise levels by means of a Wireless Acoustic Sensor Network (WASN) through the collection of measurements in real-operation environments. In the framework of the LIFE DYNAMAP project, two WASNs have been deployed in two pilot areas: one in the city of Milan, as an urban environment, and another around the city of Rome in a suburban location. For a correct evaluation of the noise level generated by road infrastructures, all Anomalous Noise Events (ANE) unrelated to regular road-traffic noise (e.g., sirens, horns, speech, etc.) should be removed before updating corresponding noise maps. This work presents the production and analysis of a real-operation environmental audio database collected through the 19-node WASN of a suburban area. A total of 156 h and 20 min of labeled audio data has been obtained differentiating among road-traffic noise and ANEs (classified in 16 subcategories). After delimiting their boundaries manually, the acoustic salience of the ANE samples is automatically computed as a contextual Signal-to-Noise Ratio (SNR) together with its impact on the A-weighted equivalent level (ΔL_{Aeq}). The analysis of the real-operation WASN-based environmental database is evaluated with these metrics, and we conclude that the 19 locations of the network present substantial differences in the occurrences of the subcategories of ANE, with a clear predominance of the noise of sirens, trains, and thunder.

Keywords: road-traffic noise; anomalous noise event; acoustic dataset; noise monitoring; smartcity; WASN; SNR; impact; L_{Aeq}; urban sound; noise maps

---

1. Introduction

Presently, cities are growing in both size and population, and the consequent increase in vehicles is making traffic noise problem more present, with a clear effect on the quality of life of their citizens [1]. Noise is one of the main environmental health concerns [2,3], and its impact on social and economic aspects has been proved [4]. To face this issue, European authorities have driven the European Noise Directive (END) [5], focused on the creation of noise-level maps to inform citizens of their exposure to noise, and aided the authorities to take appropriate action to minimize its impact.

Noise maps have been historically generated by means of costly expert measurements using certified sound-level meters, with a basis of short-term periods aimed at being sufficiently representative. This approach is presently overcome by the technological advances of the Internet of Things in the framework of smart cities, which has allowed the emergence of Wireless Acoustic Sensor Networks (WASN) [6,7]. In the literature, several different WASNs have been designed for urban sound monitoring, some of them focused on security and surveillance and others on city noise management, involving noise mapping, the development of action plans, and public awareness campaigns. For example, the SENSEable project [8] deployed a WASN to collect information from the acoustic environment by means of a set of low-cost acoustic sensors with the goal of analyzing that data together with public health information. Other similar projects are the IDEA project in Belgium [9], or the RUMEUR network in France [10] with special focus on aircraft noise, or even the Barcelona noise-monitoring network [11], whose data is integrated in the Sentilo city management platform [12]. Recently, the SONYC project has deployed 56 low-cost acoustic sensors across New York City to monitor urban noise and perform a multi-label classification of urban sound sources in real time [4], but not on site. Finally, the LIFE DYNAMAP project [13] aims to monitor the noise level generated by road infrastructures by means of two WASNs installed in two pilot areas, one within an urban environment in Milan (District 9), and another in a suburban area surrounding Rome (A90 highway). To monitor Road-Traffic Noise (RTN) levels reliably, all Anomalous Noise Events (ANE) unrelated to regular RTN (e.g., sirens, horns, speech, etc.) should be removed before updating the corresponding noise maps [13].

The deployment of these projects has shown that the WASN paradigm entails several challenges, from technical issues [14,15], to other aspects related to the WASN-based application, such as the automation of data collection and the subsequent signal processing [16–18], especially if the system intends detect acoustic events in real operation and locally in each sensor. Acoustic event detection and classification belongs to the Computational Auditory Scene Analysis (CASA) paradigm [19], and it is usually based on the segmentation of the input acoustic data into slices that represent a single occurrence of the target class, and focus on individual simultaneous events [20]. To do so, Acoustic Event Detection (AED) algorithms are typically trained with databases designed ad hoc in each of the problems to be solved, hence typically considering a finite set of predefined acoustic classes [21,22].

Therefore, the development of AED-based applications entails representative audio databases with all kinds of sounds of interest, as in the one obtained in the SONYC project [4], with data from 56 sensors deployed in different neighborhoods of New York, which considers 10 different kinds of common urban sound sources labeled in an urban soundscape. As a first attempt to create an acoustic dataset to model the acoustic environments of urban and suburban pilot areas in the framework of the DYNAMAP project, an expert-based recording campaign was conducted before the two WASNs were deployed [23]. The analyses showed the highly local, unpredictable, and diverse nature of ANEs in real acoustic environments can be far different to previous models obtained by means of synthetically generated datasets [23]. After labeling the gathered acoustic data, the dataset was used to train the AED-based algorithm designed to detect ANEs, known as Anomalous Noise Event Detector (ANED) [18]. Although that preliminary dataset collected a representative number of acoustic events of interest from both acoustic environments, it missed several key aspects, such as different RTN patterns observed during day–night, weekday–weekend and the effect of diverse weather conditions [24]. This work describes the generation of the acoustic dataset to model the Rome's acoustic environment in real-operation conditions, after deploying the 19-node WASN in its final location. The paper describes the conducted recording campaign and the subsequent labeling of ANEs in 16 different subcategories (without considering combined sounds in a sample), as well as the analysis of their occurrences, duration, Signal-to-Noise Ratio (SNR), and impact on the A-weighted equivalent noise level (L_{Aeq}) computation.

The remainder of this paper is the following. Section 2 details the most relevant previous attempts to generate environmental audio databases. Section 3 describes the generation and labeling of the real-operation conditions environmental audio database in the suburban scenario. Section 4 analyses the ANE of the dataset in terms of occurrences, duration, SNR, and impact on the L_{Aeq}. Finally, Section 5 discusses in detail the results obtained in the analysis and the future applications of the designed dataset.

2. Related Work

In the literature, several audio databases related to machine hearing (or machine listening) have been unveiled for benchmarking purposes under the umbrella of the so-called CASA, and mainly oriented to evaluate the performance of acoustic scene classification and AED. This section reviews the literature about environmental acoustic databases and the datasets designed for challenges (e.g., DCASE), and describes their characteristics and limitations.

2.1. Environmental Acoustic Databases

The environmental acoustic databases employed by the machine-hearing research community have been generally created from live recordings directly and/or synthetically generated by artificially mixing sound events with certain acoustic environments (i.e., background noise). The latter allows control of the SNR of the mixture, and dealing with data scarcity of specific audio events in real-life contexts—which is one of the key problems when trying to gather representative data from live environments [25]—while the former entails a huge effort for data collection and subsequent manual annotation to generate the labeled database or ground truth [23,26,27].

Regarding real-life environmental acoustic databases, in [28] a 1133-min audio database including 10 different acoustic environments, both indoor and outdoor was introduced. On the other hand, the MIVIA audio events dataset was designed for surveillance applications focused on the identification of glass breaking, gun shots, and screams (https://mivia.unisa.it/datasets/audio-analysis/mivia-audio-events/) [21]. The training dataset is about 20 h, while the test set is about 9 h. Moreover, the same research laboratory developed a smaller dataset of about 1 h duration also for surveillance purposes focused on road audio events, which contains sound events from tire skidding and car crashes (https://mivia.unisa.it/datasets/audio-analysis/mivia-road-audio-events-data-set/) [29].

In [23], a real-life acoustic database collected from the urban and suburban environments of the pilot areas of the Life DYNAMAP project [13] is described. The database composed of 9 h and 8 min was obtained through an in situ recording campaign. This acoustic database was developed for discriminating road-traffic noise from ANE through the ANED [18]. The ANEs, which only represented 7.5% of the annotated data, were subsequently classified in 19 different subcategories after manual inspection, showing SNR levels with respect to background noise that ranged from −10 dBm to +15 dBm and showing a high heterogeneity of intermediate SNR levels. When comparing both environments, it can be observed that the number of ANEs in the urban area is approximately four times higher than in the suburban area, also including events with larger acoustic salience. Nevertheless, it is worth mentioning that the recordings in the urban area were conducted at the street level of the preselected locations [30] within District 9 of Milan, while the recordings in the suburban area were conducted on the A90 ring-road portals surrounding Rome (see [31] for further details).

One of the main sources employed to build acoustic databases is the well-known Freesound online repository (https://www.freesound.org). For instance, *freefield1010* is a database composed of 7690 audio clips tagged as "field recording" in the metadata of the original recordings uploaded in this online repository, totaling over 21 h of audio [32]. In [33], 60 h of real field recordings uploaded in Freesound from urban environments were used to build the UrbanSound database. The database is composed of 27 h, which includes 18.5 h of verified and annotated sound event occurrences classified in 10 sound categories (i.e., air conditioner, car horn, children playing, dog bark, drilling, engine idling, gun shot, jackhammer, siren, and street music). Moreover, the authors also provide an 8.75 h subset—denoted as UrbanSound8k—designed to train sound classification algorithms and obtained after arbitrarily fixing the number of items to 1000 slices per class. In [34], a mixture of sound sources from Freesound mixed with real-life traffic noise was considered to train and evaluate an AED algorithm (considering two SNRs levels: +6 dB and +12 dB). Finally, "ESC: Dataset for Environmental Sound Classification" [35] is composed of three subsets: *(i)* ESC-50, a strictly balanced 50 classes of various environmental sounds obtained through manual annotation, *(ii)* ESC-10, as a reduced 10 classes subset of the former as a proof-of-concept dataset, and *(iii)* EC-US, which contains 250,000 recordings directly extracted from the "field recording"-tagged category of Freesound. Nevertheless, due to the uncontrolled origin of the sound sources uploaded to Freesound and similar online repositories (see [23] for further examples), involving a wide variation in the recording conditions and quality, the derived environmental acoustic databases may become unsuitable for reliable CASA- and AED-based systems evaluation purposes [27,32].

In [27], the described acoustic dataset covers both indoor and outdoor environments, including real-life recordings of predefined acoustic event sequences and individual acoustic events synthetically mixed with background recordings by considering specific SNRs levels, such as −6 dB, 0 dB, and +6 dB. In [25], the authors developed a mixed acoustic database composed of acoustic data from real-life recordings, which was subsequently extended with synthetic mixtures of extra events of interest to increase database diversity. In [21], an acoustic database for surveillance purposes that includes sound events such as screams, glass breaking, and gunshots was also artificially generated from indoor and outdoor environments considering different SNR levels (from +5 to +30 dB). Moreover, a small environmental acoustic database containing 20 scenes mixing background noise with car, bird, and car horn samples synthesized with SimScene software [36] was described in [37]. Following a similar approach, in [38], the TUT Sound Events Synthetic 2016 (TUT-SED-2016 for short) was introduced. The 566 min dataset is composed of synthetic mixtures created by mixing isolated sound events from 16 sound event classes from the original TUT database (The reader is referred to http://www.cs.tut.fi/sgn/arg/taslp2017-crnn-sed/tut-sed-synthetic-2016 for a detailed explanation).

Finally, it is worth mentioning the recent development of an open-source library for the synthesis of soundscapes named Scaper, mainly focused on SED-related applications [39]. This library provides an audio sequencer to generate synthetic soundscapes following a probabilistic approach including isolated sound events. The proposal allows control of the characteristics of the sound mixtures, such as the number of events, and their type, timing, duration, and SNR level with respect to the background noise. The authors validate their proposal through the development of the URBAN-SED database from UrbanSound8k as an example of the result of this kind of data augmentation. The synthetic generation of acoustic databases is a potential solution to address data scarcity when training Deep Neural Networks (DNN) for CASA-related problems (e.g., see [38,40]). Nevertheless, although the artificial generation of sound mixtures allows the creation of controlled training and evaluation environments, it has been also stated that these artificially generated databases could not represent the variability encountered in real-life environments accurately enough [18,26].

2.2. Challenge-Oriented Acoustic Datasets

Over the last decade, the CASA research community has provided publicly available datasets and standard metrics to evaluate the development of their investigations. One of the seminal international efforts that emerged to evaluate systems developed to model the perception of people, their activities, and interactions was the CLassification of Events, Activities and Relationships (CLEAR) competition, which included in its 2006 and 2007 editions specific data for acoustic event detection and classification mainly collected from indoor environments—specifically, from meeting rooms (see e.g., [41–43]). Although other attempts emerged to provide evaluation material for CASA-focused systems, such as DARESounds.org initiative [44] or TRECVID Multimedia Event Detection competition (focused on audiovisual and multi-modal event detection (https://www.nist.gov/itl/iad/mig/multimedia-event-detection)), neither them nor CLEAR led to the establishment of a reference evaluation challenge for the CASA research community [20].

Later, the IEEE AASP supported a new competition named Detection and Classification of Acoustic Scenes and Events (DCASE), which started in 2013 at WASPAA conference (http://www.waspaa.com/waspaa13/d-case-challenge/index.html). The database that was provided for that challenge contained both live and synthetic recordings [45]. The results of that competition can be found in [27]. Since 2016, DCASE has become an annual competition including different challenges covering acoustic scene classification, sound event detection in synthetic and in real-life audio, domestic audio tagging, or the detection of bird or rare sound events, to name a few (see [22,46,47] for further details) (http://dcase.community/events), mainly thanks to the contribution of several researchers who have made available different datasets for public evaluation.

Among them , it is worth mentioning the creation of the "TUT Database for Acoustic Scene Classification and Sound Event Detection" from real-life recordings [26]. This database allows the evaluation of automatic event or acoustic scene detection systems within 15 different real-life acoustic environments, such as lakeside beach, bus, cafe/restaurant, car, city center, forest path, grocery store, home, library, metro station, office, urban park, residential area, train, and tram. It is worth mentioning that the sound events subset, which covers both indoor and outdoor environments, it is mainly focused on surveillance and human activity monitoring. Finally, it is worth noting that in the recent announcement of the DCASE2019 competition, an Urban Sound Tagging (UST) challenge has been presented. The goal of this task is to predict whether each of 23 sources of noise pollution is present or absent in a 10-s scene. In this challenge, the audio in the dataset has been acquired with the acoustic sensor network of the SONYC project: SOunds of New York City [4], and it provides a simplified taxonomy of the sounds of the city in two levels, 8 coarse categories, and 23 fine labels. The challenge dataset includes 2351 recordings in the training split and 443 in the validation split, making a total of 2794 10-s audios. The full taxonomy and details of the SONYC project dataset can be found in [33].

3. Design of an Environmental Database

In this section, the real-operation environmental audio database recorded and built in for the suburban area of the DYNAMAP project is detailed. First, the conducted recording campaign is described. Second, the subsequent generation of the audio database is detailed, which includes the labeling and the computation of the acoustic salience of the anomalous noise events (in terms of SNR) and their impact on the L_{Aeq}. These are computed in this study and used jointly with duration, number of occurrences and other database descriptors to analyze in detailed nature of the ANEs of the suburban area.

3.1. Description of the WASN-based Suburban Recording Campaign

The main goal of any recording campaign is to collect representative samples of the acoustic environment under study through the WASN in real operation. Taking advantage of the experience gained from the preliminary recording and analysis of that acoustic environment [23], a second recording campaign was designed. The main reason for it was three-fold: *(i)* all the nodes of the WASN had been already deployed in their definitive operative location, which increased the number of recording points and changed slightly the sensor position in the portals, *(ii)* the sampling completeness, because the previous recording campaign did not include nighttime, weekend data, or different meteorological conditions, and *(iii)* the total amount of time recorded was quite short (4 h and 44 min), including only 12.2% of ANEs, which led us to the conclusion that ANEs were misrepresented in the dataset, after discarding the augmentation of the dataset by means of synthetic samples according to [23].

The WASN deployed in the suburban area of DYNAMAP project is located on the A90 highway surrounding Rome, and comprises 24 acoustic nodes, 5 of which are low-capacity sensors without enough computational resources to run the ANED algorithm. The locations of the 19 high-capacity sensors of the WASN in the Rome's suburban pilot area are shown in Figure 1. The set of basic specifications [48] that are defined to satisfy DYNAMAP requirements for each monitoring station are the following: *(i)* 40–100 dB(A) broadband linearity range, *(ii)* 35–115 dB working range with acceptable Total Harmonic Distortion (THD), and *(iii)* narrowband floor noise level. The project also requires the possibility of audio recording, as well as Virtual Private Network (VPN) connection and GPRS/3G/WiFi connection. The precision of the sensors is a key issue for the system reliability [49]. During the developing stage, all the elements that could increase the uncertainty of the measurement were taken into account following the requirements of IEC 61672 [50]. Several tests have been conducted with both the hardware and the software, using a climate chamber with different operation temperatures. Electromagnetic Compatibility tests were also conducted as well as atmospheric agent simulations over the designed equipment [51].

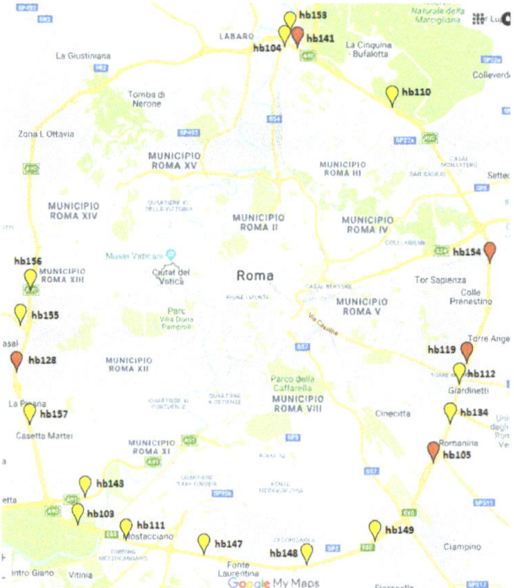

Figure 1. Map with sensors' location information within the WASN of DYNAMAP project in the suburban area of Rome, those locations also sensed during the initial recording campaign described in [23] shown in red.

Some of the recording locations are the same as the ones used in the previous recording campaign [23] (see Figure 2). However, although conducted in the same portal both the sensor and the exact location measurement differ. The microphone location is slightly different with respect to the entire structure of the portal (see Figure 2a,b).

(**a**) (**b**)

Figure 2. Location of the sensor used in the recording campaign in the portals. (**a**) Recording campaign deployment in [23]; (**b**) WASN deployment in this work (picture property of ANAS S.p.A.).

Achieving a complete and exhaustive dataset is a challenging task since the amount of available resources is limited, e.g., processing and storage capabilities, data collection using 3G modems, availability of all nodes in fully operative conditions, etc. In Figure 3, it can be observed that the L_{Aeq}

presents a diurnal variation that suggests sampling the recording differentiating day and night to obtain data from several patterns of traffic noise, and so of ANEs.

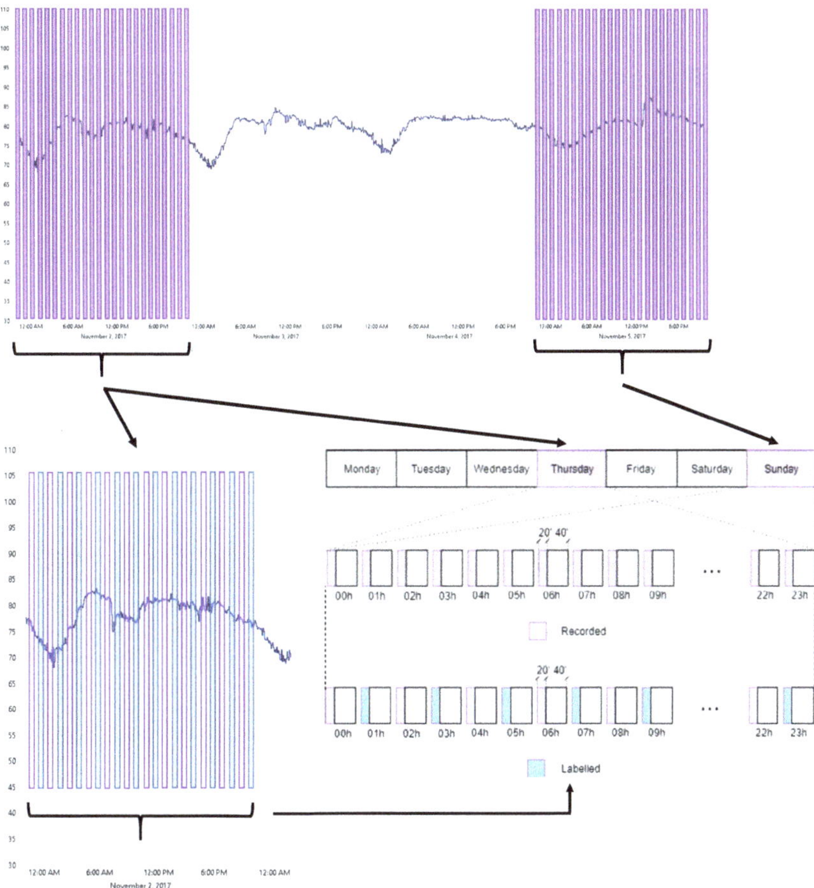

Figure 3. Daily curve of L_{Aeq} for sensor hb147 for the 2nd and the 5th of November 2017, and the recordings conducted. Diagram of the recording days and duration for each sensor, and scheme of the labeled files to build the dataset.

To this aim, one-day real-operation recordings were planned through all nodes of the suburban WASN considering different days of variating traffic conditions, and assuming a trade-off between completeness and available computation and data communication resources, which were limited by the resources available in each of the nodes of the WASN (storage capacity, throughput of data, etc.). The following data sampling approach was proposed: Thursday and Sunday were selected as representative weekday and weekend days, respectively, being the 2nd and the 5th of November 2017, specifically. From each sensor, 20 min have been recorded per hour which was limited by the storage capacity and communication resources of each of the nodes. Figure 3 shows the recordings over the values of L_{Aeq}, while Figure 3 shows a schematic diagram with the recording process during the selected weekday and weekend day. As a result, 16 h of acoustic data were collected from each sensor to cover the diversity of the acoustic environment in a workday and a weekend day in this suburban environment.

It is worth mentioning that the high-capacity sensors used to conduct the recording campaign using the WASN in real operation were low-cost acoustic sensors designed ad hoc for the DYNAMAP project [51].

3.2. Data Labeling

After recording representative acoustic data for building the suburban environmental database, a labeling process was conducted. The manual annotation of ANEs becomes particularly complex when dealing with real-operation data from raw recordings. Thus, this process must be conducted by experts, since it is very important to precisely determine the occurrence and boundaries of each event, e.g., indicating the start and end points in the mixed audio [23].

The labeling process was not exhaustive because of the excessive burden that such a task would represent if considering all recorded data, meaning that only 50% of gathered audio signals were finally labeled. This represented a total of 156 h and 20 min of labeled audio data. From the labeling process 94.8% was labeled as RTN, 1.8% as ANE and the remaining 3.4% was categorized as *others* when the audio passage was difficult to categorize in one or other class due to the complexity of the audio scene. These last passages were not included in the subsequent analyses presented throughout this paper, but they have been left for future analyses that will focus, e.g., on the impact they can have for ANED assessment.

In Figure 4, an example of the labeling process using Audacity software is shown for illustrative purposes. The example of ANE is a siren, which is a long event, recorded in sensor hb149 the 2nd of November 2017.

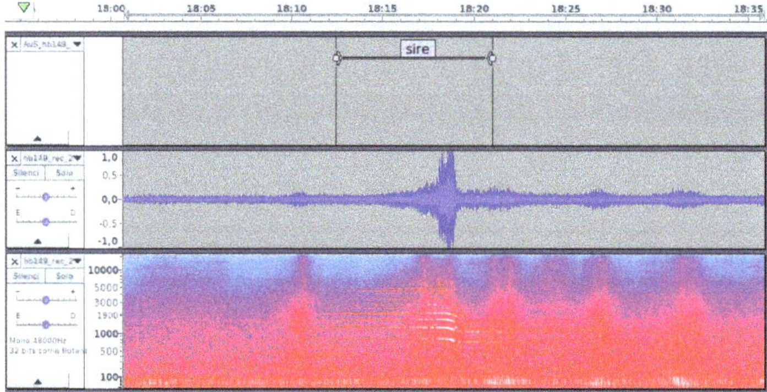

Figure 4. Example of the labeling of a siren using Audacity, with the label on top, the signal in the middle and the spectrum on the bottom.

From the labeling of all the collected data through the 19 nodes of the Rome's WASN, the following list of ANEs were observed:

- *airp*: airplanes.
- *alrm*: sounds of cars and houses alarm systems.
- *bike*: noise of bikes.
- *bird*: birdsong.
- *brak*: noise of brake or cars' trimming belt.
- *busd*: opening bus or tramway, door noise, or noise of pressurized air.
- *door*: noise of house or vehicle doors, or other object blows.
- *horn*: horn vehicles noise.
- *inte*: interfering signal from ad industry or human machine.
- *musi*: music in car or in the street.
- *rain*: sound produced by heavy rain.

- *sire*: sirens of ambulances, police, fire trucks, etc.
- *stru*: noise of portals structure derived from its vibration, typically caused by the passing-by of very large trucks.
- *thun*: thunderstorm.
- *tran*: stop, start, and pass-by of trains.
- *trck*: noise when trucks or vehicles with heavy load passed over a bump.

Another label has been used for the annotation process, the *cmplx* label that indicates that the piece of raw acoustic audio was the result of more than one subcategory of ANE or that the sound was not identifiable by the labeler. This is not considered to be a subcategory of ANE because it cannot be assigned to a single noise source.

3.3. Characterization of the ANEs

In previous works, two parameters were considered by the authors to figure out the effects of the ANEs on noise-map generation [52]. The first of the parameters is based on the classical SNR calculation, consisting of the ratio of power of the ANE in relation to the power of the surrounding RTN. The second metric determines the impact of the ANE on the equivalent noise level used to build the noise map. The calculation of the two parameters is described below.

3.3.1. SNR Calculation

As aforementioned, the SNR is calculated as the classical signal-to-noise ratio used in signal processing, considering that the ANE corresponds to the signal and the RTN is the noise. The acoustic power of the ANE and the RTN are calculated as follows:

$$P_x = \sum_{n=1}^{N} \left(\frac{x[n]^2}{N} \right) \tag{1}$$

where $x[n]$ is the recorded audio with N samples that belongs to either the ANE of the RTN.

After the power calculations of the ANE and the surrounding RTN, the SNR is calculated as:

$$SNR = 10\ log_{10} \left(\frac{P_{ANE}}{P_{RTN}} \right) \tag{2}$$

where P_{ANE} belongs to the anomalous event in question and P_{RTN} is the power of the surrounding RTN.

All the casuistry of the calculation is detailed in [52]. Finally, it is worth mentioning that the SNR of a particular ANE could be negative if the power of the surrounding RTN is higher than the power of the ANE itself. This may happen in cases where RTN masks other low-energy sounds, e.g., birds, because of the fluctuation of the road pass-bys.

3.3.2. Impact Calculation

The computation of the ANE impact consists of determining the contribution of a particular event to the equivalent noise level of the recorded audio after applying the A-weighting filter [53], i.e., L_{Aeq}. This metric has been defined to evaluate the effect of each ANE on the noise map L_{Aeq}. It is calculated as the difference between the L_{Aeq} computed with all the raw data and the L_{Aeq} after removing the ANE by means of a linear interpolation (see Equation (3)). The event should be replaced by a lower-period linear interpolation to maintain the weight of the surrounding RTN level. That way, the road-traffic noise measurement is as accurate as possible in the whole integration time.

$$\Delta L_{Aeq} = L_{Aeq,ANE} - L_{Aeq,\overline{ANE}} \tag{3}$$

where ΔL_{Aeq} is the contribution of this ANE to the L_{Aeq}, $L_{Aeq,ANE}$ is the A-weighted equivalent noise level of the raw audio given an evaluation period and $L_{Aeq,\overline{ANE}}$ is the equivalent A-weighted equivalent noise level of the same audio after removing the ANE.

In the DYNAMAP project, noise-map values are updated every 5 min, thus the contribution of ANEs on the L_{Aeq} will be evaluated in this integration time, i.e., ΔL_{A300s}. The low-level interpolations are carried out in a 1-s integration window, i.e., L_{A1s}, as it heuristically proved to be a good trade-off between representing all audible short events and not adding imperceptible changes to the equivalent sound measurement. As the goal is to obtain the equivalent noise level completely unaffected by the ANE, a span of 500 ms is left before and after the exact ANE label. More details on the casuistry of this calculation may be found in [52].

Finally, it is worth mentioning that the L_{A300s} measurement is conducted in a 5-min sliding window where the ANE is centered. This is a better approximation than using a 5-min fixed window as the mean distance between RTN samples and ANE samples is reduced. The fact that future samples are needed (plus the fact that the exact labeling can only be provided after listening and labeling the recordings) implies that this measurement can only be applied in an off-line analysis and not in the real-time operation mode of the WASN.

4. Dataset Analysis

In this section, a detailed analysis of the ANEs present in the labeled WASN-based audio database is described. Specifically, the distribution of the occurrences and duration of the ANE subcategories together with their contextual SNR distribution and the impact of each of them to the L_{Aeq} value are analyzed. First, an analysis was conducted regarding the general characteristics of the new audio dataset considering the spectral and time behavior of nodes recordings along one day. Secondly, two more detailed analyses were performed: *(i)* aggregating values of occurrence, duration, and SNR, as well as L_{Aeq} impact considering all network sensors at a whole and, *(ii)* highlighting the particularities of sensor's locations of the WASN in terms of the same ANE statistical measures.

4.1. General Characteristics

The global trend of the audio database obtained from the sensor network has been firstly analyzed through focusing mainly in RTN, disregarding specific characteristics of the ANE observed during recordings. In this regard, spectrum-time profiles [54] defined as the hourly time evolution of the mean spectrum of the audio has been computed for each sensor and day, following the same approach as in [24]. Their computation has been performed using the 48 frequency sub-bands of the Mel-Frequency Cepstral Coefficients (MFCC) features of the incoming audio signal, following the setup explained in [55], and computing the mean spectrum along the 20 min of audio gathered every hour. This mean spectrum obtained for each of the 24 h in a day conforms the spectrum-time profile of a complete measured day, which mainly includes RTN raw signal.

Figure 5 shows four examples of the measured spectrum-time profiles, for two sensors hb134 and hb156 and for the two recorded days of the campaign. As can be observed, the general trend of the spectrum-time profile during weekday (Figure 5a,b) is quite similar for these two sensors, while sensor hb156 presents a slightly higher energy at high frequencies and during a narrower daily period. During the weekend (Figure 5c,d) a very persistent ANE of hard rain was present in almost all sensors, which can be seen as a high spectrum-profile value around 14 h. Additionally, another aspect that can be highlighted comparing week spectrum-time profiles with those of the weekend is that during weekend the raising and decreasing of acoustic energy during daily period is smoother than during the week day, which can be explained by the fact that on work days there are traffic rush hours at the beginning and ending of working hours.

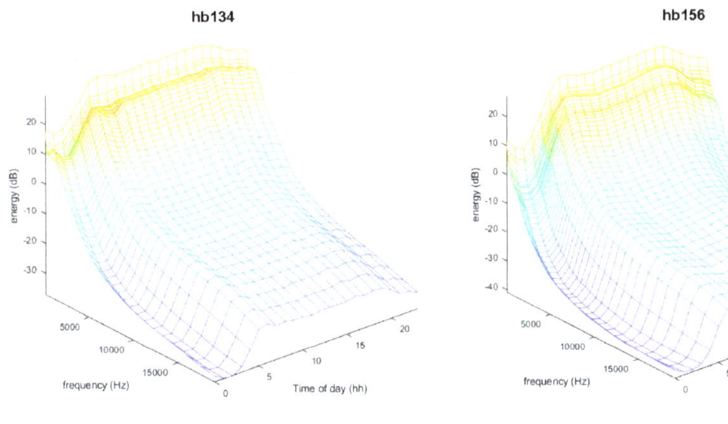

(**a**) Spectrum-time profile of sensor hb134 at week recording day.

(**b**) Spectrum-time profile of sensor hb156 at week recording day.

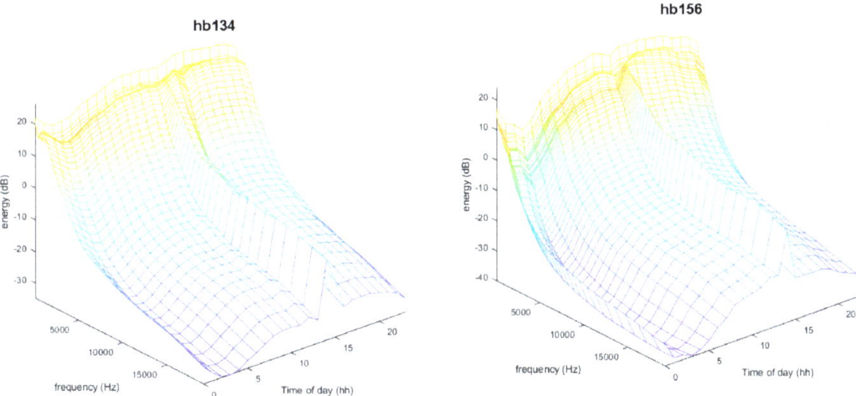

(**c**) Spectrum-time profile of sensor hb134 at weekend recording day.

(**d**) Spectrum-time profile of sensor hb156 at weekend recording day.

Figure 5. Spectrum-time profiles of sensors hb134 and hb156 at weekday (the 2nd of November 2017) and at weekend day (the 5th of November 2017).

4.2. Overall Analysis

Table 1 lists the distribution of the ANEs in terms of their number of occurrences observed within the recorded database, showing both the aggregated and segregated distributions for all sensors, respectively. The total amount of ANEs labeled in the dataset generation represents a 1.8% of all the recorded time, together with around 3.4% of raw data labeled as *cmplx*, which is the data result of the mixture of noises or with unidentifiable sound.

Table 1. Aggregated ANE occurrences and their total duration distribution per category.

Category	Number of Occurrences	Total Duration (s)
ANE	3170	10,752.9
rain	754	6413.5
brak	737	1387.2
thun	236	753.4
tran	76	655.7
trck	380	382.4
bird	482	338.9
sire	55	327.8
horn	210	165.7
alrm	5	84.3
inte	8	69.4
busd	88	56.6
stru	46	45.8
musi	3	31.6
door	83	16.3
airp	4	14.5
bike	3	9.6

Table 1 shows the distribution of total ANE duration segregated by ANE subcategory. Most frequently observed ANEs were rain (6413.5 s) and thunder (753.4 s), due do the thunderstorm episode observed during the weekend day in the city of Rome at 14:00. Non-meteorological ANEs with higher total duration values were sounds of vehicles brakes (1387.2 s), trains (655.7 s), birds (338.9 s), sirens (327.8 s), horns (165.7 s), and sounds of trucks (382.4 s). ANEs with mid-total duration values were alarm sounds (84.3 s), interfering sounds (69.4 s), structure movement sounds (45.8 s), noise of pressurized air or *busd* (56.6 s), and music (31.6 s), while ANEs with lowest presence during the recordings were door sounds (16.3 s), airplanes (14.5 s), and bikes (9.6 s).

In Figure 6, the boxplots of the ANE durations are shown. The longer ANE correspond to *musi* and *alrm*, while the shorter are the *door*, *bird* and *busd*.

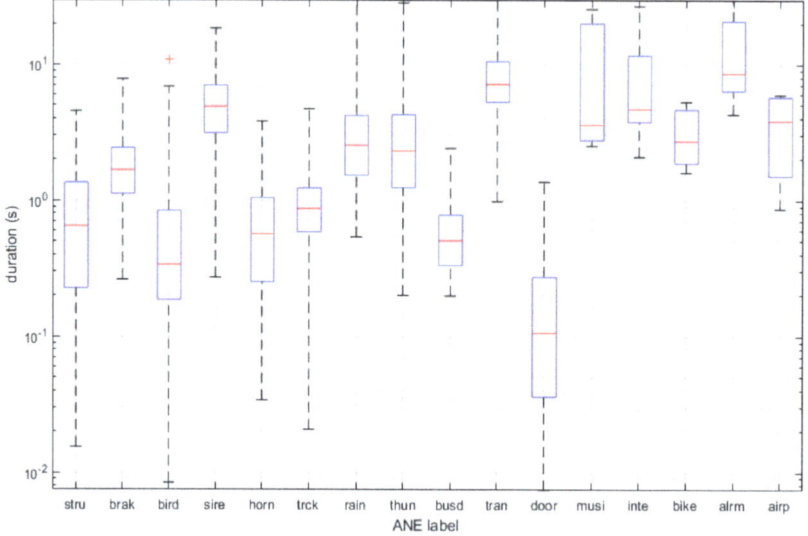

Figure 6. Boxplots of duration for each ANE category. Logarithmic axis of duration is used for a better observation of the duration values.

To determine the salience of the ANEs, the SNR measure is calculated by following the steps in Section 3.3.1. Figure 7 shows the SNR distributions for each ANE category. It can be observed that ANEs with higher SNR are sounds of trains, bikes, sirens, trucks, and horns.

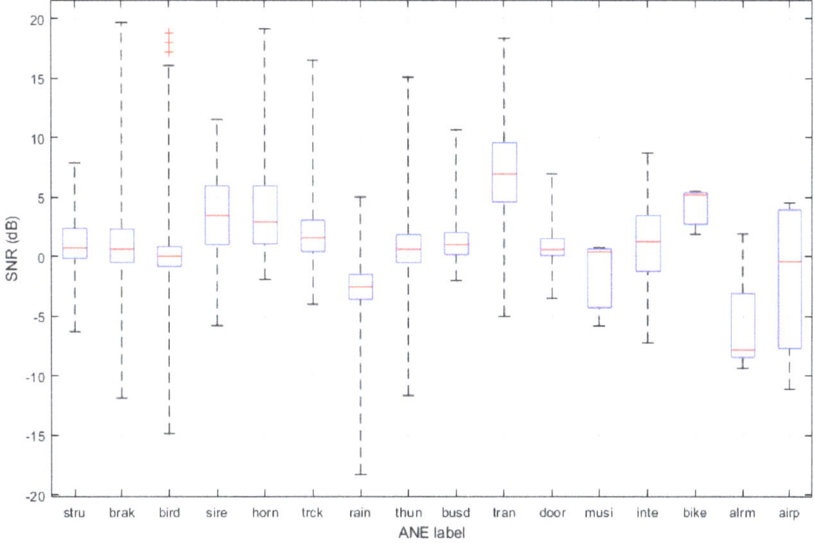

Figure 7. Boxplots of SNR for each ANE subcategory.

4.3. ANEs' Impact on the L_{Aeq}

The ANE contribution in the noise map can be analyzed in several ways. In this study, each ANE has been characterized with the two variables analyzed in the previous section: duration and SNR. Nevertheless, the impact of each ANE to the noise map should be quantified, and preliminary studies with a smaller dataset show that has a strong dependence on the SNR and the duration values [52]. The impact consists of the L_{A300s} measurement of the raw audio minus the same measurement after removing the noise event, which allows discovery of the final contribution of the event to the noise map (more details are given in Section 3.3.2).

To find a first approximation to the ANE subcategories and impact analysis, all the recorded ANEs are depicted according to their characteristics in Figure 8. The SNR is plotted on the vertical axis and the duration on the horizontal axis in a logarithmic scale, for illustration purposes. Also, the size of the marker represents the ANE impact, in a scale indicated in the legend. Besides, the class of the event is depicted in a color scale detailed in the legend, designed to distinguish events with similar parameters easily (the reader is referred to Section 3.2 for a list of all ANE subcategories).

In Figure 8 the reader may appreciate that the class distribution of the events is not uniform, as seen in Table 1. Events shorter than 1 s does not appear to have a significant impact on the L_{A300s} when evaluated individually, as they represent an impact near 0 dB. However, events that last more than 4 s and have a positive SNR may contribute to the L_{A300s} level with impacts of more than 3 dB. Among these significant ANEs, train pass-bys and sirens are the events presenting a higher overall impact score and presence, as also depicted in Figure 7 (where bike noise also appears to have a high median impact because it has only three occurrences). The events presenting a higher impact on the L_{A300s} are trains and sirens, mainly, being only the trains the ANE that surpass the 3 dB level.

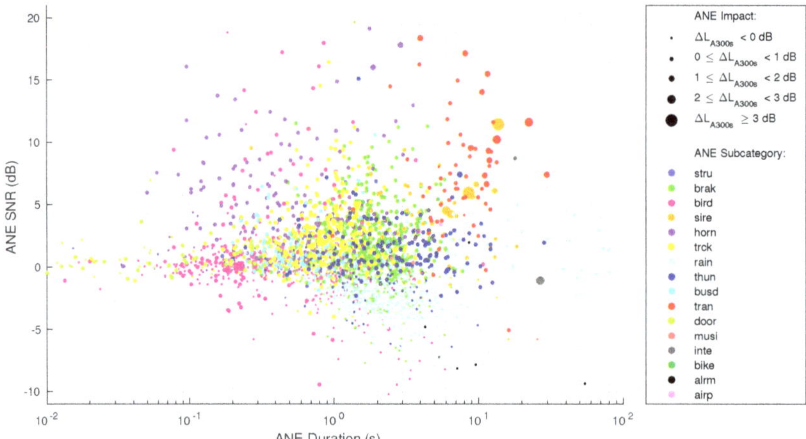

Figure 8. Scatterplot of all ANE parameters separated in subcategories (recorded and labeled in the 2nd and the 5th of November 2017).

It is worth mentioning that in some long events, the SNR and the impact on the L_{A300s} have computation problems. Hence, in Figure 8, only ANEs with both a quantifiable SNR and impact are depicted. Of the total of 2014 events, only three have a duration of more than 300 s, hence, making the impact on the L_{A300s} impossible to calculate. In addition, 61 events give SNR calculation problems, mainly due to high ANE density segments where the interpolation between RTN labels is not possible. All those events have been discarded from these representations.

4.4. Node-Based Analysis

The main upgrade of the dataset detailed in this work, apart from the time distribution of the recordings and the total amount of time labeled, is the fact that the data gathered corresponds to 19 different locations and nodes in a WASN. This leads us to detail a spatial study of the collected data, assuming that not all nodes will observe the same subcategories of ANEs and, of course, the same number of occurrences. This is a key study for the final usage of this dataset, which is the training of the ANED, to detect the ANE in all the sensors of the WASN. A first approach to the homogeneity of this network will be given by the results of the cross analysis between ANE subcategory and sensor Id.

Figure 9a shows the ANE occurrences distribution segregated by sensors Id and ANE subcategory as an image, where it can be appreciated that birds in sensor hb143 are the ones more frequently observed, as birds produce short noise bursts and they can be very repetitive in certain locations and hours. Otherwise, the rain episode during the weekend day of the recording campaign is the second mostly observed event. From the same figure, it can be observed that noise events with quite uniform distribution across all the sensors network are some which are more related to traffic (horns, brakes, and sounds of trucks), and meteorological sounds during the weekend (rain and thunder). The rest of ANEs show a more irregular distribution across the sensor network during the recording campaign.

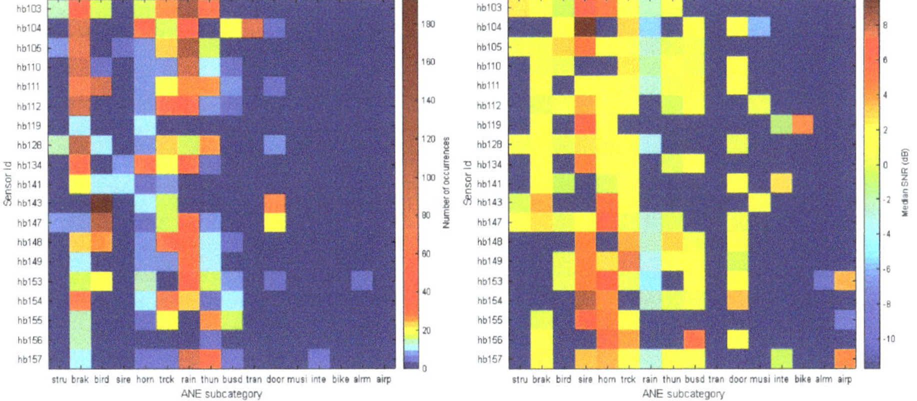

(**a**) ANE occurrences distribution for each ANE subcategory and sensor Id.

(**b**) ANE SNR distribution. Median SNR value is depicted for each ANE subcategory and sensor Id.

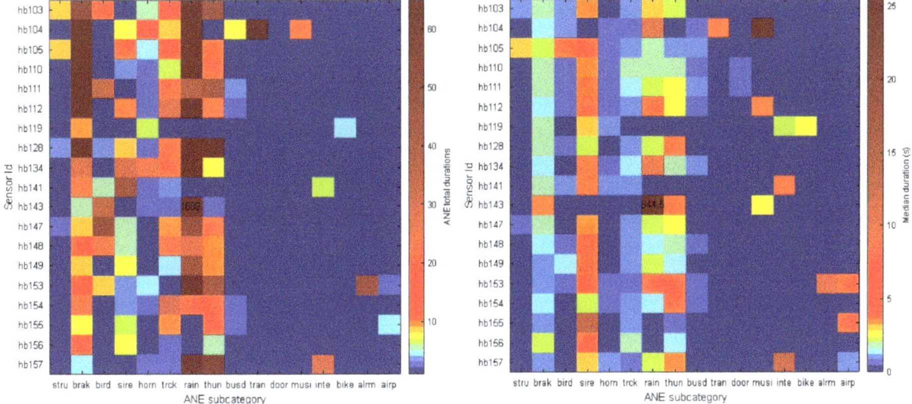

(c) ANE total duration distribution. The median duration value is depicted for each ANE subcategory and sensor Id.

(**d**) ANE duration distribution. The median duration value is depicted for each ANE subcategory and sensor Id.

Figure 9. ANE parameters distribution per subcategory for each WASN node.

Figure 9b shows the median SNR values segregated by sensors and ANE subcategory. There are several nodes and ANE subcategories that exhibit positive SNR values, and their maximum values are attained for sirens, horns, trucks, and *buses* while medium SNRs are observed for brakes, birds, thunder, doors, and airplanes. The negative median SNR values of *rain* can be basically explained by the fact that the used computation methodology of SNR (see [23] for further details) is imprecise when the ANE duration is too high because of the underlying stationary assumption of the RTN assumed within this method.

Figure 9c shows the total duration of ANE recorded for each ANE type and sensor Id. The colormap scale leaves the maximum value as an outlier, depicted as a numeric value 1689 for rain in sensor hb143. Regarding the other values, it can be appreciated that *brak*, *sire*, *horn*, *trck*, *rain* and *thun* are the ANEs with more regular presence across the entire sensor network, while other ANEs like *stru*, *tran*, *musi*, *inte*, *bike*, and *alrm* are more irregularly observed. Figure 9d shows the ANE mean duration distribution per category for each of the WASN sensors, i.e., the median ANE duration statistic has been computed for each cell of the depicted matrix. The color bar legend also presents an outlier, which is reached by sensor hb143 and rain ANE category. The following maximum mean duration is related to sensor hb104 and *musi*.

As a conclusion of this WASN-based analysis, *siren* is the subcategory of ANEs with longest duration and with presence in most of the sensors, and so are *horns*, but the latter have shorter duration; both ANEs present high values of SNR. Another two subcategories are present in most of the sensors in this WASN-based recording campaign; *rain* and *thunder* have a wide presence in the recording, due to the fact that on the 5th of November nearly all the WASN suffered heavy rain in the afternoon. The main difference between them is that while *rain* presents mainly low SNR values, *thunder* presents mid SNR values. *birds* present values of high occurrence and duration for several sensors, but the SNR associated with the birdsong is moderate or low. Nevertheless, the main output of this analysis is that there are few events with uniform appearance in all the sensors, and that most of the subcategories labeled in this work correspond to recordings of fewer groups of sensors. This leads us to the conclusion that the WASN-based recording considering all the sensors in the network was a requirement to observe the variety of the distribution of the events occurring in the entire network.

5. Discussion and Conclusions

In this section, several key aspects of this work are discussed and concluded after the recording, labeling, and analysis of the WASN-based suburban scenario audio samples.

5.1. WASN-Based Dataset Analysis vs. Expert-Based Dataset

As previously stated, a preliminary suburban environment dataset in Rome was published in [23], which consisted of a set of expert-based recordings of limited duration and scope. The work presented in this article is based on the knowledge acquired from that first dataset. It responds to the need for increasing the coverage of the RTN and the ANEs at all hours of the day and night, the weekend, and even when elements external to the noise appear in the measurements, such as adverse weather conditions, i.e., in real-operation conditions.

The WASN-based dataset presented in this work has been enriched with seven classes in comparison to the previous one: *rain, thun, tran, bird, alrm, inte* and *airp*. On the contrary, it has not been possible to record the noise of people talking (*peop*) as in the previous dataset caused by the presence of workers in the portals. All the rest of the ANEs were already part of the expert-based dataset, but with fewer occurrences because the dataset was much smaller. The preliminary dataset contained 3.2% of ANE of the total recorded time, and this new dataset contains 1.8%. A possible explanation to these differences is that the expert-based dataset recording was centered in daytime and this WASN-based dataset has recorded day and night, where night shows low presence of ANE with respect to the day.

The longest ANEs have been found within the *sire* subcategory, while the shortest ones are found in the *door* subcategory, as also happened in the expert-based recording. Moreover, the ANEs labeled as *horn* and *sire* present the highest SNR in both datasets. However, it is worth noting that in this new dataset there are samples of *tran* and *inte* subcategories that also entail high SNRs in many occurrences, a characteristic that was not found in [23].

From this comparative analysis, it can be deduced that the data captured in the expert-based dataset was suitable enough for the first characterization of the suburban soundscape. Nevertheless, the WASN-based recording campaign has shown that there were several noise subcategories that in the preliminary recording campaign had not been recorded and labeled, which present critical characteristics in terms of SNR and duration.

5.2. WASN-Based Dataset and Node Homogeneity

The final use of the presented dataset is the training of the ANED algorithm for a precise detection of the ANEs in all the nodes of the WASN already deployed in the Rome pilot area. Although the developed dataset contains a significantly larger sample of both RTN and ANEs—around 19 times the data gathered in the preliminary expert-based dataset [23]—the analysis of the acoustic data confirms the heterogeneous distribution of the ANEs subcategories in real-life environments already observed in [23]. This heterogeneity has also been observed across the nodes as detailed in Section 4.4. Although some of the subcategories occurred in most of the nodes (e.g., *brak*, *rain*, *truck* and *horn*), there are others found particular in some sensors of the network (e.g., *airp*, *bike* and *train*).

The design of a WASN-based dataset raises the hypothesis of homogeneity of the raw data captured in each of the nodes of the network. This hypothesis was analyzed by means of the distribution of the ANEs in the previous recording campaign [56], with an analysis of the five recording locations. To ensure that the ANED will operate properly in all the nodes, their acoustic environments should present a certain homogeneity in terms of frequency distribution. To collect the data in similar conditions, all nodes have been installed maintaining the same distances and orientations to the road. Nevertheless, a study to evaluate the homogeneity should be carried out taking into account both the locations of the nodes and the occurrences of the ANEs with the final goal of a generalist training of ANED for all the network.

5.3. Impact on the L_{A300s} of the ANE Subcategories

In this paper, the impact on the computation of L_{A300s} has been evaluated for every individual ANE every 5 min. The analysis presents interesting results to be discussed considering the SNR and duration of individual ANEs. The results in Section 4 show the existence of several ANEs with high impact, which present high SNR and long duration. After the individual analysis of the impact of ANEs on the L_{A300s} computation, future work will also take into account the fact that dynamic acoustic mapping in real-life conditions face a more complex operating scenario. In a real-operation scenario, several ANEs can occur in a predefined integration time, so the ANE impact must be evaluated in an aggregated way for each period.

Another relevant result of the individual ANE analysis is the presence of ANEs with negative SNR. As detailed in Section 3.3.1, the SNR is evaluated taking into account each ANE in relation to its surrounding RTN signal level. In certain cases, the RTN decreases as the ANE occurs, so a negative SNR is obtained. Therefore, only those events with positive SNR should be removed from the L_{Aeq} computation, as also concluded in [52].

From this work, it can be concluded that working with data recorded in a real operating scenario is crucial to obtain a reliable modelling of the nodes' acoustic environment, according to the differences observed between the expert- and WASN-based datasets. Moreover, the analysis of the WASN-based collected data again shows the important role played by SNR and duration of individual ANEs in their impact on the L_{Aeq} computation, obtaining ANEs that should be considered for their high impact on the equivalent level.

Author Contributions: R.M.A.-P. has written part of the paper and has participated in the recording and labeling of the dataset. F.O. implemented the SNR and impact calculations, conducted the part of the dataset analysis and participated in the audio labeling and writing of the paper. F.A. has written the related work and reviewed the entire paper, with a special support on the contributions. J.C.S. has worked in the technical part, the recording, the labeling, and the analysis of the dataset.

Funding: The research presented in this work has been partially supported by the LIFE DYNAMAP project (LIFE13 ENV/IT/001254). Francesc Alías thanks the Obra Social La Caixa for grant ref. 2018-URL-IR2nQ-029. Rosa Ma Alsina-Pagès thanks the Obra Social La Caixa for grant ref. 2018-URL-IR2nQ-038. Ferran Orga thanks the support of the European Social Fund and the Secretaria d'Universitats i Recerca del Departament d'Economia i Coneixement of the Catalan Government for the pre-doctoral FI grant No. 2019_FI_B2_00168.

Acknowledgments: The authors would like to thank Marc Hermosilla, Ester Vidaña, Alejandro González and Sergi Barqué for helping in the audio labeling. The authors would like to thank ANAS S.p.A. for the picture of the sensor installed in the portal and Bluewave for the recording of the audio files in the sensors of the WASN.

Conflicts of Interest: The authors declare no conflict of interest.

Abbreviations

The following abbreviations are used in this manuscript:

AED	Audio Event Detector
ANE	Anomalous Noise Event
ANED	Anomalous Noise Event Detection
BCK	Background Noise
CASA	Computational Scene and Event Analysis
CLEAR	Classification of Events, Activities and Relationships
DNN	Deep Neural Networks
DYNAMAP	Dynamic Noise Mapping
END	European Noise Directive
GTCC	Gammatone Cepstral Coefficients
MFCC	Mel-Frequency Cepstral Coefficients
RTN	Road-Traffic Noise
SED	Sound Event Detector
SONYC	Sounds of New York City
SNR	Signal-to-Noise Ratio
VPN	Virtual Private Network
WASN	Wireless Acoustic Sensor Network

References

1. Goines, L.; Hagler, L. Noise Pollution: A Modern Plague. *South. Med. J. Birm. Ala.* **2007**, *100*, 287–294. [CrossRef]
2. Guite, H.; Clark, C.; Ackrill, G. The impact of the physical and urban environment on mental well-being. *Public Health* **2006**, *120*, 1117–1126. [CrossRef]
3. Hänninen, O.; Knol, A.B.; Jantunen, M.; Lim, T.A.; Conrad, A.; Rappolder, M.; Carrer, P.; Fanetti, A.C.; Kim, R.; Buekers, J.; et al. Environmental burden of disease in Europe: Assessing nine risk factors in six countries. *Environ. Health Perspect.* **2014**, *122*, 439. [CrossRef] [PubMed]
4. Bello, J.P.; Silva, C. SONYC: A System for Monitoring, Analyzing, and Mitigating Urban Noise Pollution. *Commun. ACM* **2019**, *62*, 68–77. [CrossRef]
5. EU. Directive 2002/49/EC of the European Parliament and the Council of 25 June 2002 relating to the assessment and management of environmental noise. *Off. J. Eur. Commun.* **2002**, *L189*, 12–25.
6. Manvell, D. Utilising the Strengths of Different Sound Sensor Networks in Smart City Noise Management. In Proceedings of the EuroNoise 2015, Maastrich, The Netherlands, 1–3 June 2015; EAA-NAG-ABAV: Maastrich, The Netherlands, 2015; pp. 2305–2308.
7. Alías, F.; Alsina-Pagès, R.M. Review of Wireless Acoustic Sensor Networks for Environmental Noise Monitoring in Smart Cities. *J. Sens.* **2019**. [CrossRef]
8. Nencini, L.; De Rosa, P.; Ascari, E.; Vinci, B.; Alexeeva, N. SENSEable Pisa: A wireless sensor network for real-time noise mapping. In Proceedings of the EuroNoise 2012, Prague, Czech Republic, 10–13 June 2012; pp. 10–13.
9. Botteldooren, D.; De Coensel, B.; Oldoni, D.; Van Renterghem, T.; Dauwe, S. Sound monitoring networks new style. In Proceedings of the Acoustics 2011, Gold Coast, Australia, 2–4 November 2011; Mee, D., Hillock, I.D., Eds.; Australian Acoustical Society: Queensland, Australia, 2011; pp. 93:1–93:5.
10. Mietlicki, F.; Mietlicki, C.; Sineau, M. An innovative approach for long-term environmental noise measurement: RUMEUR network. In Proceedings of the EuroNoise 2015, Maastrich, The Netherlands, 1–3 June 2015; EAA-NAG-ABAV: Maastrich, The Netherlands, 2015; pp. 2309–2314.

11. Camps-Farrés, J.; Casado-Novas, J. Issues and challenges to improve the Barcelona Noise Monitoring Network. In Proceedings of the EuroNoise 2018, Crete, Greece, 27–31 May 2018; EAA—HELINA: Heraklion, Crete, Greece, 2018; pp. 693–698.
12. Bain, M. SENTILO—Sensor and Actuator Platform for Smart Cities. 2014. Available online: https://joinup.ec.europa.eu/document/sentilo-sensor-and-actuator-platform-smart-cities (accessed on 25 June 2018).
13. Sevillano, X.; Socoró, J.C.; Alías, F.; Bellucci, P.; Peruzzi, L.; Radaelli, S.; Coppi, P.; Nencini, L.; Cerniglia, A.; Bisceglie, A.; et al. DYNAMAP—Development of low cost sensors networks for real time noise mapping. *Noise Mapp.* **2016**, *3*, 172–189. [CrossRef]
14. De la Piedra, A.; Benitez-Capistros, F.; Dominguez, F.; Touhafi, A. Wireless sensor networks for environmental research: A survey on limitations and challenges. In Proceedings of the 2013 IEEE EUROCON, Zagreb, Croatia, 1–4 July 2013; pp. 267–274.
15. Rawat, P.; Singh, K.D.; Chaouchi, H.; Bonnin, J.M. Wireless Sensor Networks: A Survey on Recent Developments and Potential Synergies. *J. Supercomput.* **2014**, *68*, 1–48. [CrossRef]
16. Bertrand, A. Applications and trends in wireless acoustic sensor networks: A signal processing perspective. In Proceedings of the 18th IEEE Symposium on Communications and Vehicular Technology in the Benelux (SCVT), Ghent, Belgium, 22–23 November 2011; pp. 1–6.
17. Griffin, A.; Alexandridis, A.; Pavlidiand, D.; Mastorakis, Y.; Mouchtaris, A. Localizing multiple audio sources in a wireless acoustic sensor network. *Signal Process.* **2015**, *107*, 54–67. [CrossRef]
18. Socoró, J.C.; Alías, F.; Alsina-Pagès, R.M. An Anomalous Noise Events Detector for Dynamic Road Traffic Noise Mapping in Real-Life Urban and Suburban Environments. *Sensors* **2017**, *17*, 2323. [CrossRef] [PubMed]
19. Wang, D.; Brown, G.J. *Computational Auditory Scene Analysis: Principles, Algorithms, and Applications*; Wiley-IEEE Press: New York, NY, USA, 2006.
20. Giannoulis, D.; Benetos, E.; Stowell, D.; Rossignol, M.; Lagrange, M.; Plumbley, M.D. Detection and classification of acoustic scenes and events: An IEEE AASP challenge. In Proceedings of the 2013 IEEE Workshop on Applications of Signal Processing to Audio and Acoustics, New Paltz, NY, USA, 20–23 October 2013; pp. 1–4.
21. Foggia, P.; Petkov, N.; Saggese, A.; Strisciuglio, N.; Vento, M. Reliable detection of audio events in highly noisy environments. *Pattern Recognit. Lett.* **2015**, *65*, 22–28. [CrossRef]
22. Mesaros, A.; Diment, A.; Elizalde, B.; Heittola, T.; Vincent, E.; Raj, B.; Virtanen, T. Sound event detection in the DCASE 2017 Challenge. *IEEE/ACM Trans. Audio Speech Language Process.* **2019**, *27*, 992–1006. [CrossRef]
23. Alías, F.; Socoró, J.C. Description of anomalous noise events for reliable dynamic traffic noise mapping in real-life urban and suburban soundscapes. *Appl. Sci.* **2017**, *7*, 146. [CrossRef]
24. Socoró, J.C.; Alsina-Pagès, R.M.; Alías, F.; Orga, F. Adapting an Anomalous Noise Events Detector for Real-Life Operation in the Rome Suburban Pilot Area of the DYNAMAP's Project. In Proceedings of the EuroNoise, Crete, Greece, 27–31 May 2018.
25. Nakajima, Y.; Sunohara, M.; Naito, T.; Sunago, N.; Ohshima, T.; Ono, N. DNN-based Environmental Sound Recognition with Real-recorded and Artificially-mixed Training Data. In Proceedings of the 45th International Congress and Exposition on Noise Control Engineering (InterNoise 2016), Hamburg, Germany, 21–24 August 2016; German Acoustical Society (DEGA): Hamburg, Germany, 2016; pp. 3164–3173.
26. Mesaros, A.; Heittola, T.; Virtanen, T. TUT database for acoustic scene classification and sound event detection. In Proceedings of the 24th European Signal Processing Conference (EUSIPCO 2016), Budapest, Hungary, 28 August–2 September 2016; Volume 2016, pp. 1128–1132.
27. Stowell, D.; Giannoulis, D.; Benetos, E.; Lagrange, M.; Plumbley, M.D. Detection and Classification of Acoustic Scenes and Events. *IEEE Trans. Multimed.* **2015**, *17*, 1733–1746. [CrossRef]
28. Heittola, T.; Mesaros, A.; Eronen, A.; Virtanen, T. Context-dependent sound event detection. *EURASIP J. Audio Speech Music Process.* **2013**, *2013*, 1–13. [CrossRef]
29. Foggia, P.; Petkov, N.; Saggese, A.; Strisciuglio, N.; Vento, M. Audio Surveillance of Roads: A System for Detecting Anomalous Sounds. *IEEE Trans. Intell. Transp. Syst.* **2016**, *17*, 279–288. [CrossRef]
30. Zambon, G.; Benocci, R.; Bisceglie, A.; Roman, H.E.; Bellucci, P. The LIFE DYNAMAP project: Towards a procedure for dynamic noise mapping in urban areas. *Appl. Acoust.* **2017**, *124*, 52–60. [CrossRef]
31. Bellucci, P.; Peruzzi, L.; Zambon, G. LIFE DYNAMAP project: The case study of Rome. *Appl. Acoust.* **2017**, *117*, 193–206. [CrossRef]

32. Stowell, D.; Plumbley, M.D. An open dataset for research on audio field recording archives: Freefield1010. *arXiv* **2013**, arXiv:1309.5275.
33. Salamon, J.; Jacoby, C.; Bello, J.P. A dataset and taxonomy for urban sound research. In Proceedings of the 22nd ACM International Conference on Multimedia, Orlando, FL, USA, 3–7 November 2014; pp. 1041–1044.
34. Socoró, J.C.; Ribera, G.; Sevillano, X.; Alías, F. Development of an Anomalous Noise Event Detection Algorithm for dynamic road traffic noise mapping. In Proceedings of the 22nd International Congress on Sound and Vibration (ICSV22), Florence, Italy, 12–16 July; The International Institute of Acoustics and Vibration: Florence, Italy, 2015; pp. 1–8.
35. Piczak, K.J. ESC: Dataset for Environmental Sound Classification. In Proceedings of the 23rd ACM International Conference on Multimedia, Brisbane, Australia, 26–30 October 2015; pp. 1015–1018.
36. Rossignol, M.; Lafay, G.; Lagrange, M.; Misdariis, N. SimScene: A web-based acoustic scenes simulator. In Proceedings of the 1st Web Audio Conference (WAC), Paris, France, 26–28 January 2015; pp. 1–6.
37. Gloaguen, J.R.; Can, A.; Lagrange, M.; Petiot, J.F. Estimating Traffic Noise Levels using Acoustic Monitoring: A Preliminary Study. In Proceedings of the Detection and Classification of Acoustic Scenes and Events 2016 (DCASE'2016), Budapest, Hungary, 3 September 2016; pp. 40–44.
38. Çakır, E.; Parascandolo, G.; Heittola, T.; Huttunen, H.; Virtanen, T. Convolutional Recurrent Neural Networks for Polyphonic Sound Event Detection. *IEEE/ACM Trans. Audio Speech Lang. Process.* **2017**, *25*, 1291–1303. [CrossRef]
39. Salamon, J.; MacConnell, D.; Cartwright, M.; Li, P.; Bello, J.P. Scaper: A library for soundscape synthesis and augmentation. In Proceedings of the 2017 IEEE Workshop on Applications of Signal Processing to Audio and Acoustics (WASPAA), New Paltz, NY, USA, 15–18 October 2017; pp. 344–348.
40. Salamon, J.; Bello, J.P. Deep Convolutional Neural Networks and Data Augmentation for Environmental Sound Classification. *IEEE Signal Process. Lett.* **2017**, *24*, 279–283. [CrossRef]
41. Temko, A.; Malkin, R.; Zieger, C.; Macho, D.; Nadeu, C.; Omologo, M. CLEAR Evaluation of Acoustic Event Detection and Classification Systems. In *Multimodal Technologies for Perception of Humans: First International Evaluation Workshop on Classification of Events, Activities and Relationships, CLEAR 2006*; Stiefelhagen, R., Garofolo, J., Eds.; Springer: Berlin/Heidelberg, Germany, 2007; pp. 311–322.
42. Temko, A. Acoustic Event Detection and Classification. Ph.D. Thesis, Universitat Politècnica de Catalunya, Barcelona, Spain, 2007.
43. Heittola, T.; Klapuri, A. TUT Acoustic Event Detection System 2007. In *Multimodal Technologies for Perception of Humans. International Evaluation Workshops CLEAR 2007 and RT 2007*; Stiefelhagen, R., Bowers, R., Fiscus, J., Eds.; Springer: Berlin/Heidelberg, Germany, 2008; pp. 364–370.
44. Van Grootel, M.W.W.; Andringa, T.C.; Krijnders, J.D. DARES-G1: Database of Annotated Real-world Everyday Sounds. In Proceedings of the NAG/DAGA Meeting 2009, Rotterdam, The Netherlands, 23–26 March 2009.
45. Giannoulis, D.; Stowell, D.; Benetos, E.; Rossignol, M.; Lagrange, M.; Plumbley, M.D. A database and challenge for acoustic scene classification and event detection. In Proceedings of the 21st European Signal Processing Conference (EUSIPCO 2013), Marrakesh, Marocco, 9–13 September 2013; pp. 1–5.
46. Mesaros, A.; Heittola, T.K.; Benetos, E.; Foster, P.; Lagrange, M.; Virtanen, T.; Plumbley, M.D. Detection and Classification of Acoustic Scenes and Events: Outcome of the DCASE 2016 Challenge. *IEEE/ACM Trans. Audio Speech Lang. Process.* **2018**, *26*, 379 – 393. [CrossRef]
47. Mesaros, A.; Heittola, T.; Diment, A.; Elizalde, B.; Shah, A.; Vincent, E.; Raj, B.; Virtanen, T. DCASE 2017 challenge setup: Tasks, datasets and baseline system. In Proceedings of the DCASE 2017-Workshop on Detection and Classification of Acoustic Scenes and Events, Munich, Germany, 16–17 November 2017
48. Nencini, L. DYNAMAP monitoring network hardware development. In Proceedings of the 22nd International Congress on Sound and Vibration (ICSV22), Florence, Italy, 12–16 July 2015; The International Institute of Acoustics and Vibration (IIAV): Florence, Italy, 2015; pp. 1–4.
49. Nencini, L. Progetto e realizzazione del sistema di monitoraggio nell'ambito del progetto Dynamap. In Proceedings of the 44° Congress of the Italian Acoustic Association, Pavia, Italy, 7–9 June 2017.
50. International Electroacoustics Commission. *Electroacoustics-Sound Level Meters-Part 1: Specifications (IEC 61672-1)*; International Electroacoustics Commission: Geneva, Switzerland, 2013.

51. Nencini, L.; Bisceglie, A.; Bellucci, P.; Peruzzi, L. Identification of failure markers in noise measurement low cost devices. In Proceedings of the 45th International Congress and Exposition on Noise Control Engineering (InterNoise 2016), Hamburg, Germany, 21–24 August 2016; pp. 6362–6369.
52. Orga, F.; Alías, F.; Alsina-Pagès, R.M. On the Impact of Anomalous Noise Events on Road Traffic Noise Mapping in Urban and Suburban Environments. *Int. J. Environ. Res. Public Health* **2017**, *15*, 13. [CrossRef] [PubMed]
53. Pierre, R.L.S.; Maguire, D.J. The impact of A-weighting sound pressure level measurements during the evaluation of noise exposure. In Proceedings of the NOISE-CON 2004, Baltimore, MD, USA,12–14 July 2004; pp. 1–8.
54. Socoró, J.C.; Alsina-Pagès, R.M.; Alías, F.; Orga, F. Acoustic Conditions Analysis of a Multi-Sensor Network for the Adaptation of the Anomalous Noise Event Detector. *Proceedings* **2019**, *4*, 51. [CrossRef]
55. Valero, X.; Alías, F. Gammatone Cepstral Coefficients: Biologically Inspired Features for Non-Speech Audio Classification. *IEEE Trans. Multimed.* **2012**, *14*, 1684–1689. [CrossRef]
56. Orga, F.; Alsina-Pagès, R.M.; Alías, F.; Socoró, J.C.; Bellucci, P.; Peruzzi, L. Anomalous Noise Events Considerations for the Computation of Road Traffic Noise Levels in Suburban Areas: The DYNAMAP's Rome Case Study. In Proceedings of the 44° Congress of the Italian Acoustic Association, Pavia, Italy, 7–9 June 2017.

Article

Evaluation of Machine Learning Algorithms for Surface Water Extraction in a Landsat 8 Scene of Nepal †

Tri Dev Acharya [1,2,3], **Anoj Subedi** [4] **and Dong Ha Lee** [2,*]

[1] Institute of Industrial Technology, Kangwon National University, Chuncheon 24341, Korea; tridevacharya@kangwon.ac.kr
[2] Department of Civil Engineering, Kangwon National University, Chuncheon 24341, Korea
[3] School of Geomatics and Urban Spatial Information, Beijing University of Civil Engineering and Architecture, Beijing 102616, China
[4] Institute of Forestry, Pokhara Campus, Tribhuvan University, Pokhara 33700, Nepal; anojsubedi99@gmail.com
* Correspondence: geodesy@kangwon.ac.kr; Tel.: +82-33-250-6232

† This paper is an extended version of the conference paper: Acharya, T.D.; Subedi, A.; Huang, H.; Lee, D.H. Classification of Surface Water using Machine Learning Methods from Landsat Data in Nepal. In Proceedings of the 5th International Electronic Conference on Sensors and Applications, 15–30 November 2018.

Received: 31 March 2019; Accepted: 17 June 2019; Published: 20 June 2019

Abstract: With over 6000 rivers and 5358 lakes, surface water is one of the most important resources in Nepal. However, the quantity and quality of Nepal's rivers and lakes are decreasing due to human activities and climate change. Despite the advancement of remote sensing technology and the availability of open access data and tools, the monitoring and surface water extraction works has not been carried out in Nepal. Single or multiple water index methods have been applied in the extraction of surface water with satisfactory results. Extending our previous study, the authors evaluated six different machine learning algorithms: Naive Bayes (NB), recursive partitioning and regression trees (RPART), neural networks (NNET), support vector machines (SVM), random forest (RF), and gradient boosted machines (GBM) to extract surface water in Nepal. With three secondary bands, slope, NDVI and NDWI, the algorithms were evaluated for performance with the addition of extra information. As a result, all the applied machine learning algorithms, except NB and RPART, showed good performance. RF showed overall accuracy (OA) and kappa coefficient (Kappa) of 1 for the all the multiband data with the reference dataset, followed by GBM, NNET, and SVM in metrics. The performances were better in the hilly regions and flat lands, but not well in the Himalayas with ice, snow and shadows, and the addition of slope and NDWI showed improvement in the results. Adding single secondary bands is better than adding multiple in most algorithms except NNET. From current and previous studies, it is recommended to separate any study area with and without snow or low and high elevation, then apply machine learning algorithms in original Landsat data or with the addition of slopes or NDWI for better performance.

Keywords: surface water mapping; machine learning; naive Bayes; recursive partitioning and regression trees; neural networks; support vector machines; random forest; gradient boosted machines; Landsat; Nepal

---

1. Introduction

Nepal is a geographically diverse country with flats in the south and increasing hills, to the mighty Himalayas in the north. In Nepal, approximately 70% to 90% of the total annual rainfall occurs during the monsoon period resulting in high runoff and sediment discharge causing surface water area change [1]. Thus, it is rich in water resources with approximately 600 rivers [2] and 5358 lakes [3]. Due to such seasonal variation and large surface water area, it is difficult to track changes in surface water [4,5]. Furthermore, the change in stream-flows due to climate change has also been predicted [6,7]. Therefore, the monitoring and estimation of surface water is an essential task.

In such cases, remote sensing technology plays a very important role on detecting, extracting, and monitoring surface water [8,9]. Open and free access mid-resolution multi-spectral satellite images such as Landsat brings further benefits in the process [10]. Thus, the authors begin to utilize the Landsat database to extract surface water from a small case of Phewa to a Landsat scene that covers different types of surface water along with features that resemble water, such as shadows, forests, built-ups, snow and clouds. In previous studies, the authors evaluated water index methods, single and combined [11,12], along with the segmentation of the scene [13]. Our latest work showed promising results for a scene in which segmentation and the optimum threshold were manually identified based on the given set of reference dataset. As a next step, automated extraction of surface water with well-known supervised classification approaches were evaluated [14].

With recent developments in computing technology, the machines are cheaper, and the algorithms are efficient. Therefore, the abundance of these machines and machine learning algorithms have been widely applicable in almost every aspect of human life. Moreover, their optimization has outperformed the classical ones. Numerous machine learning algorithms have been applied for remotely sensed imageries [15–19]. These algorithms can be divided broadly into three categories: (a) Unsupervised learning; (b) supervised learning; and (c) reinforcement learning. Unsupervised learning groups are given an unlabeled dataset based on the implicit relationship/function. Supervised learning utilizes a certain labeled instance (training dataset) to predict a similar dataset. Reinforcement learning does not provide a precise label, rather it takes the next step based on the goal-oriented feedback available for each prediction. Reinforcement learning is still in the developing stage, and as there are no errors, it could be wrong with each positive reward. Results of unsupervised learning cannot be ascertained and can be less accurate, whereas supervised provides specific class and labels with better accuracy [20]. Some of the most common supervised algorithms are decision trees, naive Bayes (NB), neural networks (NNET), regression, support vector machines (SVM), and ensemble methods [21]. The libraries for these algorithms have been well developed and implemented in reliable ecosystems of open source tools, such as Python and R languages. Despite the availability of open access data and tools, the evaluation of such in Nepal has never been documented. Moreover, the challenge of varying conditions in a single scene is also new for testing the performance of machine learning algorithms in the extraction of surface water.

Hence, the motivation of this work is to introduce the application of the most common algorithms used by the remote sensing community in Nepal and evaluate their performance for surface water extraction. The six most common algorithms, naive Bayes (NB), recursive partitioning and regression trees (RPART), neural networks (NNET), support vector machine (SVM), random forest (RF), and gradient boosted machines (GBM) were evaluated in a Landsat 8 operational land imager (OLI) bands. Also, the slope, normalized difference vegetation index (NDWI) and normalized difference water index (NDWI) were combined one at a time and all three at once with OLI bands to evaluate whether the combination can overcome the limitations of the original bands in water extraction. In the future, such evaluation will assist in selection of proper methods to develop an effective time series database at national scale in Nepal.

2. Materials and Methods

As the study is the extension of our previous study [13], the authors utilized the same Landsat scene and reference dataset in this study. Hence, details on the study area and data can be found in Acharya et al. [13]. Pre-processing of the Landsat scene was carried out in Environment for Visualizing Images (ENVI) version 5.3 (Exelis Visual Information Solutions, Boulder, CO, USA), cartographic maps were produced in ArcGIS 10.5 (Environmental Systems Research Institute, California, CA, USA), and the machine learning process were carried using Classification And REgression Training (CARET) package in R 3.5.0 (The R Foundation, Vienna, Austria) software packages.

2.1. Cases

To better evaluate and compare the results, our previous study had six classes in the scene [13]. However, it was found that Himalayan areas with shadows and melting ice are quite challenging in the task. Hence, the authors introduced two more cases that will help better understand the results. Figure 1 shows the Landsat 8 scene in pansharpened natural color composite image with water and non-water reference dataset and cases.

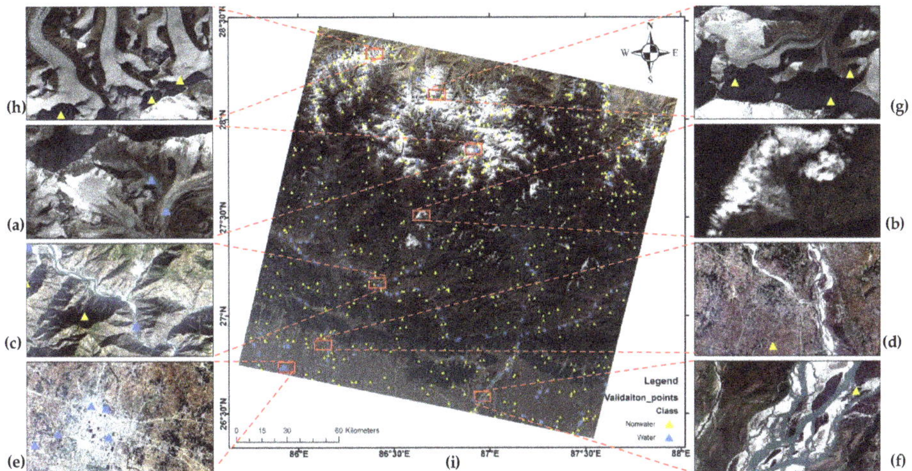

Figure 1. Pansharpened Landsat 8 natural color composite image with water and non-water reference dataset; (**i**) Each red box represents different surface water bodies and noises case in the scene (**a**–**h**).

2.2. Methods

Using the digital elevation model (DEM) and pre-possessed OLI bands (LS8), three secondary bands were created: Slope, NDVI and NDWI. Figure 2 shows the secondary bands. After which these three secondary bands were stacked one by one with LS8 such that this study has three new datasets as LS8 + Slope, LS8 + NDVI, and LS + NDWI. Finally, all three were stacked with LS8 to form another new data LS8 + Slope + NDVI + NDWI.

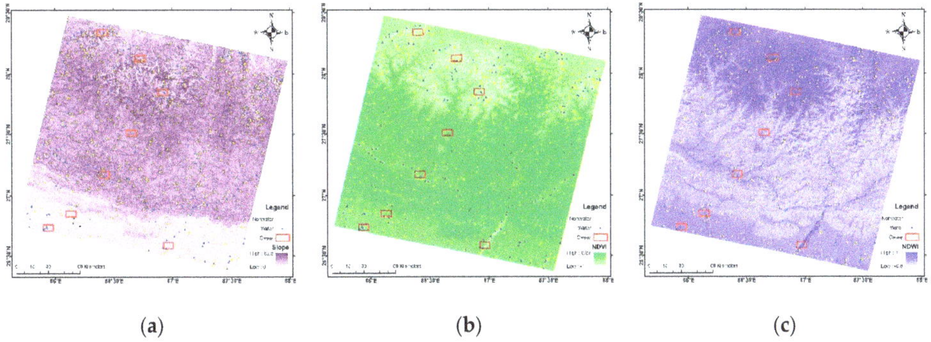

Figure 2. Secondary bands (**a**) slope; (**b**) NDVI and (**c**) NDWI.

A total of 800 reference datasets were used in the whole scene. To minimize the overfitting or selection bias in the predictive ability of machine learning algorithms, the authors used ten folds cross-validation to train the model. In the process, the data is divided into ten sets, nine sets are first used to train the model and the remaining one is used for validation purposes. The process is repeated ten times with each of the ten subsamples being used for both training and validation. Thus, each observation is used exactly once for training and validation.

After forming the dataset, six machine learning methods, NB, RPART, NNET, SVM, RF, and GBM, were implemented in the CARET package of R, and were used for training and preparing the models. The advantage of the CARET package is that it provides complete machine learning processes from data preparation, training, modelling and prediction to validation. Moreover, the train function in the caret sets up a grid of tuning parameters as required for the algorithm which fits each model and calculates a resampling based performance measure. As details on these models can be widely found in the literature, the authors will be briefly describing these algorithms.

NB is a probabilistic classifier based on Bayes' theorem with the independent assumptions between predictors. It is easy to build, with no complicated iterative parameter estimation and suited particularly when the dimensionality of the inputs is high. With few tuneable parameters and fast, they end up being very useful as a quick-and-dirty baseline for a classification problem.

RPART is a type of binary tree used for classification or regression tasks. It performs a search over all possible splits by maximizing an information measure of node impurity, selecting the covariate showing the best split.

NNET in R, is a feed-forward neural network with a single hidden layer flowing left to right. Feed-forward neural networks were the first type of artificial neural network invented, and are simpler than their counterparts, recurrent neural networks. They are called feed-forward networks because information only travels forward in the network (no loops), first through the input nodes, then through the hidden nodes (if present), and finally through the output nodes. These are primarily used for supervised learning in cases where the data to be learned are neither sequential nor time dependent.

SVM is a data classification method that separates data using a hyperplane. In other words, for the given labelled training data (supervised learning), the algorithm outputs an optimal hyperplane which separates only one type of data. The SVM technique is generally useful for data that is non-regularity, which means data whose distribution is unknown.

RF is a meta estimator that fits several decision tree classifiers on various sub-samples of the dataset and uses averaging to improve the predictive accuracy and control over-fitting.

GBM is a class of ensemble learning techniques to create a collection of shallow and weak successive trees with each tree learning and improving on the previous based on a cost function (for example, squared error).

All these six models for five different datasets were applied to the full scene to classify the image into binary water and non-water maps. After the classification, the full reference dataset was evaluated for overall accuracy (OA) and kappa coefficient (Kappa) [22].

3. Results and Discussion

In this section, first, the results of the cross-validated models were assessed, and then they were applied to the whole scene for the surface water extraction. Next, the complete reference dataset was tested against the predicted results. In the end, a detailed comparison and discussion of the different machine learning algorithms were done to evaluate the performance case by case.

Figure 3 shows the boxplot of the OA and Kappa for different models based on the cross-validated data for five different multiband data. At first glance, it can be seen that, other than RPART and ND, four algorithms showed quite high OA and Kappa in the cross-validation. In the case with only LS8 multiband data, NNET and RF were able to achieve maximum OA and Kappa. After adding the slope, NNET and RF produced the model with maximum OA and Kappa. However, adding NDVI and NDWI with the LS8 bands, only NNET achieved improved OA and Kappa. Moreover, it can be clearly seen that the OAs achieved, with all the machine learning algorithms, narrow ranges with the addition of NDWI, followed by NDVI and the slope. The mean of OA and Kappa shifted towards higher values after adding only the slope and all three secondary bands for all algorithms, except RPART.

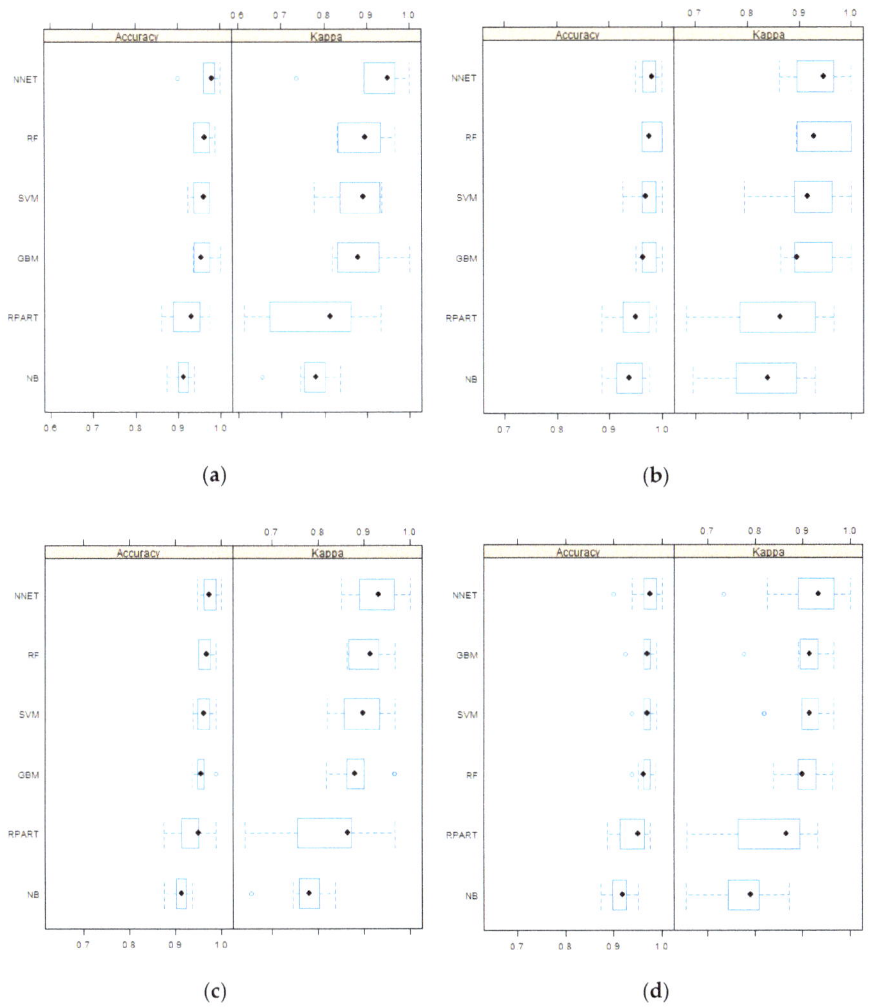

Figure 3. *Cont.*

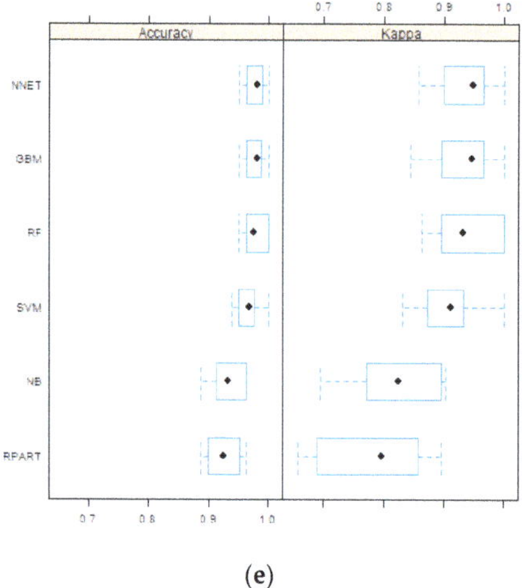

(e)

Figure 3. The boxplot of overall accuracy (OA) and the Kappa coefficient (Kappa) using different machine learning algorithms for provided multiband data: (**a**) LS8; (**b**) LS8 + Slope; (**c**) LS8 + NDVI; (**d**) LS8 + NDWI; and (**e**) LS8 + Slope + NDVI + NDWI.

In the CARET package, the train function produces the model with tuning parameters. Therefore, the produced models can be easier to apply in the prediction. Figure 4 shows the predicted result of all the algorithm models for all the multiband data. Based on the visual inspection, all the algorithms performed well in the lower flat and hilly regions compared to the Himalayas. Adding the slope made the result visually better in most algorithms, except the RPART. Similarly, adding only NDWI and adding all three secondary bands produced visually better results. In contrast, NDVI only improved with RF and GBM algorithms and others produced many misclassified surface water bodies. For further evaluation on how well all these methods have performed in the scene, eight different types of cases were carefully analyzed and compared for each algorithm in the section below.

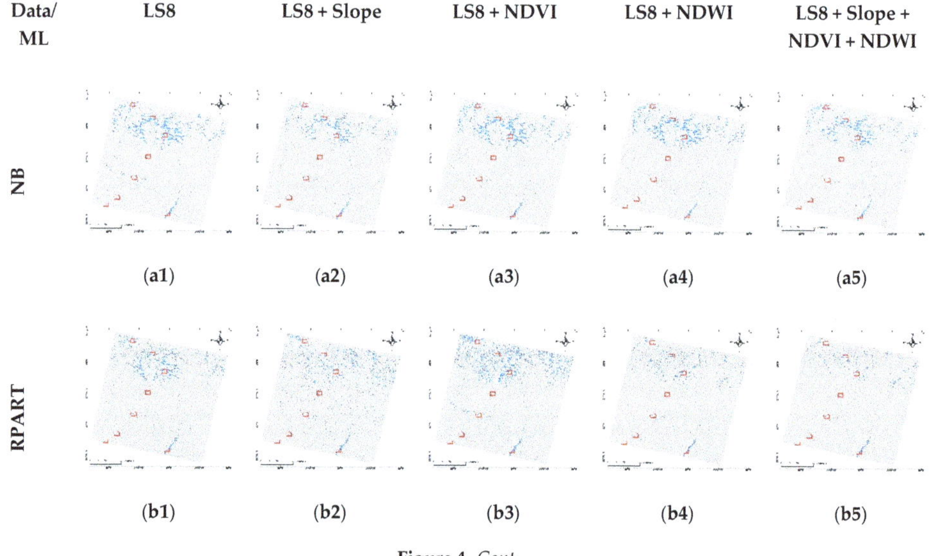

Figure 4. *Cont.*

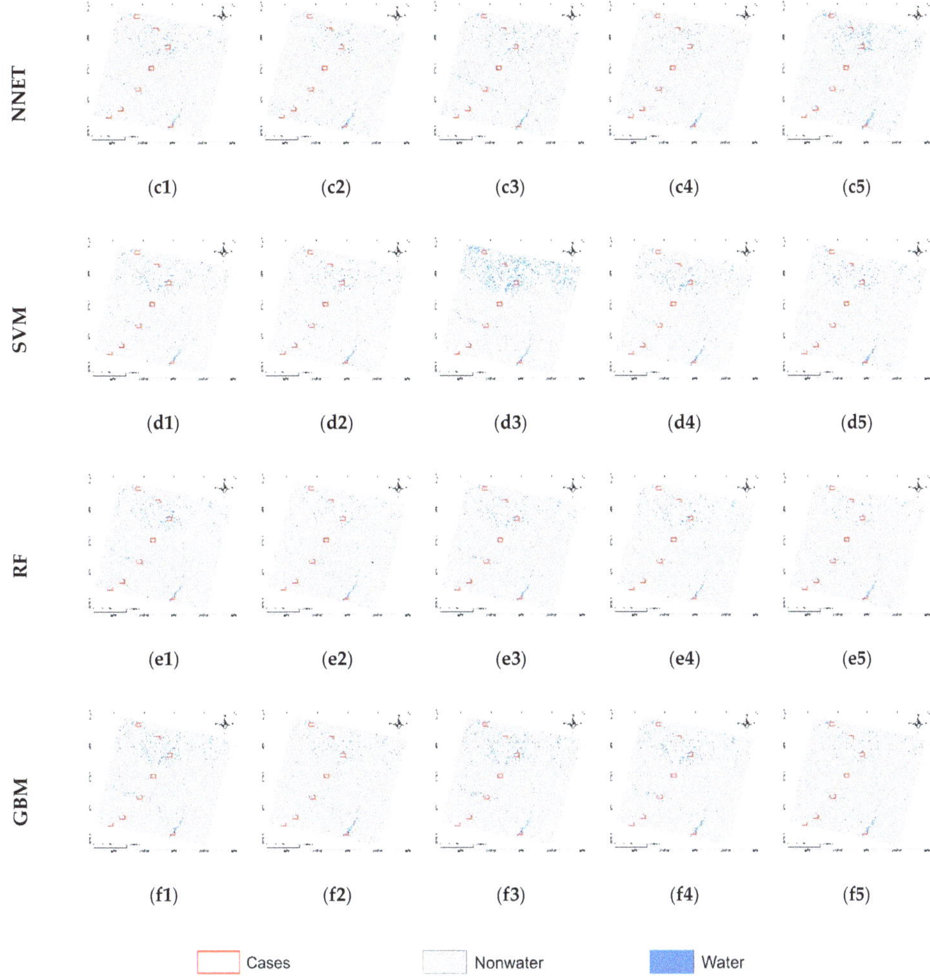

Figure 4. Results of surface water extraction using different machine learning methods (**a–f**) for provided multiband data (**1–5**).

For the quantitative evaluation, the confusion matrix-based OA and Kappa were produced for the full reference class and predicted class. In Figure 5, both OA and Kappa shows similar patterns for all the algorithms in response to the addition of the secondary band. RF shows both OA and Kappa 1 for all multiband data. Following this, GBM, NNET and SVM performed well. However, the NB and RPART performance were among the worst. In NB, LS8 and LS + NDWI data did well compared to others. The addition of NDVI or the slope or all three secondary bands decreased the performance. While in the case of RPART, the addition of one or all three bands improved the OA and Kappa. In NNET, the addition of all bands increased the performance compared to each secondary band, while it was opposite for the SVM i.e., a decrease in performance. In GBM, adding NDWI boosted the performance the highest, and the slope and all three did well, but the addition of NDVI decreased the performance.

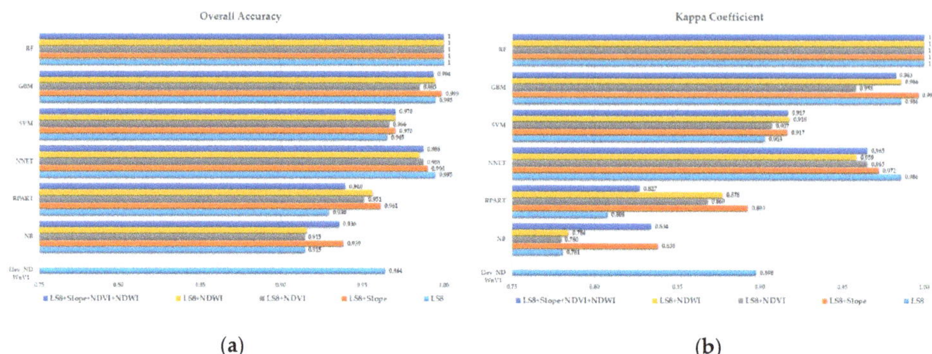

Figure 5. (**a**) Overall accuracy (OA) and (**b**) Kappa coefficient (Kappa) of all the algorithms for provided multiband data.

Except RF and NNET, similar patterns of performance were seen. NDWI followed by the slope, increases the performance of the machine learning algorithms. In contrast, NDVI or adding all bands reduced the performance. It shows that adding specific secondary bands with the original LS8 band is useful for enhancement of surface water. Similarly, the slope that ensures flat surface of water bodies is also useful if added. In NDVI, water are negative values at the same time that many other non-vegetative bodes can also have the same value, which has led to a decrease in performance. NNET and SVM are very well-known state-of-art algorithms in classification. They perform well in many cases and their performance in the original Landsat 8 scene of Nepal is satisfactory. However, the interesting result is that while NNET increased the performance with that addition of data gradually, SVM went the other side. This also shows that the addition of secondary data does not necessarily improve performance and are rather dependent on algorithms. Thus, comparative studies are necessary to check these performances and select the best one for the area under study.

Figure 6 shows the cases for NB in the entire multiband, along with the slope and results from the previous study, i.e., conditioning NDVI and NDWI with elevation (Elev_NDWnVI) [13]. Cases a, g and h were the ones representing snow and melting ice in that Himalayas, in which NB performed very bad and most of the shadowed regions were misclassified. It performed well in cases from b to f, and even did well than in the previous study in case c, i.e., narrow river channels with shadows. It was able to classify non-water shadow features well. In the case of additional band performance, all the results in the cases are similar with no distinct difference. As per the RPART performance in cases (Figure 7), it was able to remove the shadow issues in the narrow river with shadows in case b. It performed well in cases of wide shadows, ponds and wide rivers. However, it was not able to separate shadows of the Himalayas in cases a, g and h. Further, adding the slope somewhat improved these cases but failed in the case of the shadow in case b. However, adding NDVI and NDWI were unable to improve in the shadows.

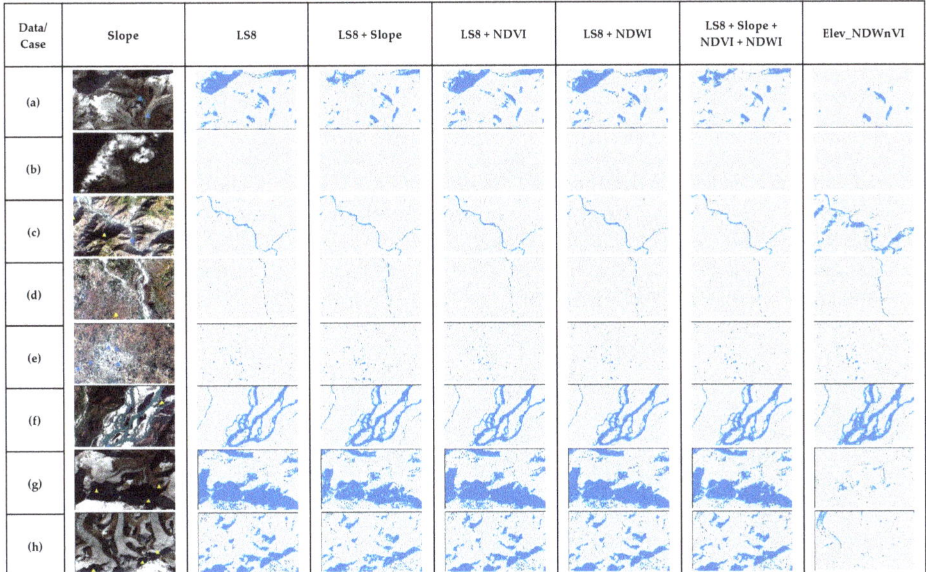

Figure 6. The comparison of NB results of special cases of surface water in the test scene (Figure 1) for different multiband data used.

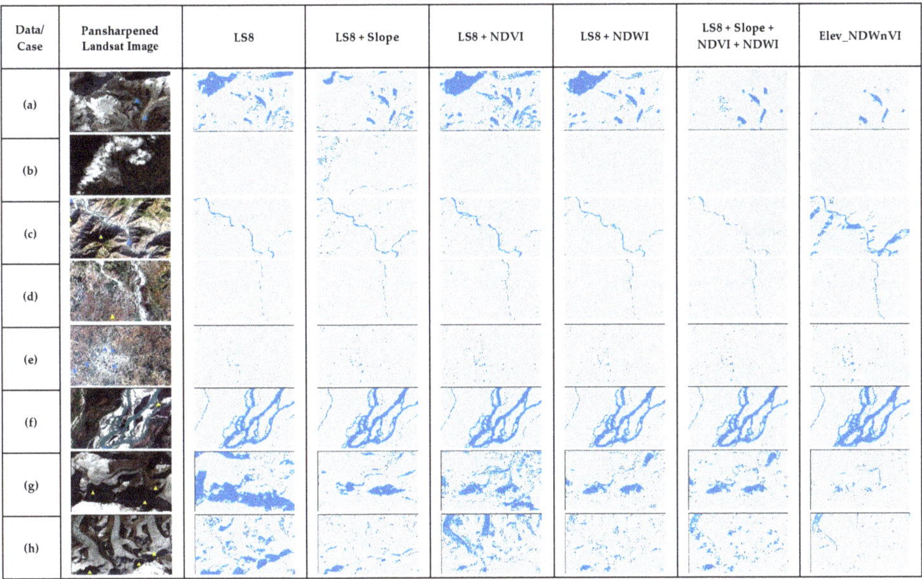

Figure 7. The comparison of RPART results of special cases of surface water in the test scene (Figure 1) for different multiband data used.

In Figure 8, NNET showed gradual improvement in identifying water bodies with the addition of secondary bands. NDWI enhanced the shadows as water compared to NDWI and all three together. For case b, the addition of NDVI misclassified few cloud shadows while others did not. For case c, all the dataset were able to solve the shadow issue from the previous shadow in the hills, except the

NDVI. For the plain area in case d with the narrow river, NNET results were good in all the dataset with even enhanced small water bodies. In case e for urban small water bodies, LS8 predicted water and some shadows but with the addition of secondary bands, the pure pixels were enhanced. With the addition of all three bands, it refined the water bodies very well. For a large river in case f, the results are similar to each other and with the previous study [13]. For cases g and h, the results are different for every dataset. The slope misclassified the melting ice, NDVI misclassified the shadows and combined, showed both misclassified. LS8 only and the addition of NDWI were somewhat better compared to others. With the SVM in Figure 9, the SVM showed similar results to NNET for most cases. However, SVM misclassified the darkest shadows as water in the whole scene. It is clearly seen in case a, b, g and h, especially with the addition of the slope. In case a, with LS8 and addition of NDVI and NDWI, even light shadows were misclassified.

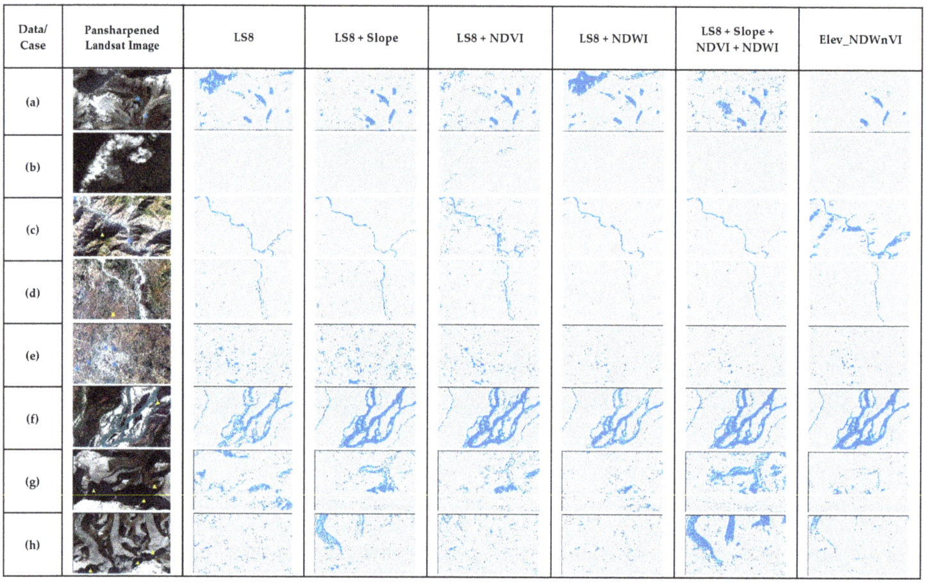

Data/ Case	Pansharpened Landsat Image	LS8	LS8 + Slope	LS8 + NDVI	LS8 + NDWI	LS8 + Slope + NDVI + NDWI	Elev_NDWnVI
(a)							
(b)							
(c)							
(d)							
(e)							
(f)							
(g)							
(h)							

Figure 8. The comparison of NNET results of special cases of surface water in the test scene (Figure 1) for different multiband data used.

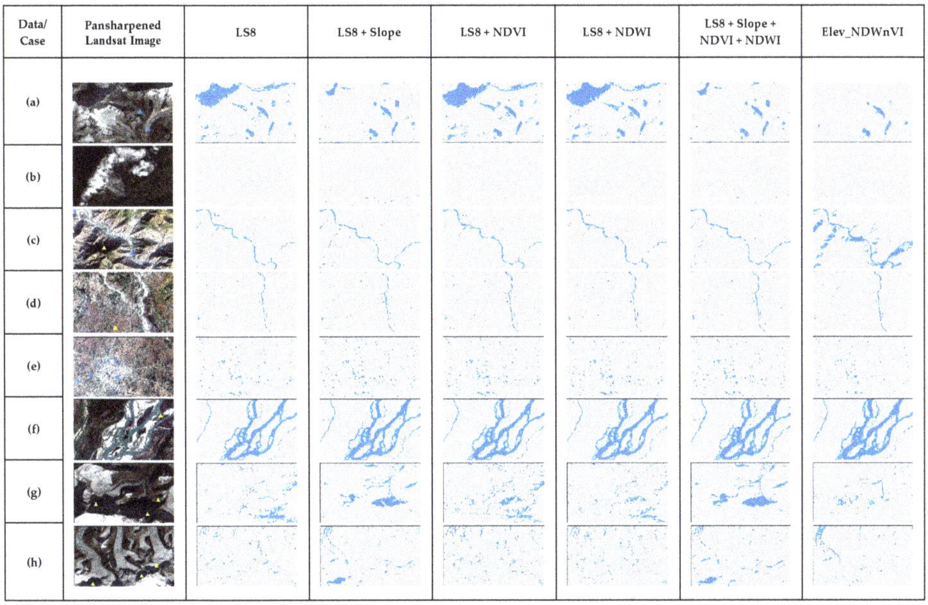

Data/ Case	Pansharpened Landsat Image	LS8	LS8 + Slope	LS8 + NDVI	LS8 + NDWI	LS8 + Slope + NDVI + NDWI	Elev_NDWnVI
(a)							
(b)							
(c)							
(d)							
(e)							
(f)							
(g)							
(h)							

Figure 9. The comparison of SVM results of special cases of surface water in the test scene (Figure 2) for different multiband data used.

As both RF and GBM are ensemble methods, their performance on the cases are quite similar in Figures 10 and 11 respectively. Both methods were successful in separating shadows in case c, but failed in cases a and g. In addition, with the addition of the slope only, the results were better than the original LS8. However, the addition of NDVI or NDWI did not perform well in the shadow areas. Only the addition of the slope as a secondary band seems to be a good choice rather than adding all for both RF and GBM algorithms in surface water mapping.

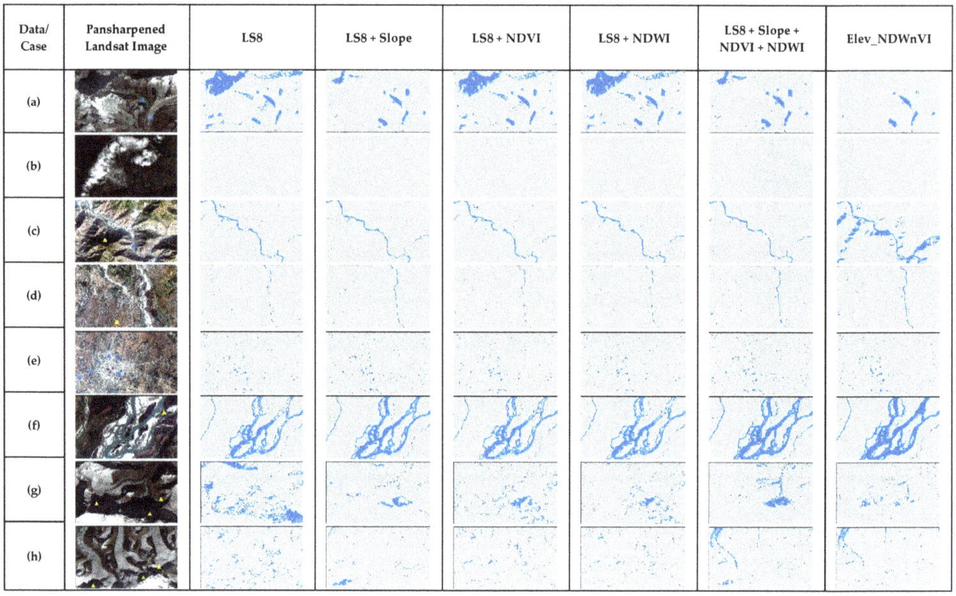

Data/ Case
Pansharpened Landsat Image
LS8
LS8 + Slope
LS8 + NDVI
LS8 + NDWI
LS8 + Slope + NDVI + NDWI
Elev_NDWnVI
(a)
(b)
(c)
(d)
(e)
(f)
(g)
(h)

Figure 10. The comparison of RF results of special cases of surface water in the test scene (Figure 1) for different multiband data used.

Data/ Case
Pansharpened Landsat image
LS8
LS8 + Slope
LS8 + NDVI
LS8 + NDWI
LS8 + Slope + NDVI + NDWI
Elev_NDWnVI
(a)
(b)
(c)
(d)
(e)
(f)
(g)
(h)

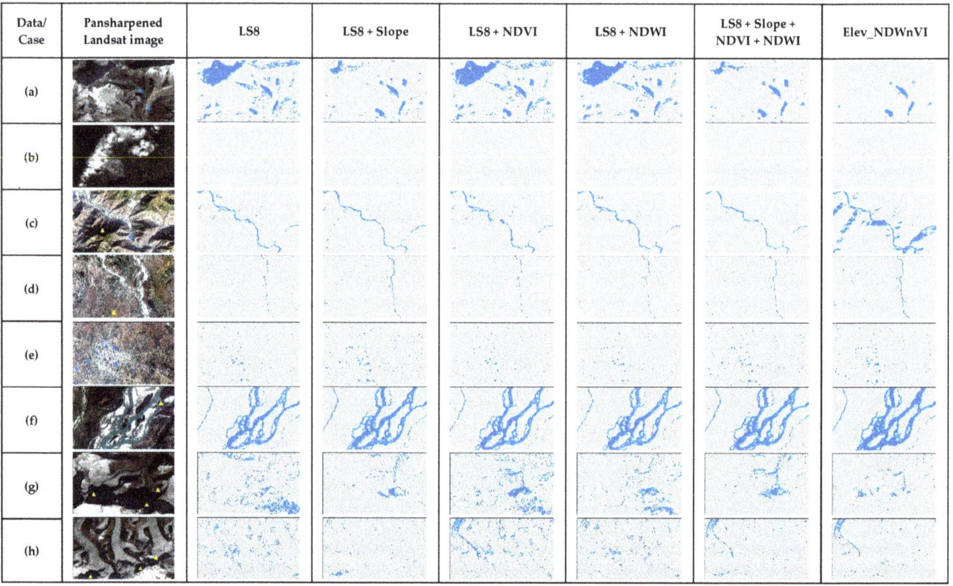

Figure 11. The comparison of GBM results of special cases of surface water in the test scene (Figure 1) for different multiband data used.

In comparison to the previous study [13], except NB and RPART, all other results are above 90% and even up to 100% as seen in the literature [16,23–25]. In addition, for a hilly test scene in the study done by Jiang et al., [15] using a multilayer perceptron neural network in Landsat 8 OLI satellite imagery, the neural network performance was similar to this work, i.e., OA 98.50 and Kappa 0.970.

Compared to our previous study [13], the results were good for hilly and lower flatlands where there is no snow. The performance of surface water detection for narrow rivers in hilly regions with shadows improved, however, they were only able to detect pure water pixels. The main issues were the ice and snow with shadows in hilly areas. Adding the slope somewhat improved cases in the Himalayas, but not the NDVI or NDWI. In a case by case evaluation also, the performance seem well but were mostly misclassified in the Himalayas. Quantitatively, machine learning algorithms were much better compared to the index methods, however, the results are not as reliable as it should be for the whole scene. In the full scene, very few areas are covered by the snow compared to the large hilly and flat lands. Nevertheless, Nepal is a mountainous country and has a quantitatively large cover, which could be a challenge in applying machine learning algorithms. Thus, the wrong predictions in those areas are less significant in the validation. A further investigation with cutting-edge machine learning technology, i.e., convolutional neural networks or deep learning could be undertaken for improvement.

4. Conclusions

In this study, we extended the previous study and applied six machine learning algorithms: NB, RPART, NNET, SVM, RF and GBM to evaluate the surface water extraction using a Landsat 8 OLI images in Nepal. Using the previous reference dataset and Landsat scene, six different models were developed using the CARET package in R software. Cross-validation was completed to minimize the overfitting then train the model to predict the surface water and validate the full reference dataset. With three secondary bands: Slope, NDVI and NDWI, the algorithms were evaluated for performance with the addition of extra information. The results were compared, case by case, and the following conclusions were drawn from the test scene and applied machine learning algorithms:

(a) All the applied machine learning algorithms showed OA above 90% but in case of Kappa except NB and RPART, it was above 90%.
(b) RF showed OA and Kappa both 1 for the all the multiband data with the reference dataset.
(c) GBM and NNET also showed good performance followed by the SVM.
(d) Machine learning algorithms were able to perform better in the hilly regions and flat lands but not well in the Himalayas with ice, snow and shadows.
(e) The addition of the slope and NDWI showed improvement in the results compared to the NDVI except NNET. Others do not improve with the addition of all three secondary bands compared with the individual addition.

It seems that machine learning methods could be very useful for the accurate automated binary classification of surface water in Nepal. The use of RF with original LS8 data or with the addition of the slope or NDWI with another algorithm can be undertaken. Based on this and previous work [13], it is recommended to segment the study area with and without snow or low and high elevation, then apply RF or GBM for better performance.

For further investigation, this study aimed to evaluate the application of convolutional neural networks or deep learning for better accuracy. In addition, individual original bands and secondary bands with the RF and GBM can be evaluated so that high accuracy can be achieved with minimum bands.

Author Contributions: Conceptualization, T.D.A.; formal analysis, T.D.A.; investigation, T.D.A. and A.S.; methodology, T.D.A.; resources, D.H.L.; software, T.D.A.; supervision, D.H.L.; validation, A.S.; visualization, A.S.; writing—original draft, T.D.A.; writing—review and editing, A.S. and D.H.L.

Funding: This work was supported by the National Research Foundation of Korea (NRF) grant funded by the Korea government (MSIT) (No. 2018R1A2B6009363).

Acknowledgments: The authors are grateful to the U.S. Geological Survey server (http://glovis.usgs.gov) for providing the Landsat data that was used in this manuscript freely. The authors would also like to thank the anonymous reviewers for their constructive comments and improving this manuscript.

Conflicts of Interest: The authors declare no conflicts of interest.

References

1. Khanal, N.R.; Chalise, S.R.; Pokhrel, A.P. Ecohydrology of River Basins of Nepal. In Proceedings of the International Conference on Ecohydrology of High Mountain Areas, ICIMOD, Kathmandu, Nepal, 24–28 March 1998; pp. 49–61.
2. WECS. *Water Resources of Nepal in the Context of Climate Change*; Water and Energy Commission Secretariat, Government of Nepal: Kathmandu, Nepal, 2011.
3. Bhuju, U.R.; Khadka, M.; Neupane, P.K.; Adhikari, R. *Lakes of Nepal: 5358—A Map Based Inventory*; National Lakes Conservation Development Committee: Kathmandu, Nepal, 2009.
4. Acharya, T.D.; Subedi, A.; Huang, H.; Lee, D.H. Application of Water Indices in Surface Water Change Detection using Landsat Imagery in Nepal. *Sens. Mater.* **2019**, *31*, 1429–1447. [CrossRef]
5. Chaubey, P.K.; Kundu, A.; Mall, R.K. A Geo-Spatial Inter-Relationship with Drainage Morphometry, Landscapes and NDVI in the Context of Climate Change: A Case Study over the Varuna River Basin (India). *Spat. Inf. Res.* **2019**. [CrossRef]
6. Nepal, S. Impacts of Climate Change on the Hydrological Regime of the Koshi River Basin in the Himalayan Region. *J. Hydro-Environ. Res.* **2016**, *10*, 76–89. [CrossRef]
7. Soncini, A.; Bocchiola, D.; Confortola, G.; Minora, U.; Vuillermoz, E.; Salerno, F.; Viviano, G.; Shrestha, D.; Senese, A.; Smiraglia, C.; et al. Future Hydrological Regimes and Glacier Cover in the Everest Region: The Case Study of the Upper Dudh Koshi Basin. *Sci. Total Environ.* **2016**, *565*, 1084–1101. [CrossRef] [PubMed]
8. Chang, H.; Yun, C.; Shiqiang, Z.; Jianping, W. Detecting, Extracting, and Monitoring Surface Water from Space using Optical Sensors: A Review. *Rev. Geophys.* **2018**, *56*, 333–360.
9. Das, R.T.; Pal, S. Exploring Geospatial Changes of Wetland in Different Hydrological Paradigms using Water Presence Frequency Approach in Barind Tract of West Bengal. *Spat. Inf. Res.* **2017**, *25*, 467–479. [CrossRef]
10. Zhu, Z.; Wulder, M.A.; Roy, D.P.; Woodcock, C.E.; Hansen, M.C.; Radeloff, V.C.; Healey, S.P.; Schaaf, C.; Hostert, P.; Strobl, P.; et al. Benefits of the Free and Open Landsat Data Policy. *Remote Sens. Environ.* **2019**, *224*, 382–385. [CrossRef]
11. Acharya, T.D.; Yang, I.T.; Subedi, A.; Lee, D.H. Change Detection of Lakes in Pokhara, Nepal using Landsat Data. In Proceedings of the 3rd International Electronic Conference on Sensors and Applications, 15–30 November 2016.
12. Acharya, T.D.; Subedi, A.; Yang, I.T.; Lee, D.H. Combining Water Indices for Water and Background Threshold in Landsat Image. *Proceedings* **2018**, *2*, 143. [CrossRef]
13. Acharya, T.D.; Subedi, A.; Lee, D.H. Evaluation of Water Indices for Surface Water Extraction in a Landsat 8 Scene of Nepal. *Sensors* **2018**, *8*, 2580. [CrossRef] [PubMed]
14. Acharya, T.D.; Subedi, A.; Huang, H.; Lee, D.H. Classification of Surface Water using Machine Learning Methods from Landsat Data in Nepal. *Proceedings* **2019**, *4*, 43. [CrossRef]
15. Jiang, W.; He, G.; Long, T.; Ni, Y.; Liu, H.; Peng, Y.; Lv, K.; Wang, G. Multilayer Perceptron Neural Network for Surface Water Extraction in Landsat 8 OLI Satellite Images. *Remote Sens.* **2018**, *10*, 755. [CrossRef]
16. Acharya, T.D.; Lee, D.H.; Yang, I.T.; Lee, J.K. Identification of Water Bodies in a Landsat 8 OLI Image using a J48 Decision Tree. *Sensors* **2016**, *16*, 1075. [CrossRef] [PubMed]
17. Karpatne, A.; Khandelwal, A.; Chen, X.; Mithal, V.; Faghmous, J.; Kumar, V. Global Monitoring of Inland Water Dynamics: State-of-the-Art, Challenges, and Opportunities. In *Computational Sustainability*; Lässig, J., Kersting, K., Morik, K., Eds.; Springer International Publishing: Cham, Switzerland, 2016; pp. 121–147.
18. Tulbure, M.G.; Broich, M. Spatiotemporal Dynamic of Surface Water Bodies using Landsat Time-Series Data from 1999 to 2011. *ISPRS J. Photogramm. Remote Sens.* **2013**, *79*, 44–52. [CrossRef]
19. Nath, R.K.; Deb, S.K. Water-Body Area Extraction from High Resolution Satellite Images-an Introduction, Review, and Comparison. *Int. J. Image Process. (IJIP)* **2010**, *3*, 353–372.
20. Guerra, L.; McGarry, L.M.; Robles, V.; Bielza, C.; Larrañaga, P.; Yuste, R. Comparison between Supervised and Unsupervised Classifications of Neuronal Cell Types: A Case Study. *Dev. Neurobiol.* **2011**, *71*, 71–82. [CrossRef] [PubMed]
21. Wu, X.; Kumar, V.; Quinlan, J.R.; Ghosh, J.; Yang, Q.; Motoda, H.; McLachlan, G.J.; Ng, A.; Liu, B.; Philip, S.Y. Top 10 Algorithms in Data Mining. *Knowl. Inf. Syst.* **2008**, *14*, 1–37. [CrossRef]

22. Congalton, R.G.; Green, K. *Assessing the Accuracy of Remotely Sensed Data: Principles and Practices*, 2nd ed.; CRC Press, Taylor & Francis Group: Boca Raton, FL, USA, 2009; p. 183. ISBN 978-1-4200-5512-2.
23. Zhou, Y.; Dong, J.; Xiao, X.; Xiao, T.; Yang, Z.; Zhao, G.; Zou, Z.; Qin, Y. Open Surface Water Mapping Algorithms: A Comparison of Water-Related Spectral Indices and Sensors. *Water* **2017**, *9*, 256. [CrossRef]
24. Yang, X.; Chen, L. Evaluation of Automated Urban Surface Water Extraction from Sentinel-2A Imagery using Different Water Indices. *J. Appl. Remote Sens.* **2017**, *11*, 026016. [CrossRef]
25. Xie, H.; Luo, X.; Xu, X.; Pan, H.; Tong, X. Evaluation of Landsat 8 OLI Imagery for Unsupervised Inland Water Extraction. *Int. J. Remote Sens.* **2016**, *37*, 1826–1844. [CrossRef]

Article

An Approach to Frequency Selectivity in an Urban Environment by Means of Multi-Path Acoustic Channel Analysis

Pau Bergadà [1,2] and Rosa Ma Alsina-Pagès [1,*]

1 Grup de recerca en Tecnologies Mèdia (GTM), La Salle, Universitat Ramon Llull, c/Quatre Camins, 30, 08022 Barcelona, Spain; pbergadac@gmail.com
2 Wavecontrol, c/Pallars, 65-71, 08018 Barcelona, Spain
* Correspondence: rosamaria.alsina@salle.url.edu; Tel.: +34-932902455

Received: 28 April 2019; Accepted: 18 June 2019; Published: 21 June 2019

Abstract: The improvement of quality of life in the framework of the smart city paradigm cannot be limited to a set of objective measures carried out over several critical parameters (e.g., noise or air pollution). Noise disturbances depend not only on the equivalent level L_{Aeq} measured, but also on the spectral distribution of the sounds perceived by people. Propagation modelling to conduct auralization can be done either with geometrical acoustics or with wave-based methods, given the fact that urban environments are acoustically complex scenarios. In this work, we present a first analysis of the acoustic spectral distribution of street noise, based on the frequency selectivity of the urban outdoor channel and its corresponding coherence bandwidth. The analysis was conducted in the framework of the data collected in the Milan pilot WASN of the DYNAMAP LIFE project, with the use of three simulated acoustic impulse responses. The results show the clear influence of the evaluated coherence bandwidth of each of the simulated channels over real-life acoustic samples, which leads us to the conclusion that all raw acoustic samples have to be considered as wide-band. The results also depict a dependence of accumulated energy at the receiver with the coherence bandwidth of the channel. We conclude that, the higher the delay spread of the channel, the narrower the coherence bandwidth and the higher the distortion suffered by acoustic signals. Moreover, the accumulated energy of the received signal along the frequency axis tends to differ from the accumulated energy of the transmitted signal when facing narrow coherence bandwidth channels; whereas the accumulated energy along the time axis diverges from the accumulated transmitted energy when facing wide coherence bandwidth channels.

Keywords: noise; propagation model; frequency selectivity; acoustic channel; wideband; auralization; smartcity; impulse response; coherence bandwidth; wireless acoustic sensor network

1. Introduction

As a result of population growth and the consequent expansion of transportation systems, including highways, railways, and airways, environmental noise pollution has been increasing. Noise pollution continues to constitute a major environmental health problem in Europe [1,2]. Among the health effects, annoyance is one of the principal environmental noise [3] issues; however, it is not merely an annoyance, as several works have detected health problems, such as sleep disorders [4], learning impairment [5], and heart diseases [6]. Thus, noise impact is one of the main environmental health concerns [7], and the harmful effects it causes on social and economic aspects have been proved [8].

The European Union reacted to this alarming increase of environmental noise pollution, especially in densely populated cities, with the Environmental Noise Directive 2002/49/EC (END) [9]. In accordance with the END, the CNOSSOS-EU methodological framework aims to improve the consistency and comparability of noise assessment results across the EU Member States [10] for its application. The main pillars of the END are the following: (i) Determining the noise exposure; (ii) updating information related to the noise available to citizens; and (iii) preventing and reducing environmental noise, where necessary.

Recent studies have showed that the effects of noise on people do not only depend on the level of noise, but also on the type of sound. In fact, in 2018, the WHO incorporated noises, such as leisure noise and wind turbine noise [11]. To accomplish the goal of measuring each type of noise source, the Anomalous Noise Event Detector (ANED) [12] was designed by this team to rule out non-road traffic noise (RTN) events from road traffic noise measurements. The ANED is an algorithm based on the spectral distribution of the different types of RTN and anomalous noise events (ANE) in order to properly identify them and, in this study, it is proven that the sound propagation and its impact on spectral behavior may change its performance [13]. Furthermore, by changing the temporal and spectral distribution of the signal, human perception may also change with respect to a reference noise measured in outdoor environmental conditions [14]. Our team has begun work on the evaluation of the perception of certain types of sound in outdoor conditions [15,16], with promising results that are still under study in the urban environment of Rome.

In an urban environment, a detailed study and simulated reproduction of the propagation of a sound—as it will be perceived by people—is a key factor in the evaluation and prediction of how people will react to the noise [3]; this approximation is called auralization [17]; virtual reality has even been used to reproduce the audio-visual environment [18]. Propagation modelling, with the final goal of auralization, has been proposed in the literature from two points of view: (i) Geometrical acoustics, and (ii) wave-based methods [19]. The analysis detailed in this work is based on the wide-band channel sounding principles which are widely used in communications [20], which demonstrate the usefulness of studying channel fading [21,22]. The techniques of channel estimation, taking into account multi-path propagation and its subsequent coherence bandwidth, have been found to be useful in acoustic propagation environments, mainly in underwater channels [23,24]. Several studies have been conducted in this area, with analysis of the scattering function, multi-path intensity profile, the coherence of an underwater acoustic channel [25], and analysing the impact of the coherence bandwidth on the transmission of pressure waves in image transmission [26].

The authors have made the assumption that the acoustic recordings are wide-band, given that the influence of the channel is in-band frequency selective [19]. The final goal of this preliminary study is to accurately determine whether the frequency selectivity of the channel changes the spectral distribution of several recorded acoustic raw signals [27]. The impulse response of the outdoor urban environment corresponds to the simulation of three different multi-path acoustic channels. We accurately describe each impulse response of the channel and its effects on the spectral distribution of real-life acoustic data collected in the framework of the DYNAMAP pilot project carried out in Milan, focusing on the coherence bandwidth. This work intends to be a first step in the analysis of the frequency selectivity applied to non-traffic-related noise, in order to evaluate the effect of the channel [28] on the spectrum-temporal vision of several real-operation raw signals collected in the Milan pilot of project DYNAMAP [29,30], with the final goal of its impact on auralization.

This paper is structured as follows. Section 2 details the methodology used to conduct this analysis. Section 3 details the conditions under which the acoustic raw data were collected in a real-operation environment. Section 4 details the mathematical models used to simulate the propagation impairments by means of pseudo-noise sequences. Section 5 details the results of the propagation of the simulated channels using real-life data and, finally, Section 6 details the conclusions and future work.

2. Overview of Methodology

The study we present in this paper uses the basis of the application of propagation modelling methods in an outdoor environment with the concept of wide-band sounding using pseudo-noise (PN) sequences, mostly used in channel soundings in communications. The acoustic communication systems are considered to be wide-band, since the channel coherence bandwidth is similar, or even narrower, than the acoustic signal bandwidth.

2.1. Outdoor Acoustic Propagation Modelling Basics

Propagation modelling with auralization purposes can be done both with geometrical acoustics and wave-based methods [19]. Urban sound propagation modelling should take into account the reflections coming from facades and other reflection surfaces, which might be specular or diffuse [31–33]. Each element has its own frequency-dependent reflection properties, which modifies the phase and the amplitude of the acoustic wave, accordingly.

Furthermore, modelling of the diffraction corresponding to the acoustic wave should also be included; in an urban environment, they often occur with multiple building edges, roofs, and corners [34]. Additionally, meteorological conditions will affect at least the long-range sound propagation; wind and turbulence also play a role in this [35]. This brief review of the most relevant elements for modelling the acoustic channel in a city demonstrates its complexity.

This work does not intend to deal with this in depth but, rather, the influence of the most basic parameters of acoustic propagation (e.g., attenuation, reflection, and influence of the medium of propagation) from a real network.

2.2. Wide-Band Channel Sounding with PN-Sequences

Any wide-band sounding focuses on the analysis of time-frequency dispersive features of the channel under study. Time dispersion [20] is the time-spread suffered by any wave when propagated in a medium. This corresponds to the interval of delay that causes that the auto-correlation function of the channel impulse response to differ from zero in the receiver. Frequency dispersion [20] describes the channel variation speed; the Doppler spread is the frequency range of the channel impulse response auto-correlation function that differs from zero [36].

2.2.1. PN-Sequence Wide-Band Analysis Proposal

There are several methods to conduct wide-band sounding and channel analysis. In this work, we have used the transmission of PN waveforms [37] with good cyclic cross-correlation characteristics, such as m-sequences [20]. The computation process for the wide-band sounding is detailed in Figure 1.

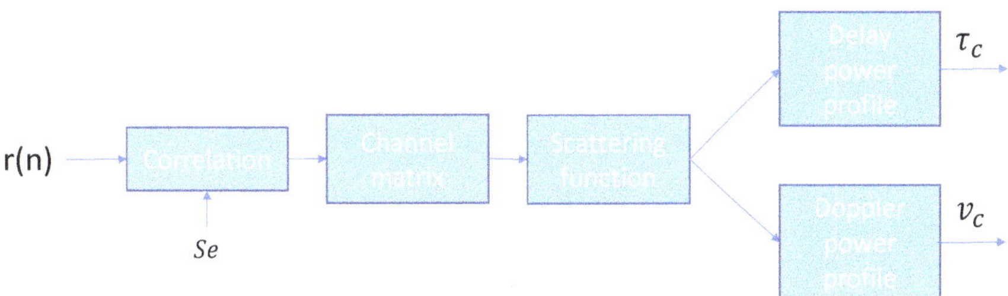

Figure 1. Diagram of the computation process for wide-band channel sounding.

The signal $r[n]$ in the receiver is correlated with the original PN sequence S_e shaped by a raised cosine filter with a determined roll-off factor, which is evaluated depending on the application. The correlation function is calculated as:

$$\phi_{r[n],S_e}[n] = \sum_{k=0}^{N_e-1} r[n+k]S_e[k], \quad (1)$$

where N_e is the length of the PN sequence and S_e is the sequence filtered with a raised cosine filter. Then, the channel impulse response $h[n, \tau]$ can be written as:

$$h[n, \tau] = \phi_{r,S_e}[nlN_c + \tau], \tag{2}$$

where τ is the variable corresponding to the delay, l is the number of chips [20] (i.e., a pulse of a PN sequence), and N_c is the number of samples per chip. From Equation (2), we can evaluate the scattering function, which leads us to the multi-path and Doppler spreads caused by the variant channel. The scattering function $R_s[\tau, v]$ is calculated as the Fourier transform of the channel impulse response [20]:

$$R_h[\xi, \tau] = \sum_{\xi} h^*[n, \tau]h[n + \xi, \tau], \tag{3}$$

$$R_s[\tau, v] = \sum_{\xi} R_h[\xi, \tau]e^{-j2\pi\xi v}. \tag{4}$$

Both the multi-path spread (τ_c) and the Doppler spread (v_c) can be computed from the scattering function $R_s[\tau, v]$ using a certain observation window, which is wider or narrower depending on the channel variations and the application under study.

2.2.2. Underwater Acoustic Channel Sounding

The acoustic communication channel may have a sparse impulse response, where physical paths act as time-varying low-pass filters and where movement introduces both Doppler spread and shift. One of the applications of PN-sequence wide-band sounding is the field of underwater acoustic channel communications. Underwater acoustic channels are usually catalogued as one of the most hostile communications systems [38]. As the bandwidth is extremely limited, an acoustic system may operate in a frequency range between 10 and 15 kHz; although the total bandwidth is low (around 5 kHz), the system is considered wide-band as its bandwidth is not negligible in terms of coherence bandwidth.

Several studies, in which scientists used PN-sequence based systems to conduct sounding in a wide-band acoustic communications underwater channel can be found in literature. In [39], the authors described an underwater sensor network, which they took advantage of to perform acoustic tomography. In [40], the authors described Hermes, an asymmetrical point-to-point underwater acoustic modem designed for short-range operations, and explored the possibility of its possible evolution into a multiple-input-multiple-output (MIMO) device. In [41], the authors conducted a survey in northern Europe, covering the continental shelf, Norwegian fjords, a sheltered bay, a channel, and the Baltic Sea. The sounding measurements were performed in various frequency bands between 2–32 kHz, in order to define a typical acoustic communications channel.

3. Real-Operation Acoustic Data Recordings in the DYNAMAP Project

In this section, we describe the real-operation conditions used to record the acoustic raw data used for the analysis. First, we describe the recording campaign in Milan by means of a WASN. Then, we detail which of the sensors of the entire WASN were chosen to analyse their data and, finally, we describe the types of noise event used for this test and our reasons for their selection.

3.1. The DYNAMAP Project

The DYNAMAP LIFE project proposed the implementation of a dynamic noise mapping system [29], able to determine the acoustic impact of road infrastructures in real-time, following the European Noise Directive 2002/49/EC. A Multi-Sensor Network collects the noise level measurements in two pilot areas: In the city of Milan, and on the A90 motorway around Rome.

Each of the sensor nodes has to accomplish a set of basic specifications [42] defined to satisfy DYNAMAP requirements for each monitoring station, which are the following: (i) A 40–100 dB(A) broadband linearity range; (ii) a 35–115 dB working range with acceptable Total Harmonic Distortion

(THD); (iii) a narrow-band floor noise level; and (iv) a sampling rate of 48 kHz. The project also requires the possibility of audio recording, as well as VPN and GPRS/3G/Wi-Fi connections. The precision of the sensors is a key issue for system reliability [43].

In the case of the city of Milan, the deployment of the network is shown in Figure 2, where a team of acoustic experts chose up to 24 low-cost sensor locations to collect the data to generate noise maps. All of these locations are a key issue for the study of the features of the recorded signal and, therefore, for the performance of the ANED [12]. For more details about the location of the sensors, the reader should refer to [44,45].

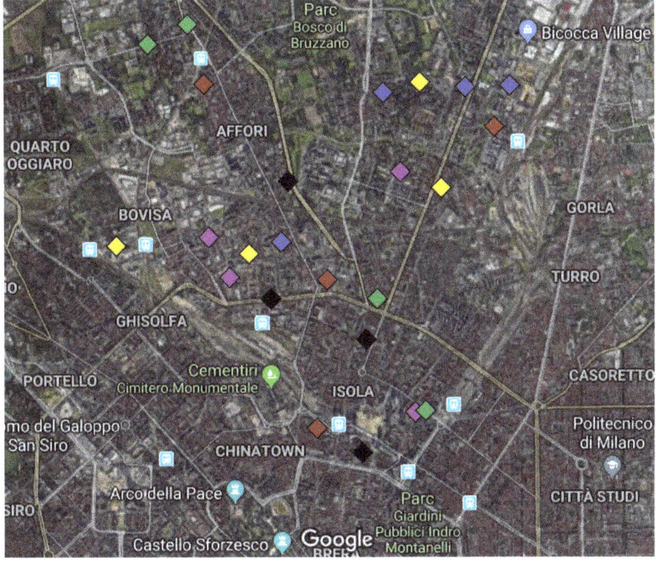

Figure 2. Map of the sensor locations in Milan. Different colors correspond to a catalog of six groups of streets [46].

Although the final goal of the DYNAMAP project is to dynamically update the noise maps in a GIS-based platform [29,47], preliminary work has been done in the sensors to evaluate the features of the raw acoustic signals collected by the network; this work enabled the recollection of up to 100 h of outdoor urban acoustic raw data in the Milan environment.

3.2. Description of the Recording Campaign

Table 1 lists the sensor nodes of the WASN, indicating their identifiers, street location, and GPS coordinates within district nine of Milan, in which they were installed. A recording campaign considering both weekends and weekdays (recording 20 min each hour every day) was performed through the multi-sensor network in real-operation conditions. The final selection of files to be labelled allowed our team to collect more than 100 h of raw urban acoustic data. Previous studies [48] showed that sampling two different days could accomplish the requirements of diversity of the urban activity during weekend days and weekdays; the different traffic flow attains both RTN and also the anomalous noise events found in an urban environment, showing the relevant differences. These previous studies took into account the experience of data set designs for other acoustic event detection projects, such as those in [8,49]. Furthermore, the recordings were exhaustive, considering the entire day period of each day, for two reasons: (i) The road traffic noise profile changes between day and night, as can be observed in the two-day equivalent noise level curve in Figure 3—noise profile curves attain a certain 24 h regularity, and this property has been taken into account in the recordings; and (ii) because the types of events occurring are substantially different between day and night, and recording with a periodic sample widens the possibility of collecting more types of noise.

Table 1. List of the nodes of the Wireless Acoustic Sensor Network (WASN) deployed in district nine in Milan.

Sensor ID	Street	GPS Coordinates
hb106	Via Litta Modignani	(45.5227587,9.1596847)
hb108	Via Piero e Alberto Pirelli	(45.5144707,9.2107111)
hb109	Viale Stelvio	(45.4929125,9.1919035)
hb114	Via Melchiorre Gioia	(45.4815058,9.1913241)
hb115	Via Fara	(45.4855843,9.1991161)
hb116	Via Moncalieri	(45.5098883,9.1968012)
hb117	Viale Fermi	(45.5089072,9.1802412)
hb120	Via Baldinucci	(45.5032677,9.1686595)
hb121	Via Piero e Alberto Pirelli	(45.5185641,9.2129266)
hb123	Via Galvani	(45.4857107,9.2005241)
hb124	Via Grivola	(45.5179185,9.1943259)
hb125	Via Abba	(45.5028072,9.179285)
hb127	Via Quadrio	(45.4839506,9.1845167)
hb129	Via Crespi	(45.4989476,9.1860456)
hb133	Via Maffucci	(45.4992223,9.1717236)
hb135	via Lambruschini	(45.5024486,9.1548883)
hb136	Via Comasina	(45.5247882,9.1655266)
hb137	via Maestri del Lavoro	(45.518893,9.1997167)
hb138	Via Novaro	(45.5187445,9.1678656)
hb139	Via Bruni	(45.5015796,9.1745067)
hb140	Viale Jenner	(45.4970863,9.1777414)
hb144	Via D'Intignano	(45.5082648,9.2027579)
hb145	Via Fratelli Grimm	(45.5184213,9.2062962)
hb151	Via Veglia	(45.4970074,9.1934109)

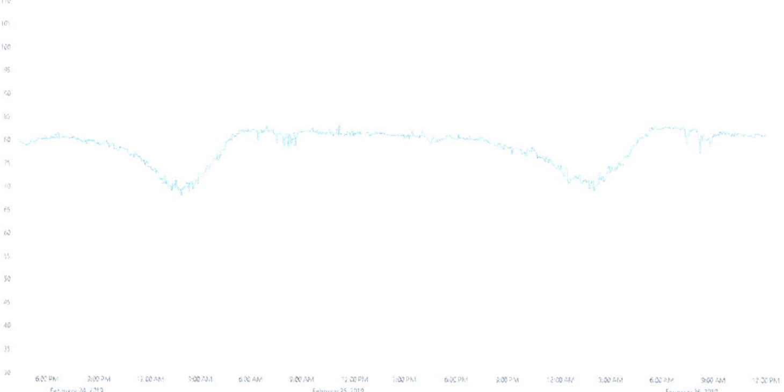

Figure 3. Example of two-day L_{Aeq300} curve of sensor hb148. The vertical axis is the measured L_{Aeq300}, in dB.

To that effect, recordings were planned for the 24-node network during the first 20 min of each hour, for each hour of the day, during two selected days: A weekday (Tuesday, 28th of November 2017) and a weekend day (Sunday, 3rd of December 2017), in order to maximize the diversity of the recorded ANEs and with a schedule intended to be exhaustive. A total of 1116 recordings of 20 min each were gathered during this recording campaign, producing an audio database of 372 h, from which our research group managed to manually listen and label up to 100 h.

3.3. Acoustic Environment of the Nodes of the WASN

There are currently 24 low-cost high-capacity sensors deployed in a WASN in district nine in Milan, as described in Table 1. Nevertheless, in the framework of the DYNAMAP project, the acoustic environment of the sensors has been studied [45] and, together with the description of several roads conducted by the colleagues of Universitá degli Studi di Milano Bicocca [46], we reached the conclusion that not all the acoustic data collected in all the sensors was suitable to conduct this kind of analysis. Under ideal conditions, the noise events under test should be recorded in an free-field environment; an unfeasible requirement, not only for the DYNAMAP project, but also for any other project that intends to collect urban noise events.

An analysis was conducted with the aim of determining which sensors could capture noise events in a similar way to this ideal free-field environment. The first item to consider was the placement of sensors in narrow streets with buildings, due to the acoustic effect over noise events, which substantially changes the frequency distribution of the recorded data. We finally discovered that two of the sensors (hb137 and hb145) were placed in a facade, but surrounded by parks. Sensor hb137 was located in a quiet area, surrounded by two parks with trees. Furthermore, the closest building to this sensor was more than 55 meters away (see Figure 4). Sensor hb145 was also in a narrow street, but was surrounded by a park; both of them were placed in public buildings, one of them being a school. The closest wall to the sensor was around 35 m away (see Figure 5). Nevertheless, in both locations, the sensors were placed on the façade, so the *façade effect* will appear for both recordings; in future work, the authors will consider working on its mitigation [50].

(**a**) Location of sensor hb137 (yellow square) (**b**) Sensor placement in the facade

Figure 4. Sensor hb137 location and picture.

(**a**) Location of sensor hb137 (blue square) (**b**) Sensor placement in the facade

Figure 5. Sensor hb145 location and picture.

Once the sensors had been chosen, an analysis of the recorded anomalous noise events was conducted. Dozens of types of anomalous noise events were found to be recorded by those two sensors, such as: Birds (609 occurrences), brake (62), truck (20), siren (15), door (131), people (169), trolley (10), dog (134), aeroplane (28), chain (21), step (45), bell (11), bike (16), glass (3), bus door (8), saw (8), and tram (16). In Figure 6, four examples can be found of anomalous noise events recorded in sensor hb137 and hb145, showing that different ANEs present very variate frequency distributions.

The analysis of the available samples of anomalous noise events was focused on the spectral diversity and minimum length of the events. The events used for this analysis, including those from both sensors (hb137 and hb145), had an overall length of 4865 s. The maximum length of the processed ANEs was 191 s and the minimum length was 0.1 s. Another of the pursued goals was to study several types of spectral distribution, both wide-band and narrow-band.

In Figure 6, the bird spectrogram in Figure 6a presents a short audio of 0.5 s and a distribution of frequencies between 5 kHz and 7.5 kHz. The siren spectrogram in Figure 6b shows a good signal-to-noise (SNR) ratio in the raw audio, which clearly presents a Doppler effect. Even more, it was a long event, with a duration of around 25 s. In Figure 6c, a good signal-to-noise ratio brake is plotted, with a clear component around 11,000 Hz. This event was around 2.5 sec length. In Figure 6d, we can observe a wide-band signal, whose spectrum includes frequencies up to around 20 kHz. Low frequencies, between 0 Hz and 1 kHz, are occupied in all the audio pieces by road traffic noise, which presents energy mainly in those frequencies.

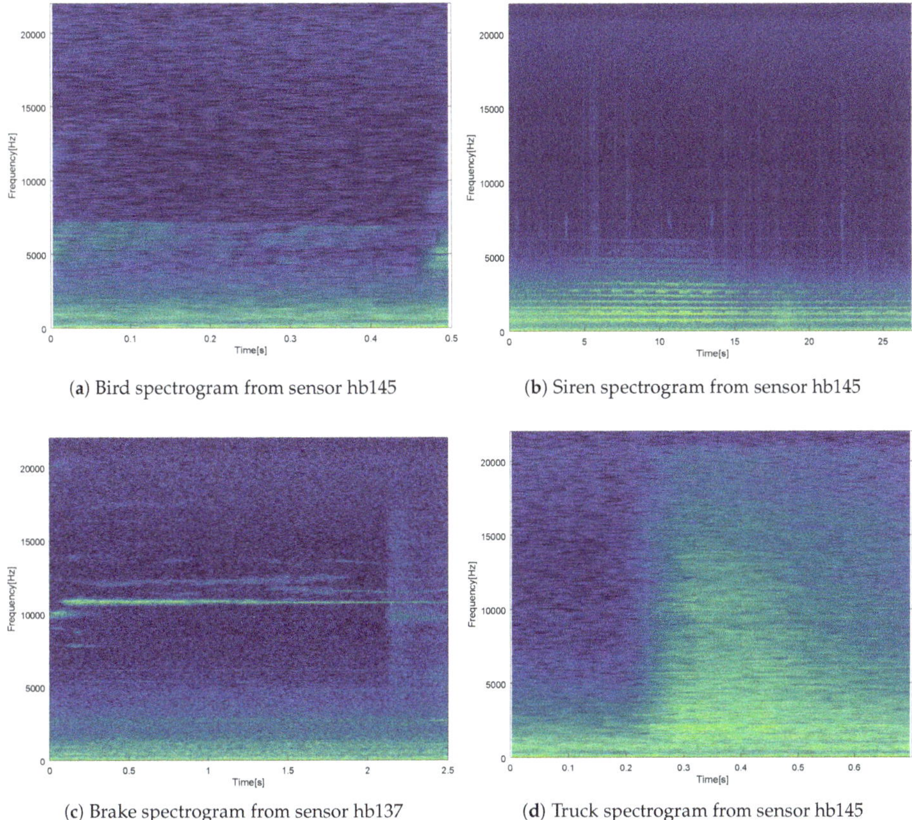

(**a**) Bird spectrogram from sensor hb145

(**b**) Siren spectrogram from sensor hb145

(**c**) Brake spectrogram from sensor hb137

(**d**) Truck spectrogram from sensor hb145

Figure 6. Examples of anomalous noise events recorded by sensors hb137 and hb145.

4. Channel Model Design

In this section, we define three urban channel models, computed as a linear combination of the input source and attenuated delayed paths. The three channel models are static (i.e., neither the transmitter nor the receiver changed their position while the sound signal was propagating) and, hence, no signal frequency spread and Doppler shift are expected. We defined three channel models which represent three different urban scenarios, with different number of paths and delay. The aim is to transmit the sound signals through each of the channel models and study the changes of time and frequency response.

Prior to applying the recorded sound signals to the three channel models, we characterised the channel in terms of time-spread by means of a PN sequence with good auto-correlation and cross-correlation properties [37]. The goal was to be able to detect each of the paths of the channel impulse response with a wide enough bandwidth to check whether the input sound signal had suffered any impairment throughout its whole bandwidth.

4.1. Outdoor Propagation Models

In this section, we describe the main characteristics of the outdoor propagation model used in this work. We consider a sound signal radiating as a spherical isotropic wave-front [51] through an homogeneous medium; that is, a medium with constant sound speed (C_0). To define the sound velocity as $c_0 = 343.23 \approx 343$ m/s, we assume the following conditions:

- Atmospheric density and pressure are assumed constant.
- The density of the air at ground level is 1.205 kg/m^3, the temperature is 20°, and the atmospheric pressure is 1 atm.
- The relative humidity is 70%.

In such a case, the acoustic free-field intensity of the radiation decreases with the inverse square of the distance and might travel to a receiver along a direct path and a number of reflected paths. Therefore, we can describe the acoustic pressure ($p(r)$) at a Euclidean distance r from the emitter in the following form [52]:

$$p(r) = \left[\frac{A(r_0)}{r_0}\right] e^{jkr_0} + \sum_{i=1}^{i=N} Q\left[\frac{A(r_i)}{r_i}\right] e^{jkr_i}, \tag{5}$$

where r_0 is the distance travelled by the direct path; r_i the distance travelled by each of the remaining N paths, which may be reflected by the ground, surrounding buildings, trees, and so on; $A(r)$ is the atmospheric attenuation; Q is the reflection coefficient; and $k = w/c_0$ is the wave number.

The atmosphere dissipates sound energy through two major mechanisms—viscous losses and relaxational processes—which have been extensively studied in the ANSI Standard S1-26:1995 [53]. The main mechanism of absorption is proportional to the square of the frequency. The relaxational processes also depend on the relaxation frequency of nitrogen and oxygen. Given the above atmospheric conditions, we can compute the attenuation suffered by an acoustic signal due to atmospheric absorption as:

$$A(r) = \frac{\alpha r}{100} \text{ [dB]}, \tag{6}$$

where r (in meters) is the distance between emitter and receiver and α is the absorption coefficient (in dB/100 m). Given the above meteorological conditions of temperature and relative humidity, we can take the absorption coefficient equal to 0.54 dB/100 m at 1000 Hz and 10.96 dB/100 m at 10,000 Hz.

The reflection coefficient is defined for spherical waves reflecting from complex plane boundary and can be approximated as:

$$Q = R_p(\theta) + B(1 - R_p(\theta))F(w), \tag{7}$$

where $R_p(\theta)$ is the plane-wave reflection coefficient, θ is the glancing angle, B is a correction term, and $F(w)$ is the boundary loss function defined by means of the numerical distance w [53]. If we assume a locally reacting ground and set $B = 1$, then:

$$R_p(\theta) = \frac{sin(\theta) - 1/Z}{sin(\theta) + 1/Z}, \tag{8}$$

where Z is the normalised acoustic impedance of the ground. We note that, for low glancing angles (θ), $R_p \to -1$ and, for high frequencies, $F(w)$ diminishes and, consequently, $Q = R_p$.

4.2. Impulse Response for the Defined Acoustic Channels

We define three different urban channel models, which are linear combinations of sets of paths received at a sensor, each with an acoustic pressure and phase shift as described by Equation (9):

$$y(n) = \sum_{i=1}^{N} x(n - \tau_i) a_i e^{j\tau_i 2\pi w}, \tag{9}$$

where $y(n)$ is the received signal at the sensor, $x(n)$ is the transmitted signal through each path whose amplitude (i.e., a_i) is computed as explained in Equation (5), N is the number of paths of each model,

τ_i is the delay suffered by each path, and $e^{j\tau_i 2*\pi w}$ is the phase delay of each path, which is proportional to the delay and frequency.

In order to study the effect of time-dispersive channels on acoustic signals, we propose three different scenarios. We first define a simple channel (model A) with just two paths, which can describe a wide street with a direct path and a reflection path on the ground. Models B and C describe two scenarios with more dense scatterers than model A: While model B depicts a scenario with a high delay spread, which may be distinctive of a wide street, model C represents a highly scattered scenario with half the delay spread of model B, which may be characteristic of a narrower street. Table 2 shows the length and delay of each path per model (where the propagation velocity was taken as 343 m/s).

Table 2. Length and delay of each path, *rms* delay spread, and coherence bandwidth of each channel model.

	Model A		Model B		Model C	
	Length [m]	**Delay [ms]**	**Length [m]**	**Delay [ms]**	**Length [m]**	**Delay [ms]**
Path 1	5.0	14.57	8.0	23.32	5.5	16.03
Path 2	6.5	18.95	11.0	32.07	6.0	17.49
Path 3	-	-	14.0	40.81	7.5	21.86
Path 4	-	-	16	46.64	8.5	24.78
τ_{rms}	17.97 ms		34.50 ms		19.74 ms	
$B_c(0.5)$	11.30 Hz		5.79 Hz		10.13 Hz	

In Figure 7, we show a block diagram of the synthesis of the channel impulse response of a N-path time-dispersive channel, as described by Equation (9). The input signal $x(n)$ is first delayed a number of samples proportional to the delay, τ_i, suffered by each path. It is then attenuated with an inverse proportion to the distance travelled and, finally, the phase is shifted proportionally to the distance travelled.

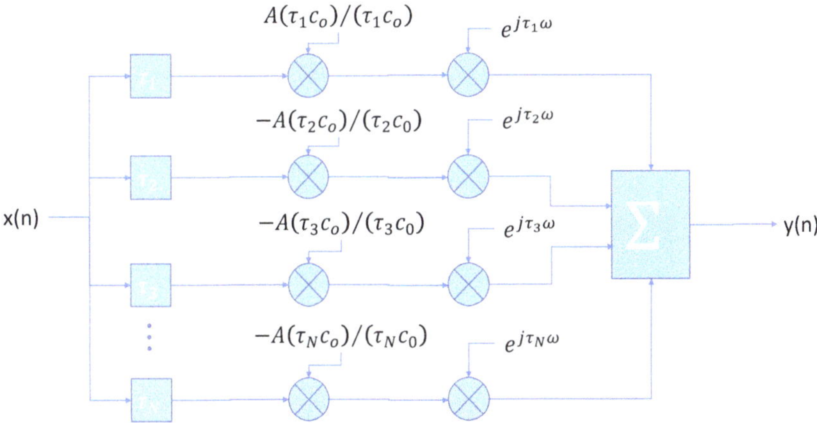

Figure 7. Block diagram for the N-path channel.

Figure 8 depicts a snapshot of the ideal impulse response of each of the three models used in this study. Each impulse response is normalised to the first path, which is the direct path and that with highest intensity at the receiver. The remaining paths are reflected paths showing negative amplitude, which denotes a value of phase between π and 2π. Model A shows two taps, whereas models B and C show four taps. The taps of model B spread in time, over a much longer period than the taps of model

A and C. Therefore, it is worth noting that model B will spread the energy of the propagated signal throughout a higher time-span than models A and C, which have closer paths.

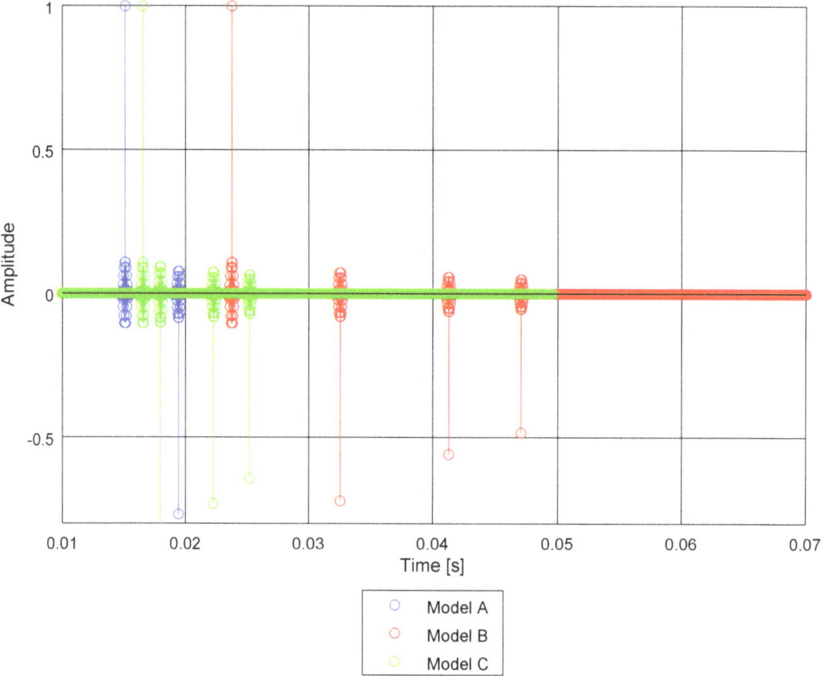

Figure 8. Channel ideal impulse response of models A, B, and C.

4.3. PN Sequence Channel Estimation

The goal of this section is to present a method to estimate the channel impulse response and frequency response of a time-dispersive and static channel (no movement in either transmitter or receiver is expected). In order to characterise the effects of this type of channel on acoustic signals, we use a PN sequence of type M, which has good cyclic cross-correlation properties [37]. The sequence is sampled at $F_s = 44.10$ kHz, with 1023 chips length and four samples per chip. Then, the chip period is $T_c = 4/F_s$ and, hence, the detection bandwidth equals $1/T_c$ and the delay resolution equals T_c. The PN sequence is also low-pass filtered with a Finite Impulse Response (FIR) root raised cosine filter (RRCOSFIR) with a roll-off factor equal to 0.9. The aim of this low-pass filtering is to limit the bandwidth of the PN sequence to 10 kHz, which is the most common band for audible signals. In Figure 9, we show the time domain (upper figure) and frequency domain (lower figure) of the PN sequence (in blue), as well as the same PN sequence after filtering (in red). It is worth noting that the time response displays a pseduo-random nature and that the bandwidth is limited after being filtered.

The analysis of the channel by means of a PN sequence is a key factor to characterise the time dispersion and, hence, to evaluate whether the signal arrived with any multi-path component at the receiver. First, the received signal $x[n]$ is correlated with the original PN sequence S. The correlation function is computed as explained in Equation (1).

From the channel impulse response (see Equation (2)), we can define the root mean square (*rms*) delay spread (τ_{rms}) as the standard deviation value of the delay reflections of the transmitted signal, weighted proportionally to the energy in the reflection waves. Table 2 shows the *rms* delay spread expected for each of the three channel models studied in this paper. Figure 10 depicts the channel impulse response of each of three models, computed by means of a PN sequence as explained in Equations (1) and (2). Model A has two paths and models B and C have four paths. Despite having different numbers of paths, models A and C showed a similar delay spread (see Table 2), as the paths were close together in channel C.

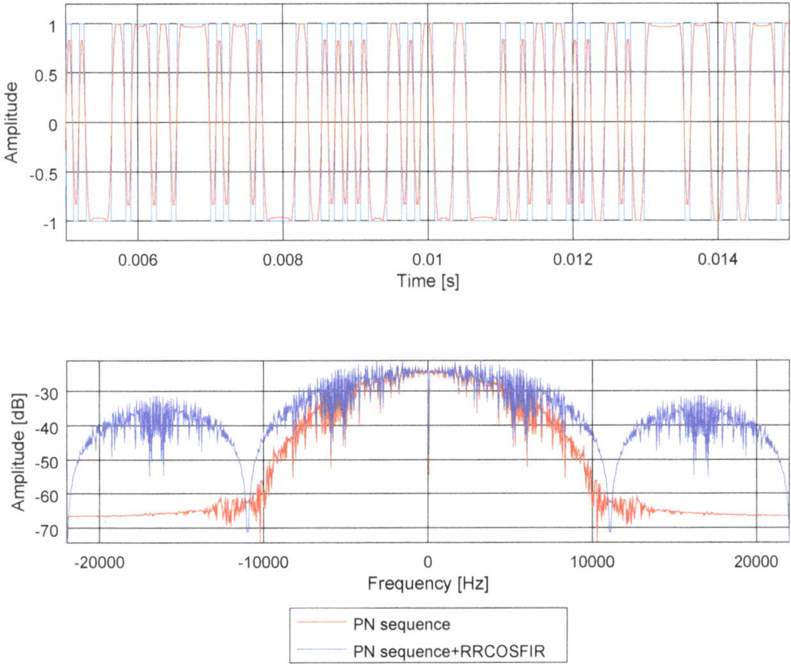

Figure 9. PN sequence time response and frequency response. The sequence without filtering is shown in blue, and after filtering (with a root raised cosine) is shown in red.

Once the channel impulse response is computed, we can derive the channel frequency transfer function as in Equation (10). A key parameter of the frequency characterisation of the channel is the coherence bandwidth. The coherence bandwidth is a statistical measure of the range of frequencies over which the channel can be considered flat (i.e., the bandwidth for which the auto co-variance of the signal amplitude at two extreme frequencies reduces from 1 to 0.5). The coherence bandwidth is inversely proportional to the *rms* delay spread of the channel (i.e., $B_c \approx 1/\tau_[rms])$ [20]; but there is no precise relationship between both. However, a widely accepted definition considers the bandwidth over which the frequency correlation is above 0.9; or a more relaxed one considers a frequency correlation of just 0.5:

$$B_c(0.9) = \frac{1}{50\tau_{rms}}, B_c(0.5) = \frac{1}{5\tau_{rms}}. \tag{10}$$

In Table 2, we show the coherence bandwidth of each channel model, with frequency correlation above 0.5. In Figure 11, we compare the frequency response of the PN sequence with the frequency response of each channel model computed with the same PN sequence, following Equations (1), (2), and (4). In Figure 11, we do not show the coherence bandwidth (for the sake of practicality), but we underline the bandwidth between consecutive fadings of the channel, which is proportional to the coherence bandwidth for each of the studied models. It is clear that, the higher the *rms* delay spread of the channel, the narrower the coherence bandwidth and, hence, the higher the probability of acoustic signal distortion.

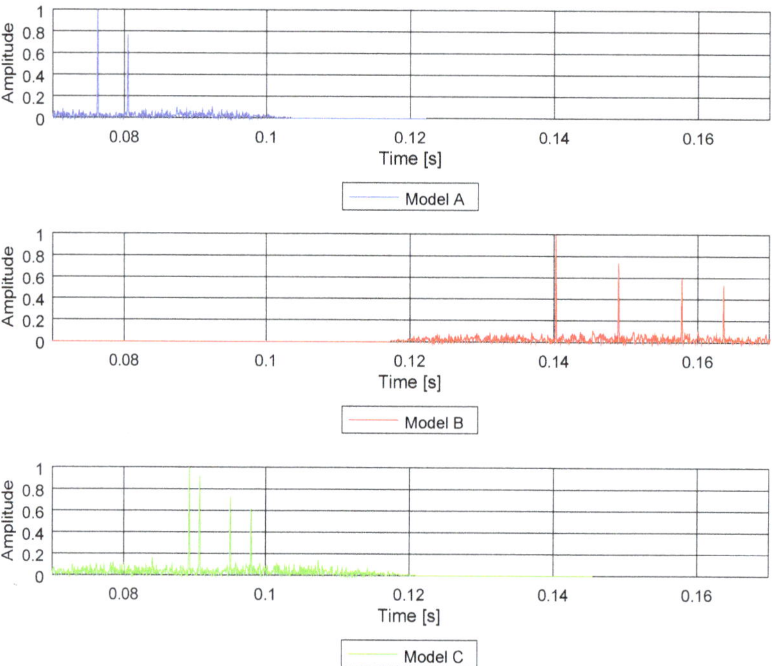

Figure 10. Channel impulse responses of models A, B, and C obtained using a PN sequence.

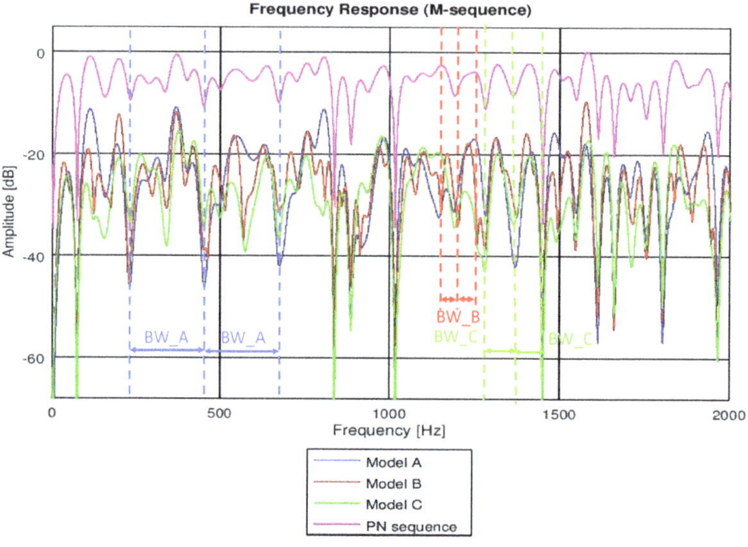

Figure 11. Channel frequency responses of models A, B, and C by means of a PN sequence. An approximate frequency range of the channel coherence bandwidth is shown for each channel model.

5. Real-Life Acoustic Recording Analysis

In this section, we show the effects of a multi-path static channel on real-life acoustic recorded signals. We, first, underline how the impulse response of a multi-path channel may be estimated by means of an acoustic signal, as if it was a PN sequence. Then, we also show how the frequency selective transfer function of a channel may distort the frequency response of an acoustic signal. Finally, we also show some examples of how the spectral and time distribution of acoustic signals may be changed because of facing different types of multi-path channels.

5.1. Propagation on Real-Life Acoustic Recordings

Among all the sample files of recorded ANEs, we have chosen some examples of those with good auto-correlation properties, in order to test whether they are able to estimate the channel response and its impairments, such as delay spread and the resultant distortion in the frequency domain. We also seek ANEs with different spectrum characteristics, in case they have different effects on the multi-path channels.

For instance, Figure 12 shows the channel impulse response, computed as in Equation (2), for the three channel models studied in this paper; however, no PN sequence was used in this case. Instead, the same ANE was used to detect the paths of the channel and, therefore, it was able to compute the rms delay spread. In Figure 12, we can see the impulse response of the three channel models computed with an ANE of a truck, with a bandwidth similar to the one shown in the spectrogram of Figure 6d. Due to the wide bandwidth of this ANE and, consequently, the high delay resolution, it was possible to detect the majority of the paths of the three models. This point can be checked if we compare the impulse response in Figure 12 with the channel impulse response computed with the PN sequence (see Figure 10). However, if the bandwidth of the ANE becomes narrow, the delay resolution dismisses and the probability of path detection reduces as well. For instance, Figure 13 depicts the impulse response computed by means of the acoustic signal of a brake, which had a narrow and discontinuous bandwidth, as shown in Figure 6c. Due to the narrow bandwidth and, hence, the poor delay resolution, it was not possible to detect all the paths of channel.

Regarding the frequency response of the three models, we found that they differed even if they were computed using the same ANE. For instance, Figure 14 plots a portion of the frequency response of the three models computed with an acoustic signal of a siren. It was a signal composed of a number of harmonics, as the spectrogram of Figure 6b shows. In Figure 14, we show three of its harmonics in the range between 800–1600 Hz and we can see that there was a maximum amplitude difference between channel frequency responses of 10 dB. This behavior is not consistent throughout the frequency axis and implies that a model does not always attenuate a frequency component with the same factor, since it computes the frequency response as a linear combination of attenuated paths with a phase that depends on the distance travelled and the frequency (see Equations (9)–(10)).

In addition to the different attenuation of frequency components in each model, we also observed a distortion phenomena which differed between models. For example, Figure 15 depicts the frequency component of an acoustic signal of a brake, for each of the three models, centered at 8 kHz. We can see that, while those models with similar delay spread (i.e., models A and C; see Table 2) had comparable frequency responses, model B had a higher delay spread, which distorted the frequency response. As explained above (see Section 4.3), the higher the delay spread, the narrower the coherence bandwidth of the channel and the higher the probability of signal distortion. The audio signal of a brake, as depicted in Figure 15, had a frequency bandwidth which happened to be flat for models A and C (wide coherence bandwidth), but not flat for model B (narrow coherence bandwidth). Therefore, the brake signal saw models A and C as non-frequency-selective channels, but model B as a frequency-selective channel.

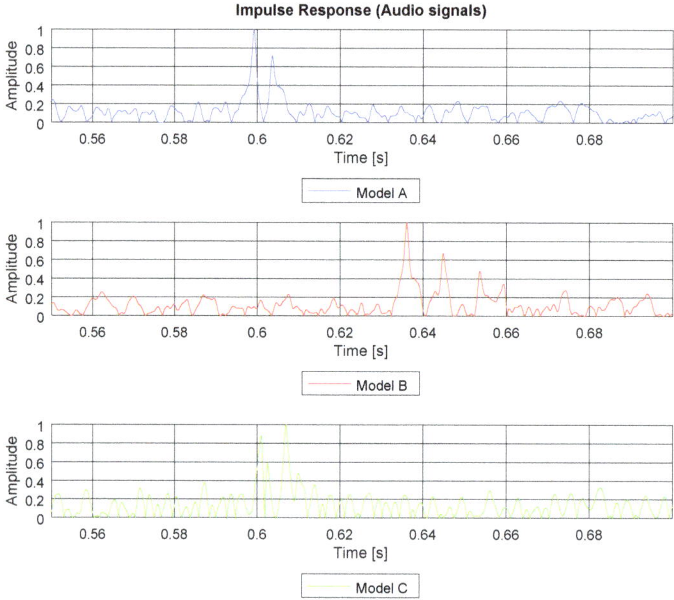

Figure 12. Channel impulse response of models A, B, and C, computed by means of an audio signal of a truck from sensor hb137.

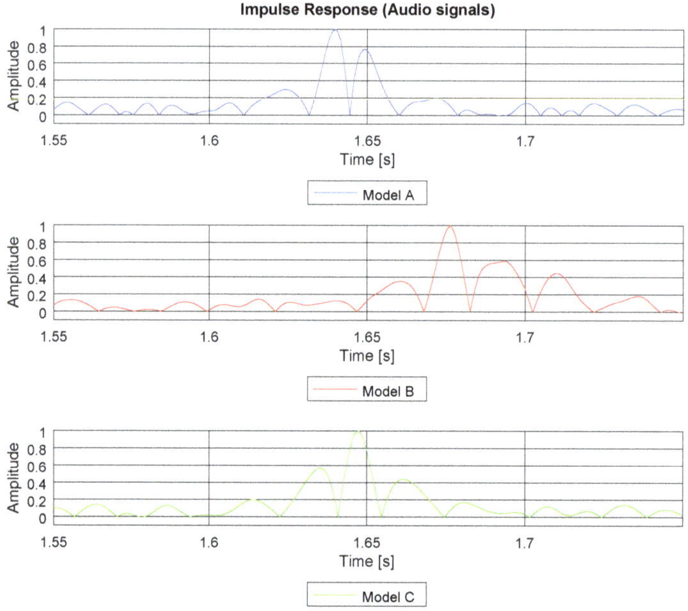

Figure 13. Channel impulse response of models A, B, and C, computed by means of an audio signal of a brake from sensor hb137.

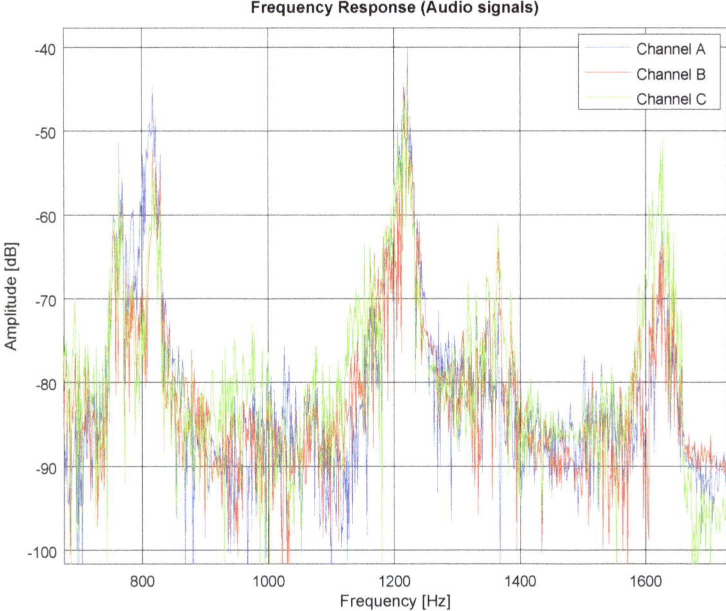

Figure 14. Frequency response of models A, B, and C, computed by means of an audio signal of a siren obtained from sensor hb145.

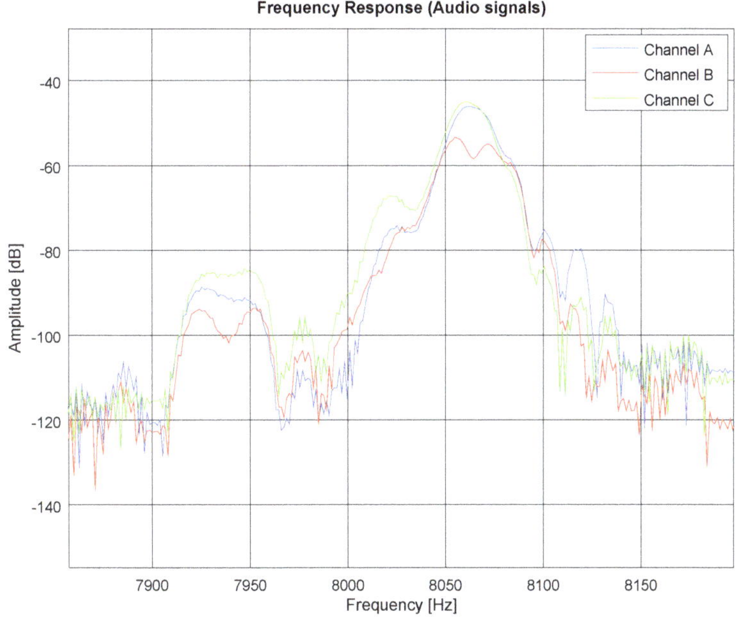

Figure 15. Frequency response of models A, B, and C, computed by means of an audio signal of a brake from sensor hb137.

5.2. Spectral Distributions over Propagation Channels

In this section, we evaluate the changes suffered in the frequency and time domain when the recorded ANEs were propagated through the three different multi-path channels (models A, B, and C, as explained in Section 4.2). Figures 16–18 show the outcomes related to different ANEs (i.e., the noise of an aeroplane, a brake, and a truck). For the sake of brevity we only show these three examples, as they are representative of the phenomena we want to outline. For each of them, we show the spectrogram of the emitted signal in the upper plot and then we divide the remaining plot into three rows, one for each model. In the first column, from row two to four, we show the spectrogram of the received signal through models A, B, and C, respectively.

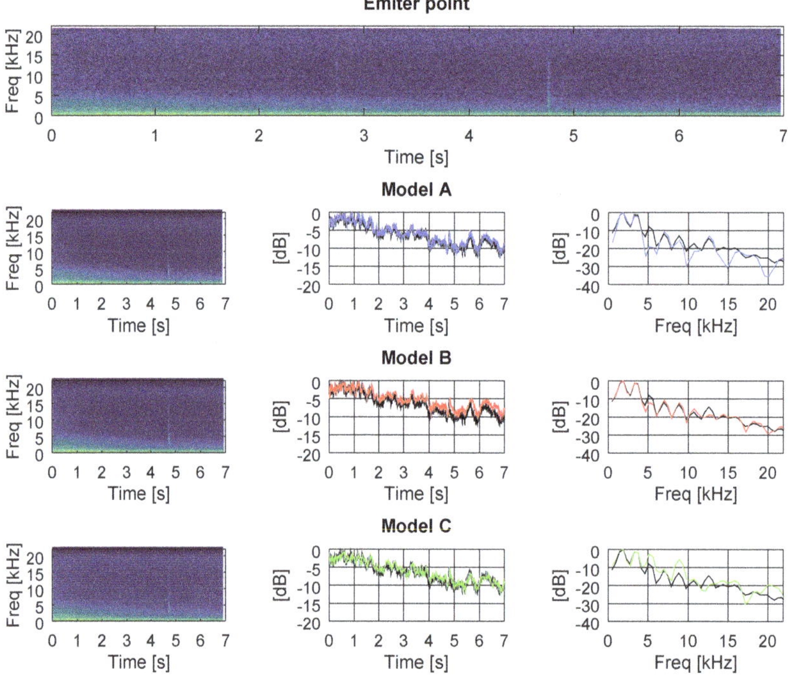

Figure 16. Accumulated energy along time and frequency, through channels A, B, and C, of an audio signal of an aeroplane at sensor hb137.

In the second column, from row two to four, we show the accumulated energy throughout the whole bandwidth for each time point, for each propagating model. Finally, in the third column, from row two to four, we show the accumulated energy throughout the whole time duration of the ANE for each frequency, for each propagating model. The spectrogram divides the signal into 40 ms-length segments, which were windowed with a Hanning window to reduce leakage and transformed into the frequency domain by means of a 2048-points Fast Fourier Transform (FFT), diplayed in logarithmic scale. Consecutive segments were overlapped by a factor of 87.5% to maximise the probability of detection. The plots of accumulated energy, in time and frequency, were normalised to the maximum energy and displayed in logarithmic scale. In these plots, we show the accumulated energy of the ANEs prior to being transmitted in black and the accumulated energy after being transmitted through models A, B, and C in blue, red, and green, respectively.

In all three figures, we can observe that the accumulated energy throughout the whole frequency band (middle column) at the emitter was more similar to the accumulated energy at the receiver when propagating through a low delay spread channel (i.e., models A and C), rather than through a higher delay spread channel (i.e., model B). The important factor that enables the spreading of energy over time is not the number of paths but the time distance between them, as well as the weight of each of them.

In all three figures, we can also underline the fact that the accumulated energy throughout the whole time span of the recorded ANEs (right column) approached the original level (i.e., prior to being propagated through the channel) when propagating through a high delay spread channel. In other words, it appears that channels with narrow coherence bandwidths managed to confine the energy into a similar range of values as the original signal. In a different manner, when facing wider coherence bandwidth channels (e.g., models A and C) the accumulated energy in the time domain tended to differ from the original one.

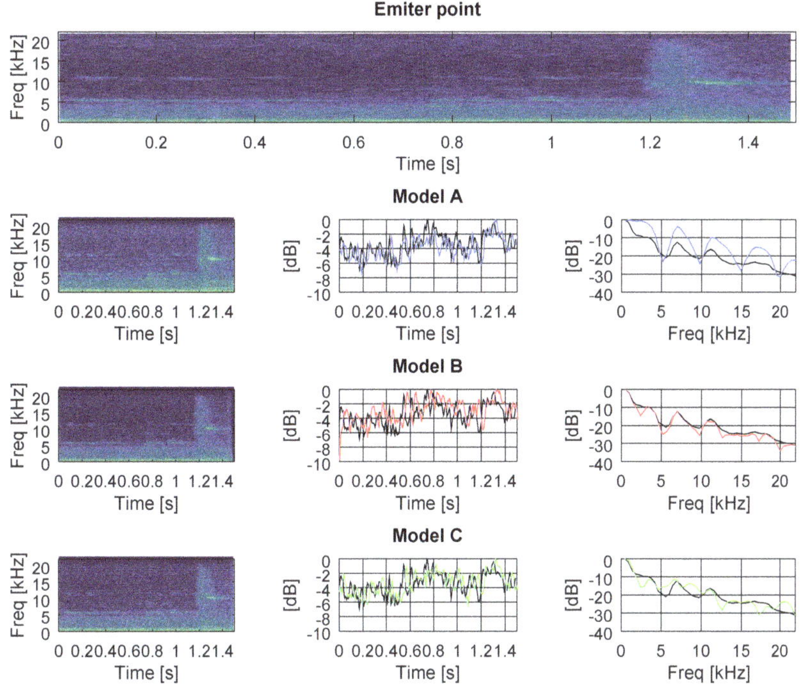

Figure 17. Accumulated energy along time and frequency, through channels A, B, and C, of an audio signal of an brake at sensor hb137.

The following step of this analysis will be to systematically analyse all ANEs from both sensors when transmitted through different types of channels (e.g., narrow streets, streets with tall buildings, wide streets with high background noise, and so on). The goal will be to find a systematic repetition of received signal characteristics related to environment factors. Another interesting point will be to study time-variant channels to check whether the time variation factor affects the received signal characteristics.

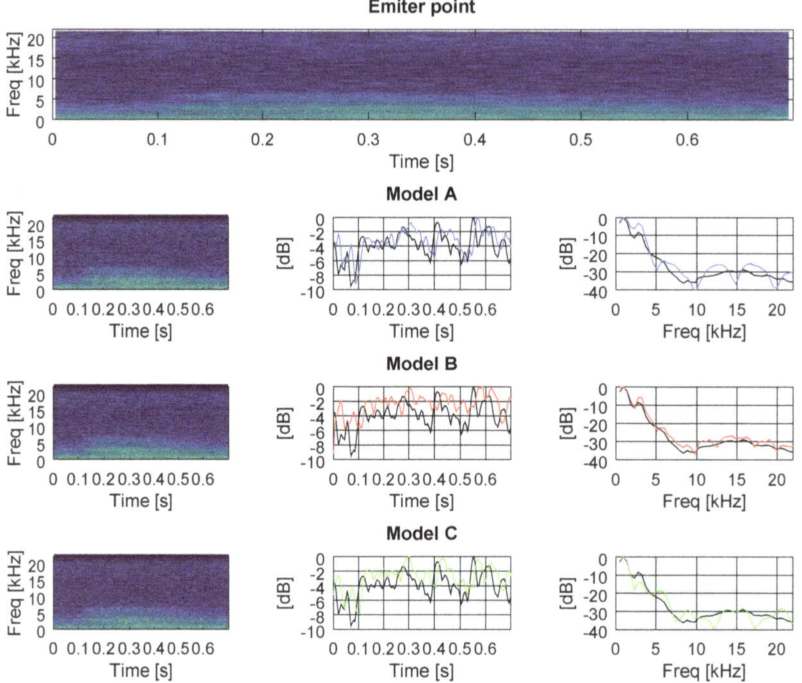

Figure 18. Accumulated energy along time and frequency, through channels A, B, and C, of an audio signal of an truck at sensor hb145.

6. Conclusions

The work presented in this paper is a preliminary study, designed to determine the spectro-temporal variations of acoustic signals in the presence of different types of propagation channels in an urban environment. It is intended to be the first detailed analysis applied to raw acoustic data of Wireless Acoustic Sensor Networks, with the aim of reaching a generalisation stage in the near future. The view of the channel over the spectrum of a raw acoustic signal in a real node of a WASN states the influence over the acoustic signal and the changes that apply.

In order to conduct the analysis, several raw acoustic signal, recorded in the framework of the DYNAMAP project, were used; the authors selected pieces of audio coming from sensors surrounded by parks, in order to have an environment similar to open air.

Throughout the research presented in this paper, we have found clear evidence that channel models with different coherence bandwidths affect the frequency response of acoustic signals in different ways. We have shown that, the higher the delay spread of the channel, the narrower the coherence bandwidth and the higher the distortion suffered by acoustic signals. Moreover, the accumulated energy along the frequency axis is more similar to the original signal when transmitted through wide coherence bandwidth channels, rather than through narrow coherence bandwidth channels. However, when computing the accumulated energy along the time axis, those channels with wide coherence bandwidths show more differences from the original signal than those channels with narrow coherence bandwidths.

On one hand, the qualitative evaluations developed in this work present substantial variations, both in terms of spectral distribution energy and in temporal variations due to delay, with a clear influence on the coherence bandwidth. These clear and severe variations in the spectro-temporal description can have severe effects on the detection of anomalous events using the ANED algorithm, as the initial hypothesis of this work stated. On the other hand, the effect of spectral-temporal noise variations on people living in these environments should also be taken into account; do these variations make the noises more annoying? Does perception change when the coefficients of spectral and temporal energy distribution are modified?

7. Future Work

The future lines of this work are to focus on the quantification of spectro-temporal variations, depending on the type of channel being used. Research into the best metrics will be pursued, in order to state whether the majority of the anomalous noise events, and even road traffic noise, is modified by the coherence bandwidth, which depends on the multi-path channel described. At the same time, the next natural step of this study will be an analysis focused on the generalisation of study of the channel performance to all available ANEs, where the impact on behavior will be determined as ANED accuracy for different types of channel. Finally, the degree of generalisation of the detection of acoustic events in time-varying propagation environments (with varying numbers and lengths of paths) will be studied, based on the sample of a narrow street with tall buildings or a wider street with more traffic noise.

Author Contributions: R.M.A.-P. has conceived the analysis and written a part of the paper. P.B. has coded the tests and written the rest of the paper.

Funding: There is no funding support.

Acknowledgments: The research presented in this work has been partially conducted with data obtained in the framework of the LIFE DYNAMAP project (LIFE13 ENV/IT/001254). The authors would like to thank Marc Arnela for his support. The authors would like to thank the 5th Electronic Conference on Sensors and Applications for the invitation to publish in Sensors regarding the Best Paper Award in the conference. Pictures of sensors hb137 and hb145 are courtesy of the colleagues of Universitá degli Studi di Milano Biccoca. The authors would like to thank 7:36 S-103 W6 14-C and 14-D for the hosting of most of this work.

Conflicts of Interest: The authors declare no conflict of interest.

Abbreviations

The following abbreviations are used in this manuscript:

ANE	Anomalous Noise Event
ANED	Anomalous Noise Event Detection
DYNAMAP	Dynamic Noise Mapping
END	European Noise Directive
FIR	Finite Impulse Response
MIMO	Multiple Input Multiple Output
PN	Pseudo-Noise
RRCOSFIR	Root Raised Cosine FIR
RTN	Road Traffic Noise
SNR	Signal-to-Noise Ratio
WASN	Wireless Acoustic Sensor Network

References

1. European Commission. *Report from the Commission to the European Parliament and the Council On the Implementation of the Environmental Noise Directive in accordance with Article 11 of Directive 2002/49/EC*; European Commission: Brussels, Belgium, 2017.
2. Recio, A.; Linares, C.; Banegas, J.R.; Díaz, J. Road traffic noise effects on cardiovascular, respiratory, and metabolic health: An integrative model of biological mechanisms. *Environ. Res.* **2016**, *146*, 359–370. [CrossRef] [PubMed]
3. Miedema, H.; Oudshoorn, C. Annoyance from transportation noise: relationships with exposure metrics DNL and DENL and their confidence intervals. *Environ. Health Perspect.* **2001**, *109*, 409. [CrossRef] [PubMed]

4. Muzet, A. Environmental noise, sleep and health. *Sleep Med. Rev.* **2007**, *11*, 135–142. [CrossRef] [PubMed]
5. Hygge, S.; Evans, G.W.; Bullinger, M. A prospective study of some effects of aircraft noise on cognitive performance in schoolchildren. *Psychol. Sci.* **2002**, *13*, 469–474. [CrossRef] [PubMed]
6. Dratva, J.; Phuleria, H.C.; Foraster, M.; Gaspoz, J.M.; Keidel, D.; Künzli, N.; Liu, L.J.S.; Pons, M.; Zemp, E.; Gerbase, M.W.; et al. Transportation noise and blood pressure in a population-based sample of adults. *Environ. Health Perspect.* **2012**, *120*, 50. [CrossRef] [PubMed]
7. Hänninen, O.; Knol, A.B.; Jantunen, M.; Lim, T.A.; Conrad, A.; Rappolder, M.; Carrer, P.; Fanetti, A.C.; Kim, R.; Buekers, J.; et al. Environmental burden of disease in Europe: Assessing nine risk factors in six countries. *Environ. Health Perspect.* **2014**, *122*, 439. [CrossRef]
8. Bello, J.P.; Silva, C.; Nov, O.; DuBois, R.L.; Arora, A.; Salamon, J.; Mydlarz, C.; Doraiswamy, H. SONYC: A System for Monitoring, Analyzing, and Mitigating Urban Noise Pollution. *Commun. ACM* **2019**, *62*, 68–77. [CrossRef]
9. Europoean Union, *Directive 2002/49/EC of the European Parliament and the Council of 25 June 2002 Relating to the Assessment and Management of Environmental Noise;* Official Journal of the European Communities: Brussels, Belgium, 2002; Volume L 189/12.
10. Kephalopoulos, S.; Paviotti, M.; Anfosso-Lédée, F. *Common Noise Assessment Methods in Europe (CNOSSOS-EU);* Report EUR 25379 EN; Publications Office of the European Union: Brussels, Belgium, 2002; pp. 1–180.
11. World Health Organization, Regional Office Europe. *Environmental Noise Guidelines for the European Region;* Technical report; World Health Organization: Geneva, Switzerland, 2018.
12. Socoró, J.C.; Alías, F.; Alsina-Pagès, R.M. An Anomalous Noise Events Detector for Dynamic Road Traffic Noise Mapping in Real-Life Urban and Suburban Environments. *Sensors* **2017**, *17*, 2323, doi:10.3390/s17102323. [CrossRef] [PubMed]
13. Alías, F.; Alsina-Pagčs, R.M.; Orga, F.; Socoró, J.C. Detection of Anomalous Noise Events for Real-Time Road-Traffic Noise Mapping: The Dynamap's project case study. *Noise Mapp.* **2018**, *5*, 71–85. [CrossRef]
14. Nilsson, M.; Forssén, J.; Lundén, P.; Peplow, A.; Hellström, B. *LISTEN Auralization of Urban Soundscapes;* Technical Report, Final Report to the Knowledge Foundation; Stockholm University: Stockholm, Sweden, 2011.
15. Orga, F.; Alías, F.; Alsina-Pagès, R.M. On the Impact of Anomalous Noise Events on Road Traffic Noise Mapping in Urban and Suburban Environments. *Int. J. Environ. Res. Public Health* **2017**, *15*, 13. [CrossRef] [PubMed]
16. Labairu-Trenchs, A.; Alsina-Pagès, R.M.; Orga, F.; Foraster, M. Noise Annoyance in Urban Life: The Citizen as a Key Point of the Directives. *Proceedings* **2019**, *6*, 1. [CrossRef]
17. Hornikx, M. *Acoustic Modelling for Indoor and Outdoor Spaces;* Taylor & Francis: Abingdon, UK, 2015.
18. Luigi, M.; Massimiliano, M.; Aniello, P.; Gennaro, R.; Virginia, P.R. On the validity of immersive virtual reality as tool for multisensory evaluation of urban spaces. *Energy Procedia* **2015**, *78*, 471–476. [CrossRef]
19. Georgiou, F. Modeling for auralization of urban environments. Ph.D. Thesis, Eindhoven University of Technology, Eindhoven, The Netherlands, 2018.
20. Proakis, J. *Digital Communications 5th Edition;* McGraw Hill: New York, NY, USA, 2007.
21. Alsina-Pagès, R.; Hervás, M.; Orga, F.; Pijoan, J.; Badia, D.; Altadill, D. Physical layer definition for a long-haul HF Antarctica to Spain radio link. *Remote Sens.* **2016**, *8*, 380. [CrossRef]
22. Hervás, M.; Alsina-Pagès, R.; Orga, F.; Altadill, D.; Pijoan, J.; Badia, D. Narrowband and wideband channel sounding of an Antarctica to Spain ionospheric radio link. *Remote Sens.* **2015**, *7*, 11712–11730. [CrossRef]
23. Stojanovic, M. Underwater acoustic communication. *Wiley Encyclopedia of Electrical and Electronics Engineering;* John Wiley: New York, NY, USA, 1999; pp. 1–12.
24. Sozer, E.M.; Stojanovic, M.; Proakis, J.G. Underwater acoustic networks. *IEEE J. Ocean. Eng.* **2000**, *25*, 72–83. [CrossRef]
25. Borowski, B. Characterization of a very shallow water acoustic communication channel. In Proceedings of the OCEANS 2009, Biloxi, MS, USA, 26–29 October 2009; pp. 1–10.
26. Kim, J.; Park, K.C.; Park, J.; Yoon, J.R. Coherence bandwidth effects on underwater image transmission in multipath channel. *Jpn. J. Appl. Phys.* **2011**, *50*, 07HG05. [CrossRef]
27. Alsina-Pagès, R.M.; Bergadà, P. The Citizen as a Key Point of the Policies: A First Approach to Auralization for the Acoustic Perception of Noise in an Urban Environment. *Multidiscip. Digit. Publ. Inst. Proc.* **2018**, *4*, 11. [CrossRef]

28. Attenborough, K.; Li, K.M.; Horoshenkov, K. *Predicting Outdoor Sound*; CRC Press: Boca Raton, FL, USA, 2014.
29. Sevillano, X.; Socoró, J.C.; Alías, F.; Bellucci, P.; Peruzzi, L.; Radaelli, S.; Coppi, P.; Nencini, L.; Cerniglia, A.; Bisceglie, A.; et al. DYNAMAP – Development of low cost sensors networks for real time noise mapping. *Noise Mapp.* **2016**, *3*, 172–189. [CrossRef]
30. Hewett, D.P. Sound Propagation in an Urban Environment. Ph.D. Thesis, Oxford University, Oxford, UK, 2010.
31. Sanchez, G.M.E.; Van Renterghem, T.; Thomas, P.; Botteldooren, D. The effect of street canyon design on traffic noise exposure along roads. *Build. Environ.* **2016**, *97*, 96–110. [CrossRef]
32. Hornikx, M.; Forssén, J. A scale model study of parallel urban canyons. *Acta Acust. United Acust.* **2008**, *94*, 265–281. [CrossRef]
33. Lyon, R.H. Role of multiple reflections and reverberation in urban noise propagation. *J. Acoust. Soc. Am.* **1974**, *55*, 493–503. [CrossRef]
34. Van Renterghem, T.; Botteldooren, D. The importance of roof shape for road traffic noise shielding in the urban environment. *J. Sound Vib.* **2010**, *329*, 1422–1434. [CrossRef]
35. Salomons, E.M. *Computational Atmospheric Acoustics*; Springer Science & Business Media: Berlin, Germany, 2012.
36. Leon-Garcia, A. *Probability and Random Processes for Electrical Engineering: Student Solutions Manual*; Pearson Education India: London, UK, 1994.
37. Golomb, S. *Shift Register Sequences*; Holden-Day: San Francisco, CA, USA, 1967.
38. Stojanovic, M.; Preisig, J. Underwater acoustic communication channels: Propagation models and statistical characterization. *IEEE Commun. Mag.* **2009**, *47*, 84–89. [CrossRef]
39. Goodney, A.; Cho, Y.H. Acoustic tomography with an underwater sensor network. In Proceedings of the 2012 Oceans, Hampton Roads, VA, USA, 14–19 October 2012; pp. 1–10.
40. Kaddouri, S.; Beaujean, P.P.J.; Bouvet, P.J.; Real, G. Least square and trended Doppler estimation in fading channel for high-frequency underwater acoustic communications. *IEEE J. Ocean. Eng.* **2014**, *39*, 179–188. [CrossRef]
41. Van Walree, P.A. Propagation and scattering effects in underwater acoustic communication channels. *IEEE J. Ocean. Eng.* **2013**, *38*, 614–631. [CrossRef]
42. Nencini, L. DYNAMAP monitoring network hardware development. In Proceedings of the 22nd International Congress on Sound and Vibration (ICSV22), Florence, Italy, 12–16 July 2015; The International Institute of Acoustics and Vibration (IIAV): Florence, Italy, 2015; pp. 1–4.
43. Nencini, L. Progetto e realizzazione del sistema di monitoraggio nell'ambito del progetto Dynamap. In *Proceedings of the AIA 2017*; AIA: Pavia, Italy, 2017.
44. Zambon, G.; Benocci, R.; Brambilla, G. Statistical road classification applied to stratified spatial sampling of road traffic noise in urban areas. *Int. J. Environ. Res.* **2016**, *10*, 411–420.
45. Zambon, G.; Benocci, R.; Bisceglie, A.; Roman, H.E.; Bellucci, P. The LIFE DYNAMAP project: Towards a procedure for dynamic noise mapping in urban areas. *Appl. Acoust.* **2017**, *124*, 52–60. [CrossRef]
46. Zambon, G.; Benocci, R.; Brambilla, G. Cluster categorization of urban roads to optimize their noise monitoring. *Environ. Monit. Assess.* **2016**, *188*, 26. [CrossRef]
47. Cerniglia, A. Development of a GIS based software for real time noise maps update. In *INTER-NOISE and NOISE-CON Congress and Conference Proceedings*; Institute of Noise Control Engineering: Hong Kong, China, 2016; Volume 253, pp. 6291–6297.
48. Socoró, J.C.; Alsina-Pagès, R.M.; Alías, F.; Orga, F. Adapting an Anomalous Noise Events Detector for Real-Life Operation in the Rome Suburban Pilot Area of the DYNAMAP's Project. In *Proceedings of EuroNoise 2018*; EAA—HELINA: Heraklion, Crete-Greece, 2018; pp. 693–698.
49. Foggia, P.; Petkov, N.; Saggese, A.; Strisciuglio, N.; Vento, M. Audio Surveillance of Roads: A System for Detecting Anomalous Sounds. *IEEE Trans. Intell. Transp. Syst.* **2016**, *17*, 279–288. [CrossRef]
50. Memoli, G.; Paviotti, M.; Kephalopoulos, S.; Licitra, G. Testing the acoustical corrections for reflections on a façade. *Appl. Acoust.* **2008**, *69*, 479–495. [CrossRef]
51. Lamancusa, J. *Noise Control, Outdoor Noise Propagation*; Penn State University: Centre County, PA, USA, 2000.

52. Attenborough, K.; Taherzadeh, S.; Bass, H.; Di, X.; Raspet, R.; Becker, G.; Gudesen, A.; Chrestman, A.; Daigle, G.; L'Esperance, A.; et al. Benchmark cases for outdoor sound propagation models. *J. Acoust. Soc. Am.* **1994**, *97*, 173–191. [CrossRef]
53. Standards Secretariat, Acoustical Society of America. *Method for Calculation of the Absorption of Sound by the Atmosphere*; Standards Secretariat, Acoustical Society of America: Melville, NY, USA, 1995.

Article

Effect of Imperfections Due to Material Heterogeneity on the Offset of Polysilicon MEMS Structures

Aldo Ghisi * and Stefano Mariani

Department of Civil and Environmental Engineering, Politecnico di Milano, Piazza Leonardo Da Vinci 32, 20133 Milano, Italy

* Correspondence: aldo.ghisi@polimi.it; Tel.: +39-0223-994-310

Received: 7 June 2019; Accepted: 22 July 2019; Published: 24 July 2019

Abstract: Monte Carlo analyses on statistical volume elements allow quantifying the effect of polycrystalline morphology, in terms of grain topology and orientation, on the scattering of the elastic properties of polysilicon springs. The results are synthesized through statistical (lognormal) distributions depending on grain size and morphology: such statistical distributions are an accurate and manageable alternative to numerically-burdensome analyses. Together with this quantification of material property uncertainties, the effect of the scattering of the over-etch on the stiffness of the supporting springs can also be accounted for, by subdividing them into domains wherein statistical fluctuations are assumed not to exist. The effectiveness of the proposed stochastic approach is checked with the problem of the quantification of the offset from the designed configuration, due to the residual stresses, for a statically-indeterminate MEMS structure made of heterogeneous (polycrystalline) material.

Keywords: polysilicon MEMS; uncertainty estimation; statistical volume element; zero offset

---

1. Introduction

Because of the microfabrication process, a polysilicon MEMS structure is never found at the exact designed place, but after the production steps, a shift from the rest position is often observed [1,2]. The difference can be so small to be inconsequential for the working conditions, but in particular cases, it can represent a significant disturbance. Since the phenomenon depends on the actual geometry and material involved, it is intrinsically of a stochastic nature. To deal with the consequences of these uncertainties quantitatively, a statistical approach should be therefore followed, by introducing a probabilistic distribution for the relevant variables involved in the manufacturing process.

This variable initial offset creates a series of uncertainties that require workarounds or specific electronics to account for them. The most common remedy is of the type exemplified by [3], where an initial offset due to process mismatch (called "zero-g offset" or "intrinsic sensor offset"), generating a variation in the initial capacitance of the sensor up to 10% and 20% for a three-axis accelerometer, is treated through a compensation circuit. Another calibration (or auto-calibration) procedure was shown in [4], to account for the varying gap geometry of MEMS featuring comb drives due to the fabrication process. More importantly, the mentioned uncertainties affect the stiffness of critical components, spreading it around the designed value: a plethora of methods, a good review of which can be found in [5], has been invented to overcome this problem. Stiffness is in fact a key parameter for defining relevant design variables, for example the resonance frequency: e.g. in [6] it was discussed how, because of the uncertainty on the geometry due to the manufacturing process, it is a challenging objective to obtain a tight sense-drive frequency separation (e.g. 2%) for an MEMS gyroscope, since the stiffness cannot be controlled finely enough.

When one considers the previous examples, it is understandable that uncertainties affecting the initial positions are however related to the ones responsible for the variability from the target status (i.e. not due to drifts or noise) during the MEMS working conditions: they are in fact both generated by the same sources during the manufacturing process. For example, in [7] a list of sensitive parameters, namely key geometrical dimensions (varying about 6–10%) or material properties (varying about 10%) (as taken from [8]) was used to explore how the uncertainty propagates in a mathematical model describing the working conditions of an MEMS energy harvester: it is easy to recognize that the same quantities would affect an offset from the designed configuration as well.

It would be therefore a useful insight to quantify the role of these uncertainties in an effective and manageable way, before moving to the working conditions, where other issues could be involved. Moreover, the corrections necessary to overcome the initial offset can become themselves too costly to be economically convenient, as shown in [3].

Going more into the details, the sources of these uncertainties can be (i) the so-called over-etch generated from the sequence of deposition, masking alignment, photolithography, and the deep reaction-ion etching process [9–11] and (ii) the effect of the material heterogeneity due to the polycrystalline morphology [12–14]: the former affects the geometry layout, while the latter influences the effective material properties. Both the causes tend to be neglected or hastily (and roughly) estimated, when the dimensions of critical MEMS components are significantly larger than a characteristic length of the microstructure, such as the average silicon grain size. However, both of these causes become relevant as far the miniaturization proceeds and the aforementioned critical dimensions shrink.

In the past, the authors devised a numerical approach to foresee the mechanical behavior of a polycrystalline ensemble by carefully representing, through an artificial reconstruction, the network of the grain and grain boundaries [15–18]. The approach is typically used to carry out a homogenization procedure and to establish the conditions necessary to build a Representative Volume Element (RVE) [19,20], but it can also be adopted to study Statistical Volume Elements (SVEs), i.e. to construct a statistical set accounting for all the desired uncertainties (such as the grain topology and the size and orientations of the elastic axis) whenever the characteristics at the micro scale become non-negligible for the mechanical quantities. This condition exactly arises when the miniaturization is put to the current technological limit and the silicon grain size becomes on the order of magnitude of the minimum dimensions of the characteristic structural parts, like the width of the slender suspension beams.

From the perspective of the quantification of the uncertainties due to material heterogeneity, in this work we aim to extract the information necessary to create quantitatively-informed stochastic analytical distributions of the elastic properties, such as the apparent Young's moduli E and G and Poisson's ratio ν, from numerical simulations of the SVEs. The latter represent only a realization of the stochastic variables involved in the definition of the mechanical properties at the upper, device-level scale, different from the more commonly-encountered RVEs that would represent the effective properties in a deterministic sense. However, the knowledge of the statistics of the aforementioned elastic quantities allows rigorously quantifying their scattering around a mean value, and therefore to foresee the mechanical uncertainty transferred to the structural behavior, e.g. the suspension spring's stiffness.

With regard to the geometrical uncertainty due to over-etch fluctuations, different from the commonly-adopted simplification of a constant value for a single device, in this work we explore the effect of a scattering from the mentioned constant value along the supporting beams. In this case, an analytical distribution is (a priori) assumed, and the contribution of variable over-etch on the beam moment of inertia is considered in addition to the previously-mentioned material-generated uncertainty.

The main purpose of our approach is to overcome the computational burden of the polycrystalline morphology analysis, thanks to the analytical distributions, which are therefore good to devise an engineering tool useful for design.

In the following Section 2, an exemplary MEMS configuration possibly leading to an offset is described. Then, the procedure exploiting (i) Monte Carlo (MC) analyses of SVEs aiming to define the elastic mechanical properties of homogenized polysilicon, (ii) the extraction of the mentioned analytical stochastic distributions from the numerical data, and (iii) their employment to build the (statistically-informed) stiffnesses of critical MEMS details is carefully detailed in Section 3. This reasoning is adopted to solve a typical offset problem in Section 4, where a discussion of the outcome is also carried out.

2. Sensitivity to Imperfections: A Simple Model for MEMS Offset

The effect of mechanical and geometrical uncertainties at the microscale can be clearly observed in the case of statically-indeterminate movable structures. As a benchmark example, inspired by the geometry of single-axis inertial MEMS devices working as shown in Figure 1 (see also [21–23]), we focus on a simplified scheme, where a proof mass is connected to the substrate through a couple of polycrystalline silicon beams or springs in series. In this case, even if the target design is to have the two stiffnesses k_1 and k_2 equal, the randomly-varying grain morphology, the over-etch defects [9], and the residual stresses arising from the manufacturing process induce an offset u away from the rest position.

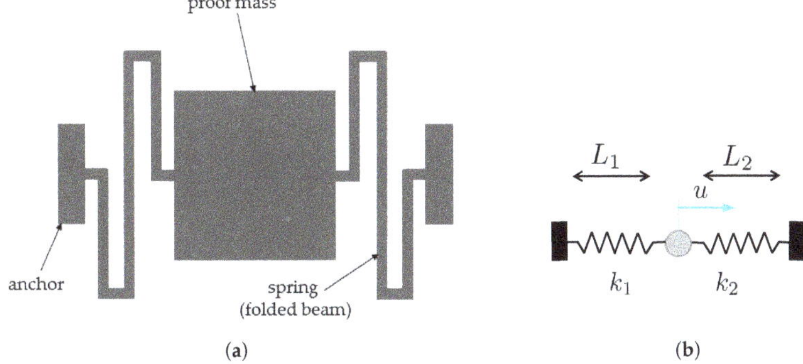

Figure 1. (**a**) Example and (**b**) structural scheme of single-axis inertial MEMS devices, featuring a proof mass anchored to the die via two springs in series.

Whatever the geometry of the two springs is, the said imperfections cause instead the stiffness values k_1 and k_2 to differ. The resulting offset u that can detrimentally affect the performance indices of the device linked, e.g. to capacitive readout, often proves negligible, but it may become relevant when the dimensions of critical structural details (like e.g. the in-plane spring width) become comparable to the average silicon grain size or to the microfabrication tolerances related to the etching stage. By assuming the proof mass to be a rigid body, in order to compute u we assume two sources that induce the (either positive or negative) elongation of the springs: an inelastic deformation ϵ_r linked to the residual stress in the polysilicon film; an elastic deformation ϵ_e induced by the constraints at the anchor points that prevent any motion at the end points. The latter effect can be formally represented by a force F acting on both the springs in series. Due to the said constraints at the anchors, the compatibility equation for this statically-indeterminate system is given by:

$$F\left(\frac{1}{k_1}+\frac{1}{k_2}\right)+\epsilon_r\,(L_1+L_2)=0. \tag{1}$$

Within the proposed frame, the sources ϵ_r and F are assumed without any dependence on the out-of-plane direction; possible effects of residual stress gradients are therefore disregarded to simplify the analysis, with a focus only on the in-plane motion of the proof mass. Moreover, the lengths L_1 and

L_2 of the springs are meant in a very general sense as kinds of effective values, in order to allow also for folded geometries like the one depicted in Figure 1a. Solving Equation (1) for F, we end up with:

$$F = -\frac{L_1 + L_2}{\frac{1}{k_1} + \frac{1}{k_2}} \epsilon_r \tag{2}$$

and the offset u thus reads:

$$u = \frac{\frac{k_1}{k_2} L_1 - L_2}{1 + \frac{k_1}{k_2}} \epsilon_r. \tag{3}$$

For the sake of simplicity, we now assume $L_1 = L_2 = L$. To avoid in the analysis any dependence on ϵ_r, which, as stated, represents an inelastic effect of the residual stresses and stands as a kind of algorithmic, model-based parameter, Equations (2) and (3) are solved for it, and the results are given next in terms of the ratio u/F, according to:

$$\frac{u}{F} = \frac{1 - \frac{k_1}{k_2}}{2\,k_1} \tag{4}$$

which depends on the stiffnesses of the springs only. Equation (4) shows that, if the two stiffnesses are equal (i.e. $k_1 = k_2$), no offset is induced. When, instead, microscale scattering leads to different stiffness values, the offset shows up. To account for such discrepancy between the two values of the spring stiffness, we assume that the scattering induces the values $k_1 = k + \Delta k_1$ and $k_2 = k + \Delta k_2$, with $\Delta k_1 \neq \Delta k_2$ (where k can be assumed as the target reference value). Equation (4) is then modified as:

$$\frac{u}{F} = \frac{1 - \frac{k+\Delta k_1}{k+\Delta k_2}}{2\,(k + \Delta k_1)} = \frac{\Delta k_2 - \Delta k_1}{2\,(k + \Delta k_1)\,(k + \Delta k_2)}. \tag{5}$$

It is noted that a difference in the stiffness values is not sufficient to get an offset: the force F arising from the aforementioned residual stress distribution is also required to trigger the proof mass displacement. This through-the-thickness distribution of the residual stresses is actually a whole problem on its own [24]: though the simple model (4) can be adopted to estimate them via MC techniques like that suggested in [25], provided that measurements are available for a number of statistically-representative devices, in this work we do not aim to address this issue, but instead a quantification of the uncertainty effects linked to the stiffness of structural components. As mentioned in the Introduction, we handle two sources of the scattering of the stiffness values: the heterogeneity of the polycrystalline material and the geometrical imperfections due to the manufacturing process. Both micromechanical sources are specifically discussed in Section 3.

3. Characterization of the Uncertainties at the Microscale

To quantify the uncertainties in the spring stiffness, the two sources linked to the polycrystalline morphology and to the etch defects are dealt with separately. In Section 3.1, an MC procedure is adopted to determine the homogenized elastic properties of the polysilicon film, in terms of microstructure-affected values of Young's modulus E, shear modulus G, and Poisson's ratio ν. Since we considered polycrystalline silicon with a columnar structure, the material was treated as transversely isotropic at the device scale. The homogenized elastic properties obtained with the MC procedure at the micro-scale refer to the plane parallel to the substrate, so that the out-of-plane direction coincides with the grain columns. Due to the FCC crystal lattice of silicon, Young's moduli obtained from the homogenization along two orthogonal directions are always equal, and deviations from isotropy prove small. Next, in Section 3.2, these scattered overall properties are used jointly with the statistics of over-etch to provide an estimate of the scattering in the overall spring stiffness according to beam bending theory.

3.1. Apparent Elastic Properties of Polysilicon Films

Let us start by allowing for the uncertainties linked to the microstructure of the polysilicon film, namely the morphology of the polycrystal. As already stated, we focus here on the in-plane motion of the structure; therefore, a two-dimensional geometry was considered. To assess the impact of the microstructure on the dispersion of the results, an MC procedure was adopted; within such a stochastic procedure, each realization or sample of the polysilicon film had the shape of a square of side-length ℓ. This length-scale ℓ was assumed to be linked to the characteristic size of the already mentioned critical details of the geometry of the movable structure. In the simple geometry of Figure 1, as the proof mass was assumed rigid, ℓ had to be driven by the springs' features: out of their length and in-plane width, the latter obviously drives scattering at the material level and is therefore considered in what follows. As far as the length was instead concerned, the way to handle it will be detailed in Section 3.2.

The microstructure inside each square sample was obtained via a (regularized) artificial Voronoi tessellation, by imposing an average grain size $g = 1$ μm [17] measured as the average distance between the centroids of the grains in the tessellation. Due to the columnar morphology of the polysilicon film [26], with a texture almost aligned with the out-of-plane axis for all the grains, the randomness of crystal lattice was assumed to hold in-plane only [19]. Because of the small ℓ/g ratio, the length-scale separation assumption of homogenization did not hold true, and relevant asymptotic bounds on the elastic properties of the RVEs did not actually cover the full range of values of microstructure-affected apparent elastic properties: outliers therefore happened to fall out of the bilateral limits, as observed in [14]. The sampling of values out of the range provided by asymptotic, analytical bounds was assumed to undermine the assumption of representativeness, and the results thus had to be interpreted in a statistical context. To understand the above reasoning, in Figure 2 a few examples of the microstructural samples are shown, reporting also the orientation of the in-plane axes of elasticity for each grain.

In the numerical, Finite Element (FE) procedure feeding the MC analysis, the whole sample was meshed with quadratic triangular elements featuring a characteristic size of 125 nm, to assure accuracy in the results. This element size was already proven in [17,27] to lead to mesh-independent results, in terms of the homogenized elastic property of the polycrystalline samples. To decrease the computational burden of the simulations, especially in view of the number of samples to be handled in each MC analysis, some authors proposed the use of (grain) boundary element formulations; see e.g. [28–31], wherein only the grain boundary network has to be discretized to reduce the dimensionality of the problem.

Figure 2. The 2×2 μm^2 SVE examples of the in-plane polycrystalline morphology adopted in the MC analysis.

The reasons for relying on a numerical strategy with numerical simulations to feed an MC analysis were already discussed in [27]. If one is interested in the reference or mean values of the mechanical properties of the polysilicon film, asymptotic approaches will suffice; if the scattering induced by micromechanical features is instead the main focus of the analysis, a proper statistical description of all the stochastic variables and a proper sampling out of the relevant statistical distributions are indeed necessary. The aforementioned need is strongly enhanced in the considered case of ℓ/g values on the order of 2–3. To this end an SVE, as handled in the analysis, was therefore defined as follows [32,33]: it is a domain whose dimension is smaller than a conventional RVE, but larger than the characteristic length scale (i.e. the grain size), so that the elastic properties obtained from the homogenization procedure are not to be considered as the effective ones to be adopted for a homogeneous deterministic device-level analysis, but as representative of the actual microstructure randomness. These elastic properties have to be considered as the "apparent" ones, according to the definition provided in [32]: i.e. the apparent Young's modulus obtained from each SVE is simply a realization of a random variable, which accounts for the relevant microstructure, and it should not be used in a deterministic analysis at the device-level scale.

For the MC analysis, two different sets of SVEs were handled with $\ell = 2, 3$ μm. These two values were selected in accordance with our previous studies [12–14]: to emphasize the microstructural effects, ℓ/g has to be minimized in compliance with the microfabrication constraints. This rationale justified the value $\ell = 2$ μm, whereas $\ell = 3$ μm was selected on the basis of the dimension of beams in the structure described in [34].

The apparent elastic moduli of each sample were obtained according to the procedure discussed in [17]. A bilateral bounding scheme was adopted by imposing either uniform stress or uniform strain Boundary Conditions (BCs) all over the boundary of the SVE. Periodic BCs were not adopted, since a periodic microstructure can never be observed in real polysilicon films. For each type of BCs, three analyses were run, inducing in turn only one non-zero component of the macroscopic in-plane stress or strain; the results of these three analyses then allowed estimating the microstructure-affected values of E, G, and ν. The latter values turned out to depend on the film morphology due to the local intensification of the stress and strain fields caused by the misorientation of adjacent grains. Additional details, which are beyond the scope of this work, can be found in [17]; see also [27].

The outcomes of this statistical analysis are reported with continuous lines in Figures 3–5, in terms of the Cumulative Distribution Functions (CDFs) of E, G, and ν, as obtained with the two types of BCs and for the two SVE sizes. As is known, the uniform strain BCs lead to an upper bound on the actual value of E and G and a lower bound on the actual value of ν; vice versa, the uniform stress BCs provide a lower bound for E and G and an upper bound for the actual ν. The plots allow also to compare easily the 2×2 μm^2 and the 3×3 μm^2 cases and to quantify the effect of a smaller device-level size ℓ with respect to the average grain size: it marginally affects the mean values of the elastic moduli, while the scattering around them is strongly modified. The same results are also collectively reported in Figure 6 as box-whisker plots, to show in a compact way the mean values and the dispersion around them. No limits were assumed to exclude possible outliers in the data, so the full range of values of the results was included in the plots.

The mean value of the apparent Young's modulus was slightly dependent on the BCs (as expected, see [32] and see Table 1): for the 2×2 μm^2 case, it amounted to E = 149.98 GPa for uniform strain BCs, while it read E = 148.11 GPa for uniform stress BCs. It was almost negligibly affected (on the order of 1%) by the dimension of the SVE (ℓ = 2 or 3 μm) for the assumed grain size $g = 1$ μm. The results showed instead a larger influence of the SVE size on the standard deviation, as qualitatively shown in Figure 3 by the less steep CDF curve for the 2×2 μm^2 case with respect to the 3×3 μm^2 case. Quantitatively, the standard deviation for the uniform strain BCs increased from 3.34 GPa to 5.47 GPa when the SVE size decreased from $\ell = 3$ μm to $\ell = 2$ μm: the correspondent coefficients of variations (i.e. the ratio between the standard deviation and the mean value) were around 2.2% and 3.5%, respectively. We can interpret this larger scattering as the quantification of the effect of the grain morphology on the statistical distribution of E, as the grain size becomes comparable with the SVE size.

Similar considerations can be made for the apparent shear modulus G; see Table 2: the scattering between the mean values was small (on the order of 2%) when the BCs and/or the SVE size were varied; the scattering became more significant if the standard deviation was considered, as evidenced by the values of the coefficients of variations, ranging from 3.9% (3×3 µm^2) to 6.2% (2×2 µm^2).

For the apparent Poisson's ratio (see Table 3), the mean values did not change significantly with the BCs, but they varied with the reduction of the SVE size of about 9%. The corresponding standard deviation instead showed a more meaningful change, varying from about 10% for the 3×3 µm^2 case to about 17% for the 2×2 µm^2 case.

As in any MC analysis, results relevant to a sufficiently high number of samples have to be collected to guarantee the convergence in the statistics of the apparent properties. With the focus on the homogenized Young's modulus of the polycrystalline film, some exemplary results are collected in Figure 7 for the 2×2 µm^2 SVE (top row), in terms of the convergence of the mean value and of the standard deviation with an increasing number n of samples. Besides some small fluctuations, visible only because of the reduced scale of the graphs, outcomes tended to converge and became stable even before $n = 100$. For the 3×3 µm^2 SVE shown in Figure 7 (bottom row), the convergence was attained even faster, as expected for a larger SVE. Such an evolution of the graphs was not directly related to some specific SVE morphologies, since they remained unchanged (within the limits of the statistical representation) by shuffling the order of the SVEs handled in the MC analysis. Similar results were observed for any elastic property of the film, and for any SVE size; the only slight difference at varying SVE size was the different value at convergence for the standard deviation.

A further interesting feature of the mechanical response of the polysilicon film was its in-plane degree of anisotropy. In Figure 8, box-whisker plots are reported, again at varying BCs and SVE sizes, for the ratio G/G_{iso}, where $G_{\mathrm{iso}} = E/(2(1+\nu))$ would be the shear modulus of a perfectly in-plane isotropic material, whose Young's modulus E and Poisson's ratio ν were those obtained by means of the same homogenization procedure. This measure was already adopted in [17] to assess the material-dependent isotropy of microstructured materials and, to some extent, the representativeness of the SVE size. Since silicon, due to its FCC crystal structure, is a moderately-anisotropic material, the results related to aggregates of randomly-oriented grains are supposed to be characterized by a ratio $G_{\mathrm{iso}} = E/(2(1+\nu))$ tightly bonded around isotropy, i.e. around $G/G_{\mathrm{iso}} = 1$. This ratio in the case of single crystal silicon, which has a low degree of anisotropy, turns out to be 1.57; it can be therefore appreciated that, in the case of a polycrystalline material, the ratio tends towards a unitary value, even for small SVEs as the ones considered in this work (see Figure 2).

To deal with the analysis of the imperfection sensitivity at the spring scale, the numerical statistical distributions were fitted with analytical lognormal CDFs (represented by the dashed lines in the graphs of Figures 3–5):

$$\mathcal{C}(\xi) = \frac{1}{2} + \frac{1}{2}\,\mathrm{erf}\left(\frac{\ln \xi - \mu}{\sqrt{2}\omega}\right) \tag{6}$$

where μ is the mean value and ω is the standard deviation of the natural logarithm of the random variable ξ (i.e. alternatively E, G, or ν); erf$(\cdot)$ is the normalized Gaussian function (also known as the "error function"). These distributions statistically satisfied the constraint for E and G to be always positive, with no artificial procedures necessary to avoid handling negative elastic moduli in the calculation of the overall spring stiffness; see also [35]. Moreover, an analytical, but SVE-informed CDF had the advantage of the ease of handling in further theoretical developments, as we will show for the spring stiffness in Section 3.2.

In Tables 1–3, the discussed mean and the standard deviation values obtained with the homogenization procedure are shown, together with the parameters μ and ω of Equation (6) as identified through a nonlinear fitting. These latter parameters made the analytical lognormal distribution ready to be used for the estimate of the stiffnesses of MEMS springs whose polysilicon morphology showed a grain size on the order of $g = 1$ μm. The quality of the adopted fitting via the lognormal distributions was quantified by the very small standard deviations shown in the tables for the parameters μ and for ω in all the considered cases: for the mean and standard deviation values corresponding to E, G, and ν, the scattering was even smaller and not affecting the least significant digit there reported.

The choice of the probability distribution for the fitting of the results of the MC analysis was a relevant decision. A lognormal probability distribution was assumed for each elastic modulus, but we assumed independence among E, G, and ν. This latter hypothesis does not prove always correct: in this way, in fact, one neglects the influence of the microstructure on the correlation among them. However, since in this work, we used the outcomes of the stochastic analysis to compute the stiffness of slender beams depending on the Young's modulus only, the hypothesis did not detrimentally affect the results. This would not be the case for more complex inferences, i.e. in the case of the stiffness evaluation for a shorter beam, where also G and shear deformations play a role. In these cases, a multivariate probability distribution should be considered, and the correlations between the variables should be evaluated, e.g. through marginal distributions. Assuming that the dependency structure of the multivariate probability distribution would be static, i.e. the stochastic process is stationary, copulae could be used to evaluate the said marginal probability distributions. For time-evolving stochastic processes, other approaches such as spectral analysis would be necessary. Both copulae and spectral analysis are out of the scope of this paper; interested readers can find information on them, e.g. in [36,37], respectively.

Table 1. Young's modulus E: apparent values of the mean and standard deviation obtained with the MC analysis and parameters for the lognormal CDFs in Equation (6).

Case	Mean (GPa)	Std Deviation (GPa)	μ	ω
2×2 μm^2, uniform strain BCs	149.98	5.47	5.010 ± 0.003	0.037 ± 0.002
2×2 μm^2, uniform stress BCs	148.11	5.40	4.997 ± 0.003	0.037 ± 0.002
3×3 μm^2, uniform strain BCs	149.91	3.34	5.010 ± 0.002	0.022 ± 0.001
3×3 μm^2, uniform stress BCs	148.37	3.32	4.997 ± 0.002	0.023 ± 0.001

Table 2. Shear modulus G: apparent values of the mean and standard deviation obtained with the MC analysis and parameters for the lognormal CDFs in Equation (6).

Case	Mean (GPa)	Std Deviation (GPa)	μ	ω
2×2 μm^2, uniform strain BCs	63.93	3.98	4.156 ± 0.004	0.062 ± 0.003
2×2 μm^2, uniform stress BCs	62.58	3.91	4.134 ± 0.004	0.062 ± 0.003
3×3 μm^2, uniform strain BCs	63.51	2.46	4.150 ± 0.003	0.038 ± 0.002
3×3 μm^2, uniform stress BCs	62.41	2.38	4.133 ± 0.003	0.038 ± 0.002

Table 3. Poisson's ratio ν: apparent values of the mean and standard deviation obtained with the MC analysis and parameters for the lognormal CDFs in Equation (6).

Case	Mean	Std Deviation	μ	ω
2×2 μm^2, uniform strain BCs	0.180	0.030	-1.790 ∓ 0.013	0.184 ± 0.009
2×2 μm^2, uniform stress BCs	0.170	0.030	-1.728 ∓ 0.012	0.173 ± 0.008
3×3 μm^2, uniform strain BCs	0.179	0.018	-1.777 ∓ 0.008	0.108 ± 0.005
3×3 μm^2, uniform stress BCs	0.170	0.018	-1.727 ∓ 0.007	0.103 ± 0.005

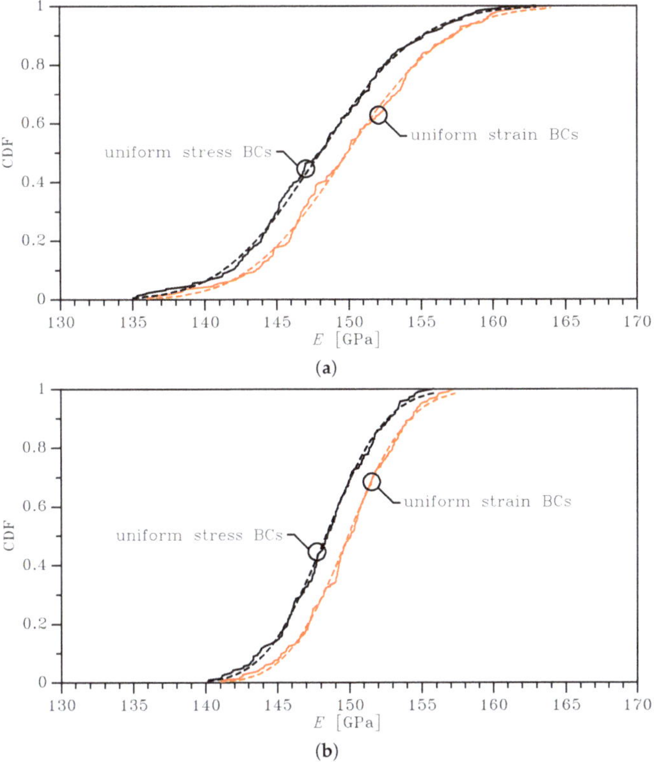

Figure 3. Effects of SVE size and BCs on the cumulative distribution function of the homogenized Young's modulus: (**a**) 2 × 2 μm^2 SVE and (**b**) 3 × 3 μm^2 SVE. Continuous lines represent the results of MC analyses, whereas dashed lines are the relevant lognormal interpolants.

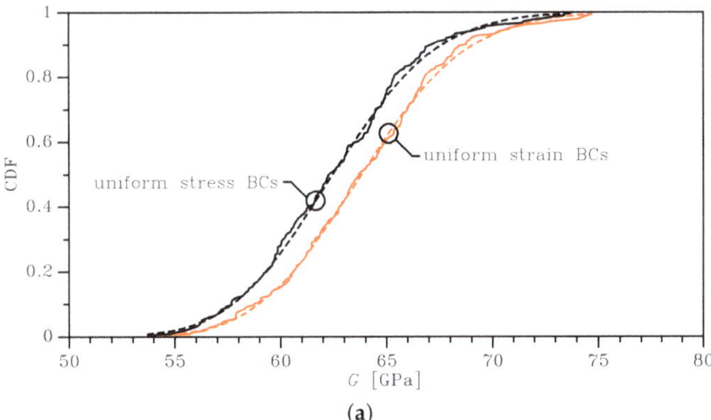

(**a**)

Figure 4. *Cont.*

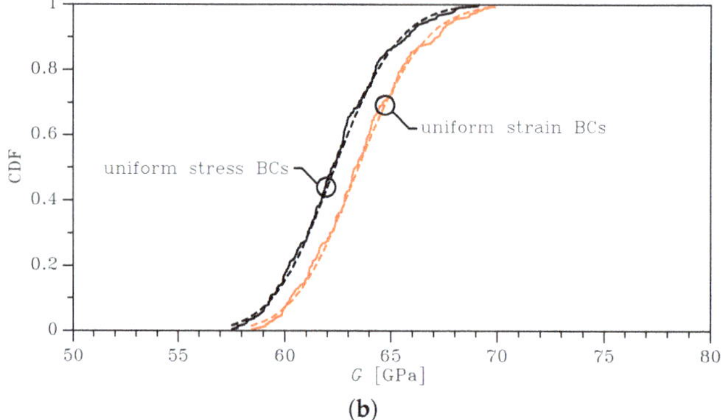

(**b**)

Figure 4. Effects of SVE size and BCs on the cumulative distribution function of the homogenized shear modulus G: (**a**) 2×2 μm^2 SVE and (**b**) 3×3 μm^2 SVE. Continuous lines represent the results of MC analyses, whereas dashed lines are the relevant lognormal interpolants.

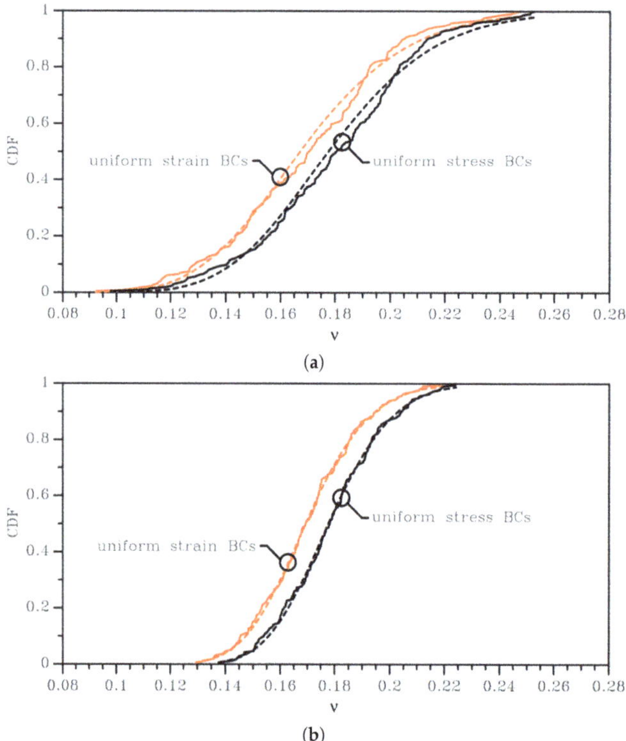

(**b**)

Figure 5. Effects of SVE size and BCs on the cumulative distribution function of the homogenized Poisson's ratio ν: (**a**) 2×2 μm^2 SVE and (**b**) 3×3 μm^2 SVE. Continuous lines represent the results of MC analyses, whereas dashed lines are the relevant lognormal interpolants.

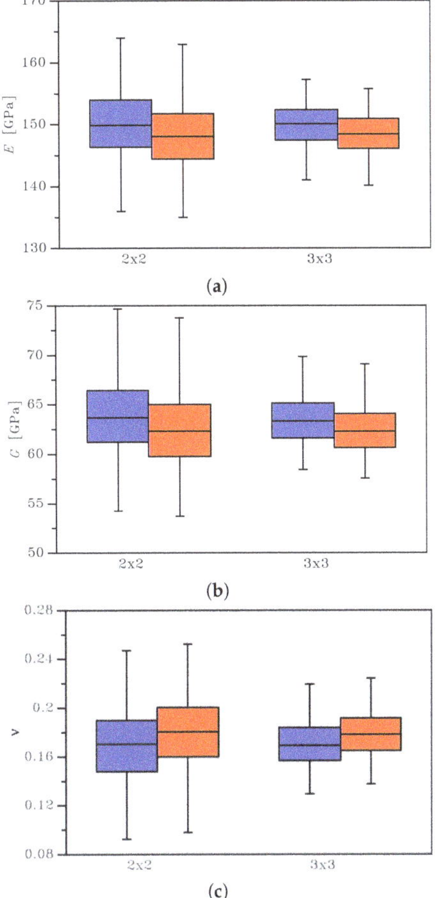

Figure 6. Box-whisker plots of the homogenized elastic properties of the SVE: effects of the BCs (blue: uniform strain BCs; orange: uniform stress BCs) on the scattering of (**a**) Young's modulus, (**b**) shear modulus, and (**c**) Poisson's ratio.

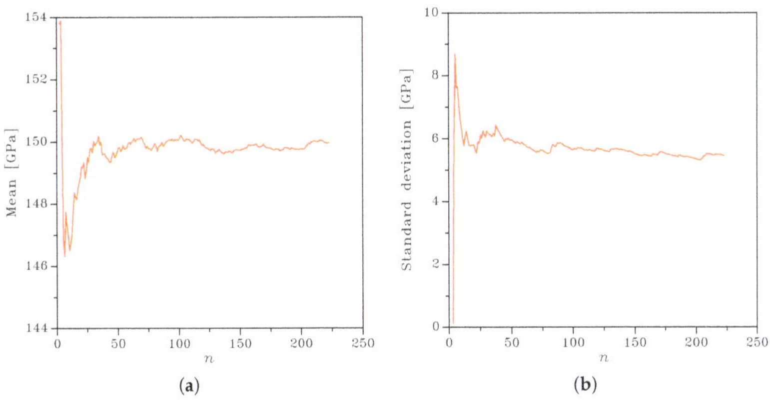

Figure 7. *Cont.*

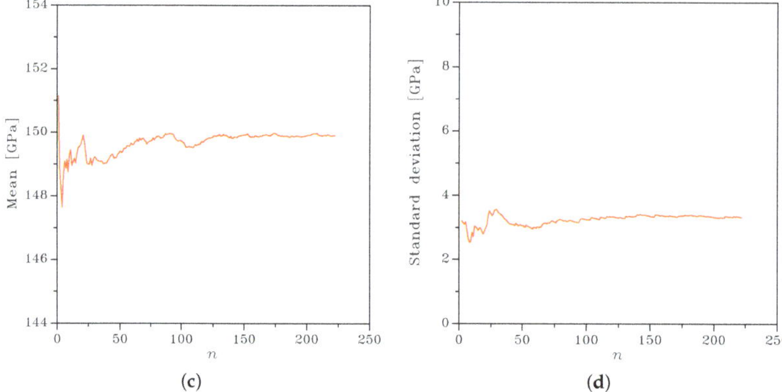

Figure 7. Effect of the number n of samples in the Monte Carlo analysis on the convergence of the (**a**,**c**) mean value and (**b**,**d**) standard deviation of in-plane Young's modulus E, for uniform strain BCs. Top: $2 \times 2\ \mu m^2$ SVE. Bottom: $3 \times 3\ \mu m^2$ SVE.

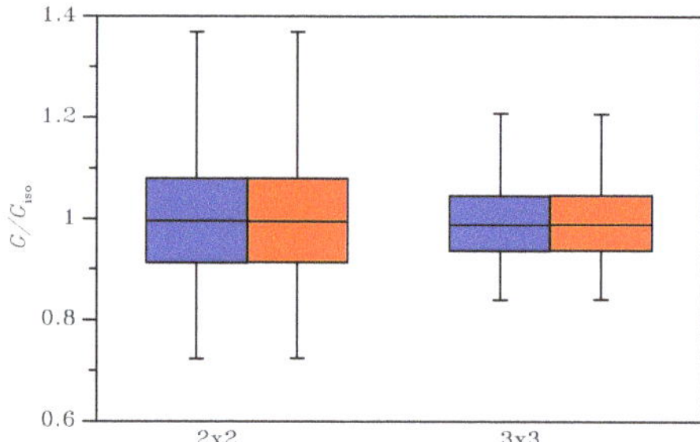

Figure 8. Box-whisker plot for the ratio between G/G_{iso} (blue: uniform strain BCs; orange: uniform stress BCs).

3.2. Overall Spring Stiffness Calculation

For the suspension spring configuration shown in Figure 9, the overall stiffness entering into play in Equation (5) with the relevant scattering can be computed as follows. Results can be easily generalized to the case of folded beams, since the stiffness scales inversely proportional to the number of folds. Stiffness values reported for homogeneous beams (see [38]) cannot be adopted in the present statistical analysis due the local heterogeneity of the material that governs the problem.

We assumed the beam to be subdivided into N disjoint subdomains, like in domain decomposition procedures [39–41]: within each subdomain, Young's modulus E was assumed constant; as far as the moment of inertia was instead concerned, we assumed that, for the sake of simplicity, its value varied for each subdomain according to a Gaussian distribution accounting for the scattering of the over-etch. The mentioned value of E was sampled out of the lognormal fitting of the statistical distributions obtained from the analysis at the polycrystalline material level. In the analysis to follow, we will consider springs characterized by high values of the ratio between length and in-plane width, namely very slender or thin geometries: Young's modulus is then the only parameter affecting the solution, according to the mathematical procedure described in the Appendix. In the case of thick springs, shear stresses gain importance in the solution, and also the distribution of G should be accounted for; in such a case, the correlation between the distributions of E and G has to be also allowed for in the analysis.

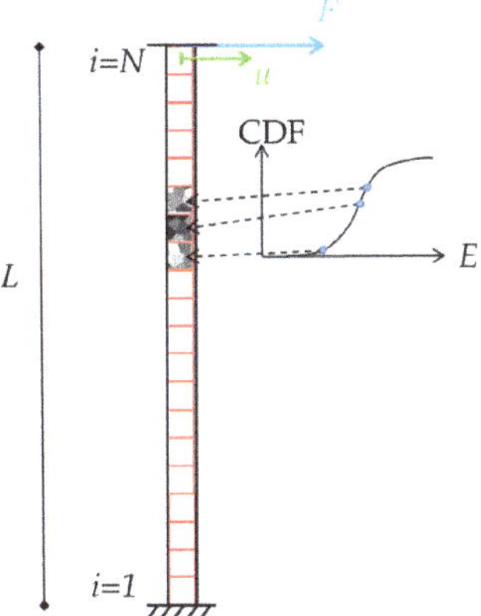

Figure 9. Scheme adopted for the calculation of the scattered spring stiffness with the spring itself subdivided into N subsets, each one with a Young's modulus sampled from the relevant size-dependent CDF and a moment of inertia dependent on the over-etch.

The adopted procedure is schematically reported in Figure 9: a beam has a fixed constraint at one end, where displacements and rotations are prevented, and a slider at the opposite end, where the motion is in the direction perpendicular to the beam axis only, without any rotation. The latter end is supposed to be the link to the proof mass, which moves as a rigid body. By means of the principle of virtual work and allowing for beam slenderness to neglect shear strains, the force F and the corresponding displacement u can be related to provide the ratio F/u as the beam stiffness k.

The outcome of this reasoning, described in detail in Appendix A, depends on the number of subdivisions N along the beam length, and it turns out to provide the spring stiffness as:

$$k = \frac{F}{u} = \frac{1}{L^3}\left(\sum_{i=1}^{N}\frac{\psi_i}{E_i\, I_i}\right)^{-1} \tag{7}$$

where L is the beam length, i is the index running over the subsets handled, I_i is the moment of inertia of the ith subset, E_i is the apparent Young's modulus at the ith subset, and ψ_i is a corrective factor dependent on the placement of the ith subset and defined in Equation (A10). Since the overall stiffness is a function of the random variables (E, I), then it is a random variable as well.

When the moment of inertia is assumed constant along the beam, the expression of the stiffness simplifies to:

$$k = \frac{F}{u} = \frac{I}{L^3}\left(\frac{1}{\sum_{i=1}^{N}\frac{\psi_i}{E_i}}\right). \tag{8}$$

4. Evaluation of the Offset at Rest

4.1. Effect of Material Uncertainties Only

In this section, the CDFs of the offset of the proof mass from its rest position are reported as obtained from Equation (5), in which u was taken as unknown, while the stiffnesses were calculated with two alternative choices: (i) from the lognormal CDFs as described in Section 3.2 or (ii) directly from FE simulations of the two-dimensional $L \times w$ (e.g. 200×2 μm^2) beam with the geometry shown in Figure 9. In these latter FE analyses, a random silicon grain morphology was considered for the whole domain: it is worthwhile to emphasize that the computational burden of these analyses was very high, since hundreds of grains were involved. To characterize the statistics of the stiffness via FE analyses of the whole beam, 200 simulations were carried out by varying in each case the whole silicon grain morphology.

As for the CDF of F in Equation (5) resulting from the production process, we considered the experimental data shown in [42] measured in a 0.3 µm-thick polysilicon layer: a residual strain constant through the thickness $\epsilon_r = 360 \pm 5 \cdot 10^{-6}$ was reported. In absence of further information, we assumed a Gaussian distribution for ϵ_r with the mentioned mean and standard deviation, and we simply computed $F = \tilde{E}\,\epsilon_r\,A$, $\tilde{E} = 150$ GPa being a reference Young's modulus and A the beam cross-section area. Therefore, the distribution of F was Gaussian as well.

In Figure 10, the offset CDFs derived from the stiffnesses obtained through the sampling of the apparent Young's modulus from the (analytical) lognormal distributions are shown with orange lines, while the CDFs derived from the stiffnesses obtained from the FE simulations are instead reported with black lines. An almost zero offset mean value was common to the two cases, so the main difference between the two solutions was related to the variance of E and, therefore, of k. Looking at the results, we can conclude that the semi-analytical approach, which was far less expensive than the FE one, correctly represented the information obtained from the FE analyses. Around the zero mean value, the offset became positive or negative depending on the difference between the stiffnesses of the right and left springs (see Figure 1). It is noted that the offset was generated by the scattering of the values of the spring stiffnesses around the mean; therefore, it depended on the standard deviations of the CDFs, and not on the mean values. Any stochastic method providing an offset estimate should then address also the quantification of the variance of the random variables associated with the elastic properties, not only their mean values.

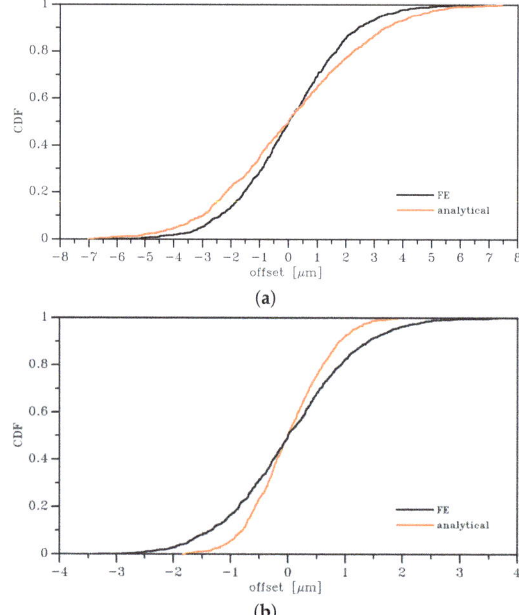

Figure 10. Offset from the rest position for length × SVE size cases: (**a**) 200 × 2 µm^2 and (**b**) 200 × 3 µm^2.

An alternative representation is provided in Figure 11 for the 200 µm × 2 µm case. The histogram represents the percentage of a given realization with a certain value of offset from the rest position. Though not further investigated in this work, it can be seen that the distribution of the offset was not unimodal.

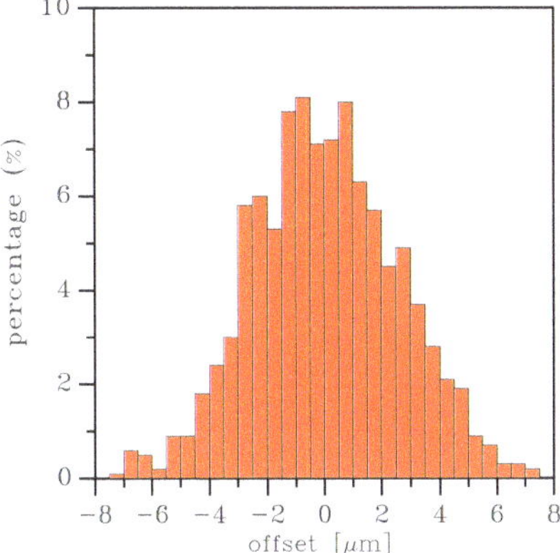

Figure 11. Histogram of the offset from the rest position for the 200 × 2 μm^2 case (uniform strain BCs).

4.2. Effect of Variable Over-Etch and Material Property Uncertainties

In Figure 12, in addition to the effect of a variable Young's modulus according to the rationale described above, we took into account the scattering of the geometry due to the over-etch variation along the beam length. We assumed that the over-etch o along the beam was distributed as a Gaussian variable with a zero mean value and a standard deviation equal to 50 nm; for the ith domain along the beam, the corresponding moment of inertia was thus computed as $I_i = t\,(w - 2\,o_i)^3/12$, where o_i is the value of the over-etch sampled out of the aforementioned Gaussian distribution. $o = 0$ corresponds to the commonly-assumed over-etch reference value in the engineering practice: a constant value for a single MEMS structure or even for MEMS distributed at different positions on a wafer. The graphs in Figure 12 include, as a reference, the previous curves for FE and analytical results related to the effects due to the scattering of Young's modulus only: the scattering due to a variable moment of inertia was shown to be far larger, as evidenced by the lower slope of the dashed orange curve and by the corresponding standard deviation, which increased from 2.5–8.2 μm, while the mean value was about zero in all cases. Two additional curves, in blue lines, were also added to the graph: they represent the scattering of the offset, due to the uncertainty on the material only, when the two springs in Figure 1 differed in width exactly by a value of 150 nm, which is representative of the 95% confidence interval from the target during the manufacturing process [12,43]. It can be seen that most of the offsets fell within the range dictated by the two blue curves.

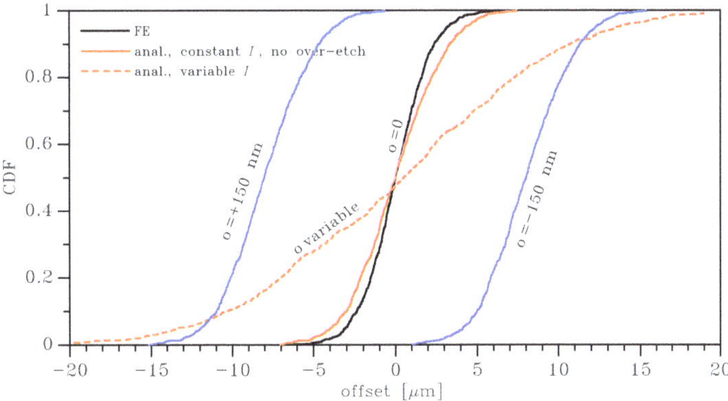

Figure 12. Effect on the offset of the variable (over-etch affected) moment of inertia I and Young's modulus along the beam length (200 × 2 μm^2 case, uniform strain BCs). Blue lines refer to fixed values for the over-etch equal to ±150 nm.

5. Conclusions

We proposed a stochastic method to evaluate the offset from the rest position for a statically-indeterminate MEMS typology, featuring a proof mass supported by two opposite springs. When the polysilicon production process was pushed to the current limit, the uncertainties about the stiffnesses of critical MEMS parts, such as nominally-identical supporting beams, generated an offset from the designed position when a residual stress distribution was also present inside the structure. Both material heterogeneity and fluctuations in critical geometry dimensions, i.e. the beam width, could be responsible for such an offset. For the first cause, we addressed the definition of the elastic properties accounting for the polycrystalline morphology by an artificially-reconstructed Voronoi tessellation in terms of statistically-representative elements, whose dimensions were equal to a characteristic dimension of the spring, namely their width, and were comparable with the silicon grain. For the second cause, we considered the effect of the over-etch on the moment of inertia of the MEMS supporting beams.

The classical homogenization procedure, under the hypothesis of statistical independence of the variables, provided a stochastic insight into the relevant mechanical properties, in particular the elastic moduli and Poisson's ratio, through numerical cumulated distribution functions that were subsequently approximated by an analytical, lognormal distribution. The stiffness of each beam supporting the moving mass was then estimated via the analytical lognormal distribution according to a simple reasoning based on the principle of virtual work applied to an exemplary beam.

By considering a statistically-indeterminate beam-moving mass-beam chain, we assumed as known the internal force due to residual stresses: it generated an offset from the rest position even in the presence of a stiffness varying only according to the polycrystalline material. A variable over-etch along the supporting beams, however, was responsible for a higher offset than the one induced by uncertainties on the material properties only. An important conclusion was that the evaluation of the offset depended on the correct characterization of the variance of Young's modulus and the moment of inertia, not on their mean value.

The approach shown here for a very simple class of structures could be extended to other statically-indeterminate MEMS configurations, provided that they are modeled as beam systems, in case the heterogeneity of the polysilicon is a matter of concern.

Author Contributions: The authors contributed equally to this work.

Funding: We gratefully acknowledge financial support through the project MaRe (Material Reliability) by STMicroelectronics.

Acknowledgments: The authors are also indebted to Marco Geninazzi for the results related to the stochastic homogenization procedure.

Conflicts of Interest: The authors declare no conflict of interest.

Abbreviations

The following abbreviations are used in this manuscript:

BC	Boundary Condition
CDF	Cumulative Distribution Function
FE	Finite Element
MC	Monte Carlo
RVE	Representative Volume Element
SVE	Statistical Volume Element

Appendix A

To quantify the stiffness of the supporting springs in the reference configuration of Figure 9, we relied upon the principle of virtual work for slender beams with a Bernoulli–Euler kinematics. When a force F acts perpendicularly to the longitudinal axis of the beam, it causes an internal bending moment $\mathcal{M}$. At equilibrium, the external work W_e equals the internal work W_i, which read:

$$W_e = F\,u \tag{A1}$$

$$W_i = \int_0^L \mathcal{M}(\zeta)\,\chi(\zeta)\,\mathrm{d}\zeta \tag{A2}$$

where u is the virtual displacement of force F in the same direction (hence, perpendicular to the beam axis), χ is the virtual curvature of the beam axis, and ζ is the axis aligned with the longitudinal direction of the spring.

For a statically-indeterminate structure, if both the elastic modulus E and the moment of inertia I are constant all along the beam axis, the solution to $W_e = W_i$ can be found in any textbook of structural mechanics. In the present case, where both E and I can vary stochastically as functions of ζ, $\mathcal{M}(\zeta)$ and $\chi(\zeta)$ need to be determined. The variation of the bending moment along the spring axis reads $\mathcal{M} = M - F\zeta$, where M is the value of the top end of the beam, as induced by the local constraint that prevents any in-plane rotation. Hence, in compliance with the mentioned constraint, we have:

$$0 = \int_0^L \frac{M - F\zeta}{E(\zeta)\,I(\zeta)}\,\mathrm{d}\zeta. \tag{A3}$$

The integral was solved numerically, by subdividing the beam into N disjoint domains of the same longitudinal length $\Delta\zeta = L/N$; within each of them, we assumed that the flexural rigidity was locally constant, i.e. stochastic fluctuations were locally neglected. Accordingly, we arrive at:

$$\sum_{i=1}^{N} \frac{1}{E_i\,I_i} \int_{\zeta_i}^{\zeta_i+\Delta\zeta} (M - F\zeta)\,\mathrm{d}\zeta = 0 \tag{A4}$$

where E_i and I_i are the Young's modulus and the moment of inertia for the ith domain, respectively. By solving for M, we obtain:

$$M = \left[\frac{\sum_{i=1}^{N} \frac{\Delta\zeta + 2\zeta_i}{2\,E_i\,I_i}}{\sum_{i=1}^{N} \frac{1}{E_i\,I_i}}\right] F. \tag{A5}$$

In compliance with the adopted subdivision of the beam, we also have $\zeta_i = (i-1)\Delta\zeta$, and Equation (A5) becomes:

$$M = \frac{F\,L}{2}\,\gamma \tag{A6}$$

where:

$$\gamma = \frac{\sum_{i=1}^{N} \frac{2i-1}{N\,(E_i\,I_i)}}{\sum_{i=1}^{N} \frac{1}{E_i\,I_i}} \tag{A7}$$

is a factor that allows for the non-uniform value of the flexural rigidity of the beam and modifies the value $M = F\,L/2$ relevant to the uniform beam case.

Now that the internal moments are fully given as a function of the external force F, we can solve $W_e = W_i$ for the displacement u to obtain:

$$u = \int_0^L \frac{F\left(\frac{L}{2}\gamma - \zeta\right)^2}{E(\zeta)I(\zeta)}\,\mathrm{d}\zeta \tag{A8}$$

being $\chi = \frac{M}{E(\zeta)\,I(\zeta)}$. With the help of the same subdivision of the beam into N domains adopted to compute M, we finally obtain Equation (7), here rewritten for convenience:

$$k = \frac{F}{u} = \frac{1}{L^3}\left(\sum_{i=1}^{N}\frac{\psi_i}{E_i\,I_i}\right)^{-1} \tag{A9}$$

where:

$$\psi_i = \left(\frac{i-1}{N} - \frac{\gamma}{2}\right)^2\frac{1}{N} + \left(\frac{i-1}{N} - \frac{\gamma}{2}\right)\frac{1}{N^2} + \frac{1}{3N^3}. \tag{A10}$$

References

1. Gaura, E.; Newman, R. *Smart MEMS and Sensors Systems;* Imperial College Press: London, UK, 2006; Chapter 4.
2. Choudhary, V.; Iniewski, K. *MEMS: Fundamental Technology and Applications;* CRC Press: Boca Raton, FL, USA, 2013; p. 189.
3. Yeh, C.Y.; Huang, J.T.; Tseng, S.H.; Wu, P.C.; Tsai, H.H.; Juang, Y.Z. A low-power monolithic three-axis accelerometer with automatically sensor offset compensated and interface circuit. *Microelectron. J.* **2019**, *86*, 150–160. [CrossRef]
4. Li, F.; Clark, J. Self-calibration for MEMS with comb drives: Measurement of gap. *J. Microelectromech. Syst.* **2012**, *21*, 1019–1021. [CrossRef]
5. De Laat, M.; Pérez Garza, H.; Herder, J.; Ghatkesar, M. A review on in situ stiffness adjustment methods in MEMS. *J. Micromech. Microeng.* **2016**, *26*, 063001. [CrossRef]
6. Weinberg, M.S.; Kourepenis, A. Error sources in in-plane silicon tuning-fork MEMS gyroscopes. *J. Microelectromech. Syst.* **2006**, *15*, 479–491. [CrossRef]
7. Madinei, H.; Khodaparast, H.H.; Friswell, M.; Adhikari, S. Minimising the effects of manufacturing uncertainties in MEMS Energy harvesters. *Energy* **2018**, *149*, 990–999. [CrossRef]
8. Alexeenko, A.; Chigullapalli, S.; Zeng, J.; Guo, X.; Kovacs, A.; Peroulis, D. Uncertainty in microscale gas damping: Implications on dynamics of capacitive MEMS switches. *Reliab. Eng. Syst. Saf.* **2011**, *96*, 1171–1183. [CrossRef]
9. Williams, K.R.; Gupta, K.; Wasilik, M. Etch rates for micromachining processing—Part II. *J. Microelectromech. Syst.* **2003**, *12*, 761–778. [CrossRef]
10. Uhl, T.; Martowicz, A.; Codreanu, I.; Klepka, A. Analysis of uncertainties in MEMS and their influence on dynamic properties. *Arch. Mech.* **2009**, *61*, 349–370.
11. Gennat, M.; Meinig, M.; Shaporin, A.; Kurth, S.; Rembe, C.; Tibken, B. Determination of parameters with uncertainties for quality control in MEMS fabrication. *J. Microelectromech. Syst.* **2013**, *22*, 613–624. [CrossRef]
12. Mirzazadeh, R.; Mariani, S. Uncertainty quantification of microstructure-governed properties of polysilicon MEMS. *Micromachines* **2017**, *8*, 248. [CrossRef]
13. Mirzazadeh, R.; Eftekhar Azam, S.; Mariani, S. Micromechanical Characterization of Polysilicon Films through On-Chip Tests. *Sensors* **2016**, *16*, 1191. [CrossRef] [PubMed]
14. Mirzazadeh, R.; Ghisi, A.; Mariani, S. Statistical investigation of the mechanical and geometrical properties of polysilicon films through on-chip tests. *Micromachines* **2018**, *9*, 53. [CrossRef] [PubMed]
15. Mariani, S.; Ghisi, A.; Corigliano, A.; Zerbini, S. Multi-scale Analysis of MEMS Sensors Subject to Drop Impacts. *Sensors* **2007**, *7*, 1817–1833. [CrossRef] [PubMed]
16. Mariani, S.; Ghisi, A.; Corigliano, A.; Zerbini, S. Modeling impact-induced failure of polysilicon MEMS: A multi-scale approach. *Sensors* **2009**, *9*, 556–567. [CrossRef] [PubMed]
17. Mariani, S.; Martini, R.; Ghisi, A.; Corigliano, A.; Beghi, M. Overall elastic properties of polysilicon films: A statistical investigation of the effects of polycrystal morphology. *Int. J. Multiscale Comput. Eng.* **2011**, *9*, 327–346. [CrossRef]
18. Mariani, S.; Martini, R.; Ghisi, A.; Corigliano, A.; Simoni, B. Monte Carlo simulation of micro-cracking in polysilicon MEMS exposed to shocks. *Int. J. Fract.* **2011**, *167*, 83–101. [CrossRef]
19. Ballarini, R.; Mullen, R.; Heuer, A. The effects of heterogeneity and anisotropy on the size effect in cracked polycrystalline films. In *Fracture Scaling;* Springer: Dordrecht, The Netherlands, 1999; pp. 19–39.

20. Cho, S.; Chasiotis, I. Elastic properties and representative volume element of polycrystalline silicon for MEMS. *Exp. Mech.* **2007**, *47*, 37–49. [CrossRef]
21. Gupta, R.K. Electronically probed measurements of MEMS geometries. *J. Microelectromech. Syst.* **2000**, *9*, 380–389. [CrossRef]
22. Young, D.J.; Pehlivanoğlu, İ.E.; Zorman, C.A. Silicon carbide MEMS-resonator-based oscillator. *J. Micromech. Microeng.* **2009**, *19*, 115027. [CrossRef]
23. Chang, C.-C.; Yang, H.-T.; Su, Y.-F.; Hong, Y.-T.; Chiang, K.-N. A method to compensate packaging effects on three-axis MEMS accelerometer. In Proceedings of the 2016 15th IEEE Intersociety Conference on Thermal and Thermomechanical Phenomena in Electronic Systems (ITherm), Las Vegas, NV, USA, 31 May–3 June 2016; pp. 536–538. [CrossRef]
24. Freund, L.B.; Suresh, S. *Thin Film Materials: Stress, Defect Formation and Surface Evolution*; Cambridge University Press: New York, NY, USA, 2004.
25. Mirzazadeh, R.; Eftekhar Azam, S.; Mariani, S. Mechanical Characterization of Polysilicon MEMS: A Hybrid TMCMC/POD-Kriging Approach. *Sensors* **2018**, *18*, 1234. [CrossRef]
26. Corigliano, A.; De Masi, B.; Frangi, A.; Comi, C.; Villa, A.; Marchi, M. Mechanical characterization of polysilicon through on-chip tensile tests. *J. Microelectromech. Syst.* **2004**, *13*, 200–219. [CrossRef]
27. Mariani, S.; Martini, R.; Corigliano, A.; Beghi, M. Overall elastic domain of thin polysilicon films. *Comput. Mater. Sci.* **2011**, *50*, 2993–3004. [CrossRef]
28. Sfantos, G.K.; Aliabadi, M.H. A boundary cohesive grain element formulation for modelling intergranular microfracture in polycrystalline brittle materials. *Int. J. Numer. Methods Eng.* **2007**, *69*, 1590–1626. [CrossRef]
29. Geraci, G.; Aliabadi, M. Micromechanical boundary element modelling of transgranular and intergranular cohesive cracking in polycrystalline materials. *Eng. Fract. Mech.* **2017**, *176*, 351–374. [CrossRef]
30. Gulizzi, V. A Computational Framework for Microstructural Modelling of Polycrystalline Materials with Damage and Failure. Ph.D. Thesis, University of Palermo, Palermo, Italy, 2018.
31. Galvis, A.; Sollero, P. Boundary Element Analysis of Crack Problems in Polycrystalline Materials. *Procedia Mater. Sci.* **2014**, *3*, 1928–1933. [CrossRef]
32. Huet, C. Coupled size and boundary-condition effects in viscoelastic heterogeneous and composite bodies. *Mech. Mater.* **1999**, *31*, 787–829. [CrossRef]
33. Yin, X.; Chen, W.; To, A.; McVeigh, C.; Liu, W.K. Statistical volume element method for predicting microstructure-constitutive property relations. *Comput. Methods Appl. Mech. Eng.* **2008**, *197*, 3516–3529. [CrossRef]
34. Corigliano, A.; Ghisi, A.; Langfelder, G.; Longoni, A.; Zaraga, F.; Merassi, A. A microsystem for the fracture characterization of polysilicon at the micro-scale. *Eur. J. Mech. A Solids* **2011**, *30*, 127–136. [CrossRef]
35. Bagherinia, M.; Bruggi, M.; Corigliano, A.; Mariani, S.; Horsley, D.A.; Li, M.; Lasalandra, E. An Efficient Earth Magnetic Field MEMS Sensor: Modeling, Experimental Results, and Optimization. *J. Microelectromech. Syst.* **2015**, *24*, 887–895. [CrossRef]
36. Jaworski, P. *Copula Theory and Its Applications, Proceedings of the Workshop Held in Warsaw, 25–26 September 2009*; Springer: Berlin/Heidelberg, Germany, 2010.
37. Hamilton, J.D. *Time Series Analysis*; Princeton University Press: Princeton, NJ, USA, 1994.
38. Corigliano, A.; Ardito, R.; Comi, C.; Frangi, A.; Ghisi, A.; Mariani, S. *Mechanics of Microsystems*; John Wiley and Sons, Ltd.: Hoboken, NJ, USA, 2018.
39. Corigliano, A.; Dossi, M.; Mariani, S. Domain decomposition and model order reduction methods applied to the simulation of multi-physics problems in MEMS. *Comput. Struct.* **2013**, *122*, 113–127. [CrossRef]
40. Confalonieri, F.; Ghisi, A.; Cocchetti, G.; Corigliano, A. A domain decomposition approach for the simulation of fracture phenomena in polycrystalline microsystems. *Comput. Methods Appl. Mech. Eng.* **2014**, *277*, 180–218. [CrossRef]
41. Corigliano, A.; Dossi, M.; Mariani, S. Model Order Reduction and domain decomposition strategies for the solution of the dynamic elastic–plastic structural problem. *Comput. Methods Appl. Mech. Eng.* **2015**, *290*, 127–155. [CrossRef]

42. Song, J.; Huang, Q.A.; Li, M.; Tang, J.Y. Influence of environmental temperature on the dynamic properties of a die attached MEMS device. *Mycrosyst. Technol.* **2009**, *15*, 925–932. [CrossRef]
43. Bagherinia, M.; Mariani, S. Stochastic Effects on the Dynamics of the Resonant Structure of a Lorentz Force MEMS Magnetometer. *Actuators* **2019**, *8*, 36. [CrossRef]

Article

Design and Experimental Validation of a LoRaWAN Fog Computing Based Architecture for IoT Enabled Smart Campus Applications †

Paula Fraga-Lamas [1,*], Mikel Celaya-Echarri [2], Peio Lopez-Iturri [3,4], Luis Castedo [1], Leyre Azpilicueta [2], Erik Aguirre [3], Manuel Suárez-Albela [1], Francisco Falcone [3,4] and Tiago M. Fernández-Caramés [1,*]

1 Department of Computer Engineering, Faculty of Computer Science, Centro de investigación CITIC, Universidade da Coruña, 15071 A Coruña, Spain
2 School of Engineering and Sciences, Tecnologico de Monterrey, 64849 Monterrey, NL, Mexico
3 Department of Electric, Electronic and Communication Engineering, Public University of Navarre, 31006 Pamplona, Spain
4 Institute for Smart Cities, Public University of Navarre, 31006 Pamplona, Spain
* Correspondence: paula.fraga@udc.es (P.F.-L.); tiago.fernandez@udc.es (T.M.F.-C.); Tel.: +34-981-167-000 (ext. 6051) (P.F.-L.)

† This paper is an extended version of our paper published in Fraga-Lamas, P.; Celaya-Echarri, M.; Lopez-Iturri, P.; Fernández-Caramés, T.M.; Azpilicueta, L.; Aguirre, E.; Suárez-Albela, M.; Falcone, F.; Castedo, L. Analysis, Design and Empirical Validation of a Smart Campus based on LoRaWAN. In Proceedings of the 5th International Electronic Conference on Sensors and Applications, 15–30 November 2018.

Received: 28 June 2019; Accepted: 23 July 2019; Published: 26 July 2019

Abstract: A smart campus is an intelligent infrastructure where smart sensors and actuators collaborate to collect information and interact with the machines, tools, and users of a university campus. As in a smart city, a smart campus represents a challenging scenario for Internet of Things (IoT) networks, especially in terms of cost, coverage, availability, latency, power consumption, and scalability. The technologies employed so far to cope with such a scenario are not yet able to manage simultaneously all the previously mentioned demanding requirements. Nevertheless, recent paradigms such as fog computing, which extends cloud computing to the edge of a network, make possible low-latency and location-aware IoT applications. Moreover, technologies such as Low-Power Wide-Area Networks (LPWANs) have emerged as a promising solution to provide low-cost and low-power consumption connectivity to nodes spread throughout a wide area. Specifically, the Long-Range Wide-Area Network (LoRaWAN) standard is one of the most recent developments, receiving attention both from industry and academia. In this article, the use of a LoRaWAN fog computing-based architecture is proposed for providing connectivity to IoT nodes deployed in a campus of the University of A Coruña (UDC), Spain. To validate the proposed system, the smart campus has been recreated realistically through an in-house developed 3D Ray-Launching radio-planning simulator that is able to take into consideration even small details, such as traffic lights, vehicles, people, buildings, urban furniture, or vegetation. The developed tool can provide accurate radio propagation estimations within the smart campus scenario in terms of coverage, capacity, and energy efficiency of the network. The results obtained with the planning simulator can then be compared with empirical measurements to assess the operating conditions and the system accuracy. Specifically, this article presents experiments that show the accurate results obtained by the planning simulator in the largest scenario ever built for it (a campus that covers an area of 26,000 m^2), which are corroborated with empirical measurements. Then, how the tool can be used to design the deployment of LoRaWAN infrastructure for three smart campus outdoor applications is explained: a mobility pattern detection system, a smart irrigation solution, and a smart traffic-monitoring deployment. Consequently, the presented results provide guidelines to smart campus designers and developers, and for easing LoRaWAN network deployment and research in other smart campuses and large environments such as smart cities.

Keywords: IoT; smart campus; sustainability; fog computing; outdoor applications; LPWAN; LoRaWAN; 3D Ray-Launching; smart cities; Wireless Sensor Networks (WSN)

1. Introduction

A smart campus is an infrastructure similar to a smart city that makes use of Internet of Things (IoT) solutions [1–6] to connect, monitor, control, optimize, and automate the systems of a university. Today, a smart campus represents a challenging scenario for IoT networks, especially in terms cost, coverage, availability, latency, security, power consumption, and scalability.

The area covered by a campus varies substantially depending on the university, its location, the financial endowment, and the year of founding. For example, Berry College (Floyd County, Georgia, United States), is often considered the largest contiguous rural campus in the world: it covers 27,000 acres (109.26 km^2) [7] of land. Other examples are the suburban/urban campuses of Duke University (Durham, NC, USA), which are deployed on 9350 acres (37.83 km^2) [8], and the campus of Stanford University (Stanford, CA, USA), which covers 8180 acres (33 km^2) [9]. Regardless of their initial surface area, it is common that campuses grow considerably as time goes by [10], hence institutions usually devise long-term sustainability plans to envision their growth in the future [11–13].

When a campus provides smart IoT services, it is necessary to provide communications connectivity to IoT nodes and gateways. Such a connectivity can be provided in a quite straightforward way indoors thanks to the use of popular technologies such as Wi-Fi, but, outdoors, technology selection becomes more complex, since it is not only necessary to provide good coverage and a cost-effective deployment, but also to decrease the communications energy consumption to maximize IoT node battery life.

To tackle such an issue in wide areas, a set of technologies grouped under the term Low-Power Wide Area Network (LPWAN) seem to be a good selection, since, in comparison to other previous technologies, they provide a wider area communications range and reduced energy consumption. In fact, LPWAN technologies have emerged as an enabling technology for IoT and Machine-to-Machine (M2M) communications [14] mainly due to their capabilities related to range, cost, power consumption, and capacity. Examples of such technologies are NB-IoT [15], SigFox [16], Ingenu [17], Weightless [18] or LoRaWAN [19] (a detailed comparison of these and other LPWAN technologies is given later in Section 2.2).

In the case of LoRaWAN, it is gaining momentum from both industry and academia [20–22]. LoRaWAN defines a communications protocol and a system architecture for the network. In addition, it uses LoRa for its physical layer [23], which is able to create long-range communications links and makes use of a Chirp Spread Spectrum (CSS) modulation that conserves the power features of Frequency Shifting Keying (FSK) while increasing its communications range. All these features make LoRaWAN a good candidate for providing wireless communications to outdoor IoT nodes in a smart campus.

Traditionally, gateways connect the IoT nodes with the cloud and among them. The cloud is basically one or more servers with large computational power, communication, and storing capabilities that receives, processes, and analyzes the data collected from the IoT nodes by performing computational-intensive tasks. Although cloud-based solutions are appropriate at a small scale, when the number of IoT nodes grows significantly and, consequently, the network traffic they generate, congestion may lead to increasing latency responses and slower data processing. Among the different alternatives to confront this challenge and to guarantee a flexible, scalable, robust, secure, and energy-efficient deployment of IoT networks, the design and implementation of a fog computing architecture was chosen. Fog computing supports physically distributed, low-latency (e.g., real-time or quasi real-time responses) and location-aware applications that decrease the network traffic and the computational load of traditional cloud computing systems by processing in the IoT nodes most of the data generated by their sensors and actuators and unburdening the higher layers from data processing [24].

Furthermore, when designing a smart campus, it is necessary to plan how LoRaWAN gateways and nodes are deployed to guarantee good IoT node coverage while minimizing the number of gateways (i.e., minimizing the smart campus communications infrastructure cost). The problem is that there are only a few examples of academic and commercial tools that create such a planning [25,26], so developers have to adapt tools previously optimized for other technologies (e.g., Wi-Fi [27]) or have to carry out tedious empirical measurements throughout the campus [28,29].

This article confronts the mentioned challenges by designing and implementing a cost-efficient, scalable, and low-power consumption LoRaWAN fog computing-based architecture for wide areas. Specifically, the system was designed with the aim of developing novel latency-sensitive IoT outdoor applications that create more sustainable and intelligent campuses. The following are the main contributions of the article, which as of writing, have not been found together in the literature:

- To establish the basics, it presents the main characteristics of the so-called smart campuses together with a detailed review of the state of the art of the main and the latest communications architectures and technologies, previous academic deployments, novel potential LPWAN applications and relevant tools for radio propagation modeling and planning.
- It thoroughly details the design, implementation, and practical evaluation of a scalable LPWAN-based communications architecture for supporting the smart campus IoT applications.
- The article presents the 3D modeling of a real 26,000 m^2 campus whose LoRaWAN wireless propagation characteristics are evaluated with an in-house developed 3D Ray-launching radio-planning simulator. The results obtained by such a simulator are validated by comparing them with empirical LoRaWAN measurements obtained throughout the campus.
- It details how the radio-planning tool can be used to design the deployment of LoRaWAN infrastructure for three smart campus applications: a mobility pattern detection system, a smart irrigation solution, and a smart traffic-monitoring deployment. Thus, it demonstrates the usefulness of the proposed tools and methodology, which are able to provide fast guidelines to smart campus designers and developers, and that can also be used for easing LoRaWAN network deployment and research in other large environments such as smart cities.

The rest of this article is structured as follows. Section 2 reviews the state of the art on smart campuses: their characteristics, technologies, architectures, previous relevant deployments, potential applications, and the previous work on modeling and planning a smart campus. Section 3 details the architecture of the proposed system and the characteristics of the LoRaWAN testbed implementation. Section 4 describes the proposed planning simulator and the analyzed scenario. Section 5 is dedicated to the experiments. Finally, Section 6 presents the main discussion on the lessons learned from these experiences, while Section 7 is devoted to the conclusions.

2. Related Work

2.1. Characteristics of a Smart Campus

It is first important to note that in the literature, some authors use the term smart campus to refer to digital online platforms to manage learning content [30,31] or to strategies or solutions to increase the smartness of the students [32–34]. In this article, the term smart campus is used for referring to the hardware infrastructure and software that provides smart services and applications to the campus users (i.e., to students and to the university staff). In this regard, a smart campus, such as a smart city, can be modeled along six different smart fields [35]:

- Smart governance. It provides users with mechanisms to participate in decision-making or in public services.
- Smart people. It deals with social issues, including the engagement in campus events and learning activities.
- Smart mobility. This field is related to the accessibility of the campus, including the use of efficient, clean, safe, and intelligent transport means.
- Smart environment. It contemplates the monitoring and protection of the environment, as well as the sustainable management of the available resources.
- Smart living. The technologies used in these fields can monitor diverse living aspects in the campus facilities, such as personal safety [36], health [37] or crowd sensing [38].
- Smart economy. It is related to the competitiveness of the campus in terms of entrepreneurship, innovation, or productivity.

These smart campus fields can be further refined to determine specific smart services and solutions that should be ideally provided by a smart campus [39]:

- Smart living services: room occupation, classroom/lab equipment access control, health monitoring and alert services, classroom attendance systems, teaching interaction services, or context-aware applications (e.g., guidance or navigation solutions).
- Smart environment solutions: they include solutions for monitoring waste, water consumption, air quality (e.g., pollution) or the status of the campus green areas.
- Smart energy systems: they control and monitor the production, distribution, and consumption of energy in a campus.

These novel smart services and solutions make use of a growing number of enabling technologies, being the most relevant represented in Figure 1.

Figure 1. Most relevant enabling technologies and applications in a smart campus.

2.2. Smart Campus Communications Architectures and Technologies

In the literature, different approaches to smart campus architectures can be found, but it seems that two main paradigms drive clearly the most popular designs: IoT and cloud computing [40]. For instance, a cloud-based smart campus architecture is presented in [41]. In such a work the authors state that they were able to build their smart campus platform within three months thanks to the use of Commercial Off-The-Shelf (COTS) hardware and Microsoft Azure cloud services. Regarding IoT, it has been suggested as a tool to be considered in the architecture of a smart campus to ease the development of learning applications, access control systems, smart grids or water management systems [42,43]. Nonetheless, cloud computing and IoT solutions are often helped by Big Data techniques and Service Oriented Architecture (SOA) architectures, since they ease the processing and analysis of the collected data [44,45].

Some authors have suggested alternative paradigms for developing smart campuses. For example, in [46] a sort of opportunistic communications architecture called Floating Content is proposed that shares data through infrastructure-less services. The idea is essentially based on the ability of each Floating Content node to produce information that is shared with the interested users within a limited physical area. Other researchers propose similar architectures, but including enhancements in aspects such as security [47].

Other proposals revolve around the application of the Edge Computing paradigm and its sub-types (e.g., Mobile Edge Computing, Fog Computing), which have been previously applied to other fields [48,49]. Essentially, Edge Computing offloads the cloud from a relevant amount of processing and communications transactions, delegating such tasks to devices that are closer to the IoT nodes. In this way, such edge devices not only offload the cloud, but are also able to reduce latency response and provide location-aware services [50]. For example, in [51] the authors propose to enhance a smart campus architecture by including Edge Computing devices to provide trustworthy content caching and bandwidth allocation services to mobile users. Similarly, the authors of [52] harness street lighting to embed Edge Computing node hardware to provide different smart campus services. The Mobile Edge Computing paradigm is used in [53], where the authors present a smart campus platform called WiCloud whose servers are accessed through mobile phone base stations or wireless access points. Furthermore, other authors propose the use of fog computing nodes to improve user experience [54].

Different wireless technologies have been used to interconnect IoT nodes with smart campus platforms. For instance, BLE and ZigBee were used in [41] to provide both short and medium range communications, although ZigBee nodes can be used as relays to cover very long distances. For this latter reason, in [55] the authors make use of a ZigBee mesh network to interconnect the nodes of their campus smart grid.

Wi-Fi has also been suggested for providing connectivity [56], although the proposed applications are usually restricted to indoor locations and nearby places. Bluetooth beacons give more freedom to certain outdoor applications [57], but they require deploying dense networks that may be difficult to manage [58].

Mobile phone communications technologies (2G/3G/4G) have also been used in the literature [59], but in most cases just for the convenience of being already embedded into smartphones. 5G is currently still being tested, but some researchers have already proposed its use for providing fast communications and low-latency responses to smart campus platforms [60].

Although 5G technologies seem to have a bright future, as of writing, LPWANs are arguably one of the best alternatives for providing long-range and low-power communications. There are different LPWAN technologies such as SigFox [16], Random Phase Multiple Access (RPMA) [17], Weightless [18], NB-IoT [15], Telensa [61] or NB-Fi [62]. Among such technologies, NB-IoT, SigFox and LoRa/LoRaWAN are currently the most popular (their main characteristics are shown in Table 1).

Table 1. Comparison of the three most popular LPWAN technologies.

Technology	Operating Frequency	Modulation	Maximum Range	Speed	Max. Payload	Bandwidth	Main Characteristics
NB-IoT	LTE in-band, guard-band	QPSK	<35 km	<250 kbi	1500 bytes	180 kHz	Low power and wide-area coverage
SigFox	868–902 MHz	DBPSK	50 km	100 kbit/s	12 bytes	0.1 kHz	Global cellular network
LoRa, LoRaWAN	Diverse UHF ISM (Industrial, Scientific, Medical) bands (e.g., 863–870 MHz and 433 MHz in Europe)	CSS	<15 km	0.25–50 k	51–222 bytes	125 kHz	Low power and wide range

There are several recent studies on the performance of LoRa/LoRaWAN technology for certain scenarios, but only a few describe real-world LoRaWAN deployments explicitly aimed at providing communications to a smart campus. For instance, Loriot et al. [63] conducted LoRaWAN measurements in a French campus both outdoors and indoors and showed that the technology can provide good performance over the major part of the campus. Another development is presented in [64], where the authors set up a LoRaWAN-based air quality system in their campus. Other interesting paper is [65], which details the design of a LoRa mesh network system within a campus. Finally, in [66] the authors briefly describe a smart campus platform that includes a LoRaWAN network to support faculty research projects.

2.3. Smart Campus Deployments

Despite the existence of many well-documented smart campus applications, there are only a few academic articles that describe in detail the deployment of real smart campuses.

For instance, in [67] an overview of the neOCampus of the Toulouse III Paul Sabatier University (France) is given. Such a smart campus runs different projects to make use of collaborative Wi-Fi, it provides an open-data platform, it fosters the reduction of the ecological footprint related to human activities and it aims for protecting the biodiversity of the campus.

Another interesting smart campus is detailed in [56], where an IoT platform deployed across different engineering schools of the Universidad Politécnica de Madrid in its Moncloa Campus (Spain) is described. Such an IoT platform is based on a cloud that provides services that follow the SOA paradigm. Two main applications are implemented: one for monitoring people flows and another for environmental monitoring.

The Sapienza smart campus (Italy) roadmap is described in [68]. Such a paper is interesting since, although it is a theoretical approach, it indicates how to structure the services to be provided by the smart campus infrastructure to scale it appropriately.

In [45] the author gives details on the Birmingham City University smart campus (United Kingdom). The smart campus platform integrates diverse business systems and smart building protocols thanks to an Enterprise Service Bus (ESB) and to the use of a SOA architecture, which provides scalability, flexibility, and service orchestration.

In the United States, an example of smart campus can be found in West Texas A&M University [66]. According to the authors, the smart campus is based on the IoT principles and covers an area of 176 acres, requiring connecting more than 42 different buildings. The described project is focused on two main tasks: to foster IoT collaboration and to provide an appropriate security framework. The proposed system has already supported diverse IoT projects, such as a LoRaWAN pilot for monitoring environmental conditions or an OpenCV-based smart parking system.

Finally, a smart campus for Wuhan University of Technology (China) is proposed in [69]. In such a paper the authors depict an architecture based on the IoT paradigm and in cloud-computing infrastructure that supports multiple applications.

2.4. Potential Smart Campus LPWAN Applications

Although a smart campus can support multiple indoor applications [70], in general, in such environments IoT nodes have access to power outlets and their communications can be usually easily handled with already common communications transceivers (e.g., Wi-Fi, Bluetooth, Ethernet). In contrast, this article focuses on the challenging environments that arise outdoors due the usual dependency on batteries to run IoT nodes and the need for exchanging data at relatively long distances (at least several hundred meters and up to 2 km), where LPWAN devices outperform other popular communications technologies.

The following are some of the most relevant outdoor applications that have already been implemented by using LPWAN technologies:

- Smart mobility and intelligent transport services. These applications require ubiquitous outdoor coverage to provide continuous data streams. For instance, in [71] researchers of Soochow University (China) propose the deployment of different smart mobility applications for their campuses, which include automatic vehicle access systems, a parking guidance service, a bus tracking system, or a bicycle rental service. Other authors also proposed similar solutions for providing campus services for smart parking [72], electric mobility [73,74], smart electric charging [75], the use of autonomous vehicles [76] or bus tracking [77].
- Smart energy and smart grid monitoring. Certain energy sources (e.g., renewable sources such as photo-voltaic panels or windmills) and smart grid components may be in remote locations, so it would be helpful to make use of LPWAN technologies to monitor them. For this reason, in recent years, special attention has been given to smart campus microgrids [78], smart grids [79] and smart energy systems [80].
- Resource consumption efficiency monitoring. These fields include waste collection [81], water management [82], energy monitoring [83], power consumption optimization [84] and sustainability [5].
- Campus user profiling. It is interesting for the campus managers to determine user patterns and behaviors to optimize the provided services. Thus, user profiling can be helpful to obtain mobility patterns, student daily walks, user activities, or social interactions, which can be obtained through opportunistic messaging apps [85], Wi-Fi monitoring [86] or on-board mobile phone sensors [87].
- Outdoor guidance and context-aware applications. This kind of systems are usually based on sensors and actuators spread throughout the campus and help people to reach their destination. There are examples in the literature of systems for guiding hearing and visually impaired people [88] or for navigating through the campus paths [89]. There are also augmented reality guidance applications [90], but it is important to note that LPWAN technologies could only help in small packet exchanges (e.g., for transmitting certain telemetry or positioning data), since the real-time multimedia content that can be demanded by augmented reality applications requires high-speed rates to preserve a good user experience.
- Classroom attendance. Some university events are carried out outdoors, what makes it difficult to control classroom attendance. To tackle such an issue, some researchers have proposed different sensor-based student monitoring systems that can be repurposed to be used outdoors [91].
- Infrastructure monitoring. It is possible to monitor remotely the status of certain assets that are scattered throughout the campus. For example, some authors presented smart campus solutions for managing campus greenhouses [92] or for monitoring high power lines with Unmanned Aerial Vehicles (UAVs) [93].
- Remote health monitoring. Smart campus technologies receive medical data in real time [94] or for measuring student stress [95]. Some researchers have even proposed to monitor the health of the campus trees [96].

2.5. Smart Campus Modeling and Planning Simulators

A smart campus is similar to an urban microenvironment where different buildings coexist with streets, open areas, parking lots, trees, benches, or people, among others. Different propagation channel models have been presented in the literature to characterize electromagnetic propagation phenomena in this type of scenarios, ranging from empirical methods to deterministic methods based on Ray Tracing (RT) or Ray-Launching (RL) approaches.

Empirical methods are based on measurements and their subsequent linear regression analysis. These methods are accurate for scenarios with the same characteristics as the real measurements, but their main disadvantages are the lack of scalability and the expensive cost in time and resources that is required to perform a measurement campaign. For example, in [97,98] empirical path loss models in a typical university campus are proposed for a frequency of 1800 MHz. The authors of [99] present a spatially consistent street-by-street path loss model for a 28 GHz channel in a micro-cell urban environment. The main drawback of these results is that the suitability of these models for path loss prediction has yet to be confirmed by the literature in other campuses.

On the other hand, deterministic methods are based on Maxwell's equations, which provide accurate propagation predictions. These approaches usually consider the three-dimensional geometry of the environment and model all propagation phenomena in the considered scenario. Their main drawback is the required computational time, which may not be afforded in large and complex scenarios. For this reason, RT or RL techniques, which are based on Geometrical Optics (GO) and the Uniform Theory of Diffraction (UTD), have been widely used for radio propagation purposes as an accurate approximation of full-wave deterministic techniques. For example, the authors of [100] modeled the dominant propagation mechanisms using advanced RT simulations in an urban microenvironment at 2.1 GHz, showing that diffuse scattering plays a key role in urban propagation. In addition, in [101] the importance of scattering when analyzing outdoor environments in the presence of trees is reported. In [102,103] the authors present propagation analyses (at 949.2 MHz and 2162.6 MHz, respectively) that make use of RT tools in a university campus with a Universal Mobile Telecommunications System (UMTS) base station placed on rooftops. Similarly, in [104], the radio-planning analysis of a Wi-Fi network in a university campus is presented using an RT approach. In addition, it has also been reported in the literature the calibration of RT simulators for millimeter-wave propagation analyses based on the measured results in a university campus at 28 GHz [105] and 38 GHz [106]. However, all the presented works are focused on micro-cellular environments and on analyzing the wireless propagation channel between a base station at a certain height in a building and a client in a pedestrian street. In this work, a radio wave propagation analysis for the connectivity of IoT LoRaWAN-based nodes by means of an in-house developed RL algorithm is reported (which is later detailed in Section 4).

2.6. Key Findings

After analyzing the state of art, it is clear that a smart campus faces similar challenges to smart cities and that they share certain use cases and communications technologies. Table 2 summarizes the characteristics of the most relevant smart campuses and related solutions, and compares them with the proposed system for the University of A Coruña.

As can be observed, the systems in the table make use of diverse short-range and long-range communications technologies, multiple sensors and actuators, different hardware and software platforms, and provide services for several practical use cases. Although some systems provide a holistic approach to a smart campus, devising potential outdoor use cases and applications, their implementations are mainly focused on environmental aspects, missing other smart fields such as the ones defined in Section 2.1.

Table 2. Comparison of the main features of the most relevant deployed smart campuses and related solutions.

Smart Campus	Area	Access Technology	Sensors and Actuators	IoT Hardware Platform	Software Platform	Use Cases	Fog Computing Capabilities	Network Planning	Sustainable Development Goals (SDGs) [107], KPIs or Results
School of STEM, University of Washington Bothell (United States) [41]	-	Zigbee, BLE, 6LowPAN	Sensor Tag 2.0 (accelerometer, magnetometer, gyroscope, light, humidity object and ambient temperature, microphone)	COTS hardware, Arduino	AWS, Microsoft Azure cloud services	-	No	No	Built in 3 months, it includes monthly cloud service bill
QA Higher Education (QAHE), University of Business and Technology, Birmingham (United Kingdom) [42]	-	-	NFC and RFID tags, QR codes	Wearables	Cisco Physical Access Control technology	Learning applications, access control systems	No	No	Deliver high quality services, protect the environment, and save costs
Tennessee State University, Nashville (United States) [43]	-	-	-	-	-	Survey on intelligent buildings, smart grid, learning environment, waste and water management and other applications	No	No	-

Table 2. *Cont.*

Smart Campus	Area	Access Technology	Sensors and Actuators	IoT Hardware Platform	Software Platform	Use Cases	Fog Computing Capabilities	Network Planning	Sustainable Development Goals (SDGs) [107], KPIs or Results
Northwestern Polytechnical University (China) [44]	-	Wi-Fi, Bluetooth	Built-in smartphone sensors	-	Android 2.1 platform, Big Data techniques and SOA	Where2Study, I-Sensing (participatory sensing), BlueShare (media sharing application)	No	No	-
Birmingham City University (United Kingdom) [45]	Two campuses of circa 18,000 and 24,000 m^2, respectively	-	-	-	Microsoft's BizTalk Server as ESB, SOA	Business systems, smart buildings	No	No	Cost savings; improved energy rating from F to B; 40% reduction in CO_2 emissions
IMDEA Networks Institute (Spain) [46]	-	Wi-Fi, Bluetooth	-	-		Mobility model	Opportunistic Floating Content (FC) communication paradigm	No	Performance of the service in terms of content persistence, availability and efficiency
University of Oradea (Romania) [47]	-	4G, Zigbee	-	RFID labels, mobile devices, sensor equipment	Private/public cloud with steganography	No. Only architecture design	No	No	-
[51]	-	-	-	Edge computing devices	Network model and bandwidth allocation scheme for mobile users	Trustworthy content caching	Edge caching reverse auction game and bandwidth allocation for multiple contents in Mobile Social Networks	No	Resource efficiency

Table 2. *Cont.*

Smart Campus	Area	Access Technology	Sensors and Actuators	IoT Hardware Platform	Software Platform	Use Cases	Fog Computing Capabilities	Network Planning	Sustainable Development Goals (SDGs) [107], KPIs or Results
[52]	-	MESH Wi-Fi	Environmental sensors, IP cameras, emergency buttons	-	Neural network learning algorithms	Street lighting	Edge Computing	No	Workload prediction accuracy, resource management dashboard
WiCloud [53]	-	Wi-Fi	-	Servers, mobile phone base stations or wireless access points	Network Functions Virtualization (NFV), Software-Defined Network (SDN)	Semantic information analysis, smart class	Mobile Edge Computing paradigm	No	Historical data
WiP [54]	-	3G/4G/5G, Wi-Fi	Smartcam, smart cards, light and temperature sensor, smartphone, tablet, smartwatch	-	-	Energy consumption savings, virtual support to students, augmented reality for museum collections	Yes	No	-
Smart CEI Moncloa, Universidad Politécnica de Madrid (Spain) [56]	5.5 km^2, 144 buildings, daily flow up to 120,000 people	Wi-Fi, Ethernet	Smart Citizen Kit (SCK)	Raspberry Pi, Arduino	Cloud, SOA paradigm	Smart emergency management and traffic restriction	No	No	Dashboard with historical data

Table 2. *Cont.*

Smart Campus	Area	Access Technology	Sensors and Actuators	IoT Hardware Platform	Software Platform	Use Cases	Fog Computing Capabilities	Network Planning	Sustainable Development Goals (SDGs) [107], KPIs or Results
West Texas A&M University (United States) [66]	176 acres (0.71 km^2) campus that connects 42 buildings and a 2393 acres (9.68 km^2) working ranch	LoRAWAN, 4G/LTE	Temperature, air pressure, relative humidity and partial concentrations	Arduino	NIST Cybersecurity Framework, standards such as COBIT and ISO	Connect cattle across the feed yard; monitor environmental conditions for network equipment; campus-wide environmental monitoring system; water irrigation; smart parking (GPS data, 800 video surveillance cameras and OpenCV-based)	No	No	-
Sapienza smart campus, University of Rome (Italy) [68]	-	N/A	N/A	N/A	Theoretical and methodological framework	Living, economy, energy, environment and mobility	No	No	Set of smart campus indicators and incidence matrix
Wuhan University of Technology (China) [69]	-	Cable, wireless, 3G/4G	Perception layer with RFID, cameras and sensors	-	Framework design, cloud computing and virtualization (Oracle 10G RAC)	Learning and living	No	No	-

Table 2. *Cont.*

Smart Campus	Area	Access Technology	Sensors and Actuators	IoT Hardware Platform	Software Platform	Use Cases	Fog Computing Capabilities	Network Planning	Sustainable Development Goals (SDGs) [107], KPIs or Results
Wisdom Campus, Soochow University (China) [71]	4058 acres (16.42 km^2), 5263 staff and more than 50,000 people	-	-	-	-	Automatic vehicle access systems, parking guidance service, bus tracking system and bicycle rental service	No	No	-
IISc campus [82]	2 km × 1 km	sub-GHz radios	Low-cost ultrasonic water level sensors, solar panels	Microcontroller TI MSP432P401R	-	Water management	No	No	RSSI and Packet Error Rate (PER) performance, power budget
Ottawa City and APEC campus [104]	-	Wi-Fi	-	-	-	-	No	RT approach	Measurements and predictions of Path Loss
Universitas Indonesia [106]	Urban area	800 MHz, 2.3 GHz, and 38 GHz	-	-	RT simulators for millimeter-wave propagation analyses based on the measured results in a university campus	-	No	RT approach and physical optic near-to-far field methods	Path Loss models
University of A Coruña (This work)	26,000 m^2	LoRaWAN	-	IoT nodes and SBCs (Raspberry Pi 3)	Simulations	Scalable architecture for multiple outdoor use cases	Yes	Yes (3D RL)	Planning simulator and empirical validation

In addition, only a few solutions consider the use of fog computing, and even less made use of network planning tools. In comparison, the solution proposed in this article is one of the few academic solutions that deploys LoRaWAN infrastructure. Moreover, the proposed smart campus is almost the only one conceived from scratch to harness the benefits of fog computing.

Regarding the efficiency in the network deployment in terms of cost, coverage, and the overall energy consumption, most of the academic papers do not give any insight regarding the heterogeneous network planning, although in some cases it is specified that it is low cost (without further details). Just a couple of systems give details for the hardware used to build the demonstrator at the level that is described later in this paper in Section 3.

3. Design and Implementation of the Smart Campus System

3.1. Architecture for Outdoor Applications

The proposed communications architecture is depicted in Figure 2. As can be observed, it comprises three different layers. The layer at the bottom consists of the different IoT LoRaWAN nodes that are deployed throughout the campus. Such nodes communicate with LoRaWAN gateways that comprise the Fog Layer, since they also act as fog computing gateways, thus providing fast location-aware responses to the LoRaWAN node requests. Every fog gateway is essentially a Single Board Computer (SBC) that embeds Ethernet, Wi-Fi, and Bluetooth interfaces besides a LoRaWAN transceiver. Finally, the top layer is the cloud, where user applications run together with data storage services.

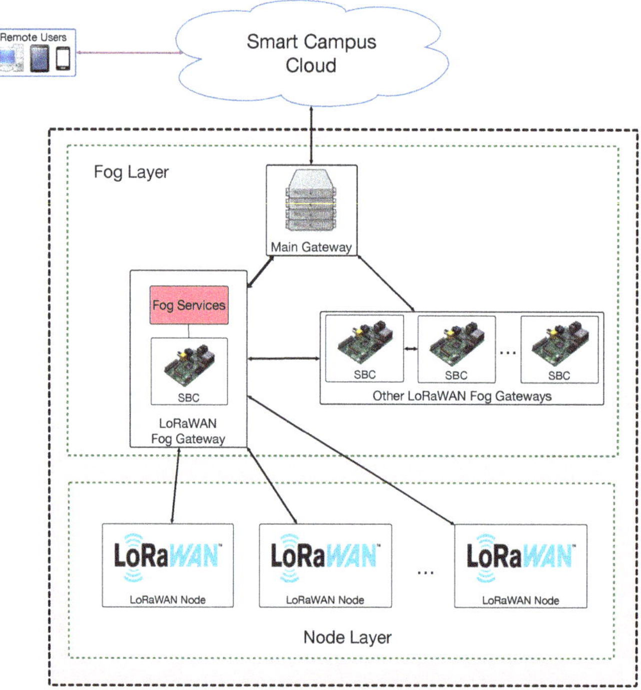

Figure 2. Proposed LoRaWAN-based smart campus architecture.

3.2. Operational Requirements for Outdoor Applications

To cover potential smart campus outdoor applications, a set of operational requirements grouped by capabilities was defined, including:

- Coverage capabilities. The coverage of the smart campus should be maximized. The typically expected coverage should be around 1 km^2 considering both Line-of-Sight (LoS) and No-Line-of-Sight (NLoS) scenarios.
- Robustness capabilities. The system should provide robustness to signal interference and/or loss of network operation. The network should provide redundancy and thus be robust against single points of failure.
- Supported services and applications. The previously mentioned applications (in Section 2.4) should be supported. Quality of Service (QoS) requirements should include support for high-peak rate demand, latency-sensitive traffic, and location-aware IoT applications. A transmission speed of up to 50 Kbps should be expected.
- Deployment features and cost. It should be expected that the deployment will depend largely on low-cost IoT nodes resource-constrained in terms of memory, battery, computing capabilities, and energy consumption.
- Network topology. The network architecture should support Point-To-Multipoint (PMP) and Point-to-Point (PtP) links. The system should be capable of establishing ad-hoc networking for specific scenarios (i.e., by using star or mesh topologies).

3.3. LoRaWAN Testbed Implementation

A LoRaWAN network consisting on a gateway and several nodes was deployed. A RisingHF RHF0M301 module [108] installed on a Raspberry Pi 3 was selected as gateway. Such a module is equipped with a dual digital radio front-end interface with a typical sensitivity of −137 dBm. The module is capable of simultaneous dual-band operation and supports Adaptive Data Rate (ADR), automatically changing between LoRa spreading factors. It can use a maximum of 10 channels: 8 multi Spreading Factor (SF) channels (SF7 to SF12 with 125 KHz of bandwidth), one FSK channel and one LoRa channel. Another interesting feature of the module is its ability to operate with negative Signal-to-Noise Ratios (SNRs), with a Co-Channel Rejection (CCR) of up to 9 dB. It also supports LoRaWAN classes A, B, and C and its maximum output power is 24.5 dBm.

A 0 dBi antenna was used for the tests. The module is installed on a Raspberry Pi 3 using the provided bridge (RHF4T002). To access the LoRaWAN RFID module, configure the node, and access the message and the transmission parameters, the following software was installed on the Raspberry Pi 3: Raspbian Stretch (Linux raspberrypi 4.14.70-v7+), LoRa Gateway v5.0.1, LoRa Packet Forwarder v4.0.1 and LoRaWAN-Server v0.6.0.

The LoRa gateway was configured as follows:

- Coding Rate: 4/5.
- RX1 delay: 1.
- RX2 delay: 2.
- Power: 14 dBm.
- RX Frequency: 869.5 MHz.

A RisingHF RHF76-052 module [109] was used by the LoRaWAN nodes (one of such nodes can be observed on the right of Figure 3). The module has a maximum sensitivity of −139 dBm with spreading factor 12 (SF12) and 125 kHz of bandwidth, channels (0–2) at 868.1, 868.3 and 868.5 MHz, and a maximum output power of 14 dBm. During the experiments presented in this paper, the module made use of the pre-installed wired antenna.

Figure 3. LoRaWAN IoT node during the empirical measurement campaign.

4. LoRaWAN Planning Simulator Setup

4.1. Planning Simulator

The in-house developed 3D Ray-Launching (3D-RL) technique is based on GO and the Uniform Theory of Diffraction (UTD). The first step of such a technique is the creation of the scenario under analysis, which should consider all the obstacles within it, such as buildings, vehicles, vegetation, or people. This design phase is essential for obtaining accurate results for the real environment. Once the scenario is properly created, the frequency of operation, number of reflections, radiation pattern of the transceivers and angular and spatial resolution can be fixed as input parameters in the algorithm for simulation. Then, the whole scenario is divided into a 3D mesh of cuboids, in which all the electromagnetic phenomena are saved during simulation, emulating the electromagnetic propagation of the real waves. A detailed description of the inner-workings of the 3D-RL tool is out of the scope of this article, but the interested reader can find further information in [110].

It is worth pointing out that smart campus ecosystems are challenging environments in terms of radio propagation analysis due to their large dimensions as well as for the multipath propagation, which is caused by the multiple obstacles within them. Hence, to achieve a good trade-off between simulation computational cost and result accuracy, it is important to determine the optimal parameters for the number of reflections and the angular and spatial resolution of the RL algorithm. For such a purpose, an analysis of the optimal input parameters for the RL tool applied in large complex environments is presented in [111]. Such previous results have been considered to obtain the simulation parameters for the proposed smart campus scenario and the final values are summarized in Table 3.

Table 3. 3D Ray-Launching parameters.

Parameter	Value
Operation frequency	868.3 MHz
Output power level	14 dBm
Permitted reflections	6
Cuboid resolution	$4\,\text{m} \times 4\,\text{m} \times 2\,\text{m}$
Launched ray resolution	1°
Antenna type and gain	Monopole, 0 dBi

4.2. Scenario under Analysis

This article presents a case study conducted in the northwest of Spain at the Campus of Elviña of the University of A Coruña. The campus covers an area of 26,000 m^2 and includes elements typically found in an urban environment, such as buildings of different heights, sidewalks, roads, green areas, trees, and cars, among others. Due to the large size of the environment to be analyzed, two different scenarios were created for the simulations (in Figure 4), which correspond to the two scenarios delimited by a green and a red rectangle in the real scenario shown in Figure 5. It must be noted that the two rectangular scenarios were selected within the campus (red on the left and green on the right) to cover eight faculties. The rectangular shape is required by the 3D Ray-Launching simulator. The points marked as *M*, *A*, *Q*, *R* correspond to different reception distances from the transmitter (*T*). To obtain accurate results, the created scenarios include the urban elements previously mentioned. Furthermore, realistic object sizes as well as material properties (permittivity and conductivity) were taken into account. For the experiments, the LoRaWAN gateway explained in Section 3.3 was placed at the spot indicated by the red dot in the two images shown in Figure 4. Please note that the location corresponds to a single position, which is in the third floor of the faculty of Computer Science (located in the intersection of the two rectangles of Figure 5, marked in such a Figure with a blue *T*).

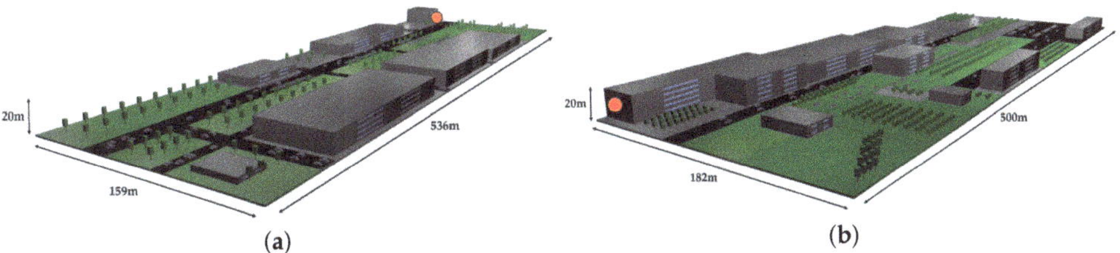

Figure 4. Simulated scenarios of the smart campus. (**a**) Red scenario; (**b**) Green scenario.

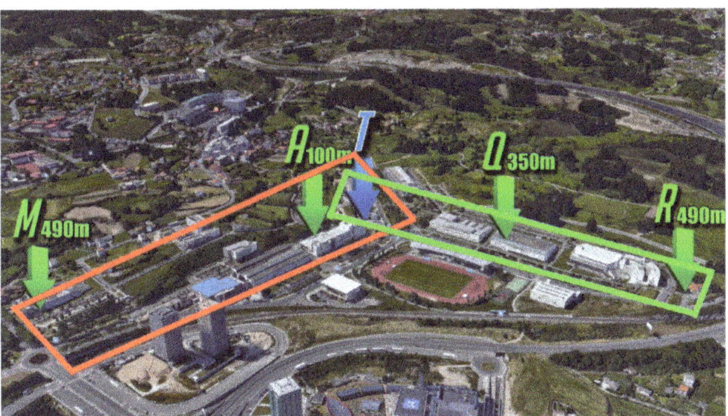

Figure 5. Aerial view of the Campus of Elviña, with the areas delimited for smart campus applications (Source: ©2019 Google).

5. Experiments

5.1. Empirical Validation: LoRaWAN Testbed

To evaluate LoRaWAN performance in a real campus, the testbed described in Section 3.3 was deployed in the Campus of Elviña. A measurement campaign was designed and carried out to validate the chosen hardware and the simulation results provided by the 3D Ray-Launching algorithm. The performed tests consisted on transmitting packets from the LoRaWAN node to the gateway from different spots throughout the campus using acknowledgment messages. Figure 6 shows such spots (in yellow) together with the LoRaWAN gateway location (in red) for one of the two parts of the evaluated campus. The transmitter was placed near a window inside a building, at a height of 3.5 m from the street ground level. In contrast, all the measurement spots were located outdoors, at a height of 0.5 m. For every of the previously mentioned spots, the Received Signal Strength Indicator (RSSI) and Signal-to-Noise Ratio (SNR) values were recorded both at the LoRaWAN gateway and at the device that acted as a node. A Debian Virtual machine was connected to the LoRaWAN-server WebSocket endpoint. Nodes were connected through an USB port and programmed to send a 6-byte payload ten times. The firmware of the LoRaWAN node uses the USB port to create a serial interface and write the SNR and RSSI of the received acknowledgment. With this setup, the LoRaWAN node was moved to different spots and at each of them the RSSI and SNR values of the gateway and the node were recorded for a total of ten packages per location.

To test for possible interference in the used Industrial, Scientific and Medical (ISM) sub-band, a USRP B210 [112] with the same 0 dBi antenna used by the LoRaWAN gateway was connected to a laptop that acted both as data-logger and spectrum analyzer (such a measurement setup is shown in Figure 3). As an example, the result of one of the analysis during the empirical measurements is shown in Figure 7, when the system was configured to monitor a central frequency of 868.3 MHz with a sampling rate of 1 MHz.

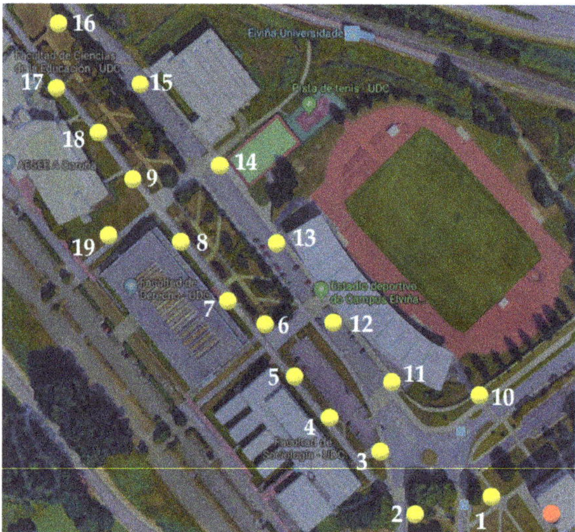

Figure 6. Empirical measurement points in the Green Scenario (Source: ©2019 Google).

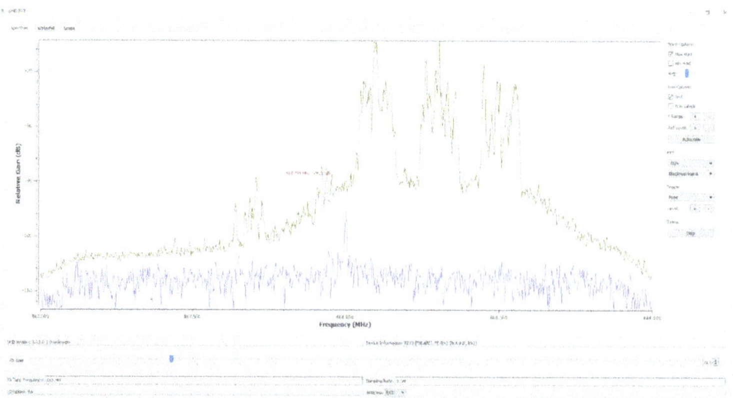

Figure 7. State of the radio spectrum during the performed empirical measurements.

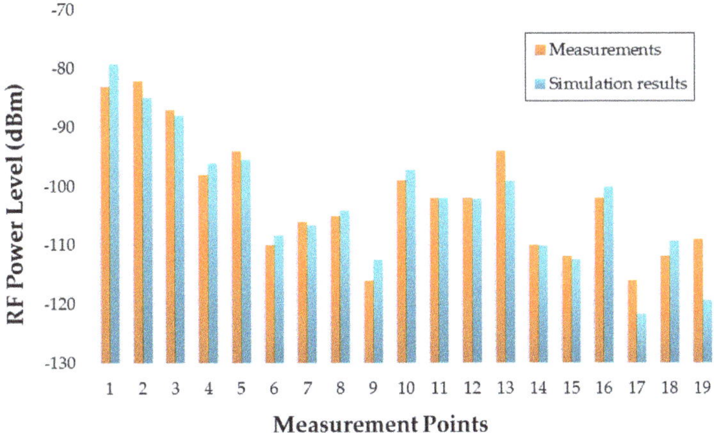

Figure 8. Comparison between empirical measurements and 3D-Ray-Launching simulation results.

After the measurement campaign within the campus, Radio Frequency (RF) power level estimations for the whole volume of the scenario were obtained with the aid of the 3D Ray-Launching simulation tool. The transmitter element was placed at the same position of the real LoRaWAN gateway (the red dot in Figure 6) and, using the simulation parameters shown in Table 3, a simulation was launched. The comparison between the measured RF power values and the simulation estimations is depicted in Figure 8. As can be seen in the Figure, the obtained estimations follow the tendency of the measured values, obtaining a mean error of 0.53 dB with a standard deviation of 3.39 dB (taking into account the 19 measurement points of Figure 6). The standard deviation is higher than the usual values provided by the 3D Ray-Launching. This effect could be due to size of the scenario (and the chosen simulation parameters such as cuboid size and launched ray resolution), since it is the largest scenario simulated so far by the developed 3D Ray-Launching tool. In addition, it must be noted the fact that measurements were based on RSSI values provided by the motes, which inherently add a received RF power level error. Nonetheless, the simulation results are accurate, and the simulation tool is validated satisfactorily. Regarding the results, it is worth noting the low RF power levels measured in several points of the scenario. The RF power level in many of these points is lower than −100 dBm, which is the typical ZigBee sensitivity. However, one of the advantages of the selected LoRaWAN devices is that their sensitivity is much lower (in the usual operating conditions, up to −137 dBm), as can it be observed in Table 4. Thus, the radio link budget for LoRaWAN has a higher margin, which means that longer communication distances can be achieved.

Table 4. Sensitivity values for LoRaWAN devices at 868 MHz.

LoRaWAN Device	Sensitivity
Seeeduino LoRaWAN	−137 dBm
Seeeduino LoRa/GPS Shield for Arduino with LoRa BEE	−148 dBm
Dragino LoRa Shield	−148 dBm
Grove—LoRa Radio	−148 dBm
DF Robot's LoRa MESH Radio Module	−148 dBm
Arduino MKR WAN 1300	−135.5 dBm
Adafruit RFM95W LoRa Radio Transceiver	−148 dBm
Adafruit Feather 32u4 RFM95 LoRa Radio	−148 dBm
Microchip LoRa Mote RN2483	−148 dBm
The Things Network TTN-UN-868	−148 dBm
The Things Network TTN-ND-868	−148 dBm

5.2. Planning of Smart Campus Use Cases

Once the presented 3D Ray-Launching simulation tool was validated by comparing the simulation results with the measurements, three different outdoor use cases were proposed for the smart campus environment, where LoRaWAN would have direct connectivity with a gateway:

- Crowdsensing/Mobility pattern detection. The purple dots depicted in Figure 9 represent the location of SBC-type devices (e.g., Raspberry Pi) that act as Bluetooth and Wi-Fi sniffers that will help to determine the mobility patterns of the users that move throughout the campus, what will optimize the deployed location-based services. In the same way, the devices could also help in crowdsensing tasks in certain areas.
- Smart irrigation. In this case, due to the location of the campus green areas, devices will be deployed only in one of the modeled scenarios. The device locations are represented by yellow dots shown in Figure 10. The aim of this system is to remotely control and automate the irrigation of green areas where the deployment of wired infrastructure to control the valves is very expensive or even unfeasible.
- Smart traffic monitoring. To detect vehicular traffic, sensors are deployed at the points represented by blue dots in Figure 11. In this way, the traffic behavior within the campus can be analyzed and the degree of parking occupancy could be inferred. Sustainability and ecological measurements to boost public transportation, to optimize routes and resources, and to adapt to real-time demand could be taken.

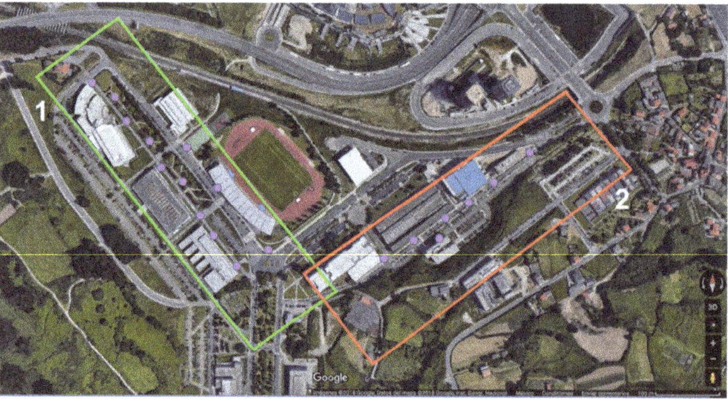

Figure 9. Aerial view of the spots monitored in the mobility pattern detection use case (Source: ©2019 Google).

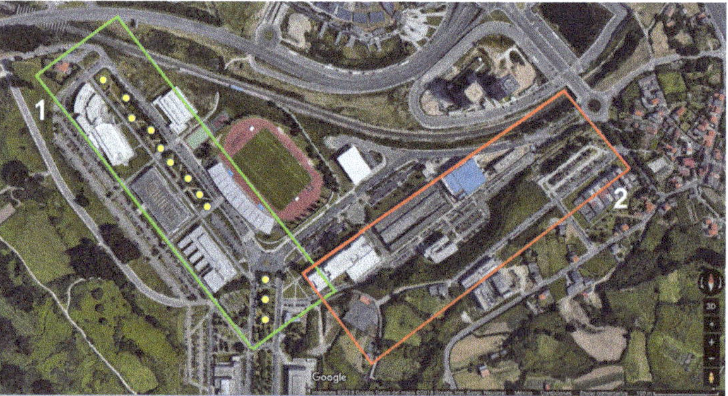

Figure 10. Aerial view of the smart irrigation monitoring spots (Source: ©2019 Google).

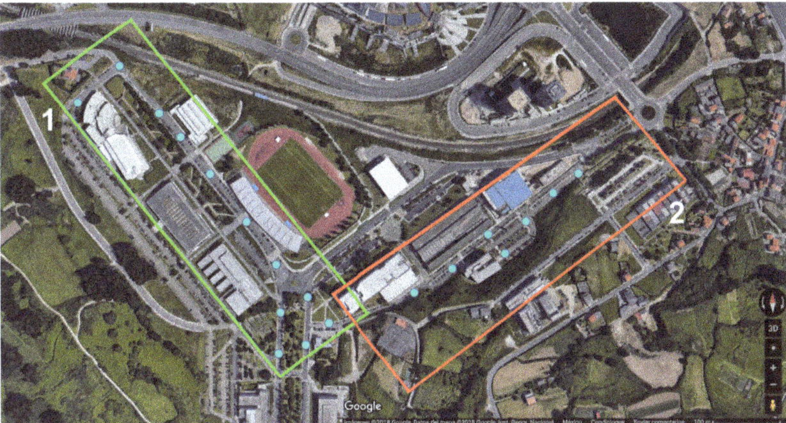

Figure 11. Aerial view of the spots monitored for the smart traffic use case (Source: ©2019 Google).

To validate the three mentioned smart campus use cases, 3D Ray-Launching simulations were launched for the proposed device locations. As an example, Figure 12 shows the estimated RF power distribution for bi-dimensional planes at two different heights (ground level and building's third floor level -at the gateway's height) for the smart irrigation use case (in Figure 12a, where the transmitter is at the center of the Green scenario) and the smart traffic-monitoring case (Figure 12b, transmitter at the furthest point of the Red Scenario). The used simulation parameters are also those shown in Table 3. As can be observed in Figure 12, the transmitter location (marked as a white circle with a T) and the morphology of the scenario (mainly the building location) greatly affect wireless signal propagation. Nevertheless, the estimated RF signal strength is quite high, taking into account the sensitivity of the employed LoRaWAN devices (i.e., −137 dBm) and the fact that the most common sensitivity value is −148 dBm (see Table 4).

To determine whether the chosen gateway location will comply with the required sensitivity for the proposed LoRaWAN node locations, 3D Ray-Launching simulation results were performed. As an example, Figure 13 summarizes the sensitivity analysis carried out for the use case illustrated in Figure 12b (i.e., for the furthest LoRaWAN node deployed for the smart traffic-monitoring use case). Specifically, Figure 13a shows the estimations obtained when the transmitter is operating at 20 dBm for different heights: ground level, third floor, and fourth floor. Figure 13b presents the same results, but for a lower transmission power (5 dBm). Finally, Figure 14 depicts the results for the sensitivity analysis based on the results obtained when transmitting at 5 dBm. This last Figure shows the areas and spots of the scenario that comply (dark blue) and do not comply (light blue) with the selected sensitivity value (in this case, the typical −148 dBm). The results show that for the case of transmitting at 20 dBm, there is no problem in terms of sensitivity threshold, but for the case of using 5 dBm, potential problems with this threshold appear within the building where the gateway is placed (this represented by the light blue surfaces at the top and left sides of the bi-dimensional planes). Therefore, a trade-off decision should be made to choose a transmission power level that ensures good sensitivity and, at the same time, the optimization of the energy consumption of the deployed motes. In fact, the results show that the gateway location could be improved by moving it from the third floor to the fourth floor. Thus, the deployment of the LoRaWAN network can be optimized by the presented 3D Ray-Launching algorithm in relation to its coverage and the overall energy consumption of the wireless communications system.

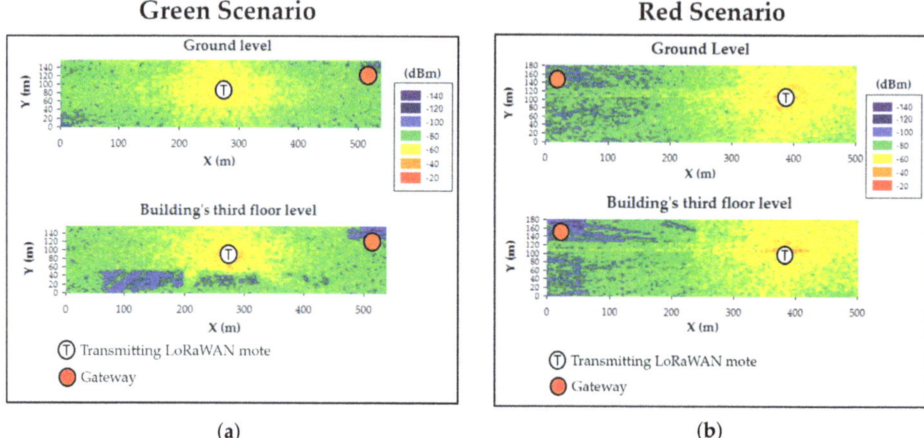

(a) (b)

Figure 12. Bi-dimensional planes of the estimated RF power distribution for two different heights. (**a**) Green scenario; (**b**) Red scenario.

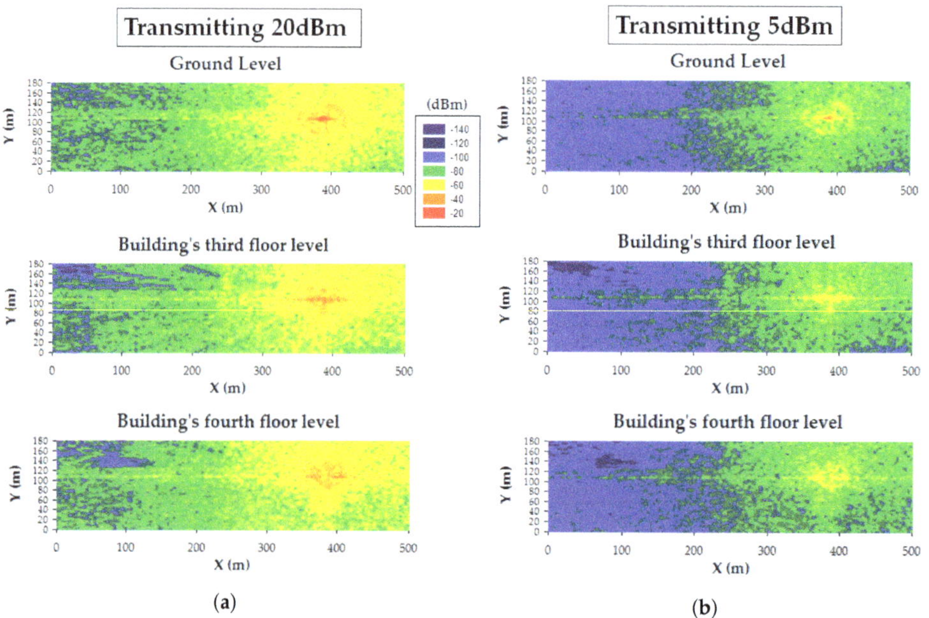

(a) (b)

Figure 13. Bi-dimensional planes of the estimated RF power distribution for two different heights. (**a**) transmission power 20 dBm, (**b**) transmission power 5 dBm.

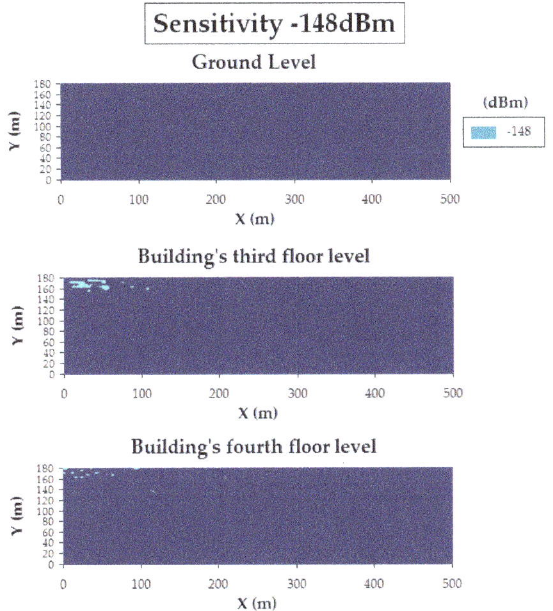

Figure 14. Bi-dimensional planes of the estimated RF power distribution for three different heights: sensitivity fulfillment (threshold = −148 dBm).

6. Discussion

The results presented in the previous section indicate the impact that the campus scenario has in radio-planning analysis and hence, in the determination of the optimal network layout. It must be first pointed out that the obtained results are hard to generalize, since the analyzed campus scenarios have particular characteristics that make them almost unique. Such characteristics include the size of the campus or the distribution of elements within it (mainly the buildings and their size and material properties), which have a great impact on radio signal propagation. Moreover, the topology of the deployed wireless network (i.e., the location of the nodes) has also a great influence on wireless communications performance. Therefore, the proposed methodology and solution have been validated in the presented paper, but it has to be noted that site-specific assessments are needed (that is, the results obtained for a specific campus environment cannot be generalized for any other campus scenario). Nonetheless, some aspects and results can be generalized up to a point (e.g., the received RF power for LoS situations), which are discussed in the following paragraphs.

One of the advantages in the use of LoRaWAN transceivers is their inherent low sensitivity values (in the range of −135.5 dBm to −148 dBm), which improve the reception range in comparison to other technologies. In all the observed simulation and measurement results, the received power levels are above −120 dBm, providing a sound margin to consider additional losses, such as the ones due to the variable fading related to user movements or to the dynamic elements within the campus.

In the specific outdoor applications considered for the smart campus, non-directional antennas provide adequate functionality, given the fact that theoretically, users and nodes can be located at any given location within the scenario. Nonetheless, in certain applications directional antennas may be helpful (e.g., in telemetry applications where the receiver is static, Yagi-Uda, helical or patch array antennas could be used), increasing received power levels, thus improving the communications range.

As can be observed in the experimental results, a coverage level of 20 dBm is appropriate for all the considered scenarios. However, it is desirable to reduce transmission power to reduce overall energy consumption as well as potential interference. For certain applications, tailored antennas may be considered during the network planning and the deployment phases.

The presented measurement results indicate that mainly due to the characteristics of the scenario, there is an appropriate coverage for a linear distance of 450 m with LoS (measurement point #16) and 330 m with NLoS (measurement point #19). The obtained results conclude that the location of the gateway is appropriate in terms of coverage estimation when there is LoS and in most situations where there is NLoS. The latter case requires in-depth analysis of the potential locations of the nodes, in order to consider effective losses related to building penetration, which on average can vary from 7 dB to over 25 dB depending on the building and wall structure [113].

In terms of capacity, LoRa/LoRaWAN provides a transmission speed of up to 50 Kbps, which is enough for a wide range of remote monitoring applications where users send small amounts of information or where nodes are polled with a moderate periodicity (i.e., several seconds) to provide information from sensors. Specific applications (e.g., real-time image monitoring) that require higher bandwidths can make use of alternative wireless technologies that can coexist together with the proposed LoRaWAN network.

7. Conclusions

After a thorough review of the state of the art on the most relevant aspects related to the deployment of smart campuses, this work detailed the design and deployment of a LoRaWAN IoT-based infrastructure able to provide novel applications in a smart campus. The architecture proposed the deployment of fog computing nodes throughout the campus to support physically distributed, low-latency, and location-aware applications that decrease the network traffic and the computational load of traditional cloud computing systems. To evaluate the proposed system, a campus of the University of A Coruña was recreated realistically through an in-house developed 3D Ray-Launching radio-planning simulator. Such a tool was tested by comparing its output with empirical measurements, showing a good accuracy. The proposed solution meets the established operational requirements:

- The provided coverage is roughly 1 km^2.
- The system provides robustness against signal losses and interference by being able to deploy redundant gateways.
- The use of fog computing nodes supports low-latency and location-aware IoT applications.
- The maximum provided transmission speed reaches 50 Kbps.
- The system has been devised to make use of low-cost resource-constrained IoT nodes.
- The network topology support both PMP and PtP links.

In addition, three practical smart campus applications were designed and evaluated with the planning simulator (a mobility pattern detection system, a smart irrigation solution, and a smart traffic-monitoring deployment), showing that the tool is able to provide useful guidelines and insights to future smart campus designers and developers.

Author Contributions: P.F.-L., P.L.-I. and T.M.F.-C. conceived and designed the experiments; T.M.F.-C., M.S.-A. and P.F.-L. performed the experiments; M.C.-E., P.L.-I. and L.A. created the scenario and performed the simulations; P.L.-I., E.A. and F.F. processed the simulation results; T.M.F.-C., P.L.-I. and P.F.-L. analyzed the data; P.F.-L., T.M.F.-C., P.L.-I., L.A. and L.C. wrote the paper.

Funding: This work has been funded by the Xunta de Galicia (ED431C 2016-045, ED431G/01), the Agencia Estatal de Investigación of Spain (TEC2016-75067-C4-1-R) and ERDF funds of the EU (AEI/FEDER, UE).

Conflicts of Interest: The authors declare no conflict of interest.

References

1. Fortes, S.; Santoyo-Ramón, J.A.; Palacios, D.; Baena, E.; Mora-García, R.; Medina, M.; Mora, P.; Barco, R. The Campus as a Smart City: University of Málaga Environmental, Learning, and Research Approaches. *Sensors* **2019**, *19*, 1349. [CrossRef] [PubMed]

2. Abuarqoub, A.; Abusaimeh, H.; Hammoudeh, M.; Uliyan, D.; Abu-Hashem, M.A.; Murad, S.; Al-Fayez, F. A survey on internet of things enabled smart campus applications. In Proceedings of the International Conference on Future Networks and Distributed Systems, Cambridge, UK, 19–20 July 2017; p. 50.
3. Hernández-Rojas, D.L.; Fernández-Caramés, T.M.; Fraga-Lamas, P.; Escudero, C.J. Design and Practical Evaluation of a Family of Lightweight Protocols for Heterogeneous Sensing through BLE Beacons in IoT Telemetry Applications. *Sensors* **2018**, *18*, 57. [CrossRef] [PubMed]
4. Sivanathan, A.; Sherratt, D.; Gharakheili, H.H.; Radford, A.; Wijenayake, C.; Vishwanath, A.; Sivaraman, V. Characterizing and classifying IoT traffic in smart cities and campuses. In Proceedings of the 2017 IEEE Conference on Computer Communications Workshops (INFOCOM WKSHPS), Atlanta, GA, USA, 1–4 May 2017; pp. 559–564.
5. Bates, O.; Friday, A. Beyond Data in the Smart City: Repurposing Existing Campus IoT. *IEEE Pervasive Comput.* **2017**, *16*, 54–60. [CrossRef]
6. Froiz-Míguez, I.; Fernández-Caramés, T.M.; Fraga-Lamas, P.; Castedo, L. Design, Implementation and Practical Evaluation of an IoT Home Automation System for Fog Computing Applications Based on MQTT and ZigBee-WiFi Sensor Nodes. *Sensors* **2018**, *18*, 2660. [CrossRef] [PubMed]
7. Berry College Official Web Page on Its Campus Extension. Available online: https://www.berry.edu/eaglecam/learn/ (accessed on 31 March 2019).
8. Duke University Official Web Page on Its Campus Extension. Available online: https://today.duke.edu/2011/11/onlinemaplaunch (accessed on 31 March 2019).
9. Stanford University Web Page on the Extension of Its Lands. Available online: https://facts.stanford.edu/about/lands/ (accessed on 31 March 2019).
10. Gumprecht, B. The campus as a public space in the American college town. *J. Hist. Geogr.* **2007**, *33*, 72–103. [CrossRef]
11. National Institute for Education Policy Reseach, Guide to the Creation of a Strategic Campus Master Plan. Available online: https://www.nier.go.jp/shisetsu/pdf/e-masterplan.pdf (accessed on 31 March 2019).
12. Princeton University Campus Plan. Available online: https://pr.princeton.edu/doc/PrincetonCampusPlan2017.pdf (accessed on 31 March 2019).
13. Wollongong Campus Master Plan. Available online: https://www.uow.edu.au/content/groups/public/@web/@pmcd/@smc/documents/doc/uow220188.pdf (accessed on 31 March 2019).
14. Sanchez-Iborra, R.; Cano, M.-D. State of the Art in LP-WAN Solutions for Industrial IoT Services. *Sensors* **2016**, *16*, 708. [CrossRef]
15. 3GPP Completed the Standardization of NB-IOT. Available online: http://www.3gpp.org/news-events/3gpp-news/1785-nb_iot_complete (accessed on 31 March 2019).
16. Sigfox Official Webpage. Available online: https://www.sigfox.com/ (accessed on 31 March 2019).
17. Ingenu Official Webpage. Available online: https://www.ingenu.com/ (accessed on 31 March 2019).
18. Weightless Official Webpage. Available online: http://www.weightless.org/ (accessed on 31 March 2019).
19. LoRa-Alliance. *LoRaWAN What is it? A Technical Overview of LoRa and LoRaWAN*; White Paper; The LoRa Alliance: San Ramon, CA, USA, 2015.
20. Adelantado, F.; Vilajosana, X.; Tuset-Peiro, P.; Martinez, B.; Melia-Segui, J.; Watteyne, T. Understanding the Limits of LoRaWAN. *IEEE Commun. Mag.* **2017**, *55*, 34–40. [CrossRef]
21. Yang, G.; Liang, H. A Smart Wireless Paging Sensor Network for Elderly Care Application Using LoRaWAN. *IEEE Sens. J.* **2018**, *22*, 9441–9448. [CrossRef]
22. Luvisotto, M.; Tramarin, F.; Vangelista, L.; Vitturi, S. On the Use of LoRaWAN for Indoor Industrial IoT Applications. *Wirel. Commun. Mob. Comput.* **2018**, *18*. [CrossRef]
23. Semtech AN 120022 LoRa Modulation Basics, May 2015. Available online: https://www.semtech.com/uploads/documents/an1200.22.pdf (accessed on 31 March 2019).
24. Suárez-Albela, M.; Fernández-Caramés, T.; Fraga-Lamas, P.; Castedo, L. A Practical Evaluation of a High-Security Energy-Efficient Gateway for IoT Fog Computing Applications. *Sensors* **2017**, *17*, 1978. [CrossRef]
25. Wang, S.; Chen, Y.-R.; Chen, T.-Y.; Chang, C.-H.; Cheng, Y.-H.; Hsu, C.-C.; Lin, Y.-B. Performance of LoRa-Based IoT Applications on Campus. In Proceedings of the 2017 IEEE 86th Vehicular Technology Conference (VTC-Fall), Toronto, ON, Canada, 24–27 September 2017; pp. 1–6.

26. Cesana, M.; Redondi, A.; Ort'ın, J. A Framework for Planning LoRaWan Networks. In Proceedings of the 2018 IEEE 29th Annual International Symposium on Personal, Indoor and Mobile Radio Communications (PIMRC), Bologna, Italy, 9–12 September 2018; pp. 1–7.
27. Sadowski, S.; Spachos, P. RSSI-Based Indoor Localization with the Internet of Things. *IEEE Access* **2018**, *6*, 30149–30161. [CrossRef]
28. Casino, F.; Azpilicueta, L.; Lopez-Iturri, P.; Aguirre, E.; Falcone, F.; Solanas, A. Optimized Wireless Channel Characterization in Large Complex Environments by Hybrid Ray Launching-Collaborative Filtering Approach. *IEEE Antennas Wirel. Propag. Lett.* **2017**, *16*, 780–783. [CrossRef]
29. Vargas, C.E.O.; Silva, M.M.; Arnez, J.J.A.; Mello, L.D.S. Initial Results of Millimeter Wave Outdoor Propagation Measurements in a Campus Environment. In Proceedings of the 2018 IEEE-APS Topical Conference on Antennas and Propagation in Wireless Communications (APWC), Cartagena de Indias, Colombia, 10–14 September 2018; pp. 901–904.
30. Chen, X.; Jin, R.; Suh, K.; Wang, B.; Wei, W. Network performance of smart mobile handhelds in a university campus Wi-Fi network. In Proceedings of the 2012 Internet Measurement Conference, Boston, MA, USA, 14–16 November 2012; pp. 315–328.
31. Atif, Y.; Mathew, S.S.; Lakas, A. Building a smart campus to support ubiquitous learning. *J. Ambient. Intell. Humaniz. Comput.* **2015**, *6*, 223–238. [CrossRef]
32. Hirsch, B.; Ng, J.W. Education beyond the cloud: Anytime-anywhere learning in a smart campus environment. In Proceedings of the 2011 IEEE International Conference for Internet Technology and Secured Transactions, Abu Dhabi, UAE, 11–14 December 2011; pp. 718–723.
33. Bakken, J.P.; Uskov, V.L.; Penumatsa, A.; Doddapaneni, A. Smart universities, smart classrooms and students with disabilities. In *Smart Education and e-Learning*; Springer: Cham, Switzerland, 2016; pp. 15–27.
34. Kwok, L.F. A vision for the development of i-campus. *Smart Learn. Environ.* **2015**, *2*, 1–12. [CrossRef]
35. Vienna University of Technology, University of Ljubljana, Delft University of Technology, Technical Report: Smart Cities, Ranking of European Medium-Size Cities, 2007. Available online: http://www.smart-cities.eu/download/smart_cities_final_report.pdf (accessed on 31 March 2019).
36. Wu, F.; Rüdiger, C.; Redouté, J.; Yuce, M.R. WE-Safe: A wearable IoT sensor node for safety applications via LoRa. In Proceedings of the 2018 IEEE 4th World Forum on Internet of Things (WF-IoT), Singapore, 5–8 February 2018; pp. 144–148.
37. Blanco-Novoa, O.; Fernández-Caramés, T.M.; Fraga-Lamas, P.; Castedo, L. A Cost-Effective IoT System for Monitoring Indoor Radon Gas Concentration. *Sensors* **2018**, *18*, 2198. [CrossRef] [PubMed]
38. Fernández-Caramés, T.M.; Fraga-Lamas, P. Design of a Fog Computing, Blockchain and IoT-Based Continuous Glucose Monitoring System for Crowdsourcing mHealth. *Proceedings* **2019**, *4*, 37. [CrossRef]
39. Pompei, L.; Mattoni, B.; Bisegna, F.; Nardecchia, F.; Fichera, A.; Gagliano, A.; Pagano, A. Composite Indicators for Smart Campus: Data Analysis Method. In Proceedings of the IEEE International Conference on Environment and Electrical Engineering and the IEEE Industrial and Commercial Power Systems Europe, Palermo, Italy, 12–15 June 2018.
40. Cao, J.; Li, Z.; Luo, Q.; Hao, Q.; Jiang, T. Research on the Construction of Smart University Campus Based on Big Data and Cloud Computing. In Proceedings of the International Conference on Engineering Simulation and Intelligent Control (ESAIC), Changsha, China, 10–11 August 2018.
41. Haghi, A.; Burney, K.; Kidd, F.S.; Valiente, L.; Peng, Y. Fast-paced development of a smart campus IoT platform. In Proceedings of the Global Internet of Things Summit, Geneva, Switzerland, 6–9 June 2017.
42. Majeed, A.; Ali, M. How Internet-of-Things (IoT) making the university campuses smart? QA higher education (QAHE) perspective. In Proceedings of the IEEE 8th Annual Computing and Communication Workshop and Conference (CCWC), Las Vegas, NV, USA, 8–10 January 2018.
43. Alghamdi, A.; Shetty, S. Survey Toward a Smart Campus Using the Internet of Things. In Proceedings of the IEEE 4th International Conference on Future Internet of Things and Cloud (FiCloud), Vienna, Austria, 22–24 August 2016.
44. Yu, Z.; Liang, Y.; Xu, B.; Yang, Y.; Guo, B. Towards a Smart Campus with Mobile Social Networking. In Proceedings of the International Conference on Internet of Things and 4th International Conference on Cyber, Physical and Social Computing, Dalian, China, 19–22 October 2011.
45. Hipwell, S. Developing smart campuses — A working model. In Proceedings of the International Conference on Intelligent Green Building and Smart Grid (IGBSG), Taipei, Taiwan, 23–25 April 2014.

46. Ali, S.; Rizzo, G.; Mancuso, V.; Marsan, M.A. Persistence and availability of floating content in a campus environment. In Proceedings of the IEEE Conference on Computer Communications, Hong Kong, China, 26 April–1 May 2015.
47. Popescu, D.E.; Prada, M.F.; Dodescu, A.; Hemanth, D.J.; Bungau, C. A secure confident cloud computing architecture solution for a smart campus. In Proceedings of the 7th International Conference on Computers Communications and Control (ICCCC), Oradea, Romania, 8–12 May 2018.
48. Marjanović, M.; Antonić, A.; Žarko, I.P. Edge Computing Architecture for Mobile Crowdsensing. *IEEE Access* **2018**, *6*, 10662–10674. [CrossRef]
49. Fernández-Caramés, T.M.; Fraga-Lamas, P.; Suárez-Albela, M.; Díaz-Bouza, M. A Fog Computing Based Cyber-Physical System for the Automation of Pipe-Related Tasks in the Industry 4.0 Shipyard. *Sensors* **2018**, *18*, 1961. [CrossRef]
50. Fernández-Caramés, T.M.; Fraga-Lamas, P.; Suárez-Albela, M.; Vilar-Montesinos, M. A Fog Computing and Cloudlet Based Augmented Reality System for the Industry 4.0 Shipyard. *Sensors* **2018**, *18*, 1798. [CrossRef]
51. Xu, Q.; Su, Z.; Wang, Y.; Dai, M. A Trustworthy Content Caching and Bandwidth Allocation Scheme with Edge Computing for Smart Campus. *IEEE Access* **2018**, *6*, 63868–63879. [CrossRef]
52. Chang, Y.; Lai, Y. Campus Edge Computing Network Based on IoT Street Lighting Nodes. *IEEE Syst. J.* **2018**, doi.10.1109/JSYST.2018.2873430. [CrossRef]
53. Liu, Y.; Shou, G.; Hu, Y.; Guo, Z.; Li, H.; Beijing, F.P.; Seah, H.S. Towards a smart campus: Innovative applications with WiCloud platform based on mobile edge computing. In Proceedings of the 12th International Conference on Computer Science and Education (ICCSE), Houston, TX, USA, 22–25 Auguat 2017.
54. Agate, V.; Concone, F.; Ferraro, P. WiP: Smart Services for an Augmented Campus. In Proceedings of the IEEE International Conference on Smart Computing (SMARTCOMP), Taormina, Italy, 18–20 July 2018.
55. Ghosh, A.; Chakraborty, N. Design of smart grid in an University Campus using ZigBee mesh networks. In Proceedings of the IEEE 1st International Conference on Power Electronics, Intelligent Control and Energy Systems (ICPEICES), Delhi, India, 4–6 July 2016.
56. Alvarez-Campana, M.; López, G.; Vázquez, E.; Villagrá, V.A.; Berrocal, J. Smart CEI Moncloa: An IoT-based Platform for People Flow and Environmental Monitoring on a Smart University Campus. *Sensors* **2017**, *17*, 12. [CrossRef] [PubMed]
57. Bai, S.; Chiu, C.; Hsu, J.; Leu, J. Campus-wide wireless indoor positioning with hybrid iBeacon and Wi-Fi system. In Proceedings of the 6th International Symposium on Next Generation Electronics (ISNE), Keelung, Taiwan, 23–24 May 2017.
58. Wang, M.; Brassil, J. Managing large scale, ultra-dense beacon deployments in smart campuses. In Proceedings of the IEEE Conference on Computer Communications Workshops (INFOCOM WKSHPS), Hong Kong, China, 26 April–1 May 2015.
59. Han, T.-D.; Cheong, C.; Ann, J.-W.; Kim, J.-Y.; Yoon, H.-M.; Lee, C.-S.; Shin, H.-G.; Lee, Y.-J.; Yook, H.-M.; Jeon, M.-H.; et al. Implementation of new services to support ubiquitous computing for campus life. In Proceedings of the Second IEEE Workshop on Software Technologies for Future Embedded and Ubiquitous Systems, Vienna, Austria, 12 May 2004.
60. Xu, X.; Li, D.; Sun, M.; Yang, S.; Yu, S.; Manogaran, G.; Mastorakis, G.; Mavromoustakis, C.X. Research on Key Technologies of Smart Campus Teaching Platform Based on 5G Network. *IEEE Access* **2019**, *7*, 20664–20675. [CrossRef]
61. Telensa Ultra Narrow Band (UNB) Official Webpage. Available online: https://www.telensa.com/technology#top (accessed on 31 March 2019).
62. Waviot NB-Fi Official Webpage. Available online: https://waviot.com/technology/what-is-nb-fi (accessed on 31 March 2019).
63. Loriot, M.; Aljer, A.; Shahrour, I. Analysis of the use of LoRaWan technology in a large-scale smart city demonstrator. In Proceedings of the 2017 Sensors Networks Smart and Emerging Technologies (SENSET), Beirut, Lebanon, 12–14 September 2017; pp. 1–4.
64. Wang, S.; Zou, J.; Chen, Y.; Hsu, C.; Cheng, Y.; Chang, C. Long-Term Performance Studies of a LoRaWAN-Based PM2.5 Application on Campus. In Proceedings of the 2018 IEEE 87th Vehicular Technology Conference (VTC Spring), Porto, Portugal, 3–6 June 2018; pp. 1–5.

65. Lee, H.; Ke, K. Monitoring of Large-Area IoT Sensors Using a LoRa Wireless Mesh Network System: Design and Evaluation. *IEEE Trans. Instrum. Meas.* **2018**, *67*, 2177–2187. [CrossRef]
66. Webb, J.; Hume, D. Campus IoT collaboration and governance using the NIST cybersecurity framework. In Proceedings of the Living in the Internet of Things: Cybersecurity of the IoT, London, UK, 28–29 March 2018.
67. Verstaevel, N.; Boes, J.; Gleizes, M. From smart campus to smart cities issues of the smart revolution. In Proceedings of the IEEE SmartWorld, Ubiquitous Intelligence & Computing, Advanced & Trusted Computed, Scalable Computing & Communications, Cloud & Big Data Computing, Internet of People and Smart City Innovation, San Francisco, CA, USA, 4–8 August 2017.
68. Pagliaro, F.; Mattoni, B.; Gugliermenti, F.; Bisegna, F.; Azzaro, B.; Tomei, F.; Catucci, S. A roadmap toward the development of Sapienza Smart Campus. In Proceedings of the IEEE 16th International Conference on Environment and Electrical Engineering (EEEIC), Florence, Italy, 7–10 June 2016.
69. Guo, M.; Zhang, Y. The research of smart campus based on Internet of Things & cloud computing. In Proceedings of the 11th International Conference on Wireless Communications, Networking and Mobile Computing (WiCOM 2015), Shanghai, China, 21–23 September 2015.
70. Muhamad, W.; Kurniawan, N.B.; Suhardi; Yazid, S. Smart campus features, technologies, and applications: A systematic literature review. In Proceedings of the International Conference on Information Technology Systems and Innovation (ICITSI), Bandung, Indonesia, 23–24 October 2017.
71. Hengliang, T.; Chuanrong, C. The Construction of Intelligent Transportation System Based on the Construction of Wisdom Campus—Take Soochow University as an Example. In Proceedings of the Eighth International Conference on Measuring Technology and Mechatronics Automation (ICMTMA), Macau, China, 11–12 March 2016.
72. Bandara, H.M.A.P.K.; Jayalath, J.D.C.; Rodrigo, A.R.S.P.; Bandaranayake, A.U.; Maraikar, Z.; Ragel, R.G. Smart campus phase one: Smart parking sensor network. In Proceedings of the Manufacturing & Industrial Engineering Symposium (MIES), Colombo, Sri Lanka, 22 October 2016.
73. Bracco, S.; Brenna, M.; Delfmo, F.; Foiadelli, F.; Longo, M. Preliminary study on electric mobility applied to a University Campus in North Italy. In Proceedings of the 6th International Conference on Clean Electrical Power (ICCEP), Santa Margherita Ligure, Italy, 27–29 June 2017.
74. Brenna, M.; Foiadelli, F.; Longo, M.; Bracco, S.; Delfino, F. Sustainable electric mobility analysis in the Savona Campus of the University of Genoa. In Proceedings of the IEEE 16th International Conference on Environment and Electrical Engineering (EEEIC), Florence, Italy, 7–10 June 2016.
75. Yagcitekin, B.; Uzunoglu, M.; Ocal, B.; Turan, E.; Tunc, A. Development of Smart Charging Strategies for Electric Vehicles in a Campus Area. In Proceedings of the European Modelling Symposium, Manchester, UK, 20–22 November 2013.
76. Básaca-Preciado, L.C.; Orozco-Garcia, N.A.; Terrazas-Gaynor, J.M.; Moreno-Partida, A.S.; Rosete-Beas,O.A.; Rizzo-Aguirre, J.; Martinez-Grijalva, L.F.; Ponce-Camacho, M.A. Intelligent Transportation Scheme for Autonomous Vehicle in Smart Campus. In Proceedings of the 44th Annual Conference of the IEEE Industrial Electronics Society, Washington, DC, USA, 21–23 October 2018.
77. Chit, S.M.; Chaw, L.Y.; Thong, C.L.; Lee, C.Y. A pilot study: Shuttle bus tracker app for campus users. In Proceedings of the International Conference on Research and Innovation in Information Systems (ICRIIS), Langkawi, Malaysia, 10 August 2017.
78. Bracco, S.; Delfino, F.; Laiolo, P.; Rossi, M. The Smart City Energy infrastructures at the Savona Campus of the University of Genoa. In Proceedings of the AEIT International Annual Conference (AEIT), Capri, Italy, 5–7 October 2016.
79. Sharma, H.; Kaur, G. Optimization and simulation of smart grid distributed generation: A case study of university campus. In Proceedings of the IEEE Smart Energy Grid Engineering (SEGE), Oshawa, ON, Canada, 21–24 August 2016.
80. Lazaroiu, G.C.; Dumbrava, V.; Costoiu, M.; Teliceanu, M.; Roscia, M. Smart campus-an energy integrated approach. In Proceedings of the International Conference on Renewable Energy Research and Applications (ICRERA), Palermo, Italy, 22–25 November 2015.

81. Pagliaro, F.; Mattoni, B.; Ponzo, V.; Corona, G.; Nardecchia, F.; Bisegna, F.; Gugliermetti, F. Sapienza smart campus: From the matrix approach to the applicative analysis of an optimized garbage collection system. In Proceedings of the IEEE International Conference on Environment and Electrical Engineering and the IEEE Industrial and Commercial Power Systems Europe, Milan, Italy, 6–9 June 2017.
82. Verma, P.; Kumar, A.; Rathod, N.; Jain, P.; Mallikarjun, S.; Subramanian, R.; Amrutur, B.; Kumar, M.S.M.; Sundaresan, R. Towards an IoT based water management system for a campus. In Proceedings of the IEEE First International Smart Cities Conference (ISC2), Guadalajara, Mexico, 25–28 October 2015.
83. Liu, R.; Kuo, C.; Yang, C.; Chen, S.; Liu, J. On Construction of an Energy Monitoring Service Using Big Data Technology for Smart Campus. In Proceedings of the 7th International Conference on Cloud Computing and Big Data (CCBD), Macau, China, 16–18 November 2016.
84. Lazaroiu, G.C.; Dumbrava, V.; Costoiu, M.; Teliceanu, M.; Roscia, M. Energy-informatic-centric smart campus. In Proceedings of the IEEE 16th International Conference on Environment and Electrical Engineering (EEEIC), Florence, Italy, 7–10 June 2016.
85. Bacanli, S.S.; Solmaz, G.; Turgut, D. Opportunistic Message Broadcasting in Campus Environments. In Proceedings of the IEEE Global Communications Conference (GLOBECOM), San Diego, CA, USA, 6–10 December 2015.
86. Zhang, S.; Li, X. Mobility patterns of human population among university campuses. In Proceedings of the IEEE Asia Pacific Conference on Circuits and Systems, Jeju, Korea, 25–28 October 2016.
87. Concone, F.; Ferraro, P.; Lo Re, G. Towards a Smart Campus Through Participatory Sensing. In Proceedings of the IEEE International Conference on Smart Computing (SMARTCOMP), Taormina, Italy, 18–20 June 2018.
88. Bilgi, S.; Ozturk, O.; Gulnerman, A.G. Navigation system for blind, hearing and visually impaired people in ITU Ayazaga campus. In Proceedings of the International Conference on Computing Networking and Informatics (ICCNI), Lagos, Nigeria, 29–31 October 2017.
89. Yong, Q.; Cheng, B.; Xing, Y. A Novel Quantum Ant Colony Algorithm Used for Campus Path. In Proceedings of the IEEE International Conference on Computational Science and Engineering (CSE) and IEEE International Conference on Embedded and Ubiquitous Computing (EUC), Guangzhou, China, 21–24 July 2017.
90. Özcan, U.; Arslan, A.; İlkyaz, M.; Karaarslan, E. An augmented reality application for smart campus urbanization: MSKU campus prototype. In Proceedings of the 5th International Istanbul Smart Grid and Cities Congress and Fair (ICSG), Istanbul, Turkey, 19–21 April 2017.
91. Sutjarittham, T.; Habibi Gharakheili, H.; Kanhere, S.S.; Sivaraman, V. Data-Driven Monitoring and Optimization of Classroom Usage in a Smart Campus. In Proceedings of the17th ACM/IEEE International Conference on Information Processing in Sensor Networks (IPSN), Porto, Portugal, 11–13 April 2018.
92. Liang, T.; Tsai, C. Application of Intelligent Monitoring System in Campus Greenhouse. In Proceedings of the International Conference on Information, Communication and Engineering (ICICE), Xiamen, China, 17–20 November 2017.
93. Lu, J.; Xu, X.; Li, X.; Li, L.; Chang, C.-C.; Feng, X.; Zhang, S. Detection of Bird's Nest in High Power Lines in the Vicinity of Remote Campus Based on Combination Features and Cascade Classifier. *IEEE Access* **2018**, *6*, 39063–39071. [CrossRef]
94. Liang, Y.; Chen, Z. Intelligent and Real-Time Data Acquisition for Medical Monitoring in Smart Campus. *IEEE Access* **2018**, 6, 74836-74846. [CrossRef]
95. Gjoreski, M.; Gjoreski, H.; Lutrek, M.; Gams, M. Automatic Detection of Perceived Stress in Campus Students Using Smartphones. In Proceedings of the International Conference on Intelligent Environments, Prague, Czech Republic, 15–17 July 2015.
96. Pérez, L.E.; Pardiñas-Mir, J.A.; Guerra, O.; de la Mora, J.; Pimienta, M.; Hernández, N.; de Atocha Lopez, M. A wireless sensor network system: For monitoring trees' health related parameters in a university campus. In Proceedings of the 12th International Joint Conference on e-Business and Telecommunications (ICETE), Colmar, France, 20–22 July 2015.
97. Popoola, S.I.; Atayero, A.A.; Popoola, O.A. Comparative assessment of data obtained using empirical models for path loss predictions in a university campus environment. *Data Brief* **2018**, *18*, 380–393. [CrossRef]
98. Popoola, S.I.; Atayero, A.A.; Arausi, O.D.; Matthews, V.O. Path loss dataset for modeling radio wave propagation in smart campus environment. *Data Brief* **2018**, *17*, 1062–1073. [CrossRef]

99. Karttunen, A.; Molisch, A.F.; Hur, S.; Park, J.; Zhang, C.J. Spatially Consistent Street-by-Street Path Loss Model for 28-GHz Channels in Micro Cell Urban Environments. *IEEE Trans. Wirel. Commun.* **2017**, *16*, 7538–7550. [CrossRef]
100. Fuschini, F.; El-Sallabi, H.; Degli-Esposti, V.; Vuokko, L.; Guiducci, D.; Vainikainen, P. Analysis of Multipath Propagation in Urban Environment Through Multidimensional Measurements and Advanced Ray Tracing Simulation. *IEEE Trans. Antennas Propag.* **2008**, *56*, 848–857. [CrossRef]
101. Mani, F.; Oestges, C. A Ray Based Method to Evaluate Scattering by Vegetation Elements. *IEEE Trans. Antennas Propag.* **2012**, *60*, 4006–4009. [CrossRef]
102. Chio, C.H.; Pang, C.K.; Ting, S.W.; Tam, K.W. Field prediction in urban environment using ray tracing. In Proceedings of the 2013 IEEE Antennas and Propagation Society International Symposium (APSURSI), Orlando, FL, USA, 7–13 July 2013; pp. 1926–1927.
103. Un, L.K.; Chio, C.; Ting, S. Mobile communication site planning in campus using ray tracing. In Proceedings of the 2014 IEEE Antennas and Propagation Society International Symposium (APSURSI), Memphis, TN, USA, 6–11 July 2014; pp. 959–960.
104. Thirumaraiselvan, P.; Jayashri, S. Modeling of Wi-Fi signal propagation under tree canopy in a college campus. In Proceedings of the 2016 International Conference on Communication and Signal Processing (ICCSP), Melmaruvathur, India, 24 November 2016; pp. 2273–2277.
105. Wei, S.; Ai, B.; He, D.; Guan, K.; Wang, L.; Zhong, Z. Calibration of ray-tracing simulator for millimeter-wave outdoor communications. In Proceedings of the 2017 IEEE International Symposium on Antennas and Propagation & USNC/URSI National Radio Science Meeting, San Diego, CA, USA, 9–14 July 2017; pp. 1907–1908.
106. Fathurrahman, S.Z.; Rahardjo, E.T. Coverage of Radio Wave Propagation at UI Campus Surrounding Using Ray Tracing and Physical Optics Near to Far Field Method. In Proceedings of the TENCON 2018—IEEE Region 10 Conference, Jeju, Korea, 28–31 October 2018; pp. 1123–1126.
107. UN Sustainable Development Goals. Available online: https://sustainabledevelopment.un.org/ (accessed on 14 July 2019).
108. RHF0M301 LoRaWAN Module Datasheet. Available online: https://www.robotshop.com/media/files/pdf/915mhz-lora-gateway-raspberry-pi-hat-datasheet1.pdf (accessed on 31 March 2019).
109. LoRaWAN Module RHF76-052 Datasheet. Available online: https://fccid.io/2AJUZ76052/User-Manual/Users-Manual-3211050.pdf (accessed on 31 March 2019).
110. Azpilicueta, L.; Rawat, M.; Rawat, K.; Ghannouchi, F.; Falcone, F. Convergence Analysis in Deterministic 3D Ray Launching Radio Channel Estimation in Complex Environments. *ACES J.* **2014**, *29*, 256–271.
111. Azpilicueta, L.; Lopez-Iturri, P.; Aguirre, E.; Vargas-Rosales, C.; León, A.; Falcone, F. Influence of Meshing Adaption in Convergence Performance of Deterministic Ray Launching Estimation in Indoor Scenarios. *J. Electromagn. Waves Appl.* **2017**, *31*, 544–559. [CrossRef]
112. Ettus Research USRP B210. Available online: https://www.ettus.com/product/details/UB210-KIT (accessed on 31 March 2019).
113. Allen, K.C.; DeMinco, N.; Hoffman, J.R.; Lo, Y.; Papazian, P.B. *Building Penetration Loss Measurements at 900 MHz, 11.4 GHz, and 28.8 GHz*; NTIA Report 94-306; U.S. Department of Commerce: Washington, DC, USA, 1994.

Article

Enabling the Internet of Mobile Crowdsourcing Health Things: A Mobile Fog Computing, Blockchain and IoT Based Continuous Glucose Monitoring System for Diabetes Mellitus Research and Care

Tiago M. Fernández-Caramés *,†, Iván Froiz-Míguez, Oscar Blanco-Novoa and Paula Fraga-Lamas *,†

Department of Computer Engineering, Faculty of Computer Science, Centro de investigación CITIC, Universidade da Coruña, 15071 A Coruña, Spain

* Correspondence: tiago.fernandez@udc.es (T.M.F.-C.); paula.fraga@udc.es (P.F.-L.); Tel.: +34-981167000 (ext. 6051) (P.F.-L.)

† This paper is an extended version of our conference paper: Fernández-Caramés, T. M., Fraga-Lamas, P. "Design of a Fog Computing, Blockchain and IoT-Based Continuous Glucose Monitoring System for Crowdsourcing mHealth", 5th International Electronic Conference on Sensors and Applications, 15–30 November 2018.

Received: 28 June 2019; Accepted: 25 July 2019; Published: 28 July 2019

Abstract: Diabetes patients suffer from abnormal blood glucose levels, which can cause diverse health disorders that affect their kidneys, heart and vision. Due to these conditions, diabetes patients have traditionally checked blood glucose levels through Self-Monitoring of Blood Glucose (SMBG) techniques, like pricking their fingers multiple times per day. Such techniques involve a number of drawbacks that can be solved by using a device called Continuous Glucose Monitor (CGM), which can measure blood glucose levels continuously throughout the day without having to prick the patient when carrying out every measurement. This article details the design and implementation of a system that enhances commercial CGMs by adding Internet of Things (IoT) capabilities to them that allow for monitoring patients remotely and, thus, warning them about potentially dangerous situations. The proposed system makes use of smartphones to collect blood glucose values from CGMs and then sends them either to a remote cloud or to distributed fog computing nodes. Moreover, in order to exchange reliable, trustworthy and cybersecure data with medical scientists, doctors and caretakers, the system includes the deployment of a decentralized storage system that receives, processes and stores the collected data. Furthermore, in order to motivate users to add new data to the system, an incentive system based on a digital cryptocurrency named GlucoCoin was devised. Such a system makes use of a blockchain that is able to execute smart contracts in order to automate CGM sensor purchases or to reward the users that contribute to the system by providing their own data. Thanks to all the previously mentioned technologies, the proposed system enables patient data crowdsourcing and the development of novel mobile health (mHealth) applications for diagnosing, monitoring, studying and taking public health actions that can help to advance in the control of the disease and raise global awareness on the increasing prevalence of diabetes.

Keywords: diabetes; glucose monitoring; CGM; fog computing; IoT; blockchain; crowdsourcing mHealth; public health; decision support; personalized medicine

---

1. Introduction

Diabetes Mellitus (DM), which is usually referred to as Diabetes, is a worldwide chronic metabolic disorder that involves abnormal blood glucose level oscillations that lead to both macrovascular alterations, which affect large blood vessels (coronary arteries, the aorta, and arteries in the brain and in the limbs) and microvascular complications, which affect the kidneys (nephropathy), nerves (neuropathy), and eyes (retinopathy). According to the World Health Organization (WHO) [1], there are three main types of DM:

- Type-1 DM (DM1): it is an autoimmune process that leads to remarkably diminished insulin levels.
- Type-2 DM (DM2): it arises from the compromised function of insulin-producing cells in the pancreas and insulin resistance in the peripheral tissues.
- Gestational Diabetes Mellitus (GDM): it occurs on pregnant women that suffer from above-normal blood glucose levels.

Due to the health issues related to DM, it is important to monitor vulnerable groups, especially children, the elderly and pregnant women. Such a monitoring has been traditionally performed by taking blood samples through Self-Monitoring of Blood Glucose (SMBG) techniques [2], which have a number of drawbacks (e.g., active involvement of the patient or his/her caretakers, infections) that can be tackled by Continuous Glucose Monitors (CGMs), which are based on a small device with a sensor that takes blood glucose readings 24 h a day [3]. Such measurements make it easier for DM patients to have more precise control over blood glucose, which allows patients the ability to make informed therapeutic decisions. Thus, CGMs may warn patients about hyperglycemia (high blood glucose level) and hypoglycemia (low blood glucose level) in order to take the appropriate preventive measures. Nonetheless, it is worth pointing out that the use of CGMs carry some inconveniences: CGMs are usually expensive (although some countries are starting to subsidize its acquisition), they read glucose concentration values with a delay between 5 and 10 min [4], some CGMs need to be calibrated several times a day by finger-pricking and their typical lifespan is short (it usually goes between 3 days and a couple of weeks, but recent CGMs keep on working up to six months).

Despite CGM current drawbacks, the concept of CGM opens the possibility of creating Internet of Things (IoT) devices that provide rapid warnings and are able to make autonomous decisions when actions must be performed as fast as possible to avoid dire outcomes [5]. Thus, an IoT CGM can make use of a remote cloud system where information is stored and where rule-based decisions can be taken (e.g., to warn a doctor when the patient's blood glucose level is above or below a specific threshold). However, traditional cloud computing architectures have certain limitations: all the information and decisions are centralized and managed, in general, by a third party; the cloud availability may be compromised by overloading or by cyber-attacks; and, due to physical distance, there may be a lag between the cloud and the patient resulting in a delay between the decision to perform an action and the communication to the patient to do so, which may be too long in some cases. Fortunately, for such scenarios where a fast response and low communications overhead are required, other paradigms have been successful by moving computing capabilities from the cloud towards the edge of the network [6]. One of such paradigms is fog computing, which transfers the cloud computational and communication capabilities close to the sensor nodes in order to minimize latency, to distribute computational and storage resources, to enhance mobility and location awareness, and to ease network scalability while providing connectivity among devices in different physical environments [7,8].

Security and trustworthiness are other problems that arise when collecting and processing the data sensed by IoT devices. Regarding security, some authors proposed energy-efficient mechanisms to make use of high-security cipher suites in IoT devices, since they are usually constrained in terms of computational resources, especially when they rely on batteries [9]. With respect to trustworthiness, it is essential when data are shared with third parties. Therefore, the collected data (e.g., for doctors or for autonomous systems that base their decision-making on the received blood glucose values) should be validated.

Doctors and researchers also need data in order to improve the existing knowledge on DM and to look for a potential cure. Such medical data are usually difficult to obtain due to different reasons (e.g., lack of access to useful data, existing laws, lack of user trust), so it is important to study new ways to automate data collection on a large scale. Crowdsourcing is one potential technique, since it is able to make use of the collective intelligence of an online community to research and to develop innovative human-centered approaches and novel products and services [10]. However, there are not many public health crowdsourcing applications [11], specially in the area of mobile health (mHealth), which harnesses smartphones and wireless communications technologies to ease the access to healthcare solutions. There are even less decentralized mobile crowdsourcing healthcare applications, since such systems usually depend on a public health central authority or on a private company, which may be a single point of failure prone to cyber-attacks.

To tackle the previously mentioned issues, this article details the design and implementation of a system that includes the following contributions, which have not been found together in the previous literature:

- A description of an IoT CGM-based system able to monitor glucose concentration values remotely and quickly inform patients and/or their care givers of a dangerous situation.
- The design of a system able to provide fast warnings by using fog computing gateways when the user is located in specific scenarios (e.g., inside the patient's home, in a hospital or in a nursing home).
- A description of a mHealth blockchain-based decentralized architecture that does not depend on any central authority.
- The proposal of a data crowdsourcing mechanism that encourages collaboration by rewarding patients.
- The evaluation of the proposed architecture to assess the performance of the described decentralized storage and blockchain.

The rest of this article is structured as follows. Section 2 reviews the previous work on CGM applications and on the use of fog computing, blockchain and crowdsourcing for healthcare applications. Section 3 details the proposed communications architecture and the designed crowdsourcing incentive mechanism, while Section 4 describes their implementation. Finally, Section 5 is devoted to the experiments and Section 7 to the conclusions.

2. Related Work

2.1. CGM Applications

During the last decade different authors studied the use of CGMs in order to apply them in different scenarios. For instance, in [12] the possibility of adding intelligence to CGM sensors is analyzed in order to issue alerts when glucose levels are not within the appropriate range (e.g., in case of a hypoglycaemia/hyperglycaemia). The same authors focused later on reviewing the state of the art on contributions related to the development of hardware and smart algorithms for CGMs [13]. Similar reviews on the evolution of CGM technology can be found in [14–16].

Other CGM features have been studied by other authors. For instance, some authors focused on analyzing off-the-shelf CGM calibration algorithms and on how the amount of glucose in blood plasma is obtained [17]. Other researchers studied how CGMs detect losses in the performance on insulin infusion set actuation [18], which may derive into prolonged hyperglycemia in Type-1 DM patients. A CGM is also able to decipher the influence of daily habits on the health of DM patients. For example, physical exercise clearly alters glucose concentration levels on Type-1 DM patients: glucose regulation problems may arise during or after exercising and even when performing certain daily activities [19].

Finally, it is worth mentioning that a CGM can collaborate with an artificial pancreas in order to provide the optimal insulin dose through an insulin pump. However, there are different parameters that influence the infusion of insulin, with physical activity being one of the most challenging as it may result in hypoglycemia in Type-1 DM patients [20].

2.2. Fog Computing, Blockchain and Crowdsourcing for Healthcare Applications

In the last years, cloud computing has achieved remarkable success thanks to its ability to offload computational-intensive tasks from clients [21]. Nonetheless, in applications where low latency responses are needed, other paradigms like fog computing have proven to be valuable [6]. Fog computing is usually considered a paradigm that is an extension of cloud computing where part of the computational and communication capabilities of the cloud are moved close to the sensor nodes [7], which derives into several advantages [22]:

- Novel IoT healthcare real-time applications can be provided thanks to decreasing latency.
  The fog allows for sharing computational and storage resources, which can be harnessed by distributed wireless sensor networks that can be deployed in hospitals or in other healthcare facilities.
- The fog can also connect multiple physical environments that are far from each other, easing device and user interaction and thus facilitating the development of new potential healthcare services.
- Since the fog can be scaled easily, it provides flexibility and ease of growth to large deployments.

Despite the mentioned benefits and its multiple applications in diverse fields [23], there are not many practical examples of the use of fog computing in healthcare applications. A good compilation of fog computing based healthcare applications can be found in [24], which describes potential applications for monitoring Chronic Obstructive Pulmonary Disease (COPD) patients [25], tools for in-home Parkinson disease treatments [26] or hospital platforms that make use of e-textiles [27] and wireless sensor networks [28]. Similarly, other researchers proposed applications for healthcare monitoring in smart homes [29] or for diagnosing and preventing outbreaks of certain viruses [30].

Another challenge faced by today's healthcare researchers and practitioners is the lack of interoperability between different technological platforms. Incompatible, unscalable and independent systems hinder the development of novel end-to-end patient-centered research and healthcare solutions. Unfortunately, it is difficult to exchange information among healthcare providers, patients and third-parties (e.g., insurance companies, governments, app developers). Records are often incomplete, fragmented or unavailable at the point of care and it is difficult to access patient's health information [31,32]. The integration complexity lies mainly in the lack of access outside a specific healthcare environment and in the use of incompatible and proprietary software and hardware. Although there are some standards (e.g., Health Level Seven (HL7) Fast Healthcare Interoperability Resources (FHIR), HL7 Clinical Document Architecture (CDA), ISO13606, openEHR, CDISC Operational Data Model (ODM)) for exchanging data between trusted parties, their implementation requires data mapping and additional interface adaptations [32].

A promising alternative to solve the aforementioned problems consists in using Distributed Ledger Technologies (DLTs). Examples of DLT platforms currently in use are Ethereum [33], Hyperledger Fabric [34] or IOTA [35]. DLTs, specifically blockchain, are predicted to be key technologies within the Industry 4.0 era, since they guarantee the exchange of information between different stakeholders and interested parties that do not necessarily trust each other [36]. Moreover, blockchain holds the promise of enhanced data transparency, trustworthiness, immutability, privacy and security. Furthermore, it enables Peer-to-Peer (P2P) transactions, decentralized Apps (DApps), operational efficiency and a high degree of automation thanks to the use of smart contracts that execute code autonomously [37].

There are recent reviews on the usage of blockchain to enhance the healthcare sector [31,32,38–42]. Such works essentially analyze promising technologies, potential applications and discuss potential challenges for their further adoption. Currently, in the literature there are only a couple of recent preliminary works that use a blockchain for diabetes research and care. An example is [43], where the authors give an overview of a blockchain-based architecture that implements data and access management. An implementation of an Ethereum IoT platform architecture to take care and monitor DM patients can be found in [44].

Finally, it is worth noting that, with the rise of IoT [45,46], crowdsourcing is gaining momentum in a wide range of sectors and tasks [47–51]. In healthcare, crowdsourcing has been mainly employed to accomplish problem solving, data processing, surveillance/monitoring, and surveying [52], but there are not many mobile crowdsourcing applications [11,53] and even less decentralized while focused on healthcare [48]. In health crowdsourcing, engagement is essential since it can transform users from mere passive recipients of information to active participants in a collaborative community, raising awareness for diseases like diabetes and helping to improve their own health as well as the health of those around them [54]. In addition, there are incentive mechanisms that enable community participation [48]. For example, the authors of [55] proposed an incentive mechanism to encourage hospitals to share high-quality data, which can then be aggregated to generate prediction models with higher accuracy rates.

3. Design of the System

3.1. Communications Architecture

The communications architecture proposed for the solution detailed in this article is illustrated in Figure 1 (on the left). As it can be observed at the bottom of the Figure, CGMs are responsible for collecting glucose concentration values from remote patients. Such values are then read by a smartphone, which periodically scans the available wireless networks looking for neighboring fog computing gateways. Thus, the smartphone may be in one of the two following scenarios:

- In an area covered by a fog computing gateway. In this scenario the smartphone joins the network created by the fog gateway and sends to it the data collected from the CGM sensor. This may be the case of homes and certain buildings (e.g., hospitals, nursing homes), where fog gateway infrastructure can be easily deployed and managed.
- Out of range of the deployed fog computing gateways. Patients may move to certain areas where fog gateways are not in range (e.g., outdoors, in certain areas of a home/building). In such scenarios the smartphone should detect the lack of fog gateway connectivity and send the collected data directly to remote servers or services on the Internet.

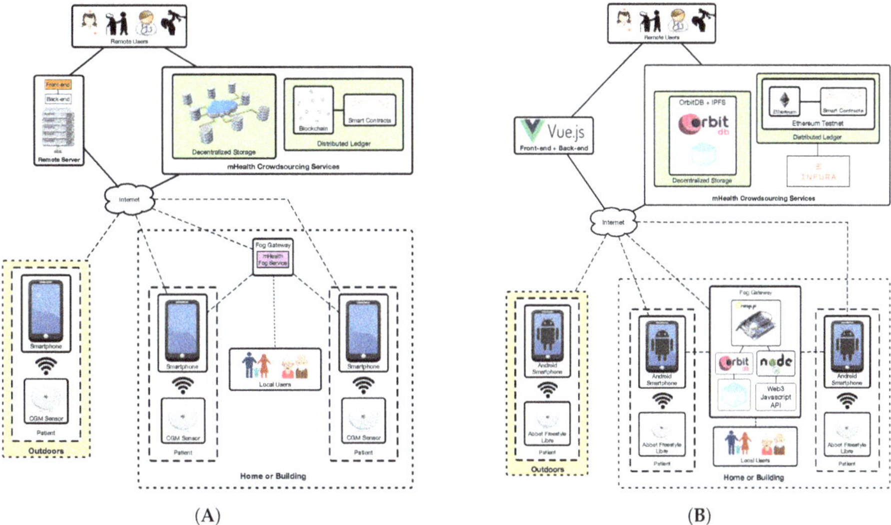

(A) (B)

Figure 1. Proposed communications architecture (**A**), and implemented architecture (**B**).

In the case of making use of a fog gateway, an mHealth fog service will be run on it. Such a service receives the sensor data from the smartphone and decides whether it is necessary to take a specific action. For instance, if the mHealth fog service detects a down trend on the glucose levels that may lead to hypoglycemia, it can warn the user or his/her caretakers (note that in Figure 1 it is considered the existence of users like relatives, doctors or nurses that may be near the patient (local users) or in remote places (remote users), which may be responsible for looking after the patient).

Besides providing fast responses to users, the mHealth fog service is also able to send the collected information to a remote server, to a decentralized storage system or to a decentralized ledger:

- Remote server: it is essentially a front-end that provides a web interface to remote users in order to allow them to access the stored information in a user-friendly way. This server also runs a back-end service that is responsible for sending notifications to remote users through SMS or instant messaging services.
- Decentralized storage system: it is able to provide redundancy, cyber-attack protection and DApp support to the stored information. This storage system is able to replicate the collected information and distribute it automatically among multiple nodes. In this way, although one storage node is not available (e.g., due to maintenance, to a malfunction or to a cyber-attack), the information can be accessed through other nodes.
- Decentralized ledger: it is included to provide transparency and decentralization, and to enhance data authenticity and security. Specifically, a blockchain can be used to preserve user anonymity and transaction privacy, so that patient data can remain private to non-authorized parties [56]. Moreover, transparency is provided to the authorized third-parties that access the stored information, which allows them to monitor and analyze it in order to determine whether it has been tampered [57]. Furthermore, a blockchain provides decentralization by distributing the data among peers and thus avoiding the involvement of middlemen [58]. Regarding data authenticity, a blockchain can be used to provide accountability, which is essential to trust the information collected from the patients. With respect to smart contracts [37], they are able to automate event triggering by monitoring when certain requirements are fulfilled. For instance, a smart contract may be used together with an oracle (i.e., an external agent that retrieves and validates certain real-world information that is later sent to a blockchain) to detect that a user is going to run out of sensors and then automate the purchase of new ones. The designed decentralized ledger also includes a cryptocurrency-based incentive mechanism (whose functioning is detailed in the next subsection) to motivate patients to share their data in the crowdsourcing platform.

3.2. Data Crowdsourcing Incentive Mechanism

One of the objectives of the proposed system is to create the basis for a mHealth data crowdsourcing ecosystem that can be harnessed by third parties in order to improve the existing knowledge and research on DM. To achieve such an objective, it is essential to motivate the patients to upload their CGM sensor data to the decentralized storage nodes so that anyone, once authorized, can access them.

The first incentive measure to ease this data collection process is its automation, which is carried out by the devised CGM-based IoT system. Thus, after setting up the system (i.e., after installing the required fog gateways and the smartphone app), the user only has to check his/her glucose levels. The rest of the processes (e.g., data processing/uploading/distribution, notifications) do not require the user intervention.

The second incentive measure is related to patients that may not be willing to share their data publicly or for free through the proposed crowdsourcing platform. In such a case, the blockchain allows for implementing a private system where patients receive some sort of reward when they contribute with their mHealth data, so the more data they add to the system, the higher the reward.

The reward system is based on the use of a metacoin, which is a digital cryptocurrency that uses an already-deployed blockchain, but which implements additional transaction logic over such a blockchain. For this article, a metacoin called GlucoCoin (₲) was created on the Ethereum blockchain. GlucoCoin requires creating a digital wallet for every participant and implements all the software operations needed by a cryptocurrency, which essentially involve getting the balance from a user and sending an amount from one user to another.

GlucoCoin can be used not only for rewarding the users for their data, but it can be used by them to purchase items or services. For instance, as it is later described in Section 4.2, patients may use their GlucoCoins to automate the purchase of new CGM sensors when they are going to expire.

4. Implementation

4.1. Implemented Architecture

In order to implement the architecture depicted on the left of Figure 1, the components indicated in Figure 1 on the right were selected. As it can be observed at the bottom of the Figure, Android smartphones are used to read the Abbot's Freestyle Libre CGM sensor. If the patient is indoors, in an area where there is fog computing gateway deployed, the smartphone can communicate with it. Such a gateway is essentially a Single-Board Computer (SBC) that runs the required software. The smartphone can also send the collected values to other decentralized nodes on the Internet or to the Ethereum blockchain, which can be accessed by remote users. In addition, such users can access the data through a front-end that shows a user-friendly web interface. The following subsections describe in detail the characteristics, software and regular functioning of every component.

4.1.1. CGM Sensor

There are a number of companies that manufacture and distribute commercial CGMs, being Dexcom [59], Medtronic [60] and Abbot [61], the most active in the field. Their most relevant and latest CGMs are compared in Table 1 together with some popular sensor-integrated insulin pump systems that are connected to specific CGM systems (e.g., MiniMed 670G). Additional systems and their main characteristics can be found in [62].

Table 1. Comparison of the characteristics of the most popular CGMs.

Characteristics \ Models	Dexcom G4 Platinum	Dexcom G5 Mobile	Dexcom G6	Medtronic Guardian Connect	Medtronic iPro 2 Prof	Medtronic MiniMed 670G	Abbot Freestyle Libre	Abbot Freestyle Libre 2	Roche Eversense XL
Duration	7 days	7 days	10 days	6 days	6 days	7 days	14 days	14 days	180 days
Daily calibration	Yes	Yes	No (manual calibration is available if the start-up calibration code is lost)	Yes	Yes	Yes	No	No	Yes
Alarms	Yes	Yes	Yes	Yes	No	Yes	No	Yes	Yes
Price	€ 1100 (Kit with a receiver, a transmitter and 4 sensors)	€ 1270 (Kit with a receiver, a transmitter and 4 sensors)	€ 1000 (Kit with a receiver, a transmitter and 3 sensors lasting 10 days each)	It varies depending on its use (ad-hoc, partial or intensive)	Not publicly available	Transmitter: $699, sensors ranging from $50–$75, depending on the number of purchased units	€ 169.90 (Kit with one reader and two sensors)	€ 169.90 (Kit with one reader and two sensors)	€ 399 for the transmitter and € 1399 for one sensor
Data transmission	Automatic, every 5 min	Automatic, every 5 min	Automatic, every 5 min	Automatic, every 5 min	Automatic, every 5 min	Automatic, every 5 min	Manual	Manual	Automatic, every 5 min
Reading range	Max. 6 m	Max. 6 m	Max. 6 m	Max. 6 m	-	-	Max. 10 cm	-	
Remote monitoring	No	Yes (iOS/Android app)	Yes (iOS/Android app)	Yes (iOS app)	Yes (iOS/Android app)	-	Yes (Android app)	-	Yes (iOS app)
Communications	Bluetooth	Bluetooth	Bluetooth	Bluetooth	-	-	NFC	NFC, BLE	Bluetooth
MARD	13%	9%	9%	13.60%	11 %	8.7%	11.40%	9.5%	11.60%

Among the analyzed CGMs, Abbott's Freestyle Libre was chosen. Although its reading range is limited (up to 10 cm) due to the use of Near Field Communication (NFC), the sensor lasts longer than others provided by Dexcom or Medtronic (14 days versus roughly a week), it is cheaper than other solutions (e.g., a kit with one reader and sensors currently costs € 169.69), finger-pricking is not necessary on a daily basis to calibrate the sensor and it is more accurate than other systems (i.e., it has a lower Mean Absolute Relative Difference (MARD)).

Note that it would be possible to make use of the second generation of Abbot's CGM (Freestyle Libre 2), which provides better MARD, but it was not available for purchase at the beginning of the work related to this article and, in practice, it is essentially the same device.

Abbott's Freestyle Libre sensor is placed on an arm by using a single-use applicator (in Figure 2, on the left). On the right of Figure 2 it is shown the sensor adhered to the optimum location, in the back of the upper arm. After activating the sensor, it takes roughly one hour before the first readings are available. Then, according to Abbot, the sensor will collect data during a maximum of 14 days (in practice, it has been observed that the sensor keeps on collecting data during almost one day more). In addition, note that each Freestyle Libre sensor also has a commercial expiration date that limits its usage time interval.

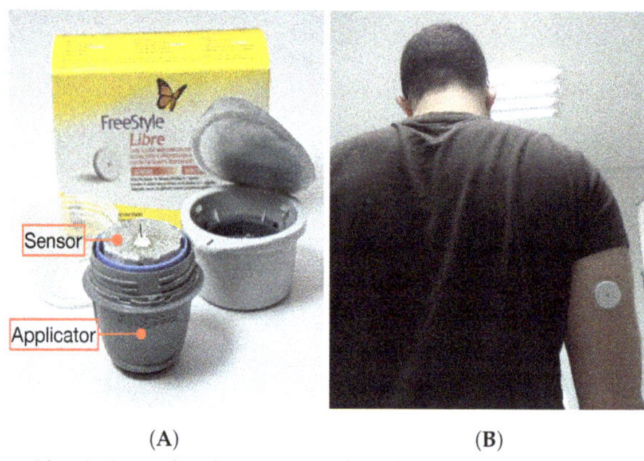

Figure 2. Abbott's Freestyle Libre sensor and applicator (**A**), and adhered sensor (**B**).

To collect readings from the sensor, the user just has to scan it with Abbot's reader or with an NFC-enabled smartphone that runs Abbot's LibreLink app. The sensor collects readings automatically every 15 min. It is necessary to scan the sensor at least once every 8 h, since the sensor stores up to 8 h of readings. Readings are stored in the internal memory of the sensor, whose structure was previously reverse engineered (at least in part) by open-source initiatives like LibreMonitor [63] or LimiTTer [64]. The most relevant parts of the sensor internal memory are illustrated in Figure 3. As it can be observed, the memory is structured into 344 bytes that are divided into a header, a footer and the main memory fields, which are the following:

- Pointer fields: bytes 27 and 28 store, respectively, the position where the last trend and historic values were written.
- Trend value fields: they are 16 fields of 6 bytes between byte 29 and byte 124 that store the latest trend measurements. Such trend measurements are obtained every minute for the last 16 min. In the position that is indicated by the Trend Pointer it is placed the current measurement, while the average of the other 15 measures is actually the last historic value. Note that the collected data are raw values that need to be calibrated after adhering sensor and initializing it with the sensor reader, therefore a researcher that develops a non-official NFC-enabled app to read the sensor should calibrate it at least once with an accurate glucose reader (or compare the collected values with the ones actually shown by Abbot's reader).

- Historic values: they are 32 values of 6 bytes that are located after the Trend values (from byte 125 to byte 316) and that represent the glucose concentration values obtained every 15 min for the last 8 h.
- Activity time: there are two bytes (317 and 318) that indicate the number of minutes since the sensor started to take measurements (i.e., since it was initialized with the official Abbot Freestyle reader), so they allow for obtaining the expiration date of the sensor.

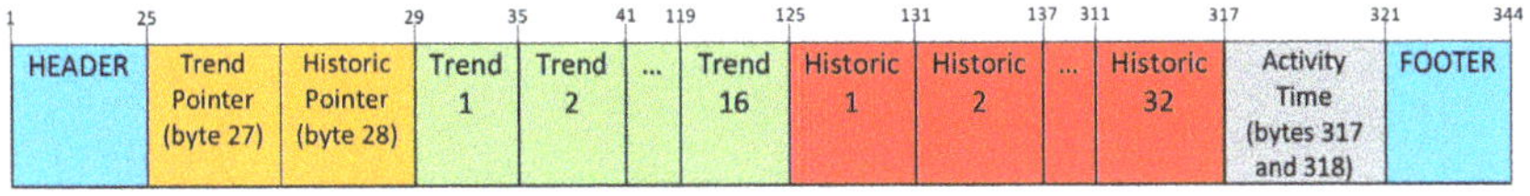

Figure 3. Structure of the internal memory (in byte) of Abbot's Freestyle Libre sensor.

4.1.2. Decentralized Storage

In order to store the collected data, the proposed system makes use of the decentralized database OrbitDB [65], which runs over InterPlanetary File System (IPFS) [66]. Specifically, OrbitDB uses IPFS for storing data and for synchronizing multiple peer databases.

OrbitDB nodes can fetch content if there are connected peers in the swarm that share the database. Obviously, in order to replicate a database, at least one node must be connected to the swarm. Nodes can use the REST API to create/open a database and then add data to them. In the system implemented for this article the database was defined as an eventlog, which creates an append-only log. Fetching and synchronizing data are automatically orchestrated by OrbitDB, so this functionality does not need to be implemented by the REST API.

During the experiments presented in this article OrbitDB nodes were running on a fog gateway and in a virtual machine on a remote cloud on the Internet. In this latter case, the virtual machine make use of two AMD Opteron 3.1 Ghz cores and 8 GB of RAM.

4.1.3. Smartphone App

The smartphone app was developed for Android, since, as of writing, nearly 80% of the market share belongs to such an operating system. Specifically, the smartphone used during the experiments for this article was an OnePlus 6T, which embeds an octa-core CPU (that runs at up to 2.8 Ghz), 6 GB of LPDDR RAM and 128 GB of internal memory.

Regarding the Android app, it was designed to provide a fast, easy and reliable way to retrieve patient's data. Figure 4, on the left, shows a screenshot of the main screen of the app, where a speedometer widget points at the current patient blood glucose concentration values (the screenshot shows real values from one of the researchers that wore the system for 14 days). The main screen also shows a graph with the evolution of the patient glucose on a daily, weekly and monthly basis and a table with the minimum, maximum and average values during the selected time period.

The smartphone app also implements the first notification layer, which is triggered after reading the values collected through NFC from the sensor and when such values are not within the predefined thresholds.

To increase the reliability of the system, data are first stored locally on the smartphone via a SQLite database, which is queried from the Android app by using Room persistence library. Then, the collected data can be saved in a fog node (in OrbitDB) using a REST Application Programming Interface (API) [67]. When the app detects that there is not a connection with a fog gateway, the REST API requests are sent to another OrbitDB nodes on the Internet (during the tests performed in this article, such nodes were in a remote cloud).

4.1.4. Back-End and Front-End

A web interface was designed and implemented to show historical values (i.e., a graph of the glucose levels) in order to ease medical supervision. Moreover, notifications are sent to the medical services whenever anomalous values of glucose are detected.

The required systems are deployed in a cloud machine that is also running an OrbitDB node and a service that exposes a REST API that obtains the data from the database, thus allowing other systems to access the decentralized data in a traditional way.

The visualization panel consists of a web interface designed as a Single Site Application using the Vue.js framework [68]. The application loads into the user's web browser and makes HTTP requests to the REST API to show patients' glucose historical data.

Along with the web application (showed in Figure 4, on the right) it is executed an event notification service. Such a service is triggered every time a user accesses the visualization interface and processes the data, thus detecting whether the glucose values are or not within certain pre-established thresholds. If the collected value is above or below such thresholds, an alert is generated and an SMS or a Telegram message [69] is sent to the medical services to warn them. For example, such messages can serve as a reminder for a nurse that a specific patient has an incoming hypoglycemia. Specifically, the notification service uses Twilio's API [70] to send SMS, while a Telegram bot was developed to warn remote users through Telegram messages. As an example, Figure 5 shows SMS and Telegram messages sent to notify hypoglycemia and hyperglycemia alerts.

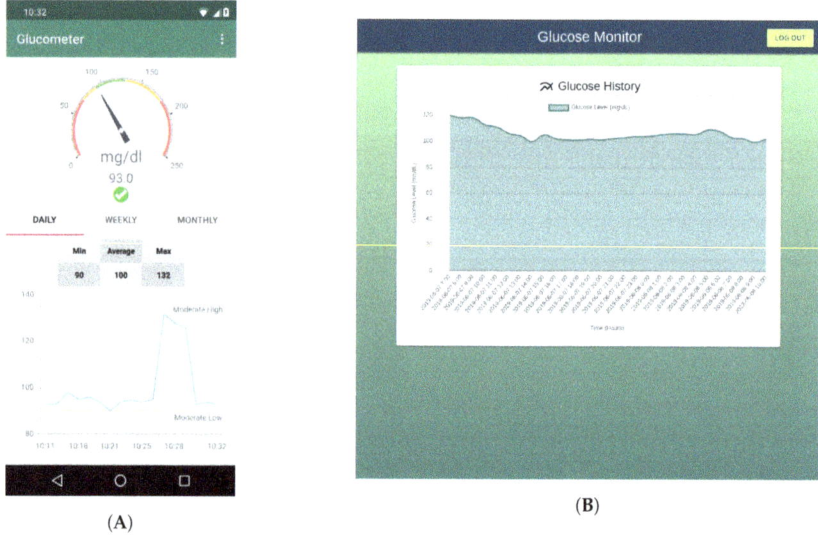

(A) (B)

Figure 4. Screenshot of the smartphone app (**A**) and of the web application (**B**).

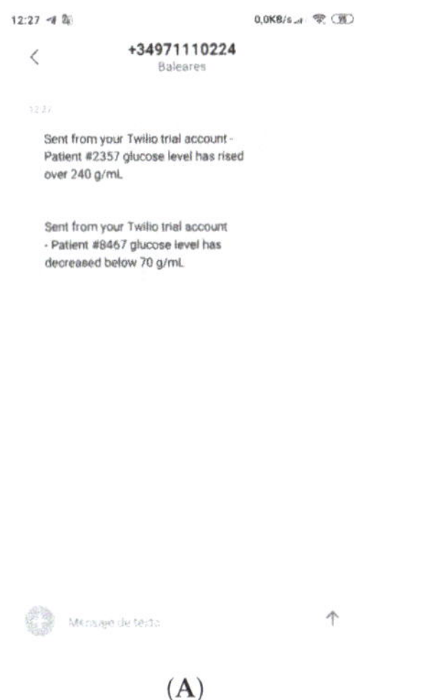

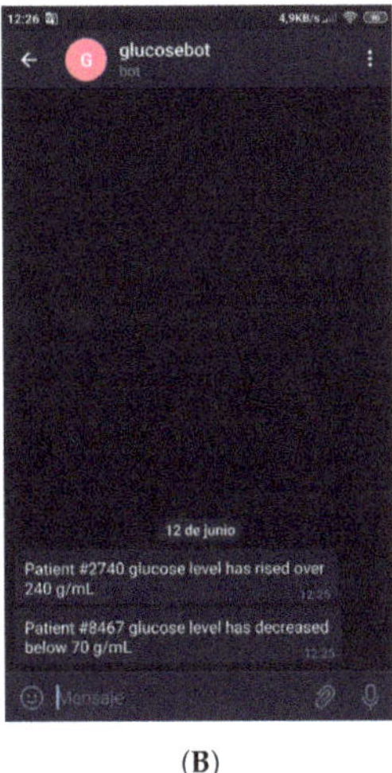

(A) (B)

Figure 5. Event notifications received through a Twilio's SMS (**A**) and a Telegram message (**B**).

4.1.5. Distributed Ledger

Ethereum was chosen as blockchain since the distributed ledger required by the proposed architecture has to be able to run smart contracts. Although the developed system can be deployed in the public Ethereum blockchain, for the experiments performed in this article, it was run on two testnets (i.e., Ethereum test networks): Rinkeby [71] and Ropsten [72]. Rinkeby and Ropsten differ in their consensus algorithm: while Rinkeby uses Proof-of-Authority (PoA) (with Clique PoA as consensus protocol), Ropsten uses Proof-of-Work (PoW). This means that, in the case of Rinkeby, there is a subset of authorized signers that assume block minting and Ether cannot be mined (it is requested through a faucet [73]). In contrast, in Ropsten Ether is mined, although it can be also requested from a faucet [74]. In practice, Rinkeby is able to create blocks on an average of 15 s, while Ropsten usually requires up to 30 s.

Smart contracts can be executed both in Rinkeby and Ropsten. For this article, the smart contracts were compiled and deployed by using Truffle [75]. The main smart contract used by the proposed system was developed to implement the meta-coin incentive mechanism detailed in Section 3.2, which initializes user account balance for the participating parties and manages the performed GlucoCoin payments. User wallets increase their balance with a fee when the owner of the wallet sends data to OrbitDB. If the user needs to buy a new glucose sensor, the account balance is decreased. The mechanism that performs the sensor purchase is implemented outside the blockchain (i.e., it is handled by an external provider). Since the expiration information is stored in OrbitDB, it can be used to trigger automatically the purchase and payment of the new sensor.

4.1.6. Fog Gateway

The fog gateway of the cture may be implemented on most current SBCs, like Raspberry Pi 3 [76], Banan Pi [77] or Odroid-XU4 [78]. Among such devices it was selected the Orange Pi Zero Plus [79], since it provides a good trade-off between features and cost (it can be currently purchased for less than $20). The Orange Pi Zero Plus features an Allwinner H2 System on Chip (SoC) that embeds a quad-core ARM Cortex-A7 microcontroller. The selected SBC also includes 256 MB of DDR RAM and includes interfaces for USB 2.0, 10/100 M Ethernet and WiFi (IEEE 802.11 b/g/n). In addition, it is worth noting that the SBC is really small (48 mm × 46 mm) and lightweight (26 g).

The fog gateway runs on ARMbian [80] and includes the following main software to implement the proposed architecture:

- OrbitDB and IPFS. Every fog gateway is actually an autonomous OrbitDB node that runs on IPFS and that synchronizes periodically with the other OrbitDB nodes deployed on the Internet.
- Web3 [81]. It is required to interact with the Ethereum blockchain. Specifically, Web3 is used as a JavaScript API to exchange requests with Infura [82], which provides an easy-to-use HTTP API that allows for accessing the Ethereum blockchain even from resource-constraint IoT devices.
- Node.js [83]. It is needed for executing JavaScript code outside of a browser, thus easing the interaction with Web3 and OrbitDB. In the case of Web3, in practice, the proposed implementation runs a Node.js instance that performs calls to the Web3 API in order to exchange requests with Infura, which interacts with Ethereum.

Finally, it is worth pointing out that every fog gateway makes use of two WiFi interfaces simultaneously. One of them is the embedded interface, while an external WiFi adapter is plugged into the Orange Pi Zero Plus to provide the second one. The reason for the use of both interfaces is the need for connecting wirelessly to the Internet through one interface, while the second one is setup in Access Point mode to broadcast a Service Set IDentifier (SSID) and a MAC address that can be identified easily by the smartphones of patients in order to connect to them fast.

4.2. Regular Functioning of the System

Three common use cases may arise during the regular operation of the system. The first one is illustrated in Figure 6 and represents the insertion process that is performed with the collected data. It must be noted that, before starting to use the system, it needs to be initialized: the decentralized storage nodes require to be synchronized through IPFS (steps 0A and 0B in Figure 6), the used smart contracts need to be compiled and deployed in a Ethereum testnet (step 0C), the GlucoCoin wallets have to be initialized (step 0D) and the CGM sensor has to be placed on the patient's arm. Moreover, the developed mobile app has to be installed on the patient's smartphone and the required Android permissions need to be granted so that the app can make use of the WiFi interface to detect and connect to the available fog gateways.

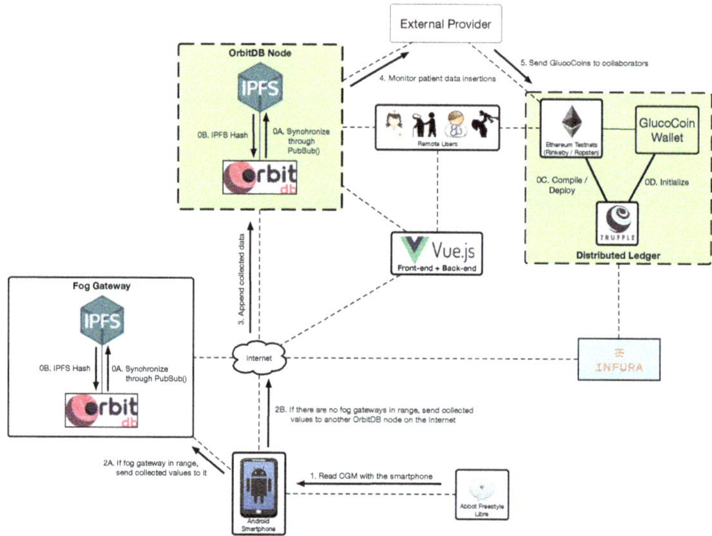

Figure 6. Insertion process for the data collected from the CGM.

Once the system is deployed and configured, the patient can read the sensor by approaching the smartphone (step 1) (note that there are amateur [63,64] and commercial [84] devices that make use of arm bands to automate this reading process). If there is a fog gateway in range, the collected glucose values are sent to it, which stores them in OrbitDB (step 2A). If there are no fog gateways in range, the smartphone sends the data to an OrbitDB node on the Internet (step 2B), which appends them to its database (step 3). Since the stored data are decentralized and therefore stored on synchronized OrbitDB nodes, an external authorized provider can access such data (step 4) and reward the patient with a number of GlucoCoins that is proportionate to his/her data contribution (step 5).

A second relevant use case is illustrated in Figure 7 and is related to the notifications performed by the system when a dangerous situation is detected (e.g., when detecting an incoming hypoglycemia). In such a case, after initializing the system (steps 0A and 0B), glucose concentration values can be read (step 1) and three notification layers may warn the user:

- Local warnings. For simple analyses, a smartphone is powerful enough to process the collected data and warn the user about a dangerous situation. Since the sensor data can be stored in the local database, the smartphone app can read such an information and then show notifications to the user (step 2).
- Fog warnings. If there is a fog gateway in range, the smartphone will send the collected values to it (step 3A) in order to store them in OrbitDB. The collected data can be accessed by the internal mHealth fog service, which can warn the patient (step 4A) and the local users (step 5A) in case of detecting a dangerous situation.
- Cloud warnings. If the smartphone detects no fog gateways, it sends the collected data to another OrbitDB node on the Internet (step 3B), which stores them (step 4B). Such data are retrieved periodically by the back-end (step 5B) in order to essentially show them to remote users through a web interface. In addition, the back-end processes the collected data (step 6B) and, if a dangerous situation is detected, it can warn local and remote users (steps 7B and 8B).

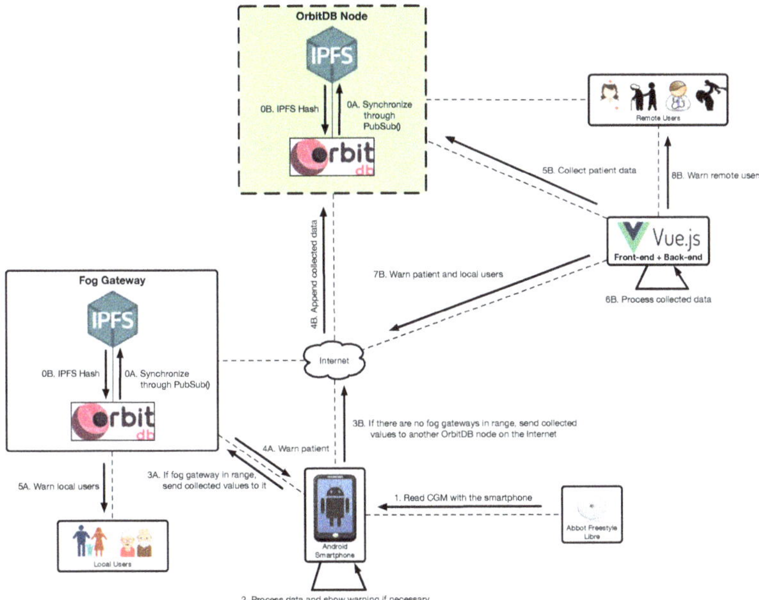

Figure 7. Patient and user notification processes.

The third relevant use case is depicted in Figure 8 and illustrates an example of smart contract execution. Specifically, this third use case refers to how a new sensor can be purchased automatically by relying on the data provided by the CGM sensor. Such a process begins after the initialization of the system (steps 0A, 0B, 0C and 0D) and requires reading the expiration date (actually, the activity time) of the sensor (step 1) and storing it either on the local fog gateway (step 2A) or on a remote OrbitDB node (steps 2B and 3). The stored data are monitored by an oracle (step 4) that feeds a smart contract (step 5) that decides whether it is time to perform the purchase of a new sensor. When such a time comes, the smart contract is executed: it is checked the patient's GlucoCoin wallet (step 6) and, if there are enough GlucoCoins, the purchase is performed (step 7). Once the sensor provider confirms the purchase (step 8), the price of the new sensor in GlucoCoins can be withdrawn from the user wallet (step 9) and the provider can send (for instance, by courier) the CGM sensor to the patient (step 10).

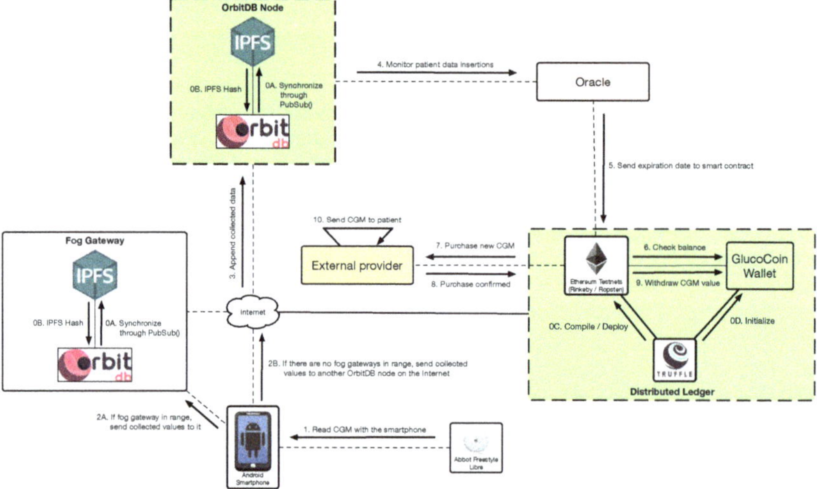

Figure 8. Process to automatically purchase CGMs due to the imminent expiration of the current one.

5. Experiments

In order to evaluate the proposed architecture, several experiments were designed to provide insightful results about three different aspects: (1) the throughput of the fog and cloud architecture, (2) the performance of the decentralized database, and (3) the performance of the blockchain.

5.1. Baseline Performance of Fog and Cloud Nodes

Before studying the performance of the proposed architecture in detail, a first preliminary test was carried out for establishing a performance baseline for the fog and cloud nodes, since their constrained hardware (in the case of fog nodes) and remote connectivity (in the case of the cloud nodes) may limit the overall performance of the proposed system. Thus, an Orange Pi Zero Plus and a cloud node were evaluated when only running a Node.js server. Specifically, each test performed 5000 connections on a fog node and 1000 on a remote node (to avoid network congestion issues) at different connection rates in order to determine the node maximum throughput up to the point where connection errors arise.

The obtained results are shown in Figure 9, where it is represented the desired request rate (the one imposed by the tests) versus the rate actually achieved during the tests. As it can be observed in the Figure, the desired and real rates are roughly the same up to a point when the node is not able to handle the requests and thus its performance decreases. Specifically, the fog node reaches its peak performance at 300 requests per second, while this point is at 200 requests per second for the cloud node. In the case of the fog node, this is due to its hardware constraints, while, in the case of the cloud node, it is related to the restrictions of its network (i.e., the load of the network and the characteristics of the devices involved in processing and routing the requests through the Internet).

In any case, these results are a useful reference when evaluating the performance analyses described in the next subsections.

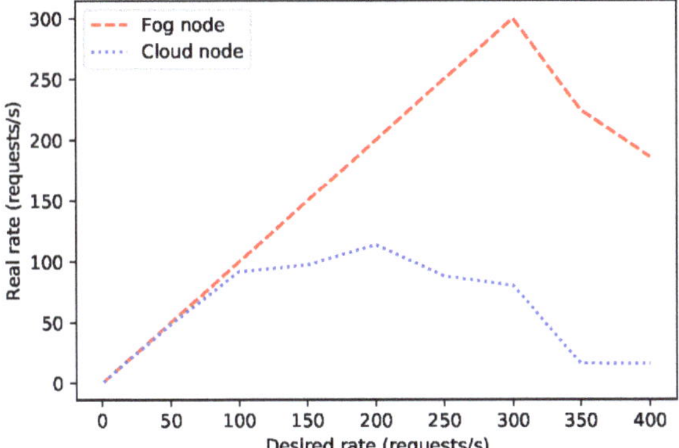

Figure 9. Desired/real request rate for 5000 connections on the fog node and 1000 on the cloud node.

5.2. Performance of the Decentralized Database

In order to estimate the throughput of the selected decentralized database, OrbitDB nodes were deployed locally (in fog nodes) and remotely (in a cloud). In such scenarios it was measured the average time required by an OrbitDB node for processing each REST API request. For the sake of fairness, tests were performed for four different payload sizes to evaluate their effect on network delay (for each payload size, the time required for processing 1000 requests was averaged).

The obtained results are shown in Figures 10 and 11. As it can be expected, the larger the payload, the larger the time response and the lower the request rate. Moreover, it can be observed that the cloud-based OrbitDB node is clearly slower in spite of being more powerful than the fog node. In the worst evaluated case (i.e., for the cloud and a 4 KB payload), the decentralized node, although it is only able to handle two requests per second, it is actually really fast, since it is able to process and to respond to each of the requests in less than half of a second, which seems to be quick enough for most glucose monitoring applications.

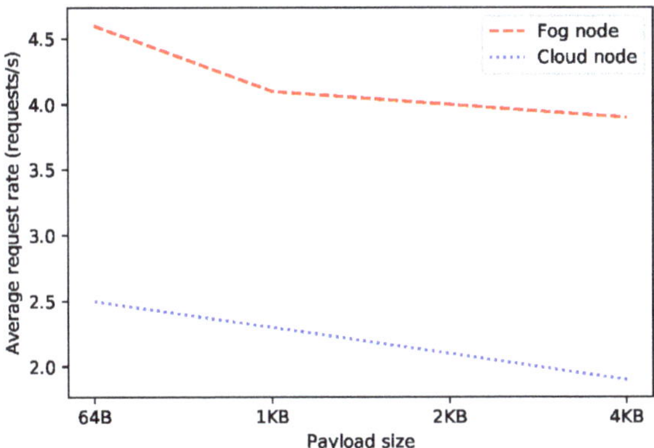

Figure 10. Average request rate for OrbitDB when running on fog and cloud nodes.

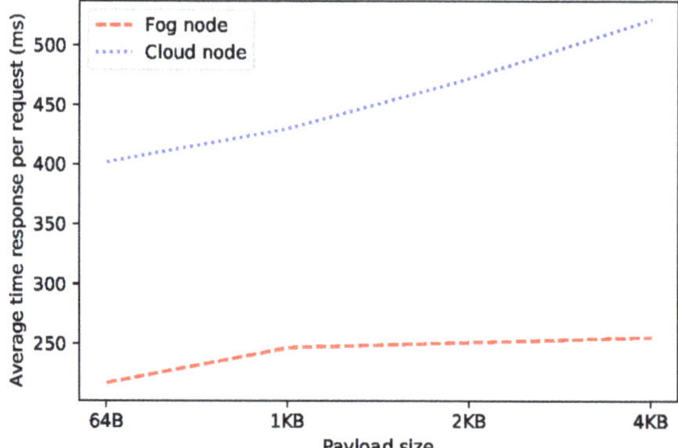

Figure 11. Average response time of OrbitDB when running it on fog and cloud nodes.

It is also worth noting that the difference in time response between the fog node and the cloud increases as payload size gets larger, mainly due to the communications transmission time through the network (i.e., although the processing time required by the OrbitDB node remains constant, the time required to exchange the request payload increases). In fact, during the performed experiments, the average round trip time (calculated using the 'ping' command in Linux, which makes use of 56-byte packets) for the fog node was 0.859 ms, while the same for the cloud was 45.920 ms, which makes a significant difference.

Figures 12 and 13 show the results of a second test, which measured the performance of the decentralized database when carrying out 2000 consecutive write operations on an OrbitDB node that ran on a fog and on a cloud node. For the sake of fairness, the data of the Figures were obtained by using the official OrbitDB benchmarking scripts [85], which obtain the average throughput of the OrbitDB node every 10 s.

While in Figure 12 it can be observed that the fog OrbitDB node throughput oscillates between 3.7 and 4.5 write requests per second, the cloud node throughput shown in Figure 13 can reach between 3.5 and 6 requests per second. This means that the fog node responds faster than the cloud node, but its hardware is less powerful than the one used by the cloud, so it is not able to process as many requests per second. However, it must be noted that both in the fog and in the cloud scenarios the average throughput oscillates continuously due to different factors, like the computational and network load (i.e., the network is actually shared with other users).

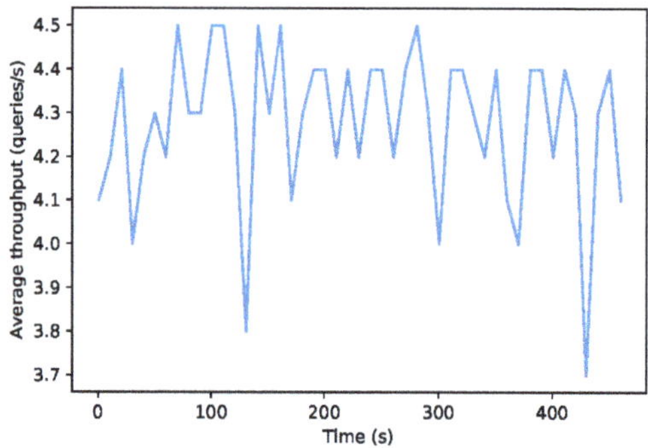

Figure 12. Performance of the fog OrbitDB node executing 2000 queries.

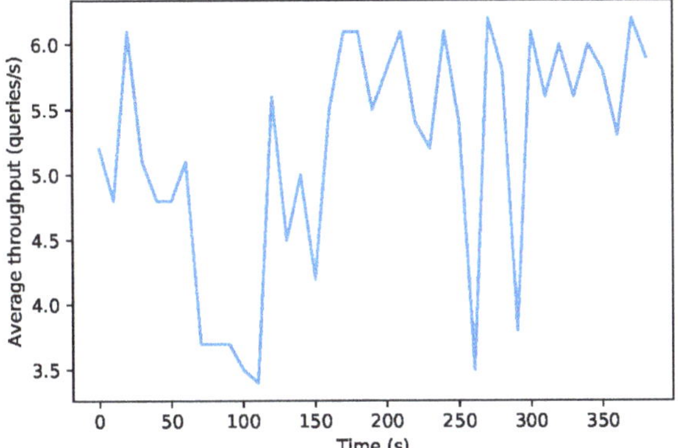

Figure 13. Performance of the cloud OrbitDB node executing 2000 queries.

A third test was carried out in order to determine the performance of an OrbitDB node when carrying out two main operations: when fetching content from connected peers in the swarm and when replicating content in other connected peers (these both operations are automatically orchestrated by OrbitDB). Figure 14 shows the average throughput for every 10 s when performing 2000 fetching and replication operations (from a fog node to a cloud node). Despite the observed oscillations, it can be concluded that the average fetching throughput is slower than the one related to a replication due to the complexity of this latter operation.

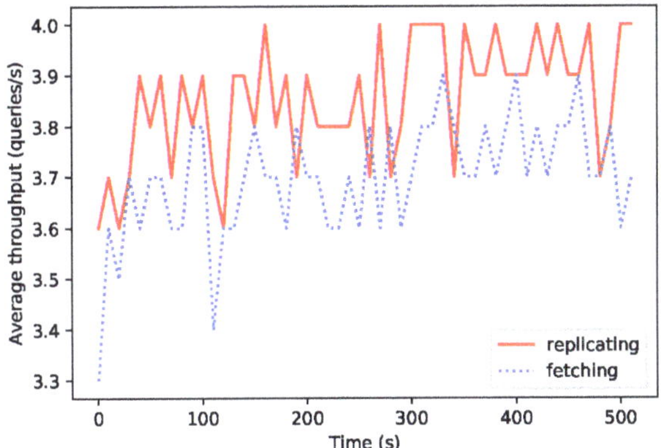

Figure 14. Performance in OrbitDB fetching and replication operations.

5.3. Time Response of Event Warnings: Local, Fog and Cloud

As it was previously indicated in Section 4.2, when a dangerous situation for the patient is detected, event notifications are issued. Depending on the target user, it can be distinguished among local, fog and cloud warnings. In order to measure the delay of these warnings, it can be averaged a number of significant requests (i.e., 500 requests for the fog node, 100 for the cloud and 20 for Android) in the three aforementioned scenarios.

In the first column of Figure 15 it is represented the average time required by the tested Android smartphone app to read the information from the local database and show a warning to the patient. As it can be observed, such a time is really small: only 11.9 ms.

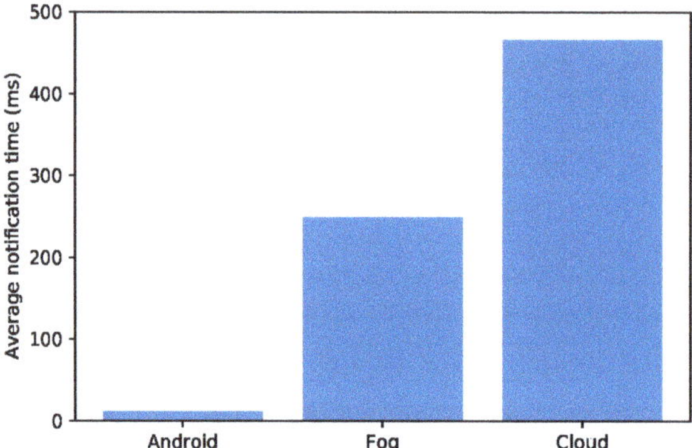

Figure 15. Average notification times in different scenarios.

The second column of Figure 15 represents the average notification time of a fog node when sending a warning. Such a time is exactly 249.5 ms, and includes the time required to send a request, to process it (thus interacting with OrbitDB) and to receive the notification from the fog gateway. The third column of Figure 15 shows the longest average notification time, which is obtained when performing the same tasks previously indicated for the fog gateway, but for an OrbitDB node that ran on the cloud. In this case, the average response time was 465.8 ms.

These results for the three selected scenarios may be used when developing applications that have to comply with strict latency requirements related to critical reaction times that might be needed to warn or to send certain information to DM patients or to health devices carried by them.

5.4. Blockchain Performance: Smart Contracts Execution Time

In order to detect potential bottlenecks in the proposed architecture, an additional experiment was performed to measure the response latency of the blockchain. This is specially interesting in the case of using Ropsten testnet, where the time to mint a block may vary noticeably as it uses PoW as consensus protocol.

Figure 16 shows the response latency of 1000 transactions in the Ropsten testnet during the execution of the smart contract related to the management of GlucoCoin (its main code is shown in Listing 1).

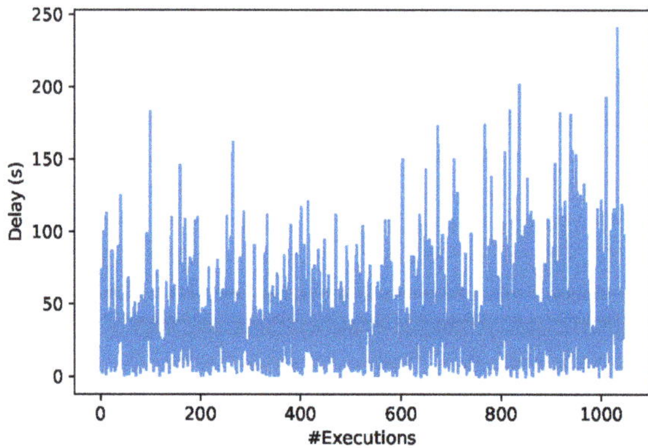

Figure 16. Ropsten testnet time response when executing smart contracts.

In Figure 16 it can be observed that the smart contract execution time oscillates continuously mainly due to the number of available miners (i.e., participants that execute the smart contract) and their computational load. Specifically, the obtained blockchain response times vary significantly from less than 20 s up to more than 240 s. Nonetheless, the average blockchain time response is 36.47 s, which is not very different from the usual average time provided by Ropsten (roughly 30 s), although the standard deviation of the performed measurements was 33.96 s. These response times can be considered normal in a blockchain based on a PoW consensus mechanism, since it includes the time for sending the transaction request, the time required by the consensus protocol to select the miner that will execute the smart contract and the time needed for executing the smart contract. As a reference, it can be indicated that, in a PoW-based cryptocurrency like Bitcoin [86], the transaction execution time in June 2019 was 9.47 min [87]. In any case, it is worth pointing out that execution delay obtained in Ropsten is noticeably larger than the time response required by OrbitDB nodes, so it will have to be taken into consideration in cases where response latency is essential.

It is worth pointing out that, thanks to the use of a blockchain, any reader can check the obtained results by browsing the wallet used for the experiments on the Ropsten testnet [88]. In addition, Figure 17 shows a screenshot of the Ethereum wallet that includes the details of some of the transactions performed on the testnet.

Listing 1: Main code of the smart contract that controls GlucoCoin.

```
pragma solidity >=0.4.25 <0.6.0;

contract GlucoCoin {
    mapping (address => uint) balances;

    event Transfer(address indexed _from, address indexed _to, uint256 _value);

    constructor() public {
            balances[tx.origin] = 10000;
    }

    function sendCoin(address receiver, uint amount) public returns(bool sufficient) {
            if (balances[msg.sender] < amount) return false;
            balances[msg.sender] -= amount;
            balances[receiver] += amount;
            emit Transfer(msg.sender, receiver, amount);
            return true;
    }

    function getBalanceInEth(address addr) public view returns(uint) {
            return ConvertLib.convert(getBalance(addr),2);
    }

    function getBalance(address addr) public view returns(uint) {
            return balances[addr];
    }
}
```

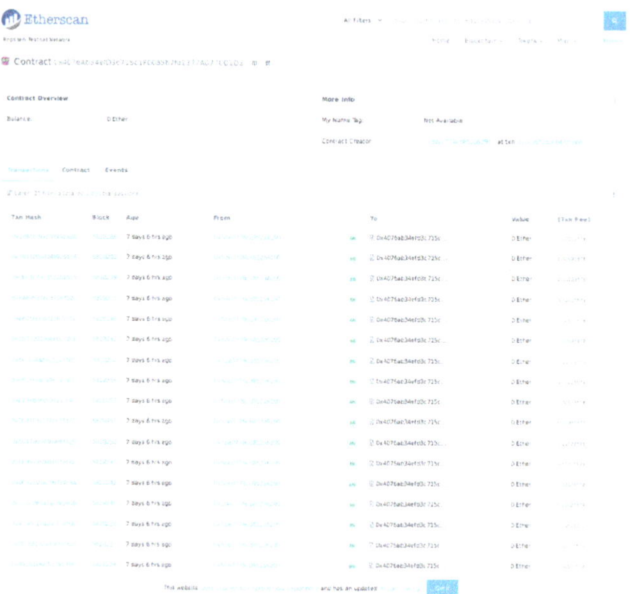

Figure 17. Ropsten testnet main view of the transactions related to the deployed smart contract.

6. Future Work

Although the tests performed on the proposed system showed its practical feasibility, there are several aspects that can be further improved:

- Automatic readings. Currently, patients have to approach their NFC-enabled smartphone to the CGM sensor to read it. Future work should be performed to design and implement solutions that automate such a process or to adapt already available amateur [63,64] and commercial [84] devices.
- Optimization of the hardware and software baseline performance. Although the obtained amounts of requests per second are high enough for the fog gateway (300 requests per second), the cloud should be improved in order to go above the achieved 200 requests per second.
- It should be further analyzed possible enhancements on the decentralized database performance, since in the experiments it was only possible to reach up to 6 write requests per second, which can be too restrictive in some practical scenarios.
- The response time of the blockchain can be improved by using faster consensus mechanisms or other new improvements for Ethereum, like sharding [89], Raiden [90] or Plasma [91].
- It would be ideal to perform trials on large sets of population and analyze the real performance of the system, the behavior of the patients regarding the incentive mechanisms and the usefulness of the collected data.

7. Conclusions

This article presented the design, implementation and evaluation of an IoT CGM-based system for mobile crowdsourcing diabetes research and care. Such a system is able to collect blood glucose levels from CGMs that can be accessed remotely. Thus, the system allows for monitoring patients and warn them in real-time in case a dangerous situation is detected. In order to create the proposed system, a fog computing system based on distributed mobile smartphones was devised to collect data from the CGMs and send them to a remote cloud and/or to a blockchain. Thanks to the blockchain and the proposed CGM-based system it is possible to provide a transparent and trustworthy blood glucose data source from a population in a rapid, flexible, scalable and low-cost way. Such mobile crowdsourced data can enable novel mHealth applications for diagnosis, patient monitoring or even public health actions that may help to advance in the control of the disease and raise global awareness on the increasing prevalence of diabetes. Furthermore, the performance of the proposed blockchain-based architecture, the decentralized database and the smart contracts were evaluated in diverse scenarios that take into account the latency demands that may be required by different potential stakeholders (e.g., patients, doctors, caretakers, scientists, government, insurance companies, governments, app developers) of the healthcare ecosystem.

Author Contributions: T.M.F.-C. and P.F.-L. conceived and designed the system; I.F.-M., T.M.F.-C., O.B.-N. and P.F.-L. performed the experiments; T.M.F.-C. and P.F.-L. tested the Freestyle Libre CGM sensor; P.F.-L. and T.M.F.-C. wrote the paper. Project administration, T.M.F.-C., and P.F.-L.; funding acquisition, T.M.F.-C., and P.F.-L.

Funding: This work has been funded by the Xunta de Galicia (ED431C 2016-045, ED431G/01), the Agencia Estatal de Investigación of Spain (TEC2016-75067-C4-1-R) and ERDF funds of the EU (AEI/FEDER, UE).

Conflicts of Interest: The authors declare no conflict of interest.

References

1. World Health Organization. Diabetes: Key Facts. Available online: http://www.who.int/news-room/fact-sheets/detail/diabetes (accessed on 31 May 2019)
2. Benjamin, E.M. Self-Monitoring of Blood Glucose: The Basics *Clin. Diabetes* **2002**, *20*, 45–47. [CrossRef]
3. Torres, I.; Baena, M.G.; Cayon, M.; Ortego-Rojo, J.; Aguilar-Diosdado, M. Use of Sensors in the Treatment and Follow-up of Patients with Diabetes Mellitus. *Sensors* **2010**, *10*, 8. [CrossRef] [PubMed]
4. Schmelzeisen-Redeker, G.; Schoemaker, M.; Kirchsteiger, H.; Freckmann, G.; Heinemann, L.; Del Re, L. Time Delay of CGM Sensors: Relevance, Causes, and Countermeasures. *J. Diabetes Sci. Technol.* **2015**, *9*, 1006–1015.
5. Fraga-Lamas, P.; Fernández-Caramés, T.M.; Noceda-Davila, D.; Díaz-Bouza, M.; Vilar-Montesinos, M.; Pena-Agras, J.D.; Castedo, L. Enabling automatic event detection for the pipe workshop of the shipyard 4.0. In Proceedings of the 2017 56th FITCE Congress, Madrid, Spain, 14–16 September 2017; pp. 20–27.
6. Markakis, E.K.; Karras, K.; Zotos, N.; Sideris, A.; Moysiadis, T.; Corsaro, A.; Alexiou, G.; Skianis, C.; Mastorakis, G.; Mavromoustakis, C.X.; et al. EXEGESIS: Extreme Edge Resource Harvesting for a Virtualized Fog Environment. *IEEE Commun. Mag.* **2017**, *55*, 7. [CrossRef]
7. Bonomi, F.; Milito, R.; Zhu, J.; Addepalli, S. Fog Computing and its Role in the Internet of Things. In Proceedings of the First Edition of the MCC Workshop on Mobile Cloud Computing, Helsinki, Finland, 17 August 2012; pp. 13–16.
8. Fernández-Caramés, T.M.; Fraga-Lamas, P.; Suárez-Albela, M.; Díaz-Bouza, M.A. A Fog Computing Based Cyber-Physical System for the Automation of Pipe-Related Tasks in the Industry 4.0 Shipyard. *Sensors* **2018**, *18*, 1961. [CrossRef] [PubMed]
9. Suárez-Albela, M.; Fernández-Caramés, T.M.; Fraga-Lamas, P.; Castedo, L. A Practical Evaluation of a High-Security Energy-Efficient Gateway for IoT Fog Computing Applications. *Sensors* **2017**, *9*, 1978. [CrossRef]
10. Sen, K.C.; Ghosh, K. Designing Effective Crowdsourcing Systems for the Healthcare Industry. In *Social Entrepreneurship: Concepts, Methodologies, Tools, and Applications*; IGI Global: Hershey, PA, USA, 2019; pp. 590–594.
11. Brabham, D.C.; Ribisl, K.M.; Kirchner, T.R.; Bernhardt, J.M. Crowdsourcing Applications for Public Health *Am. J. Prev. Med.* **2014**, *46*, 179–187.
12. Sparacino, G.; Facchinetti, A.; Cobelli, C. 'Smart' Continuous Glucose Monitoring Sensors: On-Line Signal Processing Issues. *Sensors* **2010**, *10*, 6751-6772. [CrossRef]
13. Facchinetti, A. Continuous Glucose Monitoring Sensors: Past, Present and Future Algorithmic Challenges. *Sensors* **2016**, *16*, 2093. [CrossRef]
14. Sparacino, G.; Zanon, M.; Facchinetti, A.; Zecchin, C.; Maran, A.; Cobelli, C. Italian Contributions to the Development of Continuous Glucose Monitoring Sensors for Diabetes Management. *Sensors* **2012**, *12*, 13753–13780. [CrossRef]

15. Cappon, G.; Acciaroli, G.; Vettoretti, M.; Facchinetti, A.; Sparacino, G. Wearable Continuous Glucose Monitoring Sensors: A Revolution in Diabetes Treatment. *Electronics* **2017**, *6*, 65. [CrossRef]
16. Chen, C.; Zhao, X.-L.; Li, Z.-H.; Zhu, Z.-G.; Qian, S.-H.; Flewitt, A.J. Current and Emerging Technology for Continuous Glucose Monitoring. *Sensors* **2017**, *17*, 182. [CrossRef]
17. Rosetti, P.; Bondia, J.; Vehí, J.; Fanelli, C.G. Estimating Plasma Glucose from Interstitial Glucose: The Issue of Calibration Algorithms in Commercial Continuous Glucose Monitoring Devices. *Sensors* **2010**, *10*, 10936–10952. [CrossRef]
18. Howsmon, D.P.; Cameron, F.; Baysal, N.; Ly, T.T.; Forlenza, G.P.; Maahs, D.M.; Buckingham, B.A.; Hahn, J.; Bequette, B.W. Continuous Glucose Monitoring Enables the Detection of Losses in Infusion Set Actuation (LISAs). *Sensors* **2017**, *17*, 161. [CrossRef]
19. Ding, S.; Schumacher, M. Sensor Monitoring of Physical Activity to Improve Glucose Management in Diabetic Patients: A Review. *Sensors* **2016**, *16*, 589. [CrossRef]
20. Dasanayake, I.S.; Bevier, W.C.; Castorino, K.; Pinsker, J.E.; Seborg, D.E.; Doyle, F.J.; Dassau, E. Early Detection of Physical Activity for People With Type 1 Diabetes Mellitus. *J. Diabetes Sci. Technol.* **2015**, *9*, 6. [CrossRef]
21. Peralta, G.; Iglesias-Urkia, M.; Barcelo, M.; Gomez, R.; Moran, A.; Bilbao, J. Fog computing based efficient IoT scheme for the Industry 4.0. In Proceedings of the IEEE International Workshop of Electronics, Control, Measurement, Signals and their Application to Mechatronics (ECMSM), Donostia, Spain, 24–26 May 2017.
22. Ali, O.; Shrestha, A.; Soar, J.; Wamba, S.F. Cloud computing-enabled healthcare opportunities, issues, and applications: A systematic review. *Int. J. Inf. Manag.* **2018**, *43*, 146–158. [CrossRef]
23. Froiz-Míguez, I.; Fernández-Caramés, T.M.; Fraga-Lamas, P.; Castedo, L. Design, Implementation and Practical Evaluation of an IoT Home Automation System for Fog Computing Applications Based on MQTT and ZigBee-WiFi Sensor Nodes. *Sensors* **2018**, *18*, 2660. [CrossRef]
24. Kraemer, F.A.; Braten, A.E.; Tamkittikhun, N.; Palma, D. Fog Computing in Healthcare—A Review and Discussion. *IEEE Access* **2017**, *5*, 9206–9222. [CrossRef]
25. Wac, K.; Bargh, M.S.; Van Beijnum, B.j.F.; Bults, R.G.A.; Pawar, P.; Peddemors, A. Power- and delay-awareness of health telemonitoring services: The mobihealth system case study. *IEEE J. Sel. Areas Commun.* **2009**, *27*, 4. [CrossRef]
26. Monteiro, A.; Dubey, H.; Mahler, L.; Yang, Q.; Mankodiya, K. Fit: A Fog Computing Device for Speech Tele-Treatments. In Proceedings of the IEEE International Conference on Smart Computing, St. Louis, MO, USA, 18–20 May 2016.
27. Fernández-Caramés, T.M.; Fraga-Lamas, P. Towards The Internet of Smart Clothing: A Review on IoT Wearables and Garments for Creating Intelligent Connected E-Textiles. *Electronics* **2018**, *7*, 405. [CrossRef]
28. López, G.; Custodiod, V.; Moren, J.I. LOBIN: E-Textile and Wireless-Sensor-Network-Based Platform for Healthcare Monitoring in Future Hospital Environments. *IEEE Trans. Inf. Technol. Biomed.* **2016**, *14*, 6. [CrossRef]
29. Verma, P.; Sood, S.K. Fog Assisted-IoT Enabled Patient Health Monitoring in Smart Homes. *IEEE Int. Things J.* **2018**, *5*, 3. [CrossRef]
30. Sood, S.K.; Mahajan, I. A Fog-Based Healthcare Framework for Chikungunya. *IEEE Int. Things J.* **2018**, *5*, 2. [CrossRef]
31. Khezr, S.; Moniruzzaman, M.; Yassine, A.; Benlamri, R. Blockchain Technology in Healthcare: A Comprehensive Review and Directions for Future Research. *Appl. Sci.* **2019**, *9*, 1736. [CrossRef]
32. Zhang, P.; Schmidt, D.C.; White, J.; Lenz, G. Blockchain technology use cases in healthcare. In *Advances in Computers*; Elsevier: Amsterdam, The Netherlands, 2018; Volume 111, pp. 1–41.
33. Ethereum Official Webpage. Available online: https://www.ethereum.org/ (accessed on 31 May 2019).
34. Hyperledger Official Webpage. Available online: https://www.hyperledger.org/ (accessed on 31 May 2019).
35. IOTA Official Webpage. Available online: https://www.iota.org/ (accessed on 31 May 2019).
36. Fu, Y.; Zhu, J. Big Production Enterprise Supply Chain Endogenous Risk Management Based on Blockchain. *IEEE Access* **2019**, *7*, 15310–15319. [CrossRef]
37. Christidis, K.; Devetsikiotis, M. Blockchains and smart contracts for the Internet of Things. *IEEE Access* **2016**, *4*, 2292–2303. [CrossRef]
38. Mettler, M. Blockchain technology in healthcare: The revolution starts here. In Proceedings of the IEEE 18th International Conference on e-Health Networking, Applications and Services (Healthcom), Munich, Germany, 14–17 September 2016.

39. Kuo, T.T.; Kim, H.E.; Ohno-Machado, L. Blockchain distributed ledger technologies for biomedical and health care applications. *J. Am. Med. Inf. Assoc.* **2017**, *24*, 1211–1220. [CrossRef]
40. Roman-Belmonte, J.M.; De la Corte-Rodriguez, H.; Rodriguez-Merchan, E.C.C.; la Corte-Rodriguez, H.; Carlos Rodriguez-Merchan, E. How Blockchain Technology Can Change Medicine. *Postgrad. Med.* **2018**, *130*, 420–427. [CrossRef]
41. Agbo, C.C.; Mahmoud, Q.H.; Eklund, J.M. Blockchain Technology in Healthcare: A Systematic Review. *Healthcare* **2019**, *7*, 56. [CrossRef]
42. Hölbl, M.; Kompara, M.; Kamišalić, A.; Nemec Zlatolas, L. A Systematic Review of the Use of Blockchain in Healthcare. *Symmetry* **2018**, *10*, 470. [CrossRef]
43. Cichosz, S.L.; Stausholm, M.N.; Kronborg, T.; Vestergaard, P.; Hejlesen, O. How to use blockchain for diabetes health care data and access management: An operational concept. *J. Diabetes Sci. Technol.* **2019**, *13*, 248–253. [CrossRef]
44. Azbeg, K.; Ouchetto, O.; Andaloussi, S.J.; Fetjah, L.; Sekkaki, A. Blockchain and IoT for Security and Privacy: A Platform for Diabetes Self-management. In Proceedings of the 2018 4th International Conference on Cloud Computing Technologies and Applications (Cloudtech), Brussels, Belgium, 26–28 November 2018; pp. 1–5.
45. Hernández-Rojas, D.L.; Fernández-Caramés, T.M.; Fraga-Lamas, P.; Escudero, C.J. Design and Practical Evaluation of a Family of Lightweight Protocols for Heterogeneous Sensing through BLE Beacons in IoT Telemetry Applications. *Sensors* **2018**, *18*, 57. [CrossRef]
46. Fernández-Caramés, T.M.; Fraga-Lamas, P. A Review on Human-Centered IoT-Connected Smart Labels for the Industry 4.0. *IEEE Access* **2018**, *6*, 25939–25957. [CrossRef]
47. An, J.; Gui, X.; Wang, Z.; Yang, J.; He, X. A Crowdsourcing Assignment Model Based on Mobile Crowd Sensing in the Internet of Things. *IEEE Int. Things J.* **2015**, *2*, 358–369. [CrossRef]
48. Phuttharak, J.; Loke, S.W. A Review of Mobile Crowdsourcing Architectures and Challenges: Toward Crowd-Empowered Internet-of-Things *IEEE Access* **2019**, *7*, 304–324.
49. Ren, Y.; Liu, W.; Wang, T.; Li, X.; Xiong, N.N.; Liu, A. A Collaboration Platform for Effective Task and Data Reporter Selection in Crowdsourcing Network. *IEEE Access* **2019**, *7*, 19238–19257. [CrossRef]
50. Guo, W.; Zhou, S.; Chen, Y; Wang, S.; Chu, X.; Niu, Z. Simultaneous Information and Energy Flow for IoT Relay Systems with Crowd Harvesting. *IEEE Commun. Mag.* **2016**, *54*, 143–149. [CrossRef]
51. Abualsaud, K.; Elfouly, T.M.; Khattab, T.; Yaacoub, E.; Sabry Ismail, L.; Ahmed, M.H.; Guizani, M. A Survey on Mobile Crowd-Sensing and Its Applications in the IoT Era. *IEEE Access* **2019**, *7*, 3855–3881. [CrossRef]
52. Ranard, B.L.; Ha, Y.P.; Meisel, Z.F.; Asch, D.A.; Hill, S.S.; Becker, L.B.; Seymour, A.K.; Merchant, R.M. Crowdsourcing—Harnessing the masses to advance health and medicine, a systematic review. *J. Gen. Int. Med.* **2014**, *29*, 187–203. [CrossRef]
53. Tucker, J.D.; Day, S.; Tang, W.; Bayus, B. Crowdsourcing in medical research: Concepts and applications. *PeerJ* **2019**, *7*, e6762. [CrossRef]
54. Freifeld, C.C.; Chunara, R.; Mekaru, S.R.; Chan, E.H.; Kass-Hout, T.; Iacucci, A.A.; Brownstein, J.S. Participatory epidemiology: Use of mobile phones for community-based health reporting. *PLoS Med.* **2010**, *7*, e100036. [CrossRef]
55. Xue, Q.; Chuah, M.C. Incentivising high quality crowdsourcing clinical data for disease prediction. In Proceedings of the Second IEEE/ACM International Conference on Connected Health: Applications, Systems and Engineering Technologies (CHASE), Philadelphia, PA, USA, 17–19 July 2017; pp. 185–194.
56. Fernández-Caramés, T.M.; Fraga-Lamas, P. A Review on the Use of Blockchain for the Internet of Things. *IEEE Access* **2018**, *6*, 32979–33001. [CrossRef]
57. Fraga-Lamas, P.; Fernández-Caramés, T.M. A Review on Blockchain Technologies for an Advanced and Cyber-Resilient Automotive Industry. *IEEE Access* **2019**, *7*, 17578–17598. [CrossRef]
58. Fernández-Caramés, T. M.; Fraga-Lamas, P. A Review on the Application of Blockchain for the Next Generation of Cybersecure Industry 4.0 Smart Factories. *IEEE Access* **2019**, *7*, 45201–45218. [CrossRef]
59. Dexcom Official Webpage. Available online: https://www.dexcom.com (accessed on 31 May 2019).
60. Medtronic Official Webpage. Available online: https://www.medtronic.com (accessed on 31 May 2019).
61. Abbott Official Webpage. Available online: https://www.abbott.com/ (accessed on 31 May 2019).

62. Agency for Quality and Accreditation in Health Care and Social Welfare (AAZ); Main Association of Austrian Social Security Institutions (HVB); The Norwegian Institute of Public Health (NIPHNO). Continuous Glucose Monitoring (CGM Real-Time) and Flash Glucose Monitoring (FGM) as Personal, Standalone Systems in Patients with Diabetes Mellitus Treated with Insulin; Joint Assessment; Zagreb: EUnetHTA; 27 July 2018; Report No.: OTJA08.Available online: https://www.eunethta.eu/wp-content/uploads/2018/07/OTJA08_CGM-real-time-and-FGM-aspersonal2c-standalone-systems-in-patients-with-diabetes-mellitus-treatedwith-insulin.pdf (accessed on 31 May 2019).
63. Libre Monitor Official GitHub Page. Available online: https://github.com/UPetersen/LibreMonitor (accessed on 31 May 2019)
64. Limitter Official GitHub Page. Available online: https://github.com/JoernL/LimiTTer (accessed on 31 May 2019).
65. OrbitDB: Peer-to-Peer Database for the Decentralized Web. Github Repository. Available online: https://github.com/orbitdb/orbit-db (accessed on 31 March 2019).
66. IPFS Official Webpage. Available online: https://ipfs.io/ (accessed on 31 March 2019).
67. OrbitDB HTTP API Official GitHub. Available online: https://github.com/orbitdb/orbit-db-http-api (accessed on 31 May 2019)
68. Vue.js Framework Official Website. Available online: https://vuejs.org/ (accessed on 31 May 2019)
69. Telegram Official Website. Available online: https://telegram.org/ (accessed on 31 May 2019)
70. Twilio Official Website. Available online: https://www.twilio.com/ (accessed on 31 May 2019)
71. Rinkeby's Official Website. Available online: https://www.rinkeby.io/ (accessed on 31 May 2019).
72. Ropsten's Official GitHub Page. Available online: https://github.com/ethereum/ropsten (accessed on 31 May 2019).
73. Rinkeby's Ether Faucet. Available online: https://faucet.rinkeby.io/ (accessed on 31 May 2019).
74. Metamask Ether Faucet. Available online: https://faucet.metamask.io/ (accessed on 31 May 2019).
75. Truffle Official Website. Available online: https://truffleframework.com (accessed on 31 May 2019).
76. Raspberry Pi 3 official Webpage. Available online: https://www.raspberrypi.org/products/ (accessed on 31 May 2019).
77. Banana Pi official Webpage. Available online: http://www.banana-pi.org/ (accessed on 31 May 2019).
78. Odroid-XU4 Official Webpage. Available online: https://www.odroid.co.uk/hardkernel-odroid-xu4 (accessed on 31 May 2019).
79. Orange Pi Zero Plus official Webpage. Available online: http://www.orangepi.org/OrangePiZeroPlus/ (accessed on 31 May 2019).
80. Armbian—Linux for ARM Development Boards Official Webpage. Available online: https://www.armbian.com/ (accessed on 31 May 2019).
81. Web3 Official GitHub Page. Available online: https://github.com/ethereum/web3.js/ (accessed on 31 May 2019).
82. Infura Official Website. Available online: https://infura.io/dashboard (accessed on 31 May 2019).
83. Node.js Official Website. Available online: https://nodejs.org/en/ (accessed on 31 May 2019).
84. Miaomiao Official GitHub Page. Available online: http://miaomiao.cool (accessed on 31 May 2019)
85. Official OrbitDB Benchmarking Scripts. Available online: https://github.com/orbitdb/orbit-db/tree/master/benchmarks (accessed on 31 May 2019)
86. Bitcoin Official Web Page. Available online: https://bitcoin.org/en/ (accessed on 15 July 2019).
87. Statista Statics on Bitcoin's Average Transaction Time. Available online: https://www.statista.com/statistics/793539/bitcoin-transaction-confirmation-time/ (accessed on 15 July 2019)
88. Wallet Related to the Blockchain Performance Tests of the Article. Available online: https://ropsten.etherscan.io/address/0x4076ab34efd3c715c1f0035b7fd1377a077cd1d3 (accessed on 27 June 2019)
89. Sharding Official Web Page. Available online: https://github.com/ethereum/wiki/wiki/Sharding-roadmap (accessed on 25 February 2019)
90. Raiden Official Web Page. Available online: https://raiden.network/101.html (accessed on 25 February 2019)
91. Poon, J.; Buterin, V. Plasma: Scalable autonomous smart contracts. *White Paper* **2017**, 1–47.

Article

Performance Evaluation and Interference Characterization of Wireless Sensor Networks for Complex High-Node Density Scenarios

Mikel Celaya-Echarri [1], Leyre Azpilicueta [1,*], Peio López-Iturri [2,3], Erik Aguirre [2] and Francisco Falcone [2,3]

1 School of Engineering and Sciences, Tecnologico de Monterrey, Monterrey, NL 64849, Mexico
2 Electric, Electronic and Communication Engineering Department, Public University of Navarre, Pamplona, 31006 Navarra, Spain
3 Institute of Smart Cities, Public University of Navarre, Pamplona, 31006 Navarra, Spain
* Correspondence: leyre.azpilicueta@tec.mx; Tel.: +52-818-358-2000 (ext. 5362)

Received: 1 July 2019; Accepted: 4 August 2019; Published: 11 August 2019

Abstract: The uncontainable future development of smart regions, as a set of smart cities' networks assembled, is directly associated with a growing demand of full interactive and connected ubiquitous smart environments. To achieve this global connection goal, large numbers of transceivers and multiple wireless systems will be involved to provide user services and applications anytime and anyplace, regardless the devices, networks, or systems they use. Adequate, efficient and effective radio wave propagation tools, methodologies, and analyses in complex indoor and outdoor environments are crucially required to prevent communication limitations such as coverage, capacity, speed, or channel interferences due to high-node density or channel restrictions. In this work, radio wave propagation characterization in an urban indoor and outdoor wireless sensor network environment has been assessed, at ISM 2.4 GHz and 5 GHz frequency bands. The selected scenario is an auditorium placed in an open free city area surrounded by inhomogeneous vegetation. User density within the scenario, in terms of inherent transceivers density, poses challenges in overall system operation, given by multiple node operation which increases overall interference levels. By means of an in-house developed 3D ray launching (3D-RL) algorithm with hybrid code operation, the impact of variable density wireless sensor network operation is presented, providing coverage/capacity estimations, interference estimation, device level performance and precise characterization of multipath propagation components in terms of received power levels and time domain characteristics. This analysis and the proposed simulation methodology, can lead in an adequate interference characterization extensible to a wide range of scenarios, considering conventional transceivers as well as wearables, which provide suitable information for the overall network performance in crowded indoor and outdoor complex heterogeneous environments.

Keywords: wireless sensor networks; interference characterization; performance evaluation; 3D ray launching; high-node density; smart cities

1. Introduction

The notion of a smart world, with the aid of smart devices, smartphones, smart cars, smart homes, and smart cities, the paradigm of smart everything, has been a vigorously researched topic for many years. This concept holds the view that people and the world itself will be overlaid with sensing and actuation, with the aid of the internet of things (IoT). Nowadays, IoT has been used in a large number of areas, such as government, industry, and academia [1], for different applications. For example, sensors are placed in buildings for attempting to save energy [2,3]; wireless sensor networks (WSNs) in vehicular communications trying to improve safety and transportation [4]; home automation [5];

industry [6]; or e-Health services which are relying on increased home sensing to support remote medicine and wellness [7].

Nowadays, more than half of the world's population lives in cities [8] with more than six devices per person connected to the Internet [9]. That means that billions of devices will be connected by 2020 to build the aforementioned smart city concept, which can range from end-user devices or wearables to vehicular communication systems, water and gas monitoring, smart lightning, structural monitoring, or smart healthcare systems, among others [10]. These solutions involve a high-density node environment, which in turn requires smaller outdoor and indoor cells leading to heterogeneous networks (HetNet).

Moreover, the advent of next generation 5G communication systems implies the use of ultra-dense small cells in order to increase coverage/capacity requirements inherent to the wide array of services to be offered [11]. In this context, multiple wireless communication systems can be employed in order to diversify service provision, leading to issues such as unpredictable interference sources, or idle cell interference which can greatly impact quality of service [12,13]. Interference analysis has also been considered as a potentially beneficial evaluation element that can be employed in order to enable covert communications in IoT scenarios [14].

The complexity in interference analysis is given by multiple factors, such as requested service types, heterogeneous service requirements, wireless system coordination, or user location and density. In this sense, HetNet architectures implement superimposed cell structures in order to provide adequate capacity requirements as a function of the requested service type. However, this has a serious impact in overall interference levels, requiring the use of cell coordination, such as almost blank subframe technique in order to minimize transmission times [15], employing game theory approaches to provide first order approximation to consider overall interference levels or a similar approach in order to analyze multi-user-multiple-input-multiple-output (MU-MIMO) user assignment in 5G systems [16]. Not only do user distribution and service type have an impact on interference, but also hardware constraints, such as sampling rate mismatch between end user devices and access points/base stations, also affect in the case of massive IoT deployments [17]. In this same line, interference levels can degrade operation of massively deployed transceivers, such as LoRa, owing to loss of ideal orthogonality and leading to increased packet loss [18]. In the future, these constraints can further be aggravated by the use of novel schemes, such as radio frequency (RF) energy harvesting, which in the case of IoT deployments, must consider coverage/capacity relationships in order to comply both with minimum harvested energy thresholds as well as with uplink/downlink signal to interference ratios [19]. Hence, interference analysis and control are one of the fundamental aspects to consider in the scenarios with variable quality of service (QoS) requirements and user density and location, intrinsic to IoT applications [20]. Different solutions have been recently proposed in order to control overall interference levels, such as content-aware cognitive control [21], resource split full duplex mechanisms [22], or machine learning techniques to provide adaptive transmit power control in WSNs [23].

For the successful implementation of the aforementioned dense node deployments, reliable and accurate channel models are necessary, addressing the different topologies and radio links to get reliable service, coverage, and capacity, as well as interference management. Furthermore, hot spots have a non-uniform traffic demand, so it is necessary to have three-dimensional (3D) realistic environments to achieve accurate models, which can lead to network performance improvement. The approaches followed in order to analyze interference in large areas with node density are usually based on statistical channel modelling and under certain model assumptions [13,24], providing certain consideration in relation with scenario characteristics, which can be eventually combined with measurement updates. Spatio-temporal techniques have also been proposed in order to analyze connection establishment phases in massive IoT deployment, based stochastic geometric models [25]. Measurement based interference characterization in IoT scenarios has also been proposed, assisted by supervised learning [26]. However, none of these approaches perform a complete analysis and system performance evaluation considering the whole morphology and topology of the considered scenario.

Moreover, realistic wireless system operation exhibits a complex behavior, depending on conditions such as household/office environment, wireless systems under operation within the scenario under analysis or the density of transceivers considered [27].

The analysis on wireless node density and variations within wireless channel characteristics is relevant in functionalities related with applications such as wireless cooperative location systems [28] or in passive location systems [29], in which wireless channel characteristics (line of sight-non-line of sight channel conditions as well as multipath propagation) influence ranging estimations and hence, location performance. Network synchronization is another application in which node density impacts system operation of cooperative systems, influencing the value of the cooperative dilution intensity [30], which is influenced by wireless channel conditions.

Energy analysis is a relevant aspect in the operation of wireless communication systems and particularly in wireless sensor networks, with the existence of inherent limitations given by restrictive energy sources, processing capabilities, and compact form factor requirements. In this sense, wireless sensor network energy balance analysis is compulsory in order to implement efficient system level solutions, such as scheduling algorithms for sleep/active states in order to implement wireless sensor networks operating under partial coverage conditions [31]. By studying required coverage levels (given by receiver sensitivity thresholds, determined by transmission bit rates, adaptive modulation and coding schemes, and electronic device characteristics), transceivers can be dynamically set in sleep modes, resulting in effective energy reduction and hence, enhanced operation lifetime. In this context, estimation of wireless channel behavior, in terms of coverage estimation as well as in distribution of interference sources is relevant in order to analyze overall energy consumption impact, from physical layer as well as in access control and network layer.

In this work, we present a deterministic technique to model electromagnetic propagation in high node density scenarios, specifically an in-house 3D ray-launching (3D-RL) algorithm, based on geometrical optics (GO), geometrical theory of diffraction (GTD), and its extension the uniform theory of diffraction (UTD). With the aid of the 3D-RL simulation tool, the performance evaluation and interference characterization of a dense node density scenario has been performed in order to assess the key performance indicators of the network. The contributions of this work are aimed in providing a precise tool for coverage/capacity estimation, considering relevant multipath propagation phenomena in large complex scenarios. The proposed simulation methodology employs an optimized 3D-RL code with hybrid simulation (combining 3D-RL with neural network interpolators, the electromagnetic diffusion equation for diffraction estimation, and collaborative filtering of deep learning data base algorithms), enabling the possibility to simulate large, complex scenarios. A new simulation module has been implemented in order to perform estimation of error vector magnitude (EVM) for the complete simulation volume, hence enabling further quality of service analysis as a function of the employed modulation scheme.

The remaining parts of the paper are outlined as follow: The proposed simulation technique and the scenario description are explained in Section 2. Section 3 presents the simulation results in the high node density considered scenario, at ISM 2.4 GHz and 5.8 GHz frequency bands, with the received signal strength (RSS), signal to interference noise ratio (SINR), and performance analysis in terms of constellations plots and EVM characterization considering a ZigBee system (infrastructure nodes) as well as Bluetooth transceivers (high mobility devices/users), and coverage/capacity estimations examples. In Section 4, a campaign of measurements has been presented for the same considered scenario, achieving a good match between simulation and measurement results. In addition, the comparison between the scenario full of people and without people is presented in this section. Finally, conclusions and future work are summarized in Section 5.

2. Proposed Simulation Technique

2.1. The RL Technique

The in-house developed 3D-RL simulation tool has been developed in Matlab programming environment. The detailed operating mode of the algorithm has been previously published [32] and validated in complex urban environments [33]. The principle of the RL approach is to approximate the full wave methods based on Maxwell's equations into a set of equations based on GO and UTD. The in-house implemented 3D-RL code has been optimized in order to decrease computational cost when consider large, complex scenarios, such as indoor locations, by means of hybrid simulation approach. In this sense, neural network interpolators, electromagnetic diffusion equation for diffraction estimation, and deep learning database assistance by means of collaborative filtering have been implemented within the code. The algorithm basis has three steps:

- Creation of the 3D environment.
- Simulation procedure.
- Results analysis.

The first step consists in the creation of the 3D environment under evaluation. For that purpose, all the details of the environment are considered, taking into account its real dimensions, morphology, topology, and material properties (by means of the conductivity and dielectric permittivity) for all the obstacles within the scenario at the frequency band under analysis. In the simulation procedure, a set of rays are launched from the transmitter and electromagnetic propagation phenomena such as reflection, refraction, and diffraction are considering along all the path rays. Parameters such as transmitters location, angular resolution of rays, cuboids size of the scenario, frequency of operation, and number of reflections are considered as input parameters in the algorithm. A trade-off between angular resolution of launching rays, cuboids size of the scenario, required computational time, and results accuracy must be achieved during the simulations. A convergence analysis in terms of number of reflections and launching rays of the algorithm has been performed and it is presented in [32], as well as the optimal spatial resolution for large scenarios, which is presented in [34]. These parameters are used in the simulations and are presented in Section 2.2.

Finally, the third step consists in the results analysis, where different outcomes can be obtained. The 3D-RL tool is based on a modular structure, where the user can select the results of interest. The different results that can be selected are large-scale propagation parameters such as received power or path loss analysis, or small-scale parameters such as power delay profile (PDP), delay spread, coherence bandwidth, or doppler spread, among others. In this work, interference analysis has been implemented as a new module for the network performance assessment. In this library, once the power level and signal to interference noise ratio (SINR) results have been obtained for all the spatial points of the scenario, different modulations can be assessed for different communication links, presenting evaluation of EVM within the complete volume of the scenario under analysis.

2.2. Scenario Description

The selected scenario is an auditorium placed in an open free city area surrounded by inhomogeneous vegetation. Figure 1 presents the real and schematic view of the considered scenario, which is part of the Campus of Tecnologico de Monterrey, Monterrey, Mexico. The considered scenario is a complex scenario in terms of radio wave propagation characterization as it is a combination of an outdoor and indoor environment rich in multipath trajectories due to the large quantity of obstacles and people within it. The different workspaces of the auditorium environment have been recreated in the simulation algorithm, taking into account the inhomogeneous vegetation, trees, tables, chairs, the auditorium area, the cafeteria area, corridors, and a random distribution of people, both in the outdoor and indoor areas of the auditorium. A generic human body model design created specifically to be embedded in the 3D-RL algorithm has been used to enhance a more realistic scenario, as users have a significant influence in radio wave propagation in this type of complex environments [35]. The specific details of the developed human body model and its integration with the 3D-RL tool can be found in [36].

User density within the scenario, in terms of inherent transceivers density, has been performed by simulation. The considered scenario has a user capacity of 150 persons within the auditorium, and 40 more approximately, when all the tables around the auditorium are occupied. Thus, for a high-node density within the complete scenario, it has been considered that one per four persons has a wearable that can increase overall interference levels, which lead to a sensor network of 75 wearables. A medium node density has been considered with 38 wearables (one device per eight persons), and a low node density of 19 wearables (one wearable over 16 persons). All the wearables have been considered at 1.2 and 0.8 m height, emulating smart glasses or wrist-worn devices in the case of seated persons. Figure 2 represents an aerial view of the scenario (ceiling has been removed for illustration purposes) with the wearable's location for the three different nodes density cases. For the simulations, two different frequencies have been considered, 2.4 GHz and 5.8 GHz, considering the later for infrastructure operation (i.e., not as wearable devices). For the network performance analysis, a ZigBee system has been considered at 2.4 GHz frequency band, as well as Bluetooth V4.0 transceivers within the scenario. This election is based on two modes of operation: Data gathering (from WSNs or users) and wireless transport networks (given by 5.8 GHz WLAN devices). ZigBee offset-quadrature-phase-shift-keying (O-QPSK) modulation, with a bandwidth of 3 MHz and a bit rate of 250 kbps, whilst Bluetooth V4.0 employs at higher rates, differential 8-level phase-shift keying (8-DPSK) modulation at a bit rate of 3 Mbps. Simulation parameters are summarized in Table 1.

Table 1. Simulation parameters.

Parameters	Values
Wearable TX Power	4 dBm
Frequency	2.4 GHz/5.8 GHz
Bit Rate	250 kbps/1 Mbps/3 Mbps
Antenna Radiation Pattern (RX, TX)/Gain	Omnidirectional/0 dB
3D Ray Launching: Angular Resolution/Reflections	1 degree/6
Scenario size/Unitary volume analysis	$(50 \times 37 \times 8)$ m/1 m^3 $(1 \times 1 \times 1)$ m
System/Modulation/Bandwidth	ZigBee (2.4 GHz)/O-QPSK/3 MHz Bluetooth V4.0/8-DPSK/2 MHz
Number of symbols	1000

(a)

Figure 1. *Cont.*

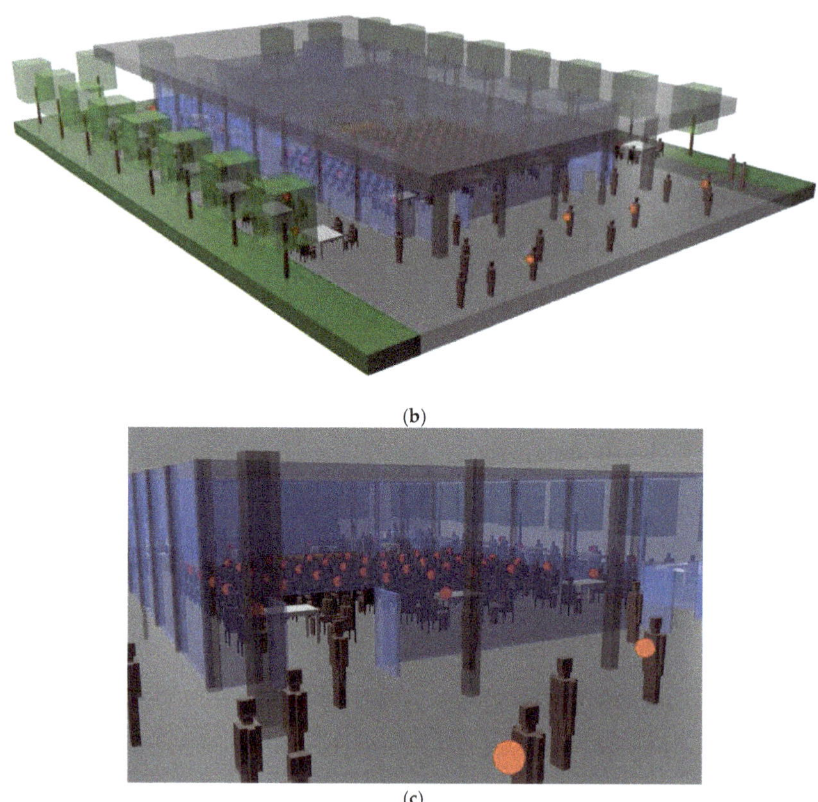

(b)

(c)

Figure 1. Urban considered scenario where, (**a**) is the real view of the scenario (retrieved from: http://www.rdlparquitectos.com/es/proyectos/pabellon-la-carreta/), (**b**) is a 3D aerial rendered view, and (**c**) detailed view of the indoor part of the scenario.

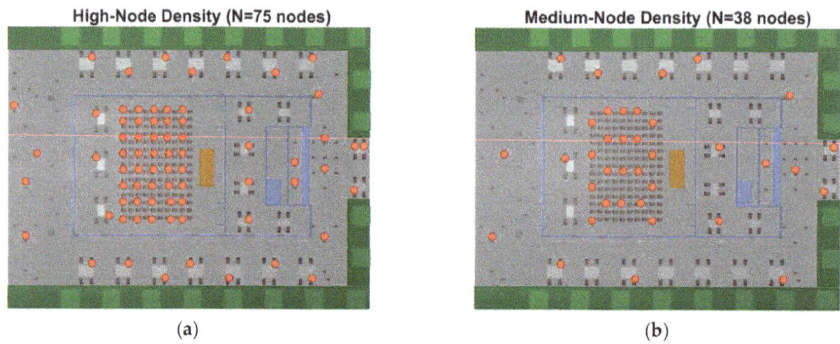

(a) (b)

Figure 2. *Cont.*

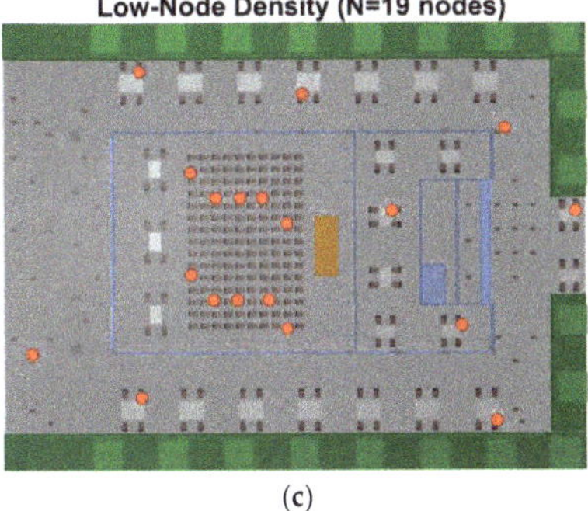

(c)

Figure 2. Aerial view of the scenario with the wearable's location for the three different node density cases. Auditorium ceiling has been removed for illustration purposes, (**a**) high-node density, (**b**) medium-node density, and (**c**) low-node density.

3. Simulation Results

3.1. Received Signal Strength

Firstly, the Received Signal Strength for the full volume of the scenario has been obtained by means of the 3D-RL simulation tool. Although the whole volume of the scenario has been analyzed in simulation, only the bi-dimensional planes of received power for a selection of relevant heights have been presented in Figures 3 and 4. The considered heights have been 1.2 m, which is the same height as the transmitters, emulating smart glasses receivers for a seated person (Figure 3), and 0.8 m height, emulating wrist-worn receivers for a seated person (Figure 4).

Figure 3 presents the RSS maps at 1.2 m height considering different node density setups for two different frequencies, 2.4 GHz and 5.8 GHz, which are the typical frequencies used in generic wearables, as well as in wireless sensor network infrastructure (i.e., 802.15.4 networks and 802.11 networks). It can be seen that the morphology and topology of the scenario, as well as node density distribution, have a great impact in radio wave propagation. It is observed a high RSS variability within the whole scenario due mainly to the presence of different scatterers such as people, furniture, walls, vegetation, and trees, among others, which cause a rich multipath propagation in the environment under analysis. It can also be remarked that RSS is around 20–30 dBs higher when high-node density is considered. Moreover, it is worth noting that there are localized regions with higher power levels, i.e., hot spots, within the scenario, caused by individual transceivers, that tends into power clusters when the number of nodes increases in a region. This is the case of the auditorium indoor area, where there is a higher concentration of wearables, and therefore it also has a higher concentration of received power (around 5–10 dB more), being more remarkable in the case of high-node density due to the higher number of wearables involved. The higher intensity in this area is also given by the body shielding effect, which enhances the signal to be concentrated in the auditorium indoor area, where more users are presented. In addition, different regions can be identified within the simulation scenario where power levels differ, such as the outdoor and indoor part of the scenario, or the cafeteria area in the right part of the scenario. These regions are delimited by different types of walls (glass in the auditorium and concrete/metal in the cafeteria), which gives rise to different associated propagation phenomena, such as the appearance of hot-spots in the aforementioned localized areas, regardless node density setups. Frequency also plays an important role showing that for all the analyzed cases, higher power levels are received at lower frequencies, as it was expected. The presence of hot-spots is also more remarkable at lower frequencies (notice that the power level scale has been maintained for both frequencies for better comparison).

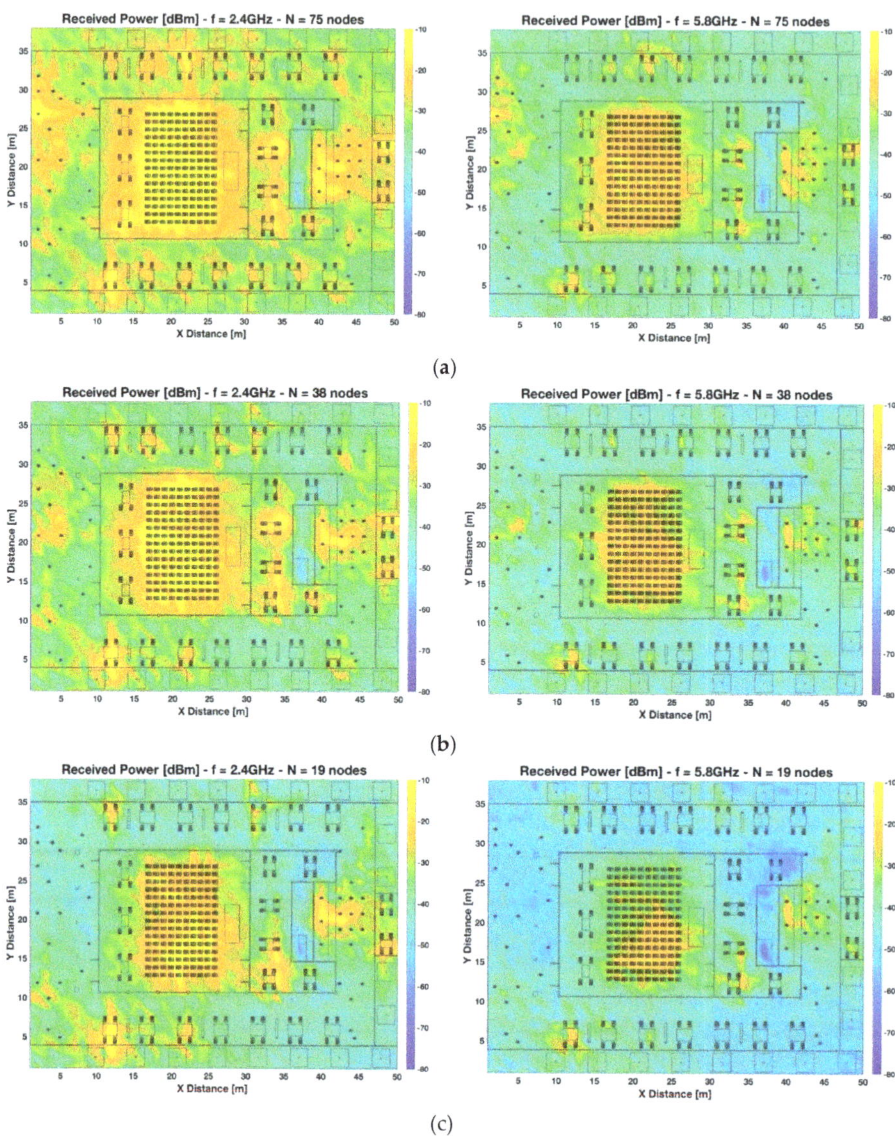

Figure 3. Bi-dimensional planes of received power [dBm] at 1.2 m height at 2.4 GHz and 5.8 GHz frequency bands. Different scenario setups have been considered: (**a**) high-node density scenario, (**b**) medium-node density scenario, and (**c**) low-node density scenario.

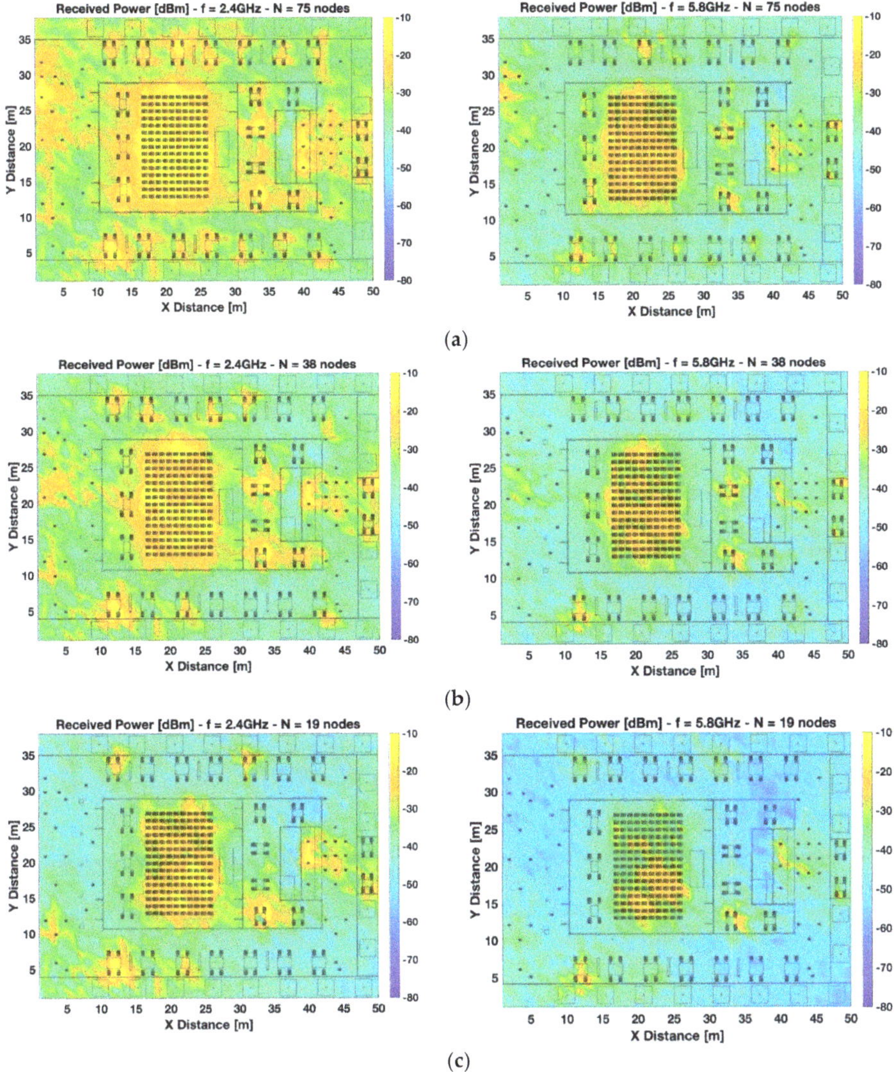

Figure 4. Bi-dimensional planes of received power [dBm] at 0.8 m height at 2.4 GHz and 5.8 GHz frequency bands. Different scenario setups have been considered: (**a**) high-node density scenario, (**b**) medium-node density scenario, and (**c**) low-node density scenario.

Figure 4 presents the same results as Figure 3, but in this case, for a 0.8 m height, emulating wearables such as smart watches operating in the wrist of a seated person. As it can be seen, a high variability of RSS is also observed in this case, showing slightly lower values of received power because the bi-dimensional plane representation is not at the same height of the transmitter's wearables. In these cut-planes, the presence of hot-spots is still visible but with lower intensity for all the different node density setups. This specific behavior is caused due to bigger link distances with the transmitters comparing with the previous cases analyzed. Besides, there are relevant differences between the operating frequencies, with more losses at the higher frequency, as it was expected.

In order to gain a better insight into the differences of received power for different heights and frequencies in the three nodes density cases, the radial distribution of RSS in the considered scenario has been assessed and it is represented in Figure 5. The considered radial line is depicted in Figure 5d, a representation of the aerial view of the scenario (ceiling has been excluded for illustration purposes), which correspond to Y = 20 m along the X-axis. As it was expected, in the comparison for the low-node density case, the trend of the obtained received power values is lower than in the other cases. Hence, the distribution of received power levels depends strongly on the node's distribution in the considered scenario. In addition, there is a slightly difference between RSS values at the different heights, showing higher values in general for the 1.2 m height (the same cut-plane as the transmitters). However, this trend is not homogeneous for all the spatial points, because as stated above, the morphology and topology of the scenario plays an important role in radio wave electromagnetic propagation for complex environments. Regarding the different frequencies, at the 5.8 GHz frequency band the losses are 5–10 dB approximately higher than the 2.4 GHz frequency band.

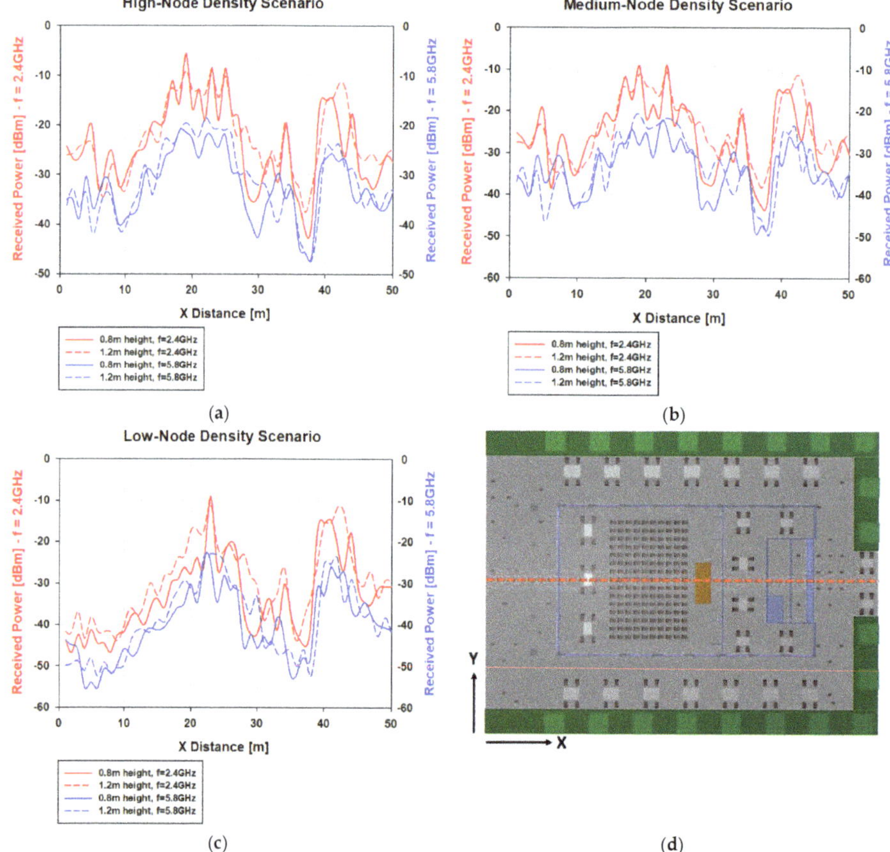

Figure 5. Linear distribution of received power [dBm] for different heights and frequency bands for Y = 20 m along the X-axis, (**a**) high-node density scenario, (**b**) medium-node density scenario, (**c**) low-node density scenario, and (**d**) radial line representation.

3.2. Signal to Interference Noise Ratio

Effective signal and interference levels are determined by received power level distributions, which in turn are determined by the number of nodes as well as their topological distribution, in relation with the surrounding environment. In order to have clear insight into the interference levels in the proposed sensor network, SINR volumetric estimations have been obtained for the whole scenario. An acceptable communication link formation in terms of quality of service is given by the fulfillment of the following condition [37]:

$$P_{RX}(\vec{d}, RX_{hw}) \geq SENS_{RX}(SINR, R_b, m_c) \quad (1)$$

where P_{RX} is the received power for each transceiver, as a function of spatial location $\vec{d}$ and receiver hardware parameters RX_{hw} (e.g., antenna gain, noise factor) and $SENS_{RX}$ is the receiver sensitivity, determined by the required $SINR$ threshold (or E_b/N_0 in the case of digital systems), transmission bit rate R_b and the modulation and coding scheme m_c. In this context, the determination of useful received power and detected interference levels provides coverage/capacity relations as a function of service requirements and density of nodes within the scenario. The scenario location of all transmitting, as well as receiving elements, is a fundamental parameter, due to the large variability in power distribution in the complex environment under consideration.

Once the wireless channel characterization has been performed for different transceiver densities, as described in previous section, interference analysis in terms of SINR can be obtained, leading to system coverage/capacity estimations. For that purpose, the worst-case conditions have been considered, in terms that SINR analysis are provided when the interconnecting device operates with in-band inter-system interference. For the different node density cases, one interconnecting device has been considered as the transmitter and the rest as in-band inter-system interference.

Figures 6 and 7 show the bi-dimensional plots at 1.2 m height of the SINR distribution for two different locations of the transmitter, indoor and outdoor, considering different sensors densities distributed non-uniformly in the scenario, as interferers (see Figure 2 to have insight into the different node distribution densities). These plots represent an upper bound in relation with quality degradation in terms of simultaneous operation of the transceivers. The provided SINR values can be mapped afterwards to E_b/N_0 ratios, where modulation scheme as well as transmission bit rate can be explicitly considered.

Figure 6 presents the SINR values considering the transmitter a wearable (i.e., smart glasses) of a seated person inside the auditorium (X = 20.35 m, Y = 22 m) (i.e., Tx 33) and the simultaneous operation of the rest of transceivers in different node densities setups. It can be observed that the highest SINR values appear in localized areas nearby the transmitter in the indoor part of the auditorium. The differences between nodes densities are pronounced, around 10–15 dB approximately between the low- and high-node density cases for the indoor area of the scenario. Besides, for all density cases, the SINR values for the outdoor region of the scenario are very low. These results show the high dependence of node density and represent an upper bound in terms of quality degradation when other transceivers are operating simultaneously. Regarding the differences between different frequency bands, it can be seen that there is not a lot of variability in the SINR values between different frequencies, but for the 5.8 GHz frequency band, SINR values are slightly lower, depending of the spatial considered point due to the morphology of the scenario.

To gain insight into the significance of considering the whole three-dimensional scenario, the same SINR analysis has been obtained for an outdoor located transmitter node, a wearable of a seated person placed in the outside tables of the auditorium, in the left down corner of the scenario (X = 11.7 m, Y = 5.8 m) (i.e., Tx 5). These results are presented in Figure 7, where a bigger area than the previous case of high values of SINR around the selected transmitter can be observed. This is caused because of the smaller number of scatterers presented in the outdoor part, which allows a higher area with high SINR values, for all different node density cases. Besides, the outdoor-indoor communication is deeply affected by the scenario boundaries, such as the auditorium walls, being the worst case for the high node density case. For all node density cases, it can be observed that the SINR values in the indoor area of the auditorium are very low, which means that the communication link will not be feasible with the indoor region of the auditorium. As in the previous case, there are not many differences in SINR values between frequencies, with slightly lower SINR values for the higher frequency band (5.8 GHz).

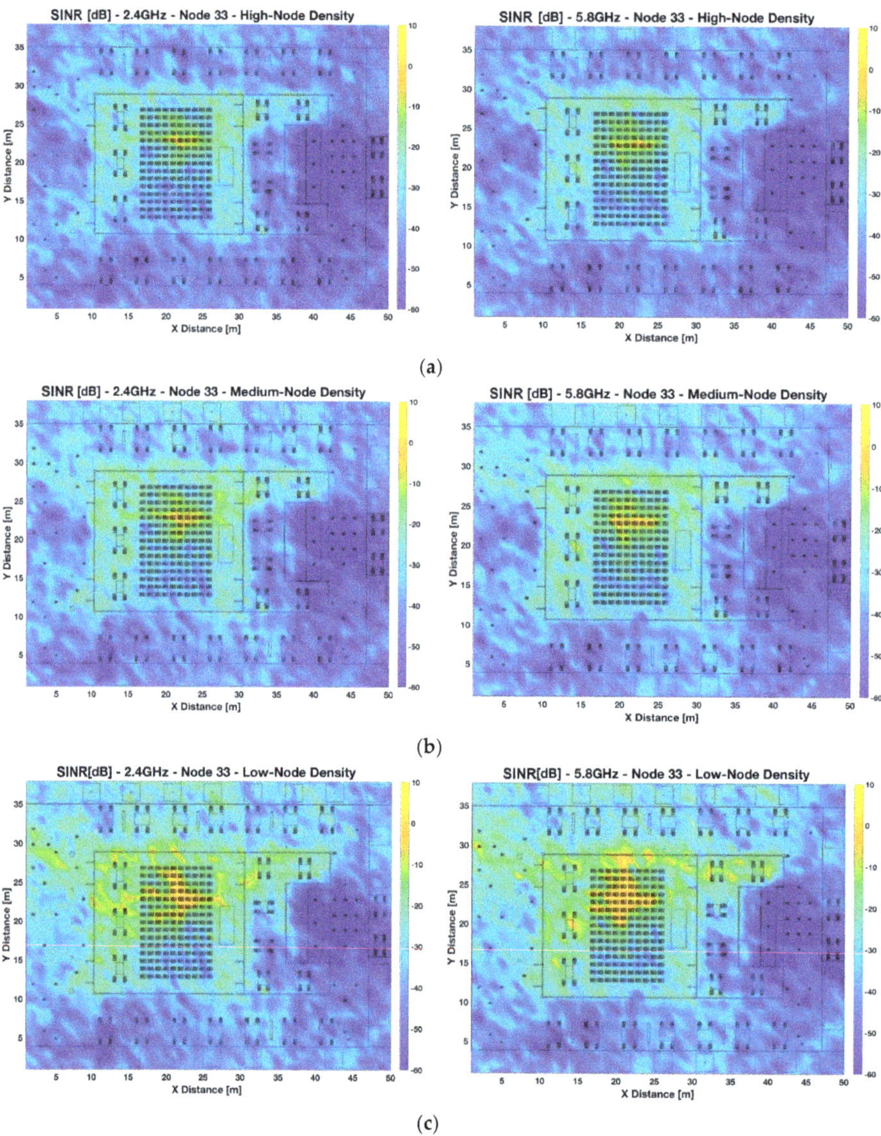

Figure 6. Bi-dimensional planes of signal to interference noise ratio (SINR) [dB] for Node 33 (wearable placed at 1.2 m height, same as a seated person in the indoor area of the auditorium, X = 20.35 m, Y = 22 m) at 2.4 GHz and 5.8 GHz frequency bands. Different scenario setups have been considered: (**a**) high-node density scenario, (**b**) medium-node density scenario, and (**c**) low-node density scenario.

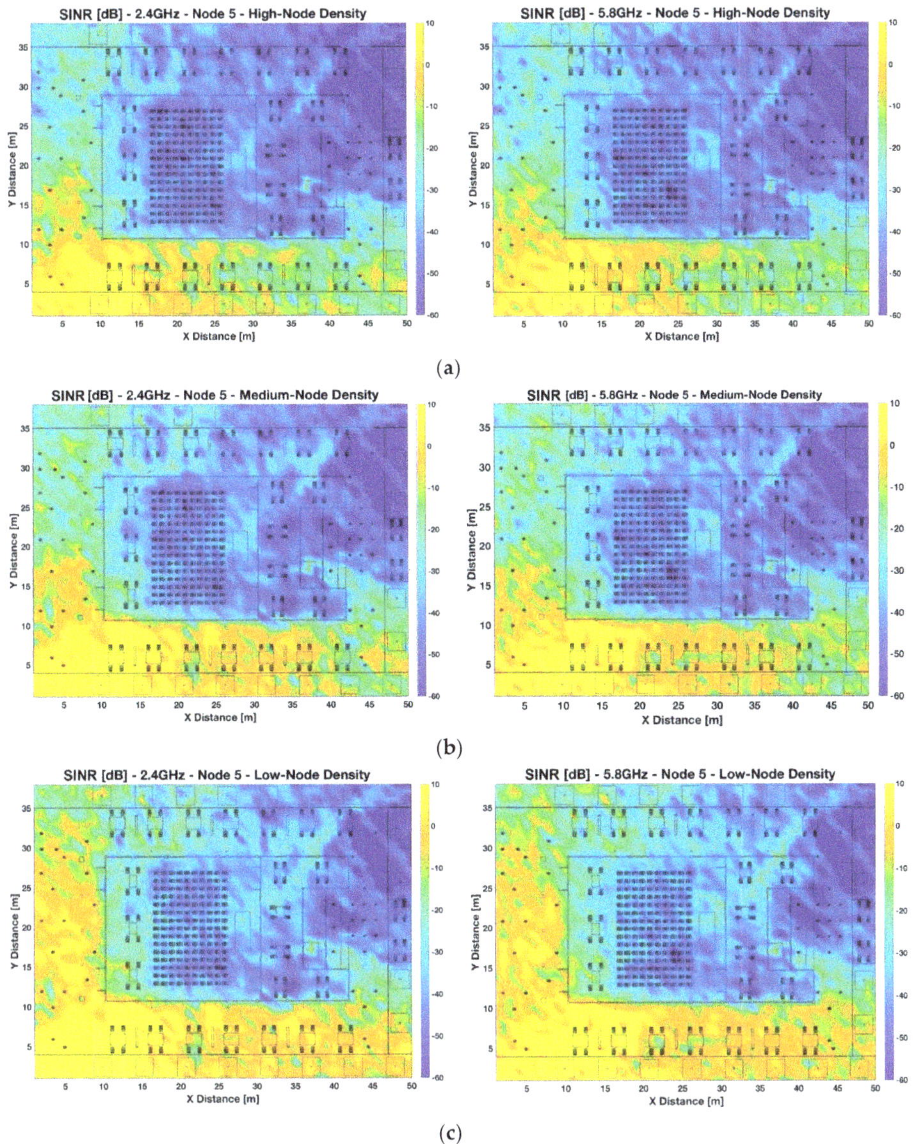

Figure 7. Bi-dimensional planes of SINR [dB] for Node 5 (wearable placed at 1.2 m height, same as a seated person in the outdoor area of the auditorium, X = 11.7 m, Y = 5.8 m) at 2.4 GHz and 5.8 GHz frequency bands. Different scenario setups have been considered: (**a**) high-node density scenario, (**b**) medium-node density scenario, and (**c**) low-node density scenario.

3.3. Performance Analysis

Coverage analysis can aid in an adequate estimation of indoor and outdoor coverage in the simulated scenario. However, since coverage and capacity are linked, it is mandatory to analyze the influence of modulation in order to determine system performance. For that purpose, a ZigBee sensor network has been considered, which uses O-QPSK modulation, with a bandwidth of 3 MHz and a bit rate of 250 kbps, as well as a Bluetooth network considering V4.0 transceivers at a bit rate of 3 Mbps. As stated before, the worst-case conditions have been considered, in terms that for the different nodes density cases, one interconnecting device has been considered as the transmitter and the rest as in-band inter-system interference.

Figure 8 presents the constellation plots at two different receiver location points placed in the indoor and outdoor area of the scenario for the ZigBee system. Two different transmitter positions have been considered, which are the same ones as the previously presented results: The first one in the indoor area of the scenario (Tx 33) and the second one in the outdoor area of the scenario (Tx 5). The considered receiver spots are placed at one-meter distance of each transmitter, in the indoor and outdoor area, respectively, considering high and low-node density. It can be seen that the symbols are more disperse in the case of the transceivers placed in the indoor part of the auditorium, for both density nodes. This is explained because simulations have been made with the auditorium full of people, which causes a high number of scatterers in the area. Thus, this coupled with the assumption of the worst-case condition, where all the nodes except the transmitter are in-band inter-system interference, causes a large dispersion in the received symbols in the indoor area of the scenario. Constellations plots for the receiver location point in the outdoor area of the scenario have less symbol dispersion, achieving the ideal constellation for the low-node density case, as the interferers in this case are not placed nearby the receiver.

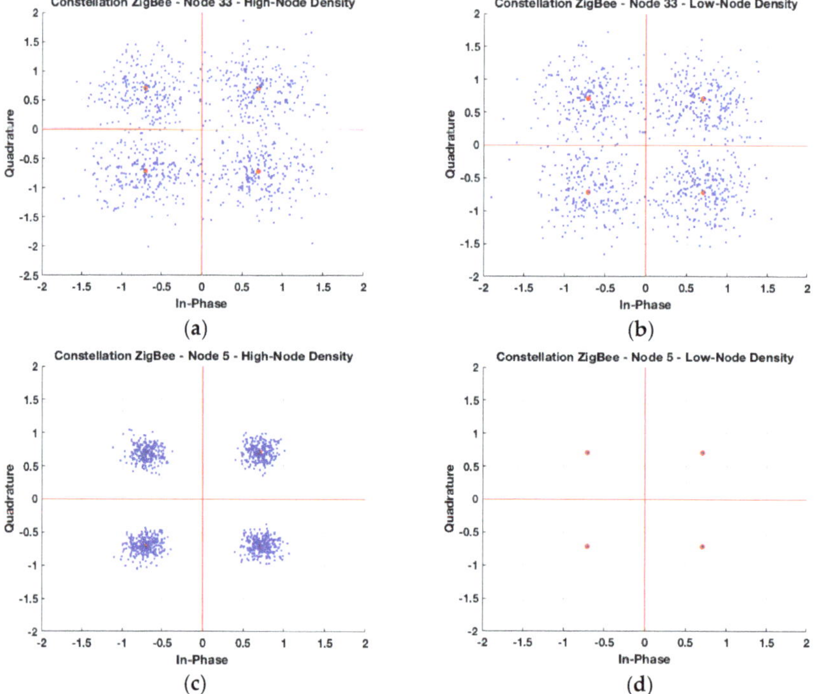

Figure 8. In Phase and Quadrature representation for two different nodes within the auditorium for offset-quadrature-phase-shift-keying (O-QPSK) modulation, (**a**) indoor node with high-node density scenario, (**b**) indoor node with low-node density scenario, (**c**) outdoor node with high-node density scenario, and (**d**) outdoor node with low-node density scenario.

To have insight into the total interference, the error vector magnitude (EVM) has been calculated for the different node-density cases. The EVM is an indicator of the modulation accuracy. To quantify the modulation error, the amplitude of the error can be calculated as:

$$Error\ Amplitude = \sqrt{(I_i - I_A)^2 + (Q_i - Q_A)^2} = \sqrt{\Delta I^2 + \Delta Q^2} \tag{2}$$

where I_i and Q_i are the In-phase and Quadrature values of an ideal signal, while the actual location of the signal is I_A and Q_A. The root mean square (RMS) value of the error amplitude for N symbols is:

$$RMS\ Error\ amplitude = \sqrt{\frac{1}{N}\left(\sum_{k=1}^{N} \Delta I_k^2 + \Delta Q_k^2\right)} \tag{3}$$

If the ideal signal amplitude is S, the EVM is defined as:

$$EVM(\%) = \frac{RMS\ Error\ amplitude}{Ideal\ Signal\ Amplitude} = \frac{\sqrt{\frac{1}{N}\left(\sum_{k=1}^{N} \Delta I_k^2 + \Delta Q_k^2\right)}}{S} \times 100 \tag{4}$$

Table 2 presents the EVM (%) for the constellations' plots presented in Figure 8, including also the medium-node density case. The EVM requirement of the IEEE 802.15.4 standard must be less than 35% [38]. According to these results, the high-node density case for the indoor link is really close to the limit, so the received signal is going to be interference limited.

Table 2. Error vector magnitude (EVM) (%) for different nodes-density cases for the ZigBee system.

RMS EVM (%)	High-Node Density	Medium-Node Density	Low-Node Density
Tx 33 (X = 20.35, Y = 22, Z = 1.2) m	32.89	29.6	11.15
Tx 5 (X = 11.7, Y = 3.8, Z = 1.2) m	7.10	3.25	0.0022

EVM analysis has also been performed in the case of considering operation of Bluetooth V4.0 transceivers within the scenario, a typical case for users or devices with high mobility. As in the previous case, variations in node density have a strong impact in EVM response, given by higher interference levels. These results are depicted in Figure 9, for the cases of node #33 and node #5, for different node densities. Table 3 presents the EVM (%) for the constellations' plots presented in Figure 9, including also the medium-node density case.

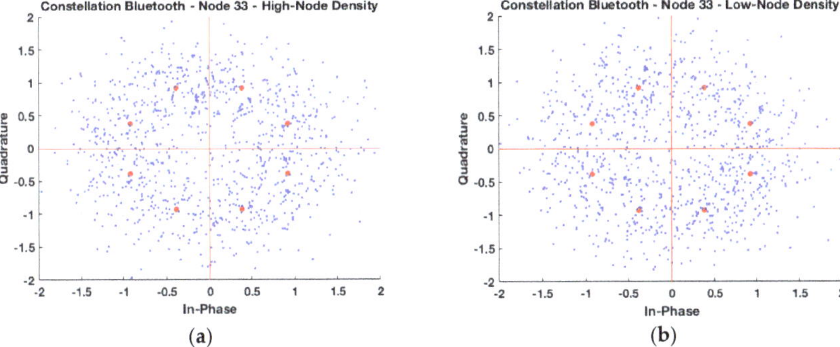

(a) (b)

Figure 9. *Cont.*

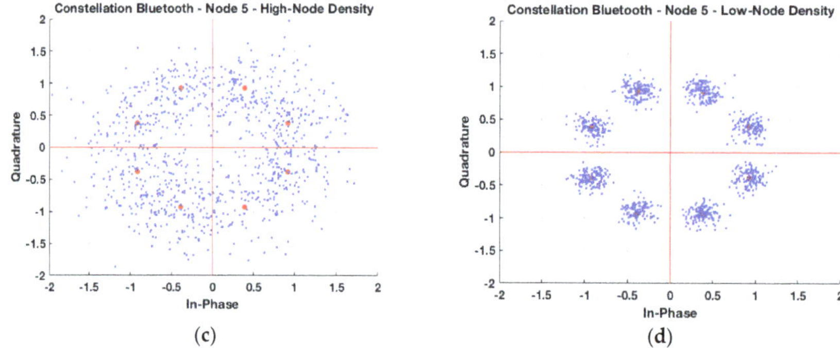

(c) (d)

Figure 9. In Phase and Quadrature representation for two different nodes within the auditorium for 8-DPSK modulation, (**a**) indoor node with high-node density scenario, (**b**) indoor node with low-node density scenario, (**c**) outdoor node with high-node density scenario, and (**d**) outdoor node with low-node density scenario.

Table 3. EVM (%) for different nodes-density cases for the Bluetooth V4.0 system.

RMS EVM (%)	High-Node Density	Medium-Node Density	Low-Node Density
Tx 33 (X = 20.35, Y = 22, Z = 1.2) m	47.61	46.39	27.24
Tx 5 (X = 11.7, Y = 3.8, Z = 1.2) m	28.66	11.19	3.06

To gain insight into the modulation error scenario characterization, the specific regions of correct operation regarding EVM for the ZigBee system have been mapped along the bi-dimensional cut planes. Although results have been obtained for the complete volume of the scenario, for the sake of clarity only the 1.2 m height has been depicted. Figure 10 presents the EVM for the different nodes-density cases and the case without interference, for a transmitter placed in the indoor area of the scenario (Tx 33) with again the worst-case conditions in terms of in-band inter-system interference. These operating regions are delimited by the configuration of the interfering network, as well as by the characteristics of the environment. It can be seen that interference levels can lead to have no service in different areas of the considered scenario, being these no-service areas bigger when high-node density is considered. It is worth noting that the low-density case considered is 19 nodes as it is presented in Figure 2c, which also implies a high interference level, as it can be seen in Figure 10c when compared with the case without interference (Figure 10d).

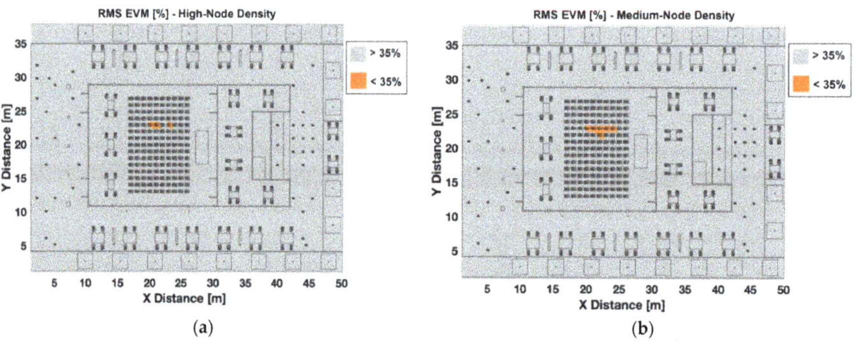

(a) (b)

Figure 10. *Cont.*

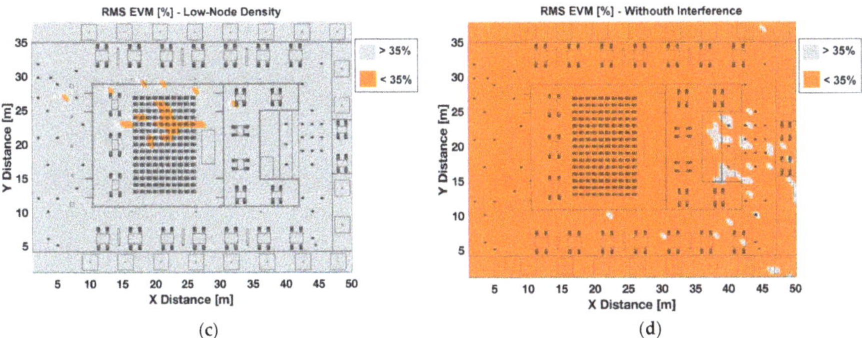

Figure 10. EVM (%) for the considered scenario for O-QPSK modulation when indoor node is transmitting, (**a**) high-node density scenario, (**b**) medium-node density scenario, (**c**) low-node density scenario, and (**d**) without interference.

Figure 11a presents the linear distribution line of EVM (%) for Y = 20 m along the X-axis in the considered scenario, for Tx 33 (indoor node X = 20.35 m, Y = 22 m), considering the three different node-density cases. There is only correct service in a small area around the transmitter because of the high interference levels which have been considered. Figure 11b presents the bit error rate (BER) for the same radial line for the three different node-density cases, showing that the BER is quite high in remote areas from the transmitter.

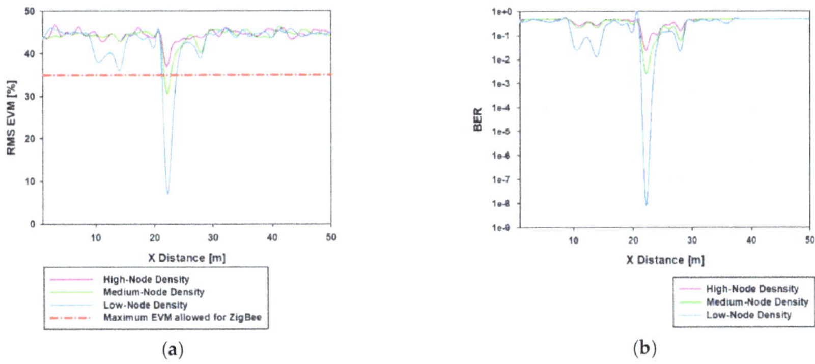

Figure 11. (**a**) Linear distribution line of EVM (%) for Y=20 m, X-axis in the considered scenario when indoor node is transmitting, (**b**) Linear distribution line of bit error rate for Y = 20 m, X-axis in the considered scenario when indoor node is transmitting.

A second transmitter position has been considered to perform the EVM analysis in terms of O-QPSK modulation. In this case, the transmitter is placed in an outdoor location (Tx 5) at 1.2 m height. Regions of correct operation can be seen in Figure 12 for the different node-density cases and the case without interference.

In comparison with the previous indoor transmitter case, the correct operation areas obtained are bigger in this case for all density cases, concentrating the valid area mostly in the outdoor part of the scenario. These results can be explained due to, as stated previously, operating regions are delimited by the boundaries of the scenario (i.e., walls) and the interfering network configuration. Besides, in the outdoor area of the scenario, there is less people, so the concentration of nodes is lower (see Figure 2 for reference), which increases correct operation area compared to the previous case.

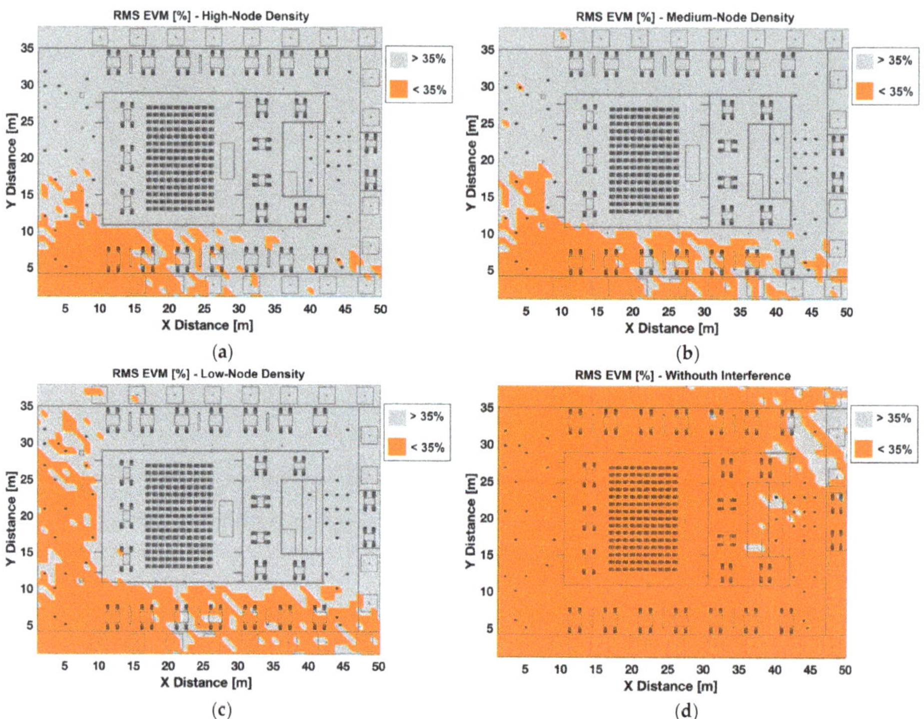

Figure 12. EVM (%) for the considered scenario for O-QPSK modulation when outdoor node is transmitting, (**a**) high-node density scenario, (**b**) medium-node density scenario, (**c**) low-node density scenario, and (**d**) without interference.

Figure 13a presents the linear distribution line of EVM (%) for Y = 5 m along the X-axis in the considered scenario for Tx 5 (outdoor node X = 11.7 m, Y = 5.8 m) considering the three different node-density cases. It can be seen that in the vicinity area of the transmitter, correct operation for the three density cases is obtained, but as we move away from the transmitter, in the medium and high-node density cases, there is no system service due to higher interference levels. Figure 13b presents the BER of the same radial line for the three different node-density cases. As it was expected, BER is quite high in remote areas from the transmitter, and lower BER is encountered close to the transmitter.

These results can aid in a better knowledge of the network performance and are relevant in terms of interference analysis as well as on the operation of mitigation schemes, when high-node density setups are presented.

Node density distributions have a direct impact in overall system performance. This can be observed in terms of coverage/capacity estimations, which depend on wireless system operating parameters (e.g., receiver sensitivity as a function of bit rate, employed adaptive modulation, and coding scheme). Coverage/capacity values can be precisely obtained by the use of 3D-RL simulation techniques, as they provide volumetric estimations of received power levels. As an example, estimations of coverage for receiver sensitivity in Bluetooth/Bluetooth Low Energy (BT/BLE) device distributions are depicted in Figure 14, as a function of transmission power levels. Two different situations are depicted: For maximum transmit power (4 dBm) and for conventional transmission power (usually set at −12 dBm). As a function of node density, variations in received power levels exhibit average received power level variations in the order of 4.8 dB, which can lead to coverage limit conditions (e.g., location at 38 m in the linear TRX radial for the conventional −12 dBm power consideration).

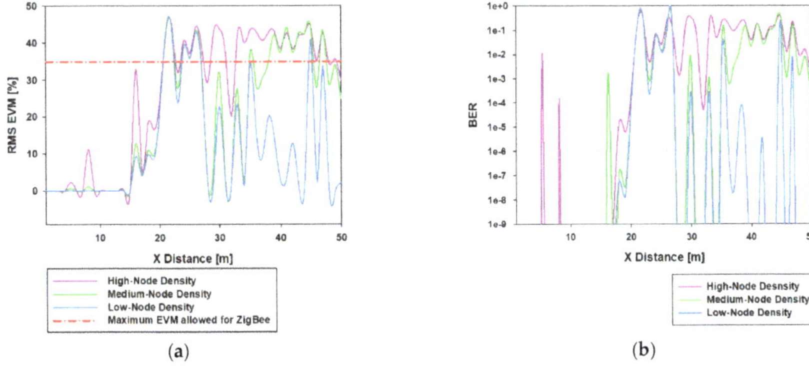

Figure 13. (**a**) Linear distribution line of EVM (%) for Y=5 m, X-axis in the considered scenario when outdoor node is transmitting, (**b**) Linear distribution line of bit error rate for Y = 5 m, X-axis in the considered scenario when outdoor node is transmitting.

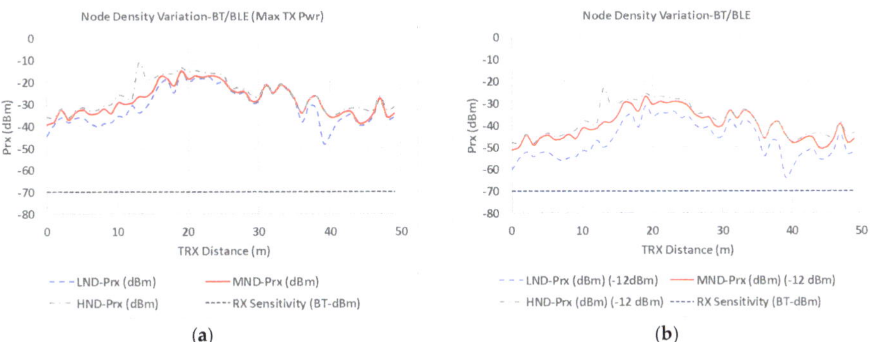

Figure 14. Coverage/capacity estimation for BT/BLE transceiver operation (**a**) max transmission power (4 dBm) (**b**) conventional transmission power (−12dBm).

From the previous results, performance is general degraded as node density increases. This is given by the fact that radio resource functionalities have not been considered, such as multiplexing strategies or dynamic frequency allocation. In this way, worst case operation conditions have been considered, as an initial bound in coverage/capacity analysis. Future work can be foreseen in the analysis of coverage/capacity relations as a function of radio resource functionality performance. Moreover, realistic operation conditions can present even further variations as a function of time dependent interfering sources, which depend on user behavior, which are specific of the scenario under operation [27].

4. Measurements Campaign

4.1. Experimental Setup

Validation of previous coverage/capacity estimations have been obtained by means of experimental characterization. A measurement campaign has been performed in the real auditorium placed at university campus at Tecnologico de Monterrey, in Monterrey, Mexico. In order to characterize the losses caused by the presence of different user densities within the auditorium, two different days have been chosen to perform the campaign of measurements: One day with an empty scenario (without people inside and outside of the auditorium) and another day with the full scenario capacity (auditorium full of people due to a conference performance). The transmitter antenna was placed in the same position for both cases, which correspond with the red circle in Figure 15 at 1.2 m height (emulating a transmitter placed in the smart glasses of a seated person). The CC2530 ZigBee development kit, from Texas Instruments, was used as the transmitter, and the N9952A Field Fox portable spectrum analyzer, from Keysight Technologies, as the receiver. The TX and RX antennas were ACA-4HSRPP-2458 from ACKme Networks, both omnidirectional, with a gain of 3.7 dB. Measurements were performed with 10 MHz bandwidth at 2.4 GHz frequency with a measurement time of 60 s at each location point. The received power at each measurement location was considered as the highest peak (Max-Hold function) of power obtained by the spectrum analyzer. Measurements were performed along the complete auditorium, as it is represented in Figure 15. It must be pointed out that the receiver antenna at each measurement location point was placed at the same height of the transmitter, 1.2 m height, which correspond with the head height of a seated person.

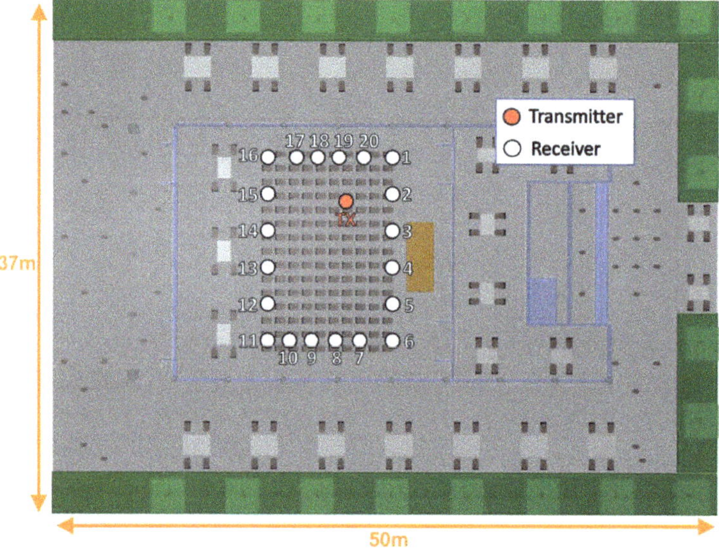

Figure 15. Transmitter position and measurement location points during the measurement campaign for the two analyzed cases: empty and full auditorium.

4.2. Measurements Results

In order to assess interference in the frequency band of interest, before starting the measurement campaign for both, empty and full scenario setups, spectrograms were obtained to verify that no other devices were transmitting at the same frequency band of interest that could disturb the measurements. It must be remarked that there were no other wireless systems interference conditioners and the noise level was around −80/−90 dBm, as it can be seen in Figure 16.

The most complex or in other words, the worst-case scenario, the auditorium at full user capacity, has been considered for comparison purposes to validate the RL simulation tool. Figure 17 presents the comparison between the 3D-RL simulation tool results and the real measurements in terms of received power for each spatial measurement point represented in Figure 15. It can be seen a good match between simulation and measurements with a mean error of 0.95 dB and standard deviation of 0.92 dB. The differences between simulation and measurement results derive from the fact that the implemented simulation scenario doesn't consider to a full extent, given by practical limitations, all of the elements within the scenario, with exact shapes, sizes, and the full material set that exists in the real world.

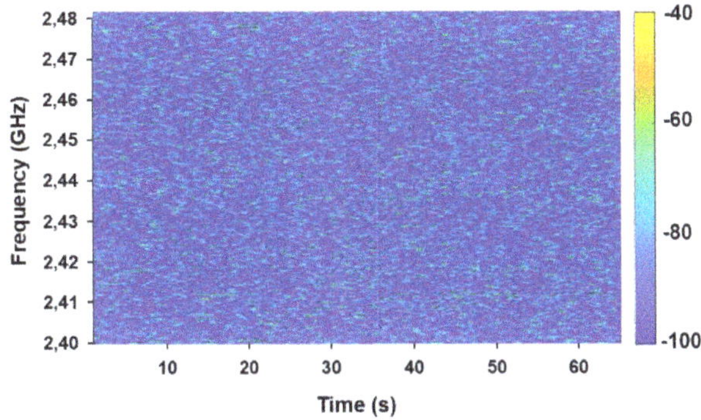

Figure 16. Spectrogram measured before starting the measurement campaign.

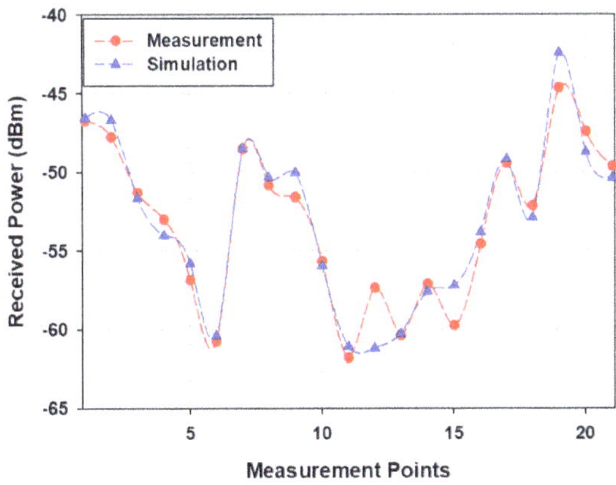

Figure 17. Received power measurements vs 3D ray-launching(3D- RL) simulation comparison.

In addition, in order to get insight into the different losses caused by users' presence within the considered scenario, Figure 18 presents the path loss comparison for the two different measurement cases, with and without people within the auditorium. The comparison shows that for almost every spatial point the path loss is higher in the full scenario case, due to the absorption caused by the users' bodies. But this phenomenon depends on the morphology and topology of the scenario, leading to some location points where the path loss is higher in the empty scenario. In average, the path loss in the full scenario (with people) is 3–5 dB bigger than in the empty scenario (without people), which is in accordance with the values reported in [39].

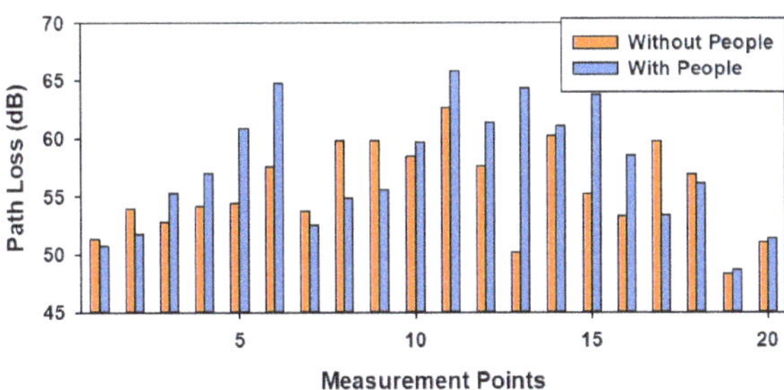

Figure 18. Measured Path Loss for the two measurement cases considered: empty and full scenario.

5. Conclusions and Future Work

In this work, the assessment of different nodes density network configurations has been addressed at 2.4 GHz and 5.8 GHz frequency bands. A complex heterogeneous indoor and outdoor environment has been selected for evaluation, which corresponds with an auditorium placed in a free open city area surrounded by inhomogeneous vegetation. The full analysis of the wireless system in terms of performance and interference characterization has been obtained by means of a deterministic in-house developed 3D-RL algorithm considering different user occupancy. Received signal power as well as SINR have been calculated, showing the effect of degradation when increasing interference in the system. A ZigBee system with O-QPSK modulation has been selected in order to obtain the network performance analysis of the full wireless system setups (infrastructure node network), as well as Bluetooth transceivers in the case of high mobility users. A new processing module has been implemented, enabling the evaluation of modulation constellations and EVM within the complete 3D volume for different indoor and outdoor links setups within the scenario, showing the correct operation regions maps considering different node density cases and distributions. As node density increases, interference values increase considering that radio resource management functionalities are not active, providing a lower bound in terms of coverage/capacity estimations. Moreover, hot-spots can be localized within the scenario, which can strongly impact overall system performance. The sensors placement, individually distributed as well as in mesh setups, is a fundamental parameter in order to assess coverage levels as well as system quality evaluation as a function of SINR. It must be pointed out, that both, coverage estimations analysis and system information provide useful knowledge of the network performance, especially when the number of sensors increases giving rise to high-nodes density scenarios. In addition, a campaign of measurements has been performed in the considered scenario, showing good agreement with simulation results. The proposed methodology makes use of in house deterministic 3D-RL code, which can consider to a high degree of accuracy elements within the scenario, in terms of shape, size and material characterization. Moreover, the 3D-RL code makes use of hybrid code simulation, employing elements such as neural network interpolators, electromagnetic diffuse scattering or collaborative filtering techniques, which reduce computational complexity and hence, enables the study of large, complex scenarios.

These analysis results and the proposed simulation methodology, can lead in an adequate interference characterization, considering conventional transceivers as well as wearables, which provide suitable information for the overall network performance in complex crowded indoor and outdoor scenarios, with no limitation in scenario definition, following a generalizable approach. Future work will consist in a deeper analysis of the network parameters as well as coverage/capacity analysis for different wireless systems. Besides, the QoS can also be characterized, as well as outage probability.

Author Contributions: M.C.-E. and L.A. conducted the simulation and analysis of the wireless propagation phenomena and the scenario impact, as well as the measurement campaign. P.L.-I., E.A. and F.F. conceived and prepared the wireless sensor network design and the performance analysis. M.C.-E. and L.A. prepared the manuscript.

Funding: This research was funded by Tecnologico de Monterrey and the Focus Research Group of Telecommunications and Networks of the School of Engineering and Sciences.

Acknowledgments: The authors would like to acknowledge the support and collaboration of the School of Engineering and Sciences and the Focus Research Group of Telecommunications and Networks at Tecnologico de Monterrey.

Conflicts of Interest: The authors declare no conflict of interest. The statements made herein are solely the responsibility of the authors.

References

1. Stankovic, J.A. Research Directions for the Internet of Things. *IEEE Internet Things J.* **2014**, *1*, 3–9. [CrossRef]
2. Bradshaw, V. *The Building Environment: Active and Passive Control Systems*; Wiley: Hoboken, NJ, USA, 2006.

3. Zeiler, W.; Houten, R.; Boxem, G.; Vissers, D.; Maaijen, R. Indoor air quality and thermal comfort strategies: The human-in-the-loop approach. In Proceedings of the 11th International Conference for Enhanced Building (ICEBO 2011), New York, NY, USA, 18–20 October 2011.
4. Burnham, G.; Seo, J.; Bekey, G.A. Identification of human driver models in car following. *IEEE Trans. Autom. Control* **1974**, *19*, 911–915. [CrossRef]
5. Control4 Home Automation and Control [Online]. Available online: http://www.control4.com (accessed on 23 January 2019).
6. M1 Security and Automation Controls [Online]. Available online: http://www.elkproducts.com/m1controls.html (accessed on 23 January 2019).
7. Dickerson, R.; Gorlin, E.; Stankovic, J. Empath: A continuous remote emotional health monitoring system for depressive illness. In Proceedings of the 2nd Conference on Wireless Health, San Diego, CA, USA, 10–13 October 2011.
8. *World's Population Increasingly Urban with More Than Half Living in Urban Areas*; United Nat: New York, NY, USA, 2014; Available online: http://www.un.org/en/development/desa/news/population/world-urbanization-prospects-2014.html (accessed on 23 January 2019).
9. Evans, D. The Internet of Things: How the Next Evolution of the Internet Is Changing Everything. 2011. Available online: http://www.cisco.com/c/dam/en_us/about/ac79/docs/innov/IoT_IBSG_0411FINAL.pdf (accessed on 23 January 2019).
10. Gharaibeh, A. Smart Cities: A Survey on Data Management, Security, and Enabling Technologies. *IEEE Commun. Surv. Tutor.* **2017**, *19*, 2456–2501. [CrossRef]
11. Al-Dulaimi, A.; Al-Rubaye, S.; Cosmas, J.; Anpalagan, A. Planning of Ultra-Dense Wireless Networks. *IEEE Netw.* **2017**, *31*, 90–96. [CrossRef]
12. Lee, S.; Kim, S.; Park, Y.; Choi, S.; Hong, D. Effect of Unpredictable Interference on MU-MIMO Systems in HetNet. *IEEE Access* **2018**, *6*, 28870–28876. [CrossRef]
13. Park, Y.; Hong, D. Theoretical Analysis of Interference Effect from Idle Cells in Ultra-Dense Small Cell Networks. *IEEE Access* **2018**, *6*, 26881–26894. [CrossRef]
14. Liu, Z.; Liu, J.; Zeng, Y.; Ma, J. Covert Wireless Communications in IoT Systems: Hiding Information in Interference. *IEEE Wirel. Commun.* **2018**, *25*, 46–52. [CrossRef]
15. Yao, K.; Wu, Q.; Xu, Y.; Jing, J. Distributed ABS-Slot Access in Dense Heterogeneous Networks: A Potential Game Approach with Generalized Interference Model. *IEEE Access* **2017**, *5*, 94–104. [CrossRef]
16. Fu, S.; Su, Z.; Jia, Y.; Zhou, H.; Jin, Y.; Ren, J.; Wu, B.; Huq, S.K.M. Interference Cooperation via Distributed Game in 5G Networks. *IEEE Internet Things J.* **2019**, *6*, 311–320. [CrossRef]
17. Zhang, L.; Ijaz, A.; Xiao, P.; Tafazolli, R. Channel Equalization and Interference Analysis for Uplink Narrowband Internet of Things (NB-IoT). *IEEE Commun. Lett.* **2017**, *21*, 2206–2210. [CrossRef]
18. Croce, D.; Gucciardo, M.; Mangione, S.; Santaromita, G.; Tinnirello, I. Impact of LoRa Imperfect Orthogonality: Analysis of Link-Level Performance. *IEEE Commun. Lett.* **2018**, *22*, 796–799. [CrossRef]
19. Kishk, M.A.; Dhillon, H.S. Joint Uplink and Downlink Coverage Analysis of Cellular-based RF-powered IoT Network. *IEEE Trans. Green Commun. Netw.* **2018**, *2*, 446–459. [CrossRef]
20. Malik, H.; Pervaiz, H.; Alam, M.M.; Le Moullec, Y.; Kuusik, A.; Imran, A.M. Radio Resource Management Schemein NB-IoT Systems. *IEEE Access* **2018**, *6*, 15051–15064. [CrossRef]
21. Baidya, S.; Levorato, M. Content-Aware Cognitive Interference Control for Urban IoT Systems. *IEEE Trans. Cogn. Commun. Netw.* **2018**, *4*, 500–512. [CrossRef]
22. Lee, H.; Park, Y.; Hong, D. Resource Split Full Duplex to Mitigate Inter-Cell Interference in Ultra-Dense Small Cell Networks. *IEEE Access* **2018**, *6*, 37653–37664. [CrossRef]
23. Lynggaard, P. Using Machine Learning for Adaptive Interference Suppression in Wireless Sensor Networks. *IEEE Sens. J.* **2018**, *18*, 8820–8826. [CrossRef]
24. Gu, Y.; Cui, Q.; Chen, Y.; Ni, W.; Tao, X.; Zhang, P. Effective Capacity Analysis in Ultra-Dense Wireless Networks with Random Interference. *IEEE Access* **2018**, *6*, 19499–19508. [CrossRef]
25. Jiang, N.; Deng, Y.; Kang, X.; Nallanathan, A. Random Access Analysis for Massive IoT Networks Under a New Spatio-Temporal Model: A Stochastic Geometry Approach. *IEEE Trans. Commun.* **2018**, *66*, 5788–5803. [CrossRef]
26. Grimaldi, S.; Mahmood, A.; Gidlund, M. Real-Time Interference Identification via Supervised Learning: Embedding Coexistence Awareness in IoT Devices. *IEEE Access* **2019**, *7*, 835–850. [CrossRef]

27. Sha, M.; Hackmann, C.G. Real-World Empirical Studies on Multi-Channel Reliability and Spectrum Usage for Home-Area Sensor Networks. *IEEE Trans Netw. Serv. Manag.* **2013**, *10*, 56–69. [CrossRef]
28. Shen, Y.; Wymeersch, H.; Moe, Z.W. Fundamental Limits of Wideband Localization—Part II: Cooperative Networks. *IEEE Trans Inf.* **2010**, *56*, 4981–5000. [CrossRef]
29. Yuan, W.; Wu, N.; Etzlinger, B.; Li, Y.; Yan, C.; Hanzo, L. Expectation Maximization-Based Passive Localization Relying on Asynchronous Receivers: Centralized versus Distributed Implementations. *IEEE Trans. Commmun.* **2019**, *67*, 668–681. [CrossRef]
30. Xiong, Y.; Wu, M.N.; Shen, Y.; Win, M.Z. Cooperative Network Synchronization: Asymptotic Analysis. *IEEE Trans. Sign. Proc.* **2018**, *66*, 757–772. [CrossRef]
31. Mostafaei, H.; Montieri, A.; Persico, V.; Pescape, A. A sleep shceduling approach based on learning automata for WSN partial coverage. *J. Netw. Comp. Appl.* **2017**, *80*, 67–78. [CrossRef]
32. Azpilicueta, L.; Rawat, M.; Rawat, K.; Ghannouchi, F.; Falcone, F. Convergence analysis in deterministic 3D ray launching radio channel estimation in complex environments. *Appl. Comput. Electromagn. Soc. J.* **2014**, *29*, 256–271.
33. Azpilicueta, L.; Vargas-Rosales, C.; Falcone, F. Deterministic Propagation Prediction in Transportation Systems. *IEEE Veh. Technol. Mag.* **2016**, *11*, 29–37. [CrossRef]
34. Azpilicueta, L.; López-Iturri, P.; Aguirre, E.; Falcone, F. An accurate UTD Extension to a Ray Launching Algorithm for the Analysis of Complex Indoor Radio Environments. *J. Electromagn. Waves Appl.* **2016**, *30*, 43–60. [CrossRef]
35. Hall, P.S.; Hao, Y. *Antennas and Propagation for Body-Centric Wireless Communications*, 2nd ed.; Artech House: Norwood, MA, USA, 2012.
36. Aguirre, E.; Arpón, J.; Azpilicueta, L.; De Miguel, S.; Ramos, V.; Falcone, F. Evaluation of Electromagnetic Dosimetry of Wireless Systems in Complex Indoor Scenarios with human body interaction. *Pier B* **2012**, *43*, 189–209. [CrossRef]
37. Lopez-Iturri, P.; Aguirre, E.; Azpilicueta, L.; Astrain, J.J.; Villadangos, J.; Falcone, F. Challenges in Wireless System Integration as Enablers for Indoor Context Aware Environments. *Sensors* **2017**, *17*, 1616. [CrossRef]
38. *IEEE 802.15.4: Wireless Medium Access Control (MAC) and Physical Layer (PHY) Specifications for Low-Rate Wireless Personal Area Networks (WPANs)*; IEEE: Piscataway, NY, USA, 2006.
39. Koutitas, G. Multiple Human Effects in Body Area Networks. *IEEE Antennas Wirel. Propag. Lett.* **2010**, *9*, 938–941. [CrossRef]

Article

A Time–Frequency Acoustic Emission-Based Technique to Assess Workpiece Surface Quality in Ceramic Grinding with PZT Transducer

Martin A. Aulestia Viera [1,*], Paulo R. Aguiar [1], Pedro Oliveira Junior [1], Felipe A. Alexandre [1], Wenderson N. Lopes [1], Eduardo C. Bianchi [2], Rosemar Batista da Silva [3], Doriana D'addona [4] and Andre Andreoli [1]

1 Department of Electrical Engineering, São Paulo State University—UNESP, Av. Eng. Luiz Edmundo Carrijo Coube, 14-01, Bauru 17033-360, Brazil
2 Department of Mechanical Engineering, São Paulo State University—UNESP, Av. Eng. Luiz Edmundo Carrijo Coube, 14-01, Bauru 17033-360, Brazil
3 School of Mechanical Engineering, Federal University of Uberlandia, Av. João Naves de Avila 2121, Uberlandia 38408-100, Brazil
4 Fraunhofer Joint Laboratory of Excellence on Advanced Production Technology (Fh-J_LEAPT Naples) Department of Chemical, Material and Industrial Production Engineering, University of Naples Federico II, Piazzale Tecchio 80, 80125 Naples, Italy
* Correspondence: martin.aulestia@unesp.br; Tel.: +55-014-99795-4453

Received: 30 July 2019; Accepted: 5 September 2019; Published: 11 September 2019

Abstract: Innovative monitoring systems based on sensor signals have emerged in recent years in view of their potential for diagnosing machining process conditions. In this context, preliminary applications of fast-response and low-cost piezoelectric diaphragms (PZT) have recently emerged in the grinding monitoring field. However, there is a lack of application regarding the grinding of ceramic materials. Thus, this work presents an analysis of the feasibility of using the acoustic emission signals obtained through the PZT diaphragm, together with digital signal processing in the time–frequency domain, in the monitoring of the surface quality of ceramic components during the surface grinding process. For comparative purpose, an acoustic emission (AE) sensor, commonly used in industry, was used as a baseline. The results obtained by the PZT diaphragm were similar to the results obtained using the AE sensor. The time–frequency analysis allowed to identify irregularities throughout the monitored process.

Keywords: piezoelectric transducer; sensor monitoring; ceramic grinding; digital signal processing; acoustic emission; short-time Fourier transform

1. Introduction

Grinding is a precision abrasive machining process used extensively for producing components with fine tolerances and high surface quality [1]. Among all the variables measured during the grinding process, the surface integrity is one of the most important parameters of any machined surface and is a decisive factor in the evaluation of a successful grinding, especially when high surface integrity requirements need to be met for some applications [2]. The monitoring of the grinding process is very complex because of the high number of influencing parameters, such as the workpiece, the grinding machine, and the process parameters. The control and monitoring of the grinding process allow the improvement of the process performance and the reduction of defects to a possible minimum to guarantee high precision and quality [3,4].

According to Teti et al. [5], the monitoring of machining operations has traditionally been categorized into two approaches: Direct and indirect. In the direct approach, the actual quantity of the variable, e.g., tool wear, is measured. On the other hand, the indirect approach uses auxiliary quantities, e.g., sensor signals (acoustic emission (AE), power, vibration, and force), to estimate a variable quantity. According to [6], among all the signals of the mentioned sensors, the acoustic emission (AE) is considered the most sensitive signal because its frequency range is beyond mechanical vibrations and electrical noises and, therefore, these noises can be easily filtered. According to Lopes et al. [7], the acoustic emission signal is defined as the transient elastic waves generated by the rapid release of energy within a material. Regarding this subject, AE techniques have been widely employed to monitor engineering applications [8–12]. For example, in He et al. [13], the AE method for detecting metallic grinding burn was presented as a nondestructive detection method. Lopes et al. [14] studied the influence of the temperature on the frequency response of an AE sensor. In Badger et al. [15], the fundamental relationships between the AE signal and the dressing variables were proposed.

Recently, piezoelectric diaphragms of lead zirconate titanate (PZT) have become popular due to the fact that it has been successfully used in many scientific applications in view of its low cost and excellent sensing capacities [16,17]. The PZT diaphragms have a simple construction consisting of a brass plate on which a ceramic disc is fixed. In the past, these acoustic components were only used for sound generation (buzzers and telephone receivers). However, the considerable potential of these transducers makes them attractive to other applications besides producing sound [18]. In this context, PZT diaphragms have been successfully used in the areas of structural health monitoring (SHM) to detect structural damages in engineering projects by signals related to the piezoelectric effect [19]. In addition, they have also been used as ultrasonic actuators [20], as the active element in acoustic position encoder [21], and as an acoustic sensor for partial discharge monitoring in power transformers [22].

The present work differs from other studies because it proposes a non-invasive and low-cost monitoring technique of the ceramic grinding process through the implementation of the PZT diaphragm and digital signal processing in the time–frequency domain, which allows a better interpretation of the results when compared to the traditional techniques. An important aspect of this research work is the use of the PZT diaphragm, which is a low-cost sensor compared to sensors such as acoustic emission, dynamometer, accelerometer, and power. PZT diaphragms can operate in both active and passive configurations and have an average cost of a few cents versus the high average cost of AE sensors, which range from hundreds to thousands of dollars [23]. In addition, they are compact, flexible, lightweight, and simple acoustic components widely used in various electronic devices to produce sound (alarm, ringing, and beep) [18,24]. Some preliminary applications of the PZT diaphragm emerged in the grinding field focusing on workpiece surface integrity and tool condition monitoring. For example, in Batista da Silva et al.'s study [25], the electromechanical impedance (EMI) method with low-cost PZT diaphragms were used to monitor the surface damage of ground steel workpieces. Microhardness and roughness measurements were compared with the results of the proposed technique. In Marchi et al. [26], two PZT diaphragms were used to detect the workpiece wear by means of the EMI method. The workpiece wear was correlated to the calculated damage statistics. Ribeiro et al. [27] proposed a new technique for monitoring the surface burning on steel workpieces using the PZT diaphragm and two types of grinding wheels (cubic boron nitride and aluminum oxide). An acoustic emission sensor was used to verify the efficiency of the PZT diaphragm. In Junior et al. [28], an approach for monitoring the dressing operation by means of PZT diaphragm-based impedance was proposed. The authors validated the proposed approach based on artificial neural networks (ANN), which selected the most damage-sensitive features based on the optimal frequency band.

The present work is an expansion of the research activities described in Viera et al. [16,29], where initial results were presented. Therefore, the scope of this research work is to illustrate a new broader approach for the real-time monitoring system of the surface quality of ground ceramics using the low-cost PZT diaphragm. In this context, the results of the time–frequency analysis, obtained through the short-time Fourier transform (STFT), are used as the basis for the computation of the ratio of power (ROP) metric, which is considered an important metric in the frequency domain.

However, in the present work, the ROP was computed in the time domain based on the study of Thomazella et al. [30], who conducted a pioneer work with the time–frequency approach along with the time domain ROP for monitoring self-vibrations on ground steel workpieces. The application of the technique in the monitoring of the ceramic grinding process in order to estimate the workpiece surface quality represents a novel approach for the manufacturing field. Moreover, this work differs from [30] because of the use of the low-cost piezoelectric diaphragm instead of the accelerometer. In addition, in order to verify the effectiveness and reliability of the proposed approach, an acoustic emission (AE) sensor, consolidated in the monitoring of manufacturing processes, was used. This paper is organized as follows: A ceramic grinding overview is presented in Section 2. Sections 3 and 4 describe the piezoelectric diaphragms and the signal processing techniques used in this study. The experimental setup is shown in Section 5. Section 6 presents the results and discussion. Finally, the conclusions are presented in Section 7.

2. Ceramic Grinding Overview

Ceramic components are susceptible to damages, such as cracks and residual stress due to their extreme hardness and high brittleness, which may affect the surface properties of the material [31]. Ceramic machining using diamond cutting tools is the primary technique for achieving specified dimensions and acceptable surface finish. Depending on the application of the component, the machining of the synthesized ceramics can represent more than 50% of the production cost, compared to 5% to 15% for metallic components. Among all machining processes, grinding represents more than 80% of all ceramic machining [32]. According to Brinksmeier et al. [33], the results of a grinding process can be subdivided into characteristics concerning the geometry and surface integrity of the workpiece. The essential macro geometric characteristics are dimension, shape, and waviness, while surface roughness is the main micro geometric characteristic. The surface integrity can be described by residual stresses, hardness, and material structure.

In Nascimento et al. [34], the viability of ceramic grinding with minimum quantity lubrication (MQL) with water was studied. A total of 45 grinding tests were performed with different MQL water concentration, depth of cut, and feed rate. The measured output variables were surface roughness, power, and scanning electron micrographs. The results show that MQL water–oil (1:1) was superior to conventional lubrication in terms of surface quality. Thomas et al. [35] proposed a new mathematical model to predict the surface roughness of ground ceramics. The effectiveness of this model was proved by the comparison of the experimental results with the predicted results. The author concluded that the optimization of the surface roughness can be done by controlling the grinding parameters. Liu et al. [36] studied the effects of the grinding parameters in the silicon nitride ceramic grinding. The influence of the grinding parameters, such as grain size, wheel speed, workpiece speed, and grinding depth, were analyzed regarding their effects on the grinding force, surface roughness, and subsurface damage. The ceramic grinding process can be optimized by the correct choice of grinding parameters. It is worth mentioning that the cited works used the surface roughness as the main variable in the evaluation of the quality of the workpiece surface.

The workpiece surface roughness is usually the most significant evaluation indicator in assessing the quality of ground surfaces. The evaluation of the competitiveness of the system can be performed through the estimation of the surface roughness, allowing productivity improvements and reducing costs [37]. The mean surface roughness (R_a) is defined as the arithmetic average of the absolute values of the deviations of the surface profile height from the mean line within the sampling length l [38]. Another tool used to evaluate the surface characteristics is the confocal microscopy. Thus, surface changes can be evaluated by reconstructing the topography of the surface from optical sections and light reflection [39]. In the ceramic grinding process, confocal microscopy can be found in the study of the cutting tool modeling [40] and the estimation of the maximum depth of cut during MQL grinding [41].

The monitoring of the ceramic grinding process by sensors has barely been studied, and only a few studies have been published. Nakai et al. [42] performed the estimation of the surface roughness of ceramic components through artificial neural networks (ANN). Four artificial models were tested with input statistics derived from the AE and power signals. The results obtained an accuracy of over 90%, which contributes to the automation of the ceramic grinding process. Feng et al. [43] monitored the cutting tool wear during ceramic micro-end grinding using AE, vibration, and force signals. The feasibility of monitoring the wear of the ceramic micro-grinding tool without knowing the machining characteristics was verified. Junior et al. [44,45] performed a feature extraction using the frequency domain spectrum and time-domain analysis of vibration signals to monitor the advanced ceramic grinding process. The statistics calculated by the application of digital filters in the chosen frequency bands were correlated with the workpiece surface roughness. Finally, the statistical analysis indicated similarity and confidence among the results presented.

3. Piezoelectric Diaphragms

The piezoelectric diaphragms have a very simple construction and are available in different sizes. This type of transducer consists of a circular piezoelectric ceramic (active element typically of barium titanate or PZT), usually ranging from 0.1 to 2 mm in thickness, mounted on a circular metal plate (diaphragm available in brass, nickel alloy, or stainless steel). The ceramic is coated with a thin metallic film (usually silver) that serves as an electrode. Figure 1 shows a typical PZT diaphragm and its parts [18]. Piezoelectric transducers can operate both as sensors and actuators due to the piezoelectric effect [46]. The piezoelectric effect consists of the generation of an electric dipole in a material that is subjected to a mechanical force, which results in an electrical output voltage. The polarization produced by the voltage creates charges and therefore an electric field. In the reverse effect, the application of an electric voltage in the piezoelectric material causes a mechanical deformation [47].

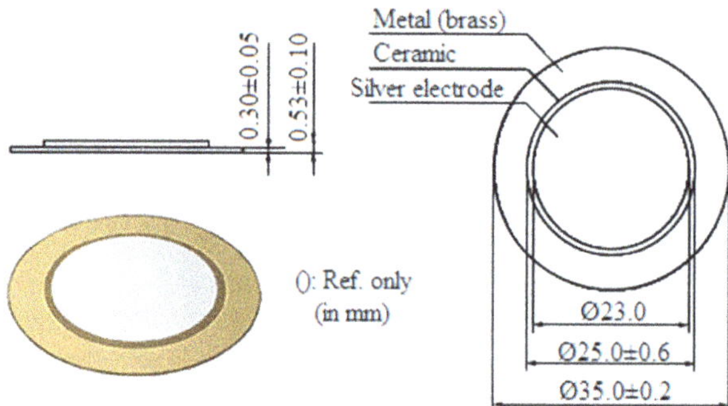

Figure 1. Murata piezoelectric sensor 7BB-35-3 used in the tests.

According to Castro et al. [22], in piezoelectric material, there is an electromechanical coupling, i.e., an electric field applied to the material generates a mechanical deformation while a mechanical change generates an electric load. Thus, piezoelectric transducers are capable of generating an electrical voltage when altered by a mechanical stress, which is generated by an acoustic wave. The application of piezoelectric sensors for different purposes can be found in the specific literature [48,49].

4. Signal Processing

The discrete Fourier transform (DFT) is a method for the analysis of frequency spectra in digital signal processing, usually implemented by the fast Fourier transform (FFT). However, the FFT is inadequate for identifying non-stationary transient information because it has no time resolution [50]. In this context, an alternative approach is to segment the sequence into a set of short subsequences, with each subsequence centered on uniform time intervals and its DFT calculated separately, thereby obtaining the short-time Fourier transform (STFT) [51]. Thus, the STFT is defined by Kim et al. [52]

$$STFT\ (t,\omega) = \int_{-\infty}^{\infty} h(u)f(t+u)e^{-j\omega u}\, du \tag{1}$$

where $f(t)$ is a given signal in the time domain, t is the time, ω the frequency, and $h(u)$ the temporal window function such as rectangular, Gaussian, Blackman, Hanning, Hamming, Kaiser, etc.

The result obtained by the STFT is a two-dimensional representation of the signal in time and frequency. However, the limitation imposed by Heisenberg's uncertainty principle requires a relation between the resolutions (time–frequency). The accuracy in time and frequency are mainly determined through the window length, which is constant for all frequencies. In this way, a longer window results in a better resolution in the frequency domain. However, to obtain a more accurate time resolution, a window of smaller length is used. Therefore, it is necessary to know the resolution in time and frequency to obtain all relevant information in both domains [53]. The resolution in time and frequency can be defined by Equations (2) and (3), respectively.

$$\Delta t = N * T_s \tag{2}$$

$$\Delta F = m\frac{F_s}{N} \tag{3}$$

where N is the window length; T_s the sampling period; F_s the sampling frequency; and m the window coefficient, and where the sampling period is equal to the inverse of the sampling frequency and the window coefficient depends on the type of window used, e.g., $m = 2$ for a rectangular window.

Many statistical parameters are applied to the signals collected during the monitoring of manufacturing processes. The most-used statistic is the root mean square (RMS) [54], however, other statistics are also applied to these signals and are able to diagnose events that occur in the monitored processes. In this context, the application of counts [7], DPO (power deviation) [55], and ROP [56] is highlighted.

According to Lin et al. [57], the ratio of power (ROP) is a statistical method for analyzing the ratio of a given frequency band with the total power spectrum. The ROP statistic can be defined as

$$ROP = \frac{\sum_{k=n1}^{n2} |x_k|^2}{\sum_{k=0}^{N-1} |x_k|^2} \tag{4}$$

where N is the block data size; $n1$ and $n2$ define the frequency range for the analysis; and x_k is the kth discrete Fourier transform [58].

The study of similarity between two or more sensor signals, statistics, or parameters is of vital importance in the validation of new techniques. The correlation study between a widely known variable in relation to a new variable in the test phase is done through several statistics and indices. In this sense, the application of correlation coefficient [59], wavelet coherence [60], and magnitude-squared coherence [61] is highlighted

According to Scannell et al. [62], the magnitude-squared coherence (MSC) lists common frequencies between two signals and evaluates their similarity. The MSC between the time domain signals x and y can be calculated by Equation (5). Results in the interval between 0 and 1 indicate the level of spectral similarity between both time-domain signals.

$$C_{xy} = \frac{\left|P_{xy}(f)\right|^2}{P_{xx}(f)P_{yy}(f)} \tag{5}$$

where $P_{xx}(f)$ and $P_{yy}(f)$ are the power spectral densities of x and y and $P_{xy}(f)$ is the cross power spectral density at frequency f.

5. Materials and Methods

5.1. Experimental Setup

Seven ceramic alumina (Al_2O_3) workpieces with 35 mm length × 8 mm width × 20 mm height and a Vickers microhardness of 1339 ± 47 HV1 (JIS R1610-1991 standard) were machined in a surface grinding machine, model RAPH 1055, from Sulmecanica. A resin-bond diamond grinding wheel, Dinser SD126MN50B2, with 15.5 mm thickness × 350 mm diameter was used in the tests. A single grinding pass was performed in the surface of each workpiece at a different depth of cut. Seven depth of cut values were tested, representing slight, moderate, and severe grinding conditions. Prior to the first test, the grinding wheel was dressed by a conglomerate-type dresser in order to ensure its best cutting performance. Coolant fluid was used in all tests. The workpieces were attached in the middle of a metal workpiece holder of 100 mm length × 35 mm width × 60 mm height by screws, which were adjusted with a torque wrench. The grinding parameters are shown in Table 1.

Table 1. Grinding parameters.

Grinding Speed	
Cutting speed (v_s)	33 m/s
Worktable speed (v_w)	58 mm/s
Depth of cut (μm)	25–35–50–105–150–210–350
Lubri-Cooling Specification	
Fluid	Shell–DMS 3200 F-1
Flow rate	27.5 L/min
Pressure	<0.7 MPa
Concentration	4% oil-water

5.2. Workpiece Surface Assessment

After grinding, the subsurface of each workpiece was evaluated by means of a three-dimensional (3D) LEICA digital confocal microscopy (DCM) with 800× magnification. In addition to the confocal analysis, the surface quality assessment was complemented with surface roughness (R_a) measurements, which were measured using a portable roughness tester Taylor Hobson 3+. A cut-off of 0.25 mm and a sampling length of 0.8 mm were considered. Each piece was divided into five equidistant points; three roughness measurements were taken from each point. The mean roughness was calculated for each workpiece, thus allowing the evaluation of the quality of the surface.

5.3. Data Acquisition

A low-cost piezoelectric diaphragm (PZT), model 7BB–35–3, from Murata Electronics, was attached to the workpiece holder with a cyanoacrylate of medium viscosity, from Tekbond. As in Viera et al. [29], the glue was evenly distributed over the entire diaphragm surface, thus maintaining a thickness of less than 1 mm. The piezoelectric diaphragm consists of (1) a bottom electrode (brass) of 35 mm diameter × 0.3 mm thickness (which is glued to the holder); (2) a piezoelectric ceramic of 25 mm diameter × 0.23 mm thickness (active element), and 3) an upper electrode (silver) of 23 mm diameter.

The study conducted by Freitas and Baptista [18] showed that the Murata piezoelectric diaphragms have a similar frequency response, determining that for most applications, the size of the diaphragm does not influence the results. Thus, the 35 mm model was chosen to validate the proposed approach. Other diaphragm models with different sizes and thicknesses can be tested in future studies. The piezoelectric diaphragm was glued to the center of the workpiece holder, thus positioning it as close as possible to the acoustic source (workpiece surface). Finally, the PZT diaphragm was covered with silicone in order to protect it from the coolant fluid. The PZT transducer was exclusively used in the passive configuration operating under d_{31} mode; in other words, the transducer was not excited by external sources; its function was to convert the acoustic waves generated during the process into electric voltage through the piezoelectric effect.

For comparison, an acoustic emission (AE) sensor was also fastened to the workpiece holder by a screw. The sensor was connected to a signal unit, model DM – 42, from Sensis manufacturer, which has an input–output gain of 3. The AE sensor consists of a metal housing of 30 mm × 20 mm × 20 mm in length, width, and height, respectively. Inside the metal housing is the piezoelectric ceramic (active element), as well as a filter and backing material, which reduces the resonance of the sensor. The AE sensor was screwed at a distance of 25 mm from the right end of the holder and at the same height as the PZT diaphragm.

The level of saturation and sensitivity of the two sensors was verified before the grinding tests. Both sensors were attached to the workpiece holder in order to reproduce an industrial application, where changes in machine cycles would be unacceptable. An oscilloscope, model DL850, manufactured by Yokogawa, collected the AE and PZT raw signals at a sample rate of 2 MS/s. The PZT output was directly connected to the oscilloscope. On the other hand, the AE output was first connected to the signal unit, which improves the signal conditions. The aliasing effect was prevented by an anti-aliasing filter, inbuilt in the oscilloscope data acquisition board, with a 65-dB cutoff centered at 80% of the sampling frequency. The experimental setup is shown in Figure 2.

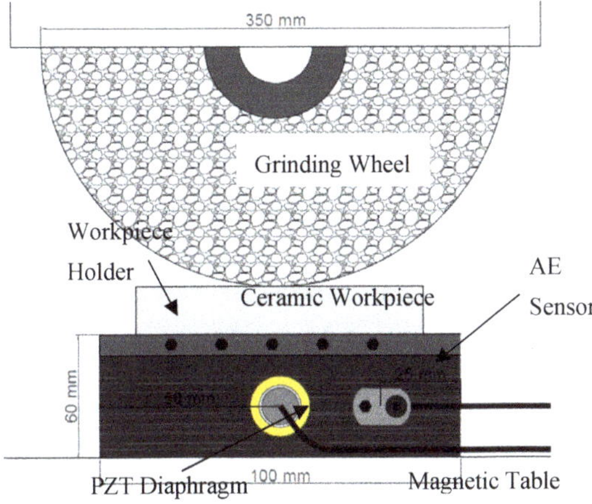

Figure 2. Experimental setup.

5.4. Signal Processing

The signals collected from the grinding tests were processed in MATLAB software. According to Ribeiro et al. [27], Moia et al. [63], and Lopes et al. [7], the AE sensor used in the tests has a frequency response of up to 300 kHz. However, as verified in the previous study of the same tests [29], the sensor presented significant frequencies up to 250 kHz. As shown by Freitas et al. [18], the PZT sensor has an efficient response of up to 200 kHz; the signal observed after this frequency has attenuations that can lead to erroneous results.

Considering the operational frequency bands of the sensors (up to 200 kHz for the PZT transducer and up to 300 kHz for the AE sensor), a digital bandpass filter (2 kHz to 300 kHz) was applied in order to eliminate noise. As there is a noise caused by the mechanical characteristics of the process and the data acquisition system, a digital Butterworth 10 order filter was also applied. The frequency of 2 kHz was chosen due to the fact that low frequencies present more information regarding noises than from the process. The signals were resampled at a rate of 1 MS/s, reducing the amount of data and respecting Nyquist's theorem. It is worth mentioning that the resample function of MATLAB applies an anti-aliasing low-pass FIR filter to the signals and compensates for the delay introduced by the filter. The STFT of each grinding test was computed with the following specifications: 5000-points rectangular window and 90% overlap. Thus, through Equations (2) and (3), the resolutions of 5 ms in time and 400 Hz in frequency were obtained.

Two frequency bands were chosen (one per sensor). The frequency bands were chosen based on the absolute values of the STFT spectrograms. Thus, the selected frequency bands presented two characteristics: (1) Frequency intervals with similar values over time (on the same spectrogram) and (2) magnitude differences (color bar) related to the process conditions (between spectrograms). Subsequently, the ROP metric was calculated for each frequency band. The absolute STFT values and the time–frequency resolution allowed the calculation of the ROP metric in the time domain. It is worth mentioning that the ROP metric is traditionally calculated in the frequency domain, so its calculation in the time domain, proposed in this paper, expands its application in real-time monitoring systems.

The magnitude-squared coherence (MSC) was then calculated for three grinding tests: 25 μm (slight), 105 μm (moderate), and 350 μm (severe). In addition, the upper envelope of each result was obtained in order to present the MSC results more clearly; the envelope was calculated by dividing the MSC results into 2048-point intervals, which correspond to about 1 ms time intervals from the total grinding pass section. Finally, the correlation between the measured surface roughness and the ROP mean values for both sensors was obtained by a linear regression. Signals were normalized in order to eliminate amplitude differences caused by the sensitivity of each sensor.

6. Results and Discussion

This section presents the results obtained from the characterization of the ground ceramic workpieces and the digital processing of the signals from both sensors. The results of the subsurface assessment by confocal microscopy are shown in Figure 3. The mean surface roughness values (R_a) measured with the portable roughness tester are presented in Table 2. Figure 4 presents the time–frequency analysis by STFT for both sensors and the seven cutting conditions. After the selection of frequency bands, the ROP parameter is shown in Figure 5. Figure 6 presents the magnitude coherence between the AE sensor and the PZT diaphragm at three cutting conditions. Finally, Figure 7 shows the correlation between the ROP metric and the surface roughness. All results were correlated with the surface quality of the ground workpieces.

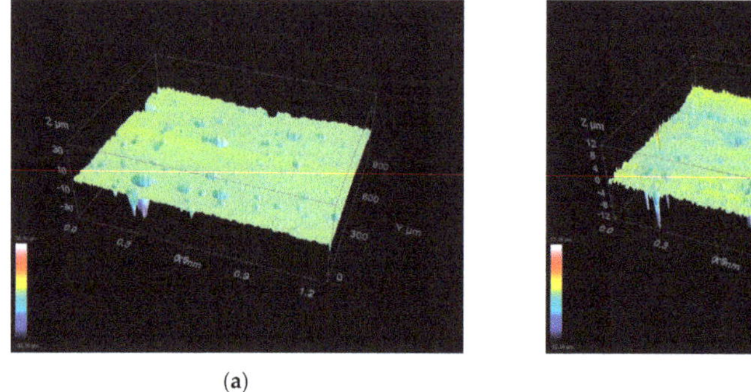

(a) (b)

Figure 3. *Cont.*

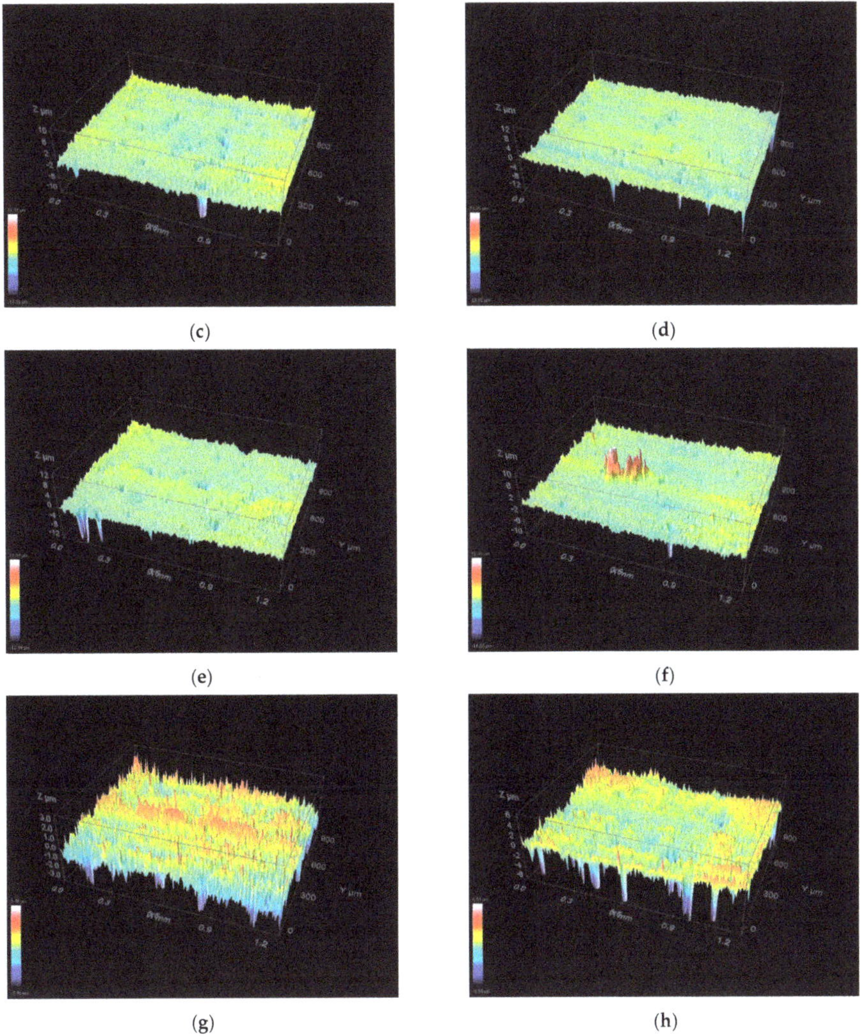

(c) (d) (e) (f) (g) (h)

Figure 3. Confocal microscopy: (**a**) Without cutting; (**b**) 25 µm; (**c**) 35 µm; (**d**) 50 µm; (**e**) 105 µm; (**f**) 150 µm; (**g**) 210 µm; and (**h**) 350 µm.

Table 2. Mean surface roughness (Ra) measured with a portable roughness tester.

Depth of Cut (a-µm)	Surface Roughness (R_a-µm)
25	0.516 ± 0.027
35	0.620 ± 0.033
50	0.647 ± 0.037
105	0.684 ± 0.040
150	0.697 ± 0.042
210	0.736 ± 0.051
350	0.793 ± 0.055

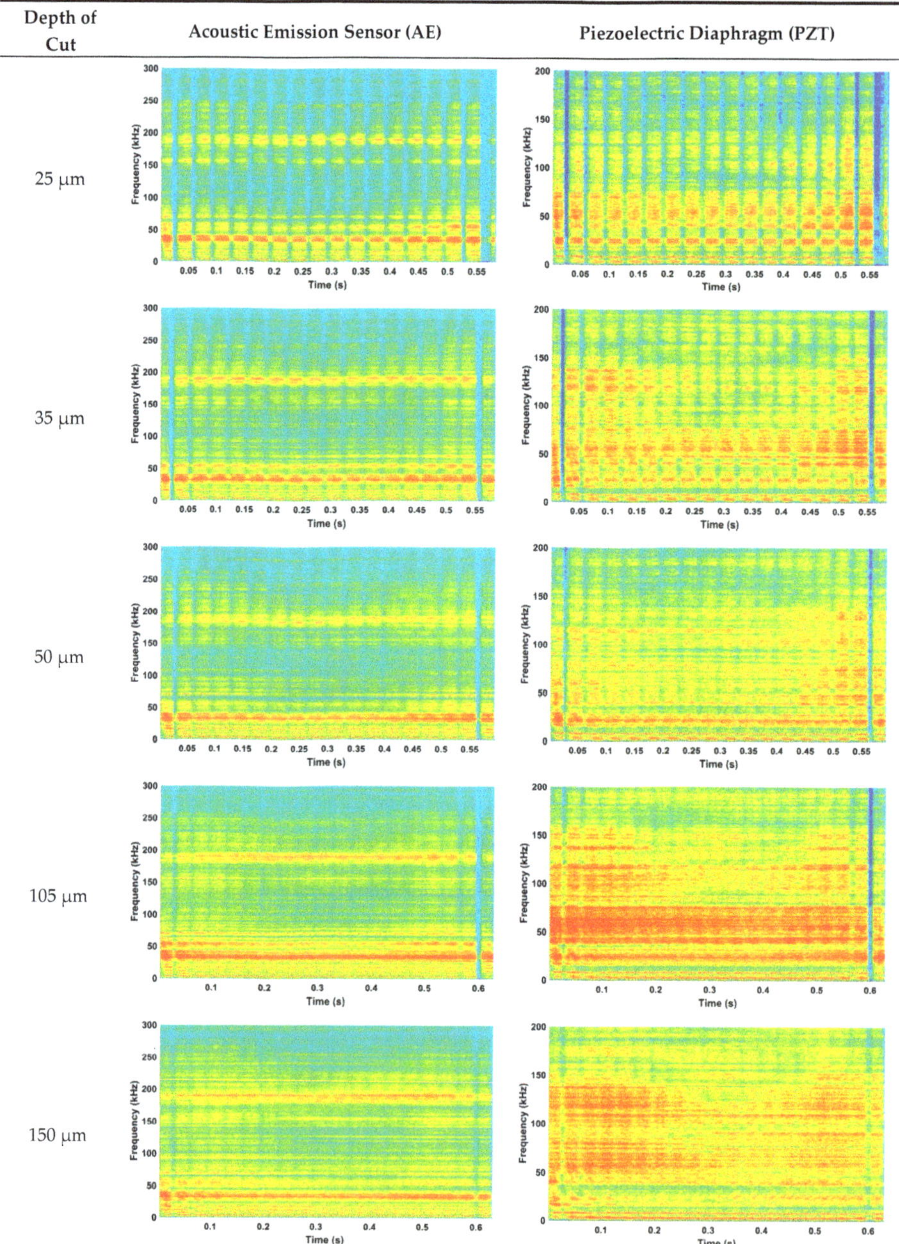

Figure 4. *Cont.*

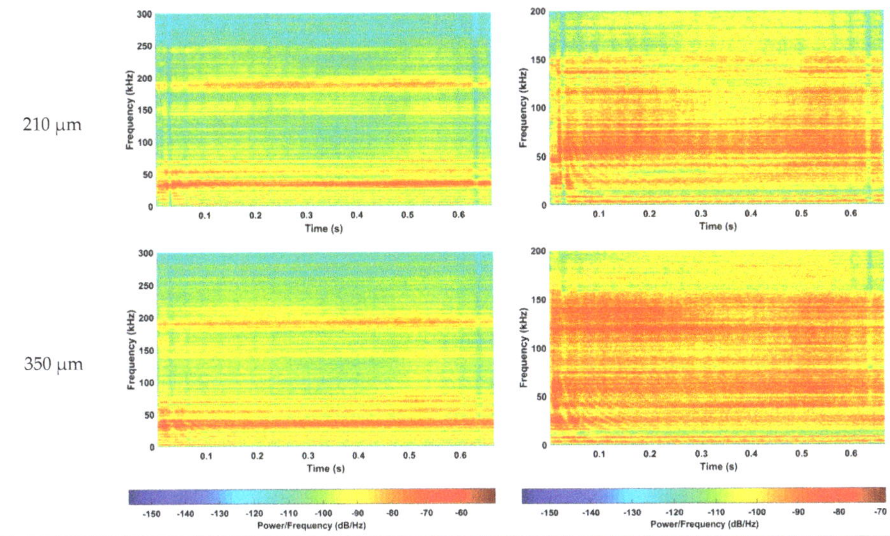

Figure 4. Short-time Fourier transform (STFT) spectrograms.

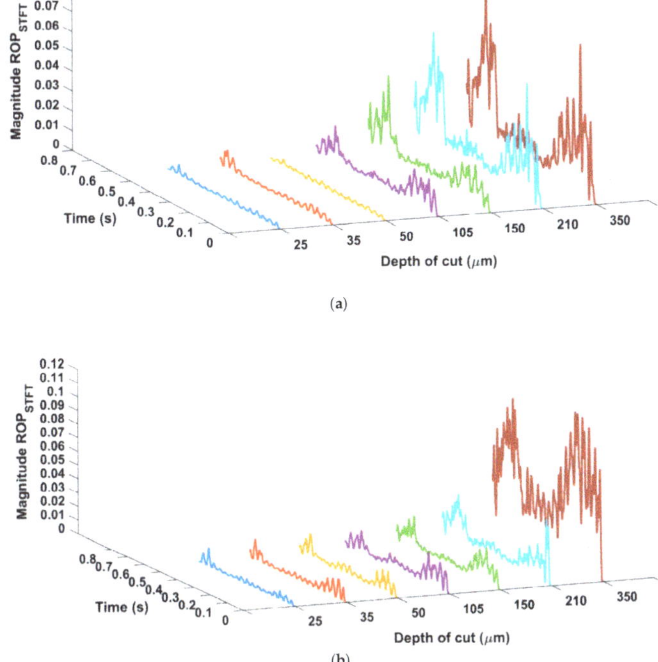

Figure 5. The ratio of power (ROP) of the (**a**) acoustic emission sensor (AE) and (**b**) piezoelectric diaphragm (PZT).

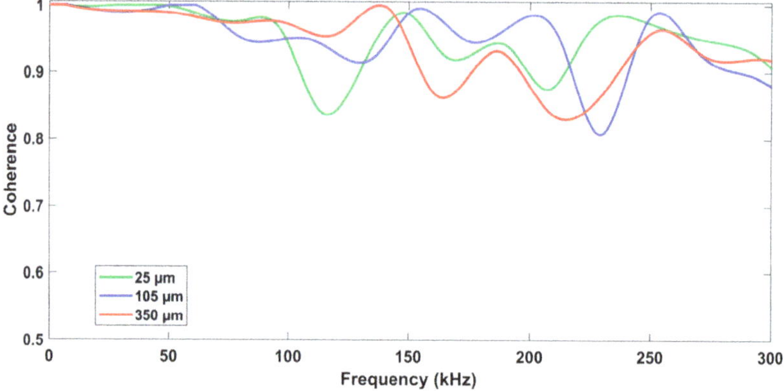

Figure 6. Magnitude-squared coherence between the AE sensor and PZT diaphragm at three grinding conditions.

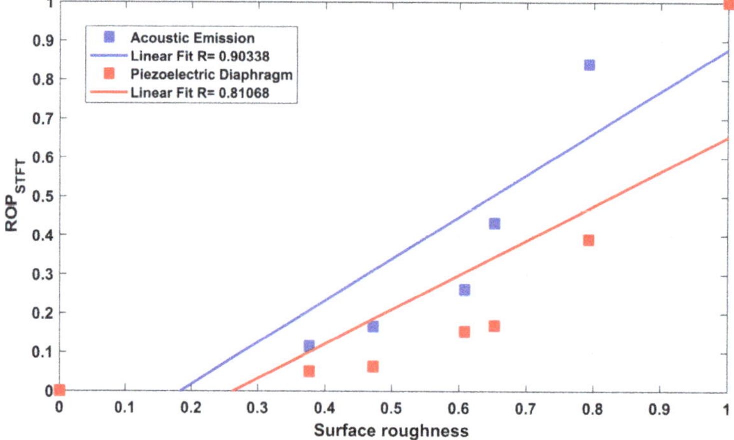

Figure 7. Correlation between ROP and surface roughness for both sensors.

6.1. Workpiece Surface Assessment

In order to detect micro-defects on the surfaces, confocal measurements were performed in a central region of each workpiece, as shown in Figure 3. As a reference, Figure 3a shows the microscopy of a workpiece prior to the grinding process (without cutting); a uniform surface can be observed with small imperfections caused by previous machining processes and short peak-to-valley distance. It can be seen in Figure 3 that the slight cutting conditions (Figure 3b–d) did not show any significant defects caused by the grinding process. However, some irregularities in the ceramic surface, such as cracks and porosity, can be observed, which resulted from previous machining processes.

The influence of the grinding process on the workpiece surface increases with the depth of cut. A bigger contact area between the workpiece surface and the abrasive grains of the grinding wheel generates an increased volume of material removed, which deteriorates surface quality and increases the number of surface cracks. In the moderate cutting conditions (Figure 3e,f), there is an increase of irregularities in the surface of the workpiece due to the increase of the cutting forces, which causes an increase in the grinding wheel wear and affects the surface quality of the workpiece. Figure 3f shows a predominant irregularity in the central-left part of the workpiece; this was caused by a crack that originated during the workpiece manufacturing process, which was enlarged by the grinding process. The other regions of the workpiece presented uniform irregularities, which are consistent with the severity of the process. The most significant effects on the workpiece surface were observed on the severe cutting conditions (Figure 3g,h); the irregularities and color differences in the images are related to the adjacency of the formed valleys and the distance between the peaks and valleys on the ground ceramic surface. The grinding effects mentioned above in the slight and moderate processes become more severe in the last two depth of cuts, where color changes can be observed throughout the workpiece surface. An increase in the peak-to-valley distance can be clearly observed, which represents a greater surface roughness.

The results of the mean surface roughness, measured with the portable roughness tester, are shown in Table 2. The results presented are the total mean of all measurements performed over the five regions that the workpiece surface was divided. As expected, a severe grinding process results in a higher mean roughness. The results presented in Figure 3 agree with those obtained in Table 2; it can be observed that the roughness values followed the same trend, rising directly with the increase of the depth of cut. According to Marinescu et al. [64], the maximum acceptable roughness value is of 1.6 μm for the steel grinding process. On the other hand, in the ceramic grinding process, a maximum roughness value of 1.0 μm is expected. Table 2 shows an increasing trend, with the lowest surface roughness value at 25 μm and the highest value at 350 μm. All roughness values are in accordance with the theory (less than 1.0 μm), which confirms that the ceramic grinding tests were successful. In addition, high standard deviation values are common for roughness measurements on ceramic components due to variations in peaks and valleys resulting from machining; these values increased as the cutting condition becomes more severe.

6.2. Signal Processing

The STFT spectrograms of both sensors for the three grinding conditions are shown in Figure 4. The color scale in the spectrograms represents the magnitude of a frequency in a given interval of time, the characteristic frequencies of the process, as well as the most influential frequencies, are shown in different shades of red colors. On the other hand, frequencies that are slightly sensitive to the grinding process are presented in shades of blue. The green, yellow, and orange colors, located in the middle of the color scale, represent frequencies sensitive to the process, but with a minor influence compared to the red ones.

In the spectrograms corresponding to the three slight grinding conditions (25 μm, 35 μm, and 50 μm), variations of magnitude can be observed over time. These variations were caused by the contact between the workpiece and the grinding wheel abrasive grains, which did not show a constant contact behavior, since many of the grains did not remove material from the workpiece surface, generating low levels of AE activity. Thus, little energy was released during the process, justifying the low harmonic content of the signals.

The two moderate grinding conditions (105 μm and 150 μm) showed higher magnitudes when compared to the slight conditions. An increase in red tones can be observed throughout the frequency spectrum, which is primarily caused by the grinding wheel wear and the process severity. The color pattern was shown to be more uniform due to the increased contact between the abrasive grains of the grinding wheel and the workpiece surface. The spectrograms representing the severe grinding processes (210 μm and 350 μm) presented uniform harmonic content over time. This behavior was caused by the continuous and severe contact between the grinding wheel grains and the ceramic component, generating more AE activity. In the three grinding conditions, a characteristic frequency was observed, with an approximate value of 35 kHz, present throughout the machining time. However, the more severe grinding condition presented higher harmonic content, assuming higher magnitude levels. The results obtained in Figure 4 are in agreement with the results of the direct measurements of Figure 3 and Table 2. As in Figure 3, it is possible to observe the color changes as the process severity increases, which is directly related to the surface roughness and surface quality of the ground workpiece. The attenuation of the PZT diaphragm response, near its maximum response region

(200 kHz), can be clearly observed in the spectrograms. There is a frequency component close to 200 kHz, which can be easily observed in the AE spectrograms, whereas in the PZT spectrograms it is difficult to observe, especially in slight cutting conditions, which generate lower acoustic levels. For this reason, it is not recommended to choose frequency bands near the maximum response region of the sensors.

The differences between the spectrograms of both sensors are related to the characteristics of the sensors, such as: The frequency response of the piezoelectric diaphragm (200 kHz) [18] and the AE sensor (300 kHz) [7,27,63], the construction of the sensors (AE sensor housing), and the signal unit. Through the spectrograms generated by the STFT, it is possible to differentiate the cutting conditions of the grinding process. In the results of Figure 4, it was possible to observe the changes in magnitude through the different cutting conditions and continuity through the grinding time. The two sensors presented similar results, showing a good sensitivity to the acoustic activity generated by the ceramic grinding process. The low-cost PZT diaphragm presented stronger color shades, which represents a greater sensitivity to the process stimuli. However, the AE sensor has a higher frequency response band and a signal unit, which filters and establishes gains for the raw signals, which makes the signal more reliable and accurate over time.

Frequency bands were chosen by means of the spectrograms shown in Figure 4. The selection criterion was based on the continuity of the frequency band over time and on the color difference between each grinding condition. As a result, two frequency bands were chosen, one for each sensor. Based on the selection criterion, the (142–147) kHz band was chosen for the piezoelectric diaphragm, while for the AE sensor, the (138–143) kHz band was selected. The frequency bands, chosen directly from the matrices containing the absolute values of each STFT, were used to calculate the ROP metric, which presented characteristics related to the process conditions.

The ROP results, applied in the selected frequency bands for both sensors, are shown in Figure 5. A higher ROP level can be clearly observed as the depth of cut increases. The results of the three slight cutting conditions presented low magnitudes, while moderate and severe conditions showed a greater magnitude and peak-to-valley distance. This behavior results from the accumulation of chip in the cutting region, which generates higher cutting force, AE activity, and surface roughness. Thus, the grinding of ceramic components at a depth of cut of 350 μm generated the highest levels of ROP, STFT, and R_a.

An increase in the ROP level at the beginning and at the end of the machining process can be clearly seen in Figure 5; all the cutting conditions showed the same characteristic. This is a typical behavior of the ceramic grinding process as it represents the first and last contact of the grinding wheel grains with the workpiece surface. The contact grinding area is smaller at the beginning and at the end of the machining process, regions in which the workpiece presents higher mechanical stress and a smaller area of energy propagation. This characteristic can be observed in Figure 2, from 0.0 to 0.2 and 1.0 to 1.2 on the x-axis, with color changes representing slight increases in surface roughness. In the middle region of the workpieces, the surface roughness presented more uniform values; these changes become clearer in the depths of cut that represent severe cutting conditions.

6.3. Correlation Analysis

The results of the magnitude-squared coherence (MSC) for three grinding conditions are observed in Figure 6. The coherence was obtained from the raw acoustic emission and piezoelectric time-domain signals. Each of the MSC values indicates how well the PZT signal corresponds to the AE signal for each frequency, with 1 representing an ideal coherence parameter and 0 representing the complete lack of relation between the signals. It can be observed that even in the severe cutting condition, where higher acoustic activity and spectral content is expected, the MSC values were higher than 80%. The successful results presented in Figure 6, directly related to the signals of both sensors under various grinding conditions, reinforce the application of the low-cost piezoelectric diaphragm in the monitoring of the grinding process of advanced ceramics. The lower values of MSC were observed in the frequency range between 200 and 300 kHz; this is due to the attenuation of the PZT sensor, which increases the difference between the signals. The low-cost PZT diaphragm demonstrated a similar response to the AE sensor, being sensitive to the stimulus caused by the grinding processes tested in this test bench.

A linear regression was performed and the coefficient of determination (R) was calculated in order to show the similarity between the measured surface roughness and the signals of each sensor, as shown in Figure 7. The coefficient of determination was 0.90338 for the AE sensor and 0.81068 for the PZT diaphragm. Although the linear fit was better for the AE sensor, there is a high coefficient for both sensors, which is also confirmed by the coherence of Figure 6. Thus, the ROP metric can contribute to the evaluation of the surface quality of the ground ceramic components, as indicated in the results obtained by the linear regression. The linear fit was close to 45°, i.e., R very close to 1, indicating a high correlation between the signals and the measured roughness, which is the focus of this study. As expected, the AE sensor showed a better response to the acoustic activity generated during the grinding process, which is related to the surface quality of the workpiece, because of the metal housing, filter, and signal unit, which improve the signal conditions. However, the signals from the PZT diaphragm, studied in this work, showed a high degree of correlation (greater than 80%), confirming the feasibility of using the low-cost PZT diaphragm in the monitoring of the surface quality of ground workpieces through the acoustic activity generated during the process. The PZT diaphragm was able to detect the same changes in acoustic activity as the AE sensor.

7. Conclusions

In this work, a flexible and low-cost piezoelectric diaphragm was applied for surface quality monitoring in grinding of ceramic materials. The PZT diaphragm was also compared to an AE sensor, which is widely used for monitoring in several machining fields. In addition, a new time-domain ROP metric, based on the short-time Fourier transform, was proposed with the objective of estimating the surface quality of the workpieces. Therefore, the following conclusions can be drawn from this work:

- A grinding process performed at a high depth of cut results in increased mean surface roughness and acoustic activity levels;
- Machining under severe conditions caused greater irregularities in the ceramic surfaces;
- The increase in ROP values was directly related to the increase in surface roughness, which was caused by the increase in process severity;
- The PZT diaphragm responded satisfactorily to the process stimuli; the results were supported by the behavior of the AE sensor;
- The coherence analysis between the responses of the low-cost PZT diaphragm and the AE sensor reinforced the results obtained, proving the viability of using the PZT diaphragm in the monitoring of the grinding process; all coherence values were higher than 80%;
- The results demonstrated the feasibility of applying the low-cost PZT diaphragms for the tested machining conditions and can be extended to other low-cost sensors and materials.

The use of the PZT diaphragm in the monitoring of the surface quality of ground ceramic components, together with time–frequency domain analysis techniques, is new, and there are several possibilities for future applications and studies. Future work can be performed with new signal analysis methods and grinding parameters. The use of neural networks of classification or even estimation can be studied further.

Author Contributions: Conceptualization, M.A.A.V. and P.R.A.; Data curation, P.O.J.; Formal analysis, F.A.A. and W.N.L.; Investigation, M.A.A.V. and P.O.J.; Methodology, F.A.A.; Supervision, P.R.A.; Writing—original draft, M.A.A.V.; Writing—review & editing, P.R.A., E.C.B., R.B.d.S., D.D. and A.A.

Funding: This research was funded by the São Paulo Research Foundation—FAPESP, grant number #2018/07292-0 and National Council for Scientific and Technological Development (CNPQ), under grant 306435/2017-9 for supporting this research work.

Acknowledgments: The authors are grateful to the São Paulo Research Foundation—FAPESP for the financial support.

Conflicts of Interest: The authors declare no conflict of interest.

References

1. Chakule, R.R.; Chaudhari, S.S.; Talmale, P.S. Evaluation of the effects of machining parameters on MQL based surface grinding process using response surface methodology. *J. Mech. Sci. Technol.* **2017**, *31*, 3907–3916. [CrossRef]
2. Onwuka, G.; Abou-El-Hossein, K. Surface Roughness in Ultra-high Precision Grinding of BK7. *Procedia CIRP* **2016**, *45*, 143–146. [CrossRef]
3. D'Addona, D.M.; Matarazzo, D.; Teti, R.; De Aguiar, P.R.; Bianchi, E.C.; Fornaro, A. Prediction of Dressing in Grinding Operation via Neural Networks. *Procedia CIRP* **2017**, *62*, 305–310. [CrossRef]
4. Karpuschewski, B.; Wehmeier, M.; Inasaki, I. Grinding monitoring system based on power and acoustic emission sensors. *CIRP Ann.-Manuf. Technol.* **2000**, *49*, 235–240. [CrossRef]
5. Teti, R.; Jemielniak, K.; O'Donnell, G.; Dornfeld, D. Advanced monitoring of machining operations. *CIRP Ann.-Manuf. Technol.* **2010**, *59*, 717–739. [CrossRef]
6. Kim, H.Y.; Kim, S.R.; Ahn, J.H. Process monitoring of centerless grinding using acoustic emission. *J. Mater. Process. Technol.* **2001**, *111*, 273–278. [CrossRef]
7. Lopes, W.N.; Ferreira, F.I.; Alexandre, F.A.; Ribeiro, D.M.S.; Junior, P.D.O.C.; de Aguiar, P.R.; Bianchi, E.C. Digital signal processing of acoustic emission signals using power spectral density and counts statistic applied to single-point dressing operation. *IET Sci. Meas. Technol.* **2017**, *11*, 631–636. [CrossRef]
8. Li, J.; Wang, X.; Shen, N.; Gao, H.; Zhao, C.; Wang, Y. Modeling of acoustic emission based on the experimental and theoretical methods and its application in face grinding. *Int. J. Adv. Manuf. Technol.* **2018**, *98*, 2335–2346. [CrossRef]
9. Boaron, A.; Weingaertner, W.L. Dynamic in-process characterization method based on acoustic emission for topographic assessment of conventional grinding wheels. *Wear* **2018**, *406–407*, 218–229. [CrossRef]
10. Holford, K.M.; Eaton, M.J.; Hensman, J.J.; Pullin, R.; Evans, S.L.; Dervilis, N.; Worden, K. A new methodology for automating acoustic emission detection of metallic fatigue fractures in highly demanding aerospace environments: An overview. *Prog. Aerosp. Sci.* **2017**, *90*, 1–11. [CrossRef]
11. Sikdar, S.; Kundu, A.; Jurek, M.; Ostachowicz, W. Nondestructive Analysis of Debonds in a Composite Structure under Variable Temperature Conditions. *Sensors* **2019**, *19*, 3454. [CrossRef] [PubMed]
12. Dias, E.A.; Pereira, F.B.; Ribeiro Filho, S.L.M.; Brandão, L.C. Monitoring of through-feed centreless grinding processes with acoustic emission signals. *Measurement* **2016**, *94*, 71–79. [CrossRef]
13. He, B.; Wei, C.; Ding, S.; Shi, Z. A survey of methods for detecting metallic grinding burn. *Measurement* **2019**, *134*, 426–439. [CrossRef]
14. Lopes, B.G.; Alexandre, F.A.; Lopes, W.N.; de Aguiar, P.R.; Bianchi, E.C.; Viera, M.A.A. Study on the effect of the temperature in Acoustic Emission Sensor by the Pencil Lead Break Test. In Proceedings of the 2018 13th IEEE International Conference on Industry Applications (INDUSCON), São Paulo, Brazil, 12–14 November 2018; IEEE: Piscataway, NJ, USA, 2018; pp. 1226–1229.
15. Badger, J.; Murphy, S.; O'Donnell, G.E. Acoustic emission in dressing of grinding wheels: AE intensity, dressing energy, and quantification of dressing sharpness and increase in diamond wear-flat size. *Int. J. Mach. Tools Manuf.* **2018**, *125*, 11–19. [CrossRef]
16. Viera, M.A.; Alexandre, F.; Lopes, W.; de Aguiar, P.; da Silva, R.B.; D'addona, D.; Andreoli, A.; Bianchi, E.; da Silva, R.; D'addona, D.; et al. A contribution to the monitoring of ceramic surface quality using a low-cost piezoelectric transducer in the grinding operation. In Proceedings of the 5th International Electronic Conference on Sensors and Applications, 15–30 November 2018; MDPI: Basel, Switzerland, 2018; Volume 4, p. 16. Available online: https://ecsa-5.sciforum.net/ (accessed on 10 September 2019).
17. Oliveira Junior, P.O.; Ferreira, F.I.; Aguiar, P.R.; Batista, F.G.; Bianchi, E.C.; Daddona, D.M. Time-domain Analysis Based on the Electromechanical Impedance Method for Monitoring of the Dressing Operation. *Procedia CIRP* **2018**, *67*, 319–324. [CrossRef]
18. De Freitas, E.S.; Baptista, F.G. Experimental analysis of the feasibility of low-cost piezoelectric diaphragms in impedance-based SHM applications. *Sens. Actuators A Phys.* **2016**, *238*, 220–228. [CrossRef]

19. Budoya, D.E.; Baptista, F.G. A Comparative Study of Impedance Measurement Techniques for Structural Health Monitoring Applications. *IEEE Trans. Instrum. Meas.* **2018**, *67*, 912–924. [CrossRef]
20. Chang, K.T.; Chiang, H.C.; Lee, C.W. Design and implementation of a piezoelectric clutch mechanism using piezoelectric buzzers. *Sens. Actuators A Phys.* **2008**, *141*, 515–522. [CrossRef]
21. Lamberti, N.; Caliano, G.; Savoia, A.S. ACUPAD: A track-pad device based on a piezoelectric bimorph. *Sens. Actuators A Phys.* **2015**, *222*, 130–139. [CrossRef]
22. Castro, B.; Clerice, G.; Ramos, C.; Andreoli, A.; Baptista, F.; Campos, F.; Ulson, J. Partial Discharge Monitoring in Power Transformers Using Low-Cost Piezoelectric Sensors. *Sensors* **2016**, *16*, 1266. [CrossRef]
23. Junior, P.; D'Addona, D.M.; Aguiar, P.R. Dressing Tool Condition Monitoring through Impedance-Based Sensors: Part 1—PZT Diaphragm Transducer Response and EMI Sensing Technique. *Sensors* **2018**, *18*, 4455. [CrossRef] [PubMed]
24. De Freitas, E.S.; Baptista, F.G.; Budoya, D.E.; De Castro, B.A. Equivalent Circuit of Piezoelectric Diaphragms for Impedance-Based Structural Health Monitoring Applications. *IEEE Sens. J.* **2017**, *17*, 5537–5546. [CrossRef]
25. Da Silva, R.B.; Ferreira, F.I.; Baptista, F.G.; de Aguiar, P.R.; de Souza Ruzzi, R.; Hubner, H.B.; Fonseca, M.D.P.C.; Bianchi, E.C. Electromechanical impedance (EMI) technique as alternative to monitor workpiece surface damages after the grinding operation. *Int. J. Adv. Manuf. Technol.* **2018**, *98*, 2429–2438. [CrossRef]
26. Marchi, M.; Baptista, F.G.; De Aguiar, P.R.; Bianchi, E.C. Grinding process monitoring based on electromechanical impedance measurements. *Meas. Sci. Technol.* **2015**, *26*, 045601. [CrossRef]
27. Ribeiro, D.M.S.; Aguiar, P.R.; Fabiano, L.F.G.; D'Addona, D.M.; Baptista, F.G.; Bianchi, E.C. Spectra Measurements Using Piezoelectric Diaphragms to Detect Burn in Grinding Process. *IEEE Trans. Instrum. Meas.* **2017**, *66*, 3052–3063. [CrossRef]
28. Junior, P.; D'Addona, D.; Aguiar, P.; Teti, R. Dressing Tool Condition Monitoring through Impedance-Based Sensors: Part 2—Neural Networks and K-Nearest Neighbor Classifier Approach. *Sensors* **2018**, *18*, 4453. [CrossRef] [PubMed]
29. Viera, M.A.A.; de Aguiar, P.R.; Junior, P.O.; da Silva, R.B.; Jackson, M.J.; Alexandre, F.A.; Bianchi, E.C. Low-Cost Piezoelectric Transducer for Ceramic Grinding Monitoring. *IEEE Sens. J.* **2019**, *19*, 7605–7612. [CrossRef]
30. Thomazella, R.; Lopes, W.N.; Aguiar, P.R.; Alexandre, F.A.; Fiocchi, A.A.; Bianchi, E.C. Digital signal processing for self-vibration monitoring in grinding: A new approach based on the time-frequency analysis of vibration signals. *Measurement* **2019**, *145*, 71–836. [CrossRef]
31. Zhang, G.; Zeng, Y.; Zhang, W. Scratching of SiC ceramics at two dimensional pre-stressing. *Int. J. Nanomanuf.* **2017**, *13*, 270. [CrossRef]
32. Sanchez, L.E.A.; Bukvic, G.; Fiocchi, A.A.; Fortulan, C.A. Allowance removal from green pieces as a method for improvement surface quality of advanced ceramics. *J. Clean. Prod.* **2018**, *186*, 10–21. [CrossRef]
33. Brinksmeier, E.; Tonshoff, H.K.; Czenkusch, C.; Heinzel, C. Modelling and optimization of grinding processes. *J. Intell. Manuf.* **1998**, *9*, 303–314. [CrossRef]
34. Do Nascimento, W.R.; Yamamoto, A.A.; de Mello, H.J.; Canarim, R.C.; de Aguiar, P.R.; Bianchi, E.C. A study on the viability of minimum quantity lubrication with water in grinding of ceramics using a hybrid-bonded diamond wheel. *Proc. Inst. Mech. Eng. Part B J. Eng. Manuf.* **2016**, *230*, 1630–1638. [CrossRef]
35. Thomas, B.; David, E.; Manu, R. Empirical modelling and parametric optimisation of surface roughness of silicon carbide advanced ceramics in surface grinding. *Int. J. Precis. Technol.* **2015**, *5*, 277–293. [CrossRef]
36. Liu, W.; Deng, Z.; Shang, Y.; Wan, L. Effects of grinding parameters on surface quality in silicon nitride grinding. *Ceram. Int.* **2017**, *43*, 1571–1577. [CrossRef]
37. Wu, C.; Li, B.; Liu, Y.; Liang, S.Y. Surface roughness modeling for grinding of Silicon Carbide ceramics considering co-existence of brittleness and ductility. *Int. J. Mech. Sci.* **2017**, *133*, 167–177. [CrossRef]
38. Khare, S.K.; Agarwal, S. Predictive Modeling of Surface Roughness in Grinding. *Procedia CIRP* **2015**, *31*, 375–380. [CrossRef]
39. Park, J.-B.; Jeon, Y.; Ko, Y. Effects of titanium brush on machined and sand-blasted/acid-etched titanium disc using confocal microscopy and contact profilometry. *Clin. Oral Implants Res.* **2015**, *26*, 130–136. [CrossRef]
40. Liao, D.; Shao, W.; Tang, J.; Li, J.; Tao, X. Numerical generation of grinding wheel surfaces based on time series method. *Int. J. Adv. Manuf. Technol.* **2018**, *94*, 561–569. [CrossRef]
41. Anand, P.S.P.; Arunachalam, N.; Vijayaraghavan, L. Study on grinding of pre-sintered zirconia using diamond wheel. *Mater. Manuf. Process.* **2018**, *33*, 634–643. [CrossRef]

42. Nakai, M.E.; Junior, H.G.; Aguiar, P.R.; Bianchi, E.C.; Spatti, D.H. Neural tool condition estimation in the grinding of advanced ceramics. *IEEE Lat. Am. Trans.* **2015**, *13*, 62–68. [CrossRef]
43. Feng, J.; Kim, B.S.; Shih, A.; Ni, J. Tool wear monitoring for micro-end grinding of ceramic materials. *J. Mater. Process. Technol.* **2009**, *209*, 5110–5116. [CrossRef]
44. Junior, P.O.C.; Aguiar, P.R.; Foschini, C.R.; França, T.V.; Ribeiro, D.M.S.; Ferreira, F.I.; Lopes, W.N.; Bianchi, E.C. Feature extraction using frequency spectrum and time domain analysis of vibration signals to monitoring advanced ceramic in grinding process. *IET Sci. Meas. Technol.* **2019**, *13*, 1–8. [CrossRef]
45. Conceicao, P.O.; Marchi, M.; Aguiar, P.R.; Bianchi, E.C.; Franca, T.V. The correlation of vibration signal features in grinding of advanced ceramics. *IEEE Lat. Am. Trans.* **2016**, *14*, 4006–4012. [CrossRef]
46. Junior, M.M.D.S.; Ulson, J.A.C.; Castro, B.A.D.; Ardila-Rey, J.A.; Campos, F.D.S.; Ferreira, L.T. Analysis of Piezoelectric Sensors in Adulteration of Bovine Milk Using the Chromatic Technique. *Proceedings* **2018**, *4*, 38. [CrossRef]
47. Castro, B.A.; Clerice, G.A.M.; Andreoli, A.L.; de Souza Campos, F.; Ulson, J.A.C. A low cost system for acoustic monitoring of partial discharge in power transformer by Piezoelectric Sensor. *IEEE Lat. Am. Trans.* **2016**, *14*, 3225–3231. [CrossRef]
48. De Almeida, V.A.D.; Baptista, F.G.; De Aguiar, P.R. Piezoelectric transducers assessed by the pencil lead break for impedance-based structural health monitoring. *IEEE Sens. J.* **2015**, *15*, 693–702. [CrossRef]
49. De Castro, B.A.; De Melo Brunini, D.; Baptista, F.G.; Andreoli, A.L.; Ulson, J.A.C. Assessment of Macro Fiber Composite Sensors for Measurement of Acoustic Partial Discharge Signals in Power Transformers. *IEEE Sens. J.* **2017**, *17*, 6090–6099. [CrossRef]
50. Yang, Z.; Yu, Z. Grinding wheel wear monitoring based on wavelet analysis and support vector machine. *Int. J. Adv. Manuf. Technol.* **2012**, *62*, 107–121. [CrossRef]
51. Zhang, S.; Yu, D.; Sheng, S. A discrete STFT processor for real-time spectrum analysis. In Proceedings of the IEEE Asia-Pacific Conference Circuits Systems (APCCAS), Singapore, 4–7 December 2006; pp. 1943–1946. [CrossRef]
52. Kim, B.S.; Lee, S.H.; Lee, M.G.; Ni, J.; Song, J.Y.; Lee, C.W. A comparative study on damage detection in speed-up and coast-down process of grinding spindle-typed rotor-bearing system. *J. Mater. Process. Technol.* **2007**, *187–188*, 30–36. [CrossRef]
53. Mangueira Lima, É.; Brito, N.S.D.; de Souza, B.A.; dos Santos, W.C.; de Andrade Fortunato, L.M. Analysis of the influence of the window used in the Short-Time Fourier Transform for High Impedance Fault detection. In Proceedings of the 2016 17th International Conference on Harmonics and Quality of Power (ICHQP), Belo Horizonte, Brazil, 16–19 October 2016; IEEE: Piscataway, NJ, USA, 2016; pp. 350–355.
54. Martins, C.H.R.; Aguiar, P.R.; Frech, A.; Bianchi, E.C. Tool condition monitoring of single-point dresser using acoustic emission and neural networks models. *IEEE Trans. Instrum. Meas.* **2014**, *63*, 667–679. [CrossRef]
55. Aguiar, P.R.; Bianchi, E.C.; Oliveira, J.F.G. A method for burning detection in grinding process using acoustic emission and effective electric power signals. *Manuf. Syst.* **2002**, *31*, 253–257.
56. Alexandre, F.A.; Lopes, W.N.; Lofrano Dotto, F.R.; Ferreira, F.I.; Aguiar, P.R.; Bianchi, E.C.; Lopes, J.C. Tool condition monitoring of aluminum oxide grinding wheel using AE and fuzzy model. *Int. J. Adv. Manuf. Technol.* **2018**, *96*, 67–79. [CrossRef]
57. Lin, Y.-K.; Wu, B.-F.; Chen, C.-M. Characterization of Grinding Wheel Condition by Acoustic Emission Signals. In Proceedings of the 2018 International Conference on System Science and Engineering (ICSSE), New Taipei, Taiwan, 28–30 June 2018; IEEE: Piscataway, NJ, USA, 2018; pp. 1–6.
58. Aguiar, P.R.; Serni, P.J.A.; Bianchi, E.C.; Dotto, F.R.L. In-process grinding monitoring by acoustic emission. In Proceedings of the 2004 IEEE International Conference Acoustics Speech, and Signal Processing, Montreal, QC, Canada, 17–21 May 2004; Volume 5, pp. 1–4. [CrossRef]
59. Guo, S.; Li, C.; Zhang, Y.; Wang, Y.; Li, B.; Yang, M.; Zhang, X.; Liu, G. Experimental evaluation of the lubrication performance of mixtures of castor oil with other vegetable oils in MQL grinding of nickel-based alloy. *J. Clean. Prod.* **2017**, *140*, 1060–1076. [CrossRef]
60. Liu, Y.; Wang, X.; Lin, J.; Zhao, W. Correlation analysis of motor current and chatter vibration in grinding using complex continuous wavelet coherence. *Meas. Sci. Technol.* **2016**, *27*, 115106. [CrossRef]
61. Gish, H.; Cochran, D. Generalized coherence (signal detection). In Proceedings of the ICASSP-88, International Conference on Acoustics, Speech, and Signal Processing, New York, NY, USA, 11–14 April 1988; IEEE: Piscataway, NJ, USA, 1988; pp. 2745–2748.

62. Scannell, E.; Carter, G. Confidence bounds for magnitude-squared coherence estimates. *IEEE Trans. Acoust.* **1978**, *26*, 475–477. [CrossRef]
63. Moia, D.F.G.; Thomazella, I.H.; Aguiar, P.R.; Bianchi, E.C.; Martins, C.H.R.; Marchi, M. Tool condition monitoring of aluminum oxide grinding wheel in dressing operation using acoustic emission and neural networks. *J. Braz. Soc. Mech. Sci. Eng.* **2015**, *37*, 627–640. [CrossRef]
64. Marinescu, I.D.; Rowe, W.B.; Dimitrov, B.; Ohmori, H. *Tribology of Abrasive Machining Processes*; Elsevier: Atlanta, GA, USA, 2013; ISBN 978-1-4377-3467-6.

Article

A Novel Ultrasound Technique Based on Piezoelectric Diaphragms Applied to Material Removal Monitoring in the Grinding Process

Felipe A. Alexandre [1,*], Paulo R. Aguiar [1], Reinaldo Götz [1], Martin Antonio Aulestia Viera [1], Thiago Glissoi Lopes [1] and Eduardo Carlos Bianchi [2]

1 Department of Electrical Engineering, São Paulo State University—UNESP, Av. Eng. Luiz Edmundo Carrijo Coube, 14-01, Bauru 17033-360, Brazil
2 Department of Mechanical Engineering, São Paulo State University—UNESP, Av. Eng. Luiz Edmundo Carrijo Coube, 14-01, Bauru 17033-360, Brazil
* Correspondence: felipe.alexandre@unesp.br; Tel.: +55-014-3234-7811

Received: 26 July 2019; Accepted: 4 September 2019; Published: 12 September 2019

Abstract: The interest of the scientific community for ultrasound techniques has increased in recent years due to its wide range of applications. A continuous effort of researchers and industries has been made in order to improve and increase the applicability of non-destructive evaluations (NDE). In this context, the monitoring of manufacturing processes, such as the grinding process, arises. This work proposes a novel technique of ultrasound monitoring (chirp-through-transmission) through low-cost piezoelectric diaphragms and digital signal processing. The proposed technique was applied to the monitoring of material removal during the grinding process. The technique is based on changes in ultrasonic waves when propagated through the material under study, with the difference that this technique does not use traditional parameters of ultrasonic techniques but digital signal processing (RMS and Counts). Furthermore, the novelty of the proposed technique is also the use of low-cost piezoelectric diaphragms in the emission and reception of ultrasonic waves, enabling the implementation of a low-cost monitoring system. The results show that the monitoring technique proposed in this work, when used in conjunction with the frequency band selection, is sensitive to the material removal in the grinding process and therefore presents an advance for monitoring the grinding processes.

Keywords: ultrasound technique; grinding monitoring; digital signal processing; piezoelectric transducers

---

1. Introduction

The manufacturing processes convert raw materials into finished products, often employing machines or machine tools [1]. Among the finishing processes, grinding is considered one of the most important processes because it is applied in the final stage of the production line, providing the final finishing of the part [2].

In the grinding process, the material removal mechanism is performed by the contact between the abrasive surface of the grinding wheel, which is composed of thousands of abrasive grains with undefined cutting geometries randomly distributed [3], with the workpiece surface, which makes grinding a highly stochastic process [4]. The monitoring of the grinding process allows a better understanding of the interaction between the grinding wheel abrasive grains and the workpiece surface [5]. In addition, a better understanding of the material removal mechanism can help in the control of the grinding wheel wear and the workpiece surface quality [6].

The monitoring methods used in manufacturing processes are usually separated into two major categories—direct and indirect [7]. Direct methods are usually employed in a laboratory environment because of their intrinsic handling limitations and are less used in industrial environments. On the other hand, indirect methods are more employed in industrial applications because of their easy handling and application, which are very important characteristics in this type of monitoring [8].

In the indirect methods, sensors are employed to measure a certain variable of interest. Among the sensors employed, the most common is the acoustic emission (AE) sensor, used by Nascimento Lopes et al. [9] in the monitoring of the dressing process; the power sensor, used by Chen et al. [10] in the study of the energy efficiency of an industrial cutting process; the vibration sensor (using accelerometers), used by Dimla [11] in the study of the cutting tool wear; and the force sensor (using dynamometers), used by Agarwal et al. [5], along with the acoustic emission sensor, in the monitoring of the grinding process.

The piezoelectric diaphragm (buzzer) is presented as a low-cost alternative to the traditional components used in structural impedance analysis. Freitas et al. [12] studied the feasibility of applying this transducer as an alternative to commonly used transducers in electromechanical impedance (EMI) methods. Budoya et al. [13] employed the piezoelectric diaphragm in the structural health monitoring system (SHM) to detect structural damages through the impedance signatures, acquired during the evaluating process. The piezoelectric diaphragm was also used as an alternative to the traditional vibration sensor (accelerometer), as presented by Lucas et al. [14], in the detection of structural anomalies in induction motors. In addition to this type of monitoring, the piezoelectric diaphragm was used in the monitoring of manufacturing processes, as presented by Marchi et al. [15], in which the electromechanical impedance method was implemented by means of piezoelectric diaphragms in the evaluation of the workpiece surface quality in the grinding process. In Ribeiro et al. [16], the piezoelectric diaphragm was employed to detect the burning phenomenon in ground workpieces, obtaining good results in the use of this transducer as an alternative to the commercial AE sensor.

This work proposes a new method for monitoring the material removal mechanism in the grinding process by means of the application of piezoelectric diaphragms, employing an alternative through-transmission technique, referred to as chirp-through-transmission ultrasound technique, along with digital signal processing techniques. The present work is an expansion of the research described in Alexandre et al. [17], where initial results were presented. In addition to the use of low-cost piezoelectric diaphragms in the emission and reception of ultrasound waves and the chirp signal as the input to the emitter transducer, what differentiates the proposed technique from others is the digital signal processing of the received signal (RMS and Counts) within a selected frequency band instead of traditional parameters of ultrasound techniques. Thus, this work paves the way to new possibilities for monitoring the material removal mechanism in the grinding process as well as for its automation. It is worth mentioning that the technique presented in this work is a novel approach to the monitoring of the grinding process, as no similar studies were found in literature.

2. Traditional Monitoring Techniques Applied to Industrial Processes

The use of sensors in the monitoring of manufacturing processes is consolidated in the manufacturing industry. Some monitoring techniques, such as the passive and active-passive, are well known. In this section, a brief review of researches that made use of such techniques is presented.

2.1. Passive Monitoring Techniques

The grinding process is a complex manufacturing process influenced by many factors, such as the workpiece, the grinding machine, the grinding wheel and the process parameters [15]. Due to the complexity of the process, grinding monitoring is very important for its automation and optimization. The use of different sensors in the passive configuration, such as the acoustic emission sensor and the piezoelectric diaphragm along with digital signal processing techniques, allows the detection of events in the grinding process and other industrial processes. Some advantages of the piezoelectric diaphragm can be herein highlighted as follow. The cost of the sensor is one of the main factors that makes the piezoelectric diaphragm more attractive than the acoustic emission sensor. This transducer has a very simple construction, consisting of a thin layer of piezoelectric material (active element) adhered to a circular metal plate (diaphragm). The ceramic is coated with a thin metallic film (usually silver) that acts as an electrode [12]. The PZT diaphragms can operate in both active and passive configurations and have an average cost of a few cents versus the high average cost of AE sensors, which range from hundreds to thousands of dollars [18]. In addition, they are compact, flexible, lightweight and simple acoustic components widely used in various electronic devices to produce sound (alarm, ringing and beep) [12,19]. Furthermore, the fastening of the PZT sensor generally requires only a thin layer of cyanoacrylate glue, while for some AE sensors it may be necessary to machine the holder in order to fit the sensor, which is difficult in some cases. In addition, the data acquisition system for the piezoelectric diaphragm has a lower cost compared to the system required for the AE sensor (this is mainly due to the lower frequency response of the PZT sensor). Finally, the PZT diaphragm is highly available as it can be easily bought by users. Although the sensitivity of the low-cost piezoelectric diaphragm is generally lower than the acoustic emission sensors, it can be considered as an alternative for monitoring several applications, as presented in Castro et al. [20], Viera et al. [21] and Ribeiro et al. [16].

The monitoring of the grinding wheel wear has been the subject of several studies since a worn grinding wheel can directly influence on the geometry and quality of the ground surfaces [18]. The most widely used sensors for grinding wheel wear monitoring (macro and micro effects) are acoustic emission, temperature, power and force sensors. The acoustic emission sensor is used for various purposes, such as monitoring the grinding wheel condition during the grinding and dressing process and detecting contact between the grinding wheel and the workpiece [22]. In this context, de Oliveira et al. [23] presented a monitoring system for the grinding process based on the analysis of the acoustic emission signal (RMS) and the construction of acoustic maps. This innovative system enables the evaluation of the dressing process, the topographic mapping of the grinding surface, as well as the interaction between the grinding wheel and the workpiece during the grinding operation.

The approach in which the PZT (buzzer) and the acoustic emission sensor are employed in conjunction with digital signal processing techniques can be found in many studies, where the objective is the monitoring of the grinding process and related processes (dressing). In order to detect the workpiece burning during the grinding process, Ribeiro et al. [16] collected the acoustic activity of the grinding process with a piezoelectric diaphragm and a commercial acoustic emission sensor. The results demonstrate that the selection of frequency bands optimized the use of metrics, such as the RMS and RMSD, in the detection of the burning phenomenon during grinding. In addition, it demonstrates that the piezoelectric diaphragm can achieve similar results to the AE sensor in detecting surface burns on ground workpieces.

A comparative study using a low-cost piezoelectric diaphragm and a commercial acoustic emission sensor, along with digital signal processing techniques in the evaluation of the surface quality of ground ceramic workpieces, was developed by Viera et al. [21,24]. The results showed that the proposed technique (statistic ROP_{STFT}) was able to detect roughness variations on the surfaces of the ground workpieces.

Castro et al. [20] employed the piezoelectric diaphragm as an alternative to the acoustic emission sensor in the detection of partial discharges in power transformers. The results showed that the low-cost sensor has low sensitivity when compared to the conventional acoustic emission sensor, however, for the application studied, the collected signals from both sensors had similar behavior, which characterizes its application in the detection of partial discharges in power transformers.

2.2. Active Monitoring with Emission-Reception Techniques

The elastic waves, when propagated in a material that presents intrinsic changes, are dispersed in all directions. The active monitoring methods employ fixed transducers at specific locations of the structure to generate and collect elastic waves after propagating through the material [25]. The active monitoring techniques typically employ ultrasonic [25] and piezoelectric transducers [26]. The following sections describe the most important piezoelectric-active techniques applied to processes monitoring.

2.2.1. Electromechanical Impedance Method (EMI) and Frequency Response Function Method (FRF)

The electromechanical impedance method (EMI) has been widely studied for the development of structural health monitoring systems [27]. According to Castro et al. [28], the EMI technique is based on the piezoelectric effect, which allows the monitoring of structural conditions through the electrical impedance of the transducer connected to a structure, since the electrical impedance of the piezoelectric transducer is related to the mechanical properties of the structure. The EMI technique can be applied to different types of materials, such as metallic materials and composites. For example, highlighting the works of Na et al. [29] on composite structure, Annamdas et al. [30] on concrete structure and Zhu et al. [31] on steel structure, thus indicating the wide range of applications. Structural damage is typically characterized using damage indexes. According to Budoya et al. [32], the most common damage indexes are based on the comparison between two electrical impedance signatures, where one of them is obtained when the structure is considered healthy. The most common indices in SHM analysis are the root mean square deviation (RMSD) and the correlation coefficient deviation metric (CCDM) [33]. The RMSD is indicated to quantify the amplitude differences between two spectrums (baseline and damage) [34], while the CCDM is appropriate to measure the variations or displacement between the frequencies of the analyzed signals [35].

In order to assess the damage present in ground 1020 steel workpieces, Marchi et al. [15] obtained the electromechanical impedance levels for the workpiece without damage and after the grinding process, which made the comparison possible. To assess the damage caused by the grinding process, the results of the RMSD and CCDM indexes were compared with the microhardness and surface roughness of the workpieces.

de Oliveira Conceição Jr. et al. [18,36] assessed the structural changes caused by wear in single-point dressing tools during their lifetime employing EMI method to ensure the reliable monitoring of the tool condition in grinding operations. Representative damage indices, such as RMSD and CCDM, obtained from impedance signatures at different frequency bands were computed for diverse tool conditions. Moreover, an intelligent system, implemented on the basis of artificial neural networks, was able to select the most damage-sensitive features based on the optimal frequency band. The authors highlighted that the EMI method was capable of effectively detect damages in the relatively small diamond tool tips showing a general overall classification error lower than 2% for the best neural models.

In contrast to the conventional EMI method, which is known for using a single transducer, the frequency response function (FRF) method uses at least two transducers, each operating separately as an actuator and as a sensor [37]. Two transducers are employed in this method, one in the emitter mode and the other one in the receiver mode. A significant change caused by incipient damage (detachment, rust, cracks, etc.) results in a difference in the FRF response for an application of the electromechanical impedance technique in an electronic circuit equipped with piezoelectric elements [38]. Liang et al. [39] employed the FRF method in the study of the loosening monitoring of threaded plumbing structures. The use of this method along with the RMSD index allowed the correlation between the RMSD values and the severity of the loosening found in the threaded structures.

2.2.2. Transmitter-Receiver Arrangements for Ultrasonic Inspection

The ultrasonic inspection of a component can be performed through two methods—active sensing and passive sensing [40]. In active sensing, a transmitter and a receiver are attached to the structure of interest. In this configuration, the transmitter is responsible for transmitting the signal and the receiver is responsible for receiving the signal. The presence of damage in the region between the transmitter and the receiver causes changes in the ultrasound signal. The damaged region can be identified by means of the analysis of the received signal [41]. The three most common configurations used in ultrasound wave analysis are shown in Figure 1.

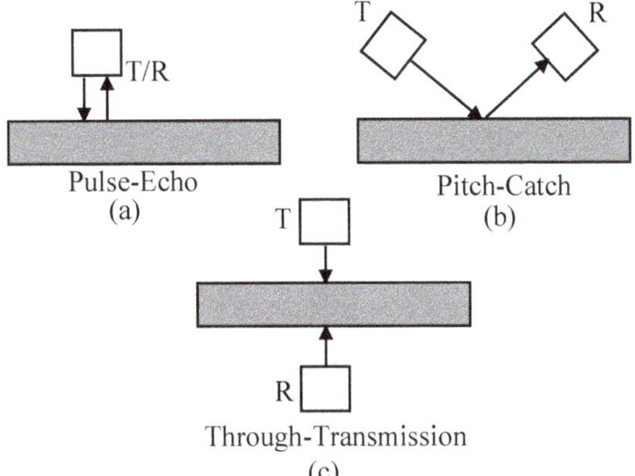

Figure 1. Three most common transmission-reception configurations: (**a**) pulse-echo; (**b**) pitch-catch; and (**c**) through-transmission [41].

Figure 1a,c show configurations based on bulk waves. This type of wave is used when analyzing defects in the internal region of the sample. Figure 1b shows the configuration in which surface waves are used, which detect anomalies on the sample surface. In this work, bulk waves are prevalent due to the transducer configuration, which is similar to the configuration of Figure 1c.

In the pulse-echo system (Figure 1a) the transducer is positioned on the surface of the workpiece. In this method, the same transducer functions as a transmitter and receiver of ultrasound waves [42]. The use of this system for the monitoring of the temperature distribution in the inner region of materials was presented by Ihara et al. [43,44]. The authors applied the proposed technique on heated steel discs at different temperatures. The results prove the feasibility of the technique, the values obtained were consistent with the values measured by thermocouples installed in the sample, thus indicating the effectiveness of the technique.

In the pitch-catch technique, the system has two transducers, the first emits the ultrasound pulses while the second transducer detects the pulses, as shown in Figure 1b. The transducers can be aligned in different ways, such as in normal incidence or oblique incidence, that is, with a certain inclination angle. The technique allows the monitoring of local damages through the emission of known input signals, adequate to diagnose a structure regarding the existence of damage [40]. A study was developed by Ihn et al. [40] employing the pitch-catch technique in the structural health monitoring of an airplane fuselage. The results showed that the use of a damage index allowed the evaluation of damage, similar to the results obtained by the Eddy current method or by the ultrasonic scanning.

The through-transmission technique, similar to the pitch-catch technique, also uses two transducers; however, the positioning of the transducers in the structure is different, as shown in Figure 1c. This technique uses two transducers positioned on opposite sides of the sample. A pulse generator is used to generate continuous electrical pulses with specific frequencies and amplitudes, which are converted into ultrasound waves by the first transducer (transmitter). The ultrasound waves, propagated through the sample, are detected by the second transducer (receiver) [42]. In Raišutis et al. [45], the through-transmission technique was used to estimate the phase velocity dispersion of ultrasound waves in plastic materials. The monitoring of lactic acid fermentation by means of the through-transmission technique was presented by Resa et al. [46]. Through the ultrasonic velocity, it was possible to detect the changes that occurred during the fermentation of carbohydrates. Thus, this non-invasive technique presents great potential in the monitoring of biological processes.

3. RMS and Counts in AE Signal Processing

One of the most used statistics in the analysis of the acoustic emission signal is the root mean square (RMS). The application of the RMS statistic in the identification of failures in the grinding and dressing processes has been extensively studied in the specific literature [47–50]. According to Webster et al. [51], the best integration interval for calculating the RMS statistic in the monitoring of the grinding process is 1 ms. The RMS statistic is defined by Equation (1).

$$X_{rms} = \sqrt{\frac{1}{N} * \sum_{i=1}^{N} x^2(i)} \tag{1}$$

where x is the raw signal and N is the number of discrete samples (i) considered in the calculation.

On the other hand, the Counts statistic was defined by Lopes et al. [9] as the number of times that the signal crosses a threshold per unit of time. The Counts statistic, calculated from the AE signal, has often been implemented in scientific works with different purposes, such as the location and measurement of acoustic emission events in snow blocks before their fracture [52] and the on-line monitoring of abrasive water cutting processes [53].

With regard to the signal analysis, the frequency domain is generally used in identifying events that are difficult to observe in the time domain. The most commonly used frequency domain techniques are the fast Fourier transform (FFT) [54], the short-time Fourier transform (STFT) [55] and the wavelet transform (WT) [56]. The FFT is an algorithm developed to improve the computational efficiency of the discrete Fourier transform (DFT), being a popular method used in spectral analysis and digital signal processing. Ahirrao et al. [57] used the FFT in the analysis of the vibration and dynamics of engines. Kang et al. [58] used the FFT in the tool condition analysis and monitoring of the micro-lens fabrication process.

The statistics presented in this section for the acoustic emission signals can also be used for signals of different natures acquired by other sensors, such as accelerometer and PZT. This work proposes the application of the RMS and Counts statistics in the signals sampled by a set of piezoelectric diaphragms in the emitter-receiver configuration for the monitoring of the material removal in the grinding process. It is worth mentioning that the implementation of the piezoelectric diaphragm in this configuration and for this purpose is unprecedented in the literature.

4. Bases of the Chirp-through-Transmission Ultrasound Technique

Nowadays, the inspection of structures by means of ultrasound techniques is widely used. A continuous effort of researchers and industries has been made in order to improve and increase the applicability of non-destructive evaluations (NDE). Thus, a new inspection technique based on ultrasound waves and low-cost piezoelectric transducers is presented in this paper. Section 4.1 presents some basic concepts of ultrasound waves, while Section 4.2 presents some common classifications of these waves. Finally, Section 4.3 presents the method proposed in this paper.

4.1. Ultrasound Waves and Their Parameters

Ultrasound signals are used to predict the behavior of materials (detecting internal defects) and to characterize them in a variety of structures. The interest of the scientific community for ultrasound techniques has increased in recent years due to its wide range of applications [41]. Immersion testing and ultrasound contact methods are very popular NDE techniques for characterizing material conditions [59]. However, this type of technique requires the removal of the structure, causing significant interruptions in the operation of the machines [60].

Ultrasound waves are composed of frequencies greater than the human's hearing range, which is typically 20 kHz [41,42,61]. The main characteristics of the ultrasound waves are length, velocity, pressure, frequency and period. The propagation of the waves in the structure results in two phenomena—velocity alteration and wave attenuation, both caused by the mechanisms of absorption and dispersion [42]. As the ultrasound waves propagate through the material, their energy is diminished and reflections are generated on the surface. These reflections are used to determine the presence and location of discontinuities and defects [61].

The wave velocity is the product between the frequency and the wavelength, so high-frequency waves have shorter wavelengths while low-frequency waves have longer wavelengths [42]. The velocity of the ultrasound waves (v) is determined by the density (ρ) and elasticity (E) of the propagation medium according to the Newton-Laplace equation.

$$v = \sqrt{\frac{E}{\rho}} \tag{2}$$

Equation (2) implies that the velocity of an ultrasound wave in a solid material is greater than in a liquid [42].

There are other ultrasound parameters that correlate the physical-chemical properties of the materials, such as the attenuation coefficient and the acoustic impedance. The attenuation is a result of the energy reduction caused by the compression and decompression of the ultrasonic waves due to the absorptions and dispersions of the propagation medium [62]. Absorption is mainly associated with homogeneous materials while dispersion is related to heterogeneous materials. In addition, attenuation is affected by viscosity, compressibility, material contours and absorption and dispersion effects [42]. It is worth mentioning that the attenuation coefficient of a given material is highly dependent on the way in which the material was manufactured, being useful for product quality control.

Acoustic impedance is the product between the density and the velocity of the wave that passes through the contours of different materials. Materials with different densities have different acoustic impedances, resulting in reflections on the contours between two materials with different acoustic impedances. The attenuation and the acoustic impedance are expressed by Equations (3) and (4), respectively.

$$A = A_o e^{-ax} \tag{3}$$

$$R = \frac{A_T}{A_t} = \frac{z_1 - z_2}{z_1 + z_2} \tag{4}$$

where A_o is the initial amplitude of the wave, x is the distance traveled; R is the ratio of the amplitude of the reflected wave (A_T) to the incident wave (A_t); z_1 and z_2 are the acoustic impedances of two materials.

4.2. Classification of Ultrasound Waves

Elastic waves (ultrasonic, sonic and subsonic) can be classified into two groups—body waves, also known as bulk waves and surface waves, also known as guided waves. Body waves propagate through the interior of the material (bulk material), while surface waves propagate along the material's surface [41].

Bulk waves are classified into two types—longitudinal waves and shear waves. According to Coramik et al. [61], longitudinal waves are also known as compression or extension waves, since the material undergoes compression and extension when the wave propagates through it. In shear waves, the motion of the particles is perpendicular to the direction of the wave propagation. Shear (transverse) waves, when compared to longitudinal waves, have shorter wavelengths and slower dispersion. In terms of elastic constants, the wave velocities of these two types of waves (longitudinal and shear) are given by Equations (5) and (6), respectively [41].

$$c_p = \sqrt{\frac{\lambda + 2\mu}{\rho}} = \sqrt{\frac{E(1-v)}{\rho(1+v)(1-2v)}} \tag{5}$$

$$c_s = \sqrt{\frac{\mu}{\rho}} = \sqrt{\frac{E}{2\rho(1+v)}} \tag{6}$$

where c_p and c_s are the longitudinal and shear velocities, respectively. ρ is the density, λ is the first Lame constant, μ is the shear modulus or second Lame constant, E is Young's modulus and v is Poisson's ratio. It should be noted that c_p is always greater than c_s, since longitudinal waves propagate faster than shear waves, longitudinal waves are called primary waves or P-waves and shear waves are called secondary waves or S-waves.

4.3. The Chirp-Through-Transmission Ultrasound Technique

As already described, the ultrasound wave transmission-reception techniques with transducers not directly coupled to the material are pulse-echo, pitch-catch and through-transmission [41]. In the pulse-echo technique, a single transducer receives and transmits the ultrasound waves. In the pitch-catch and through-transmission techniques, the transmitter and receiver are different transducers [62]. The method proposed in the present study is an alternative to the through-transmission technique, as it also uses two piezoelectric transducers at the ends of the material under study. However, a chirp signal is injected into the emitter transducer, which results in an acoustic wave that propagates through the material until it is collected by the receiver transducer. The signal is collected by a data acquisition system and processed using digital signal processing techniques in order to diagnose the condition of the structure.

The proposed method is shown in Figure 2, which shows two piezoelectric diaphragms attached at the ends of the material under study, where the right piezoelectric diaphragm was configured as emitter and the left diaphragm as ultrasonic wave receiver. The emitter is excited by a chirp signal of certain amplitude and frequency. Due to the piezoelectric effect of this transducer when excited with the chirp signal, ultrasonic bulk waves are generated and transmitted through the material under study, as illustrated in Figure 2a. The waves that propagate through the material suffer various effects, such as frequency variations, dispersion, reflection, refraction, compression, rarefaction and energy loss. These effects modify the emitted signal, creating a signature of the material under study. The frequency of the waves changes longitudinally (continuous line) and transverse (dashed line) as they propagate through the structure. Figure 2a shows the application of the technique in a healthy structure (without damage), together with the received signal and its respective frequency spectrum.

In the case of a damaged structure, as shown in Figure 2b, where the points and triangles represent, for example, cracks and material removal, the changes in the waves are different from those observed in the healthy structure and therefore its condition can be diagnosed.

Thus, the signal collected from the damaged material will differ from the signal collected from the healthy material, since changes in structure affect the propagation of ultrasound waves. After the signal acquisition, it is possible to identify the structural change of the material by comparing the signals of the healthy and damaged material. However, this identification involves digital signal processing techniques, such as spectral analysis, frequency band selection, RMS and Counts statistics, among others. Figure 2b shows the application of the technique in a damaged structure, together with the received signal and its respective frequency spectrum, both with different characteristics with respect to the healthy structure.

Thus, the proposed method is based on the changes that ultrasound waves undergo when propagating through the material under study, with the difference that this method does not use the traditional parameters commonly used in ultrasound techniques to diagnose the condition of the structure. Instead, the proposed method employs the processing of the receiver transducer signals resulting from the ultrasound waves generated by the emitting transducer. It is worth mentioning that the prevalent waves in this method are bulk waves, which are expected to allow the monitoring of defects such as microstructure alteration or material removal.

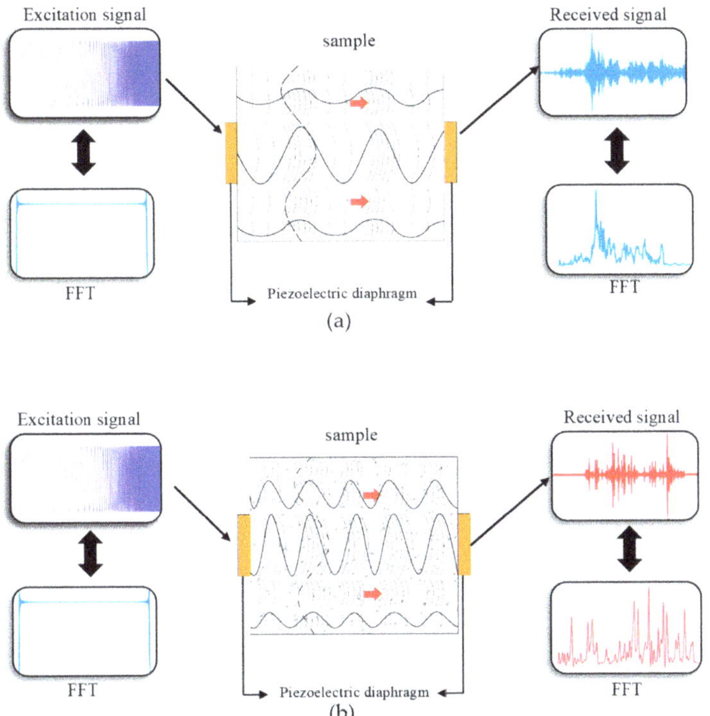

Figure 2. Chirp-Through-Transmission Technique: (**a**) health structure; (**b**) damaged structure.

The main features of the proposed method are: (1) The use of a chirp signal at the emitter transducer input instead of pulses; (2) Do not depend on the reflection of the emitted wave; (3) Not limited to specific wavelengths to identify damage; (4) Does not require the use of traditional parameters of ultrasound techniques for damage detection, such as wavelength, propagation speed and time of flight; (5) It is possible to identify the frequency bands in the received wave that best characterize the structural damage; (6) Damage detection can be performed using traditional statistics such as RMS and Counts, enabling the study of many other statistics, indexes and parameters used in process monitoring and SHM; (7) Employs low-cost piezoelectric transducers; (8) It consists of a simple experimental setup for both transducer fastening and data acquisition and signal processing system.

It is worth mentioning that in the proposed method, the changes that the ultrasound waves suffer when propagating through the structure are analyzed through digital signal processing, being necessary the comparison of two signatures, the first of the healthy structure, called baseline and the second one of the altered or damaged structure. Thus, the present method can also be classified as a structural health monitoring method.

According to Giurgiutiu et al. [63,64], the electromechanical impedance (EMI) technique considers the direct and inverse electromechanical properties of piezoelectric materials. In addition, according to Baptista et al. [65], the basic concept of this technique is to monitor structural integrity by exciting a piezoelectric transducer attached to the structure at an appropriate frequency range. Due to the piezoelectric effect, there is a relationship between the mechanical impedance of the structure (related to its integrity) and the electrical impedance of the piezoelectric transducer, obtained by the relationship between the excitation and detection signals. The proposed method, when compared to the EMI technique, does not require the measurement of the structure impedance to detect damage but the variations of the received ultrasonic waves. In addition, common digital signal processing statistics are used to detect differences between the baseline and damage signatures, the technique is not limited to the damage indices commonly used in the EMI technique [66]. Thus, the proposed technique can be expanded to the application of new indices. Finally, traditional EMI techniques use a single sensor to excite and detect the impedance variations of the structure, while in the presented method two transducers are used, analyzing only the received signal.

5. Setup, Grinding Tests and Workpiece Assessment

In order to validate the proposed ultrasound method, this section presents the experimental setup for the grinding tests, the workpiece evaluation and the digital signal processing.

5.1. Experimental Setup

The grinding tests were performed in a surface grinding machine from Sulmecânica manufacturer, model RAPH 1055 (Porto Alegre, Brazil), equipped with an aluminum oxide grinding wheel, model 38A150LVH, from NORTON company (Worcester, MA, USA). AISI 4340 steel workpieces with 150 mm length × 7 mm width × 43 mm height were machined under three different depth of cuts (a). Each test consisted of three grinding passes across the workpiece surface at a given grinding condition. The depth of cut, grinding passes, cutting speed (v_s) and workpiece speed (v_w) are shown in Table 1. The cutting fluid used in the tests was an oil-water emulsion with a concentration of 4%. Prior to test 1, the grinding wheel was dressed in order to restore its topography and cutting potential. In addition, some grinding passes were necessary to standardize the contact between the grinding wheel and the entire length of the workpiece surface. It is worth mentioning that the workpiece surface standardization operation is commonly used in the grinding process since the workpiece was undergone other manufacturing processes prior to the grinding process. These processes, such as milling and turning, hardly achieve the surface finish required by the specifications, thus making the surface of the workpiece uneven to the full contact of the grinding wheel, justifying the need of the surface standardization operation.

Table 1. Grinding test parameters.

Workpiece	Depth of Cut a (μm)	Passes	Cutting Speed v_s (m/s)	Workpiece Speed v_w (m/s)
1	10			
2	20	3	29	0.08
3	30			

Regarding the monitoring of the process, two piezoelectric diaphragms (buzzers) were attached to the workpiece holder on opposite sides. Prior to the tests, the sensors and data acquisition system were properly calibrated for good signal sensitivity and saturation. The piezoelectric diaphragms used in the tests consisted of a disc-shaped lead zirconate titanate (PZT) ceramic (active element) with a diameter of 10 mm concentrically adhered to a brass disc with a diameter of 20 mm. In addition, the transducers were placed in a metallic case that allowed for a magnetic attachment to the workpiece holder. The experimental setup is shown in Figure 3. Some sensors besides the piezoelectric diaphragm (PZT) can be seen in Figure 3, however, such sensors were not considered in this work. It can also be observed in Figure 3 the metallic case and its connections of the emitter PZT 1 and receiver PZT 2; the latter was attached on the opposite side of the workpiece holder. In order to prevent displacement, a silicone layer was applied around both sensors. Regarding the position of the sensors, both transducers were fixed at a distance of 95 mm from the end of the holder. The transducers have a capacitance of 10 nF and a resonance impedance of 350 Ω, according to the manufacturer's data. Finally, the interface between the surfaces of the transducer case and the workpiece holder was filled with oil. It is worth mentioning that a thin layer of silicon was placed around the piezoelectric transducer in order to protect it from the cutting fluid and prevent displacements during the grinding process.

The raw signal acquisition was carried out by an oscilloscope, model DL850, from Yokogawa company (Tokyo, Japan), at a sample rate of 5 MS/s. An amplifier model AD8606 (Analog Devices, Norwood, MA, USA), with simple input and gain of 50, was used to amplify the signals from the piezoelectric diaphragm set as receiver. A data acquisition system (DAQ), USB-6361 model, from National Instruments (Austin, TX, USA), was employed to generate the excitation signal of the PZT diaphragm set as emitter. The excitation signal consisted of a chirp interval of 500 ms with a frequency variation from 1 to 250 kHz. In addition, five equal packages containing the chirp signal were sent to the emitter PZT. Thus, the excitation signal was composed of five equal chirp packages sent at 250 ms intervals. The PZT excitation by various chirp signals allows the verification of the repeatability of the results of the digital signal processing. Finally, the emitter PZT was connected to the DAQ which generated the excitation signals (chirp) and the receiver PZT was connected to the signal conditioner (amplifier module). Then, the signal conditioner and the DAQ were connected to the oscilloscope in order to collect the emission and reception raw signals.

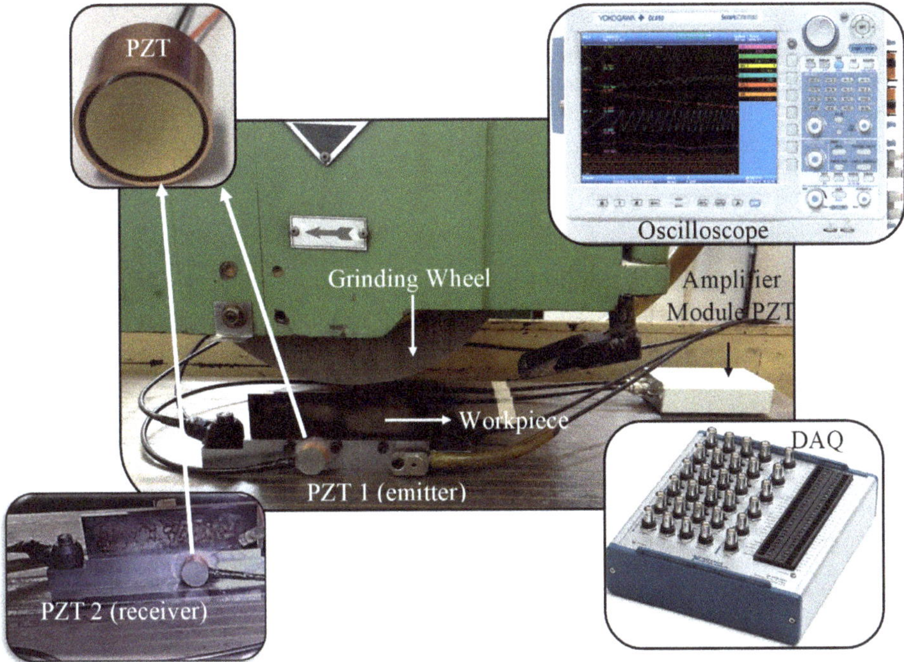

Figure 3. Experimental Setup.

5.2. Workpiece Assessment

The workpieces assessment was carried out prior to the grinding tests, that is, the workpiece mass was measured before the material removal caused by the grinding process. Thus, the data obtained shows the workpiece characteristics without material removal. After the grinding process (normalization and material removal), the workpieces were reweighed. Thus, it was possible to detect the mass variation of the workpiece machined by the grinding process. A precision scale with a scale of 500 g was used for this purpose. The workpieces were placed in the same position on the precision scale in order to avoid any changes that do not belong to the material removal caused by the grinding process.

5.3. Signal Processing and Selection of Frequency Bands

The data sets from the grinding tests were digitally processed in MATLAB software. Two distinct signals were acquired in the tests, the first signal refers to the set of chirp waves of the emitter PZT, while the second signal refers to the chirp waves collected by the receiver PZT. The emission was composed of five chirp signals sent by the emitter PZT and the reception was composed by fives signals acquired by the receiver PZT. The signals collected by the receiver, for each grinding pass and depth of cut, were divided into data vectors or packages, resulting in five packages.

The spectral analysis of the packages corresponding to the receiver signals was performed in order to find the frequency band that best characterize the workpiece conditions (without material removal and with material removal). The Fast Fourier Transform (FFT) was used to analyze the spectral content of the receiver PZT signals. The FFT magnitude was computed for each received vector and then the mean values were obtained. The criterion used for the selection of frequency bands was presented by Ribeiro et al. [16], in which the best frequency bands are those with the greatest differences in magnitudes and minimum overlap between the observed conditions.

After the selection of frequency bands, Butterworth digital filters were applied to the received packages and new vectors were obtained. Then, the RMS and Counts values were calculated for both vectors (raw unfiltered and filtered received signals). Intervals of 4096-point, corresponding to 1 ms, as reported by Webster et al. [51], were used to compute the RMS and Counts values.

Finally, the standard deviation and mean values for each received package and workpiece condition (without material removal, first grinding pass, second grinding pass and third grinding pass) were computed. It is worth mentioning that the threshold used in the Counts statistic was selected based on the noise level of the receiver signals. The flowchart presented in Figure 4 shows the digital signal processing scheme. In order to correlate the mean values of the statistics (RMS and Counts) with the volume of material removed, a linear regression was calculated. Thus, through the coefficient of determination (R) it was possible to infer the degree of correlation between the values obtained with the proposed technique and the volume of material removed. The values were normalized before the computation of the linear regression in order to eliminate the amplitude differences between the two statistics. The application of linear regression as a correlation metric can be found in several scientific works, as in de Oliveira Conceição Jr. et al. [67] and Viera et al. [21].

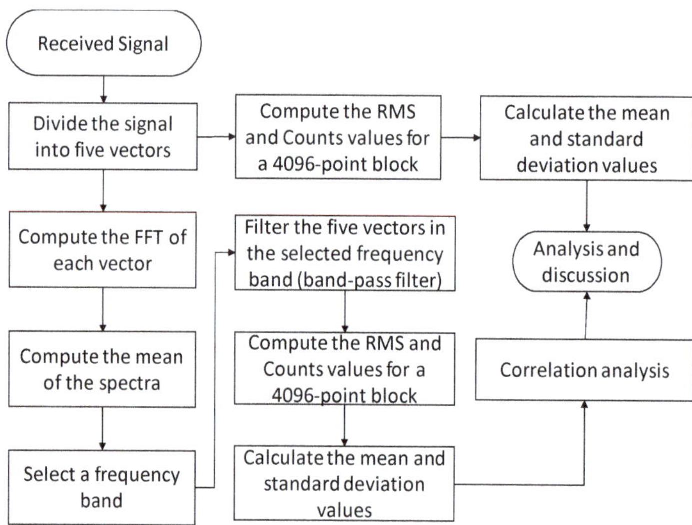

Figure 4. Flowchart of the digital signal processing scheme.

6. Results and Discussion

In this section, the results and discussion of the workpiece assessment and the digital signal processing will be presented, which focus on the correlation of the material removal of the grinding process with the statistics herein applied to the ultrasound signals.

6.1. Workpiece Assessment

The variation of masses and weights of the workpieces, measured by a precision scale, before and after the grinding passes, are shown in Table 2. Before the grinding tests, each workpiece underwent the standardization operation, which consists of grinding passes across the workpiece surface (with a depth of cut of 5 μm); the operation was repeated until uniform contact was achieved between the workpiece surface and the grinding wheel surface. After the standardization operation and the grinding passes, the workpiece mass decreases due to the material removal during the grinding process. It is worth mentioning that the mass decrease is due to the standardization and grinding passes. Thus, workpiece 2 had a smaller decrease in mass than workpiece 1 and 3 because the standardization operation was performed with fewer grinding passes than for the other workpieces (1 and 3).

Table 2. Variation of masses and weights of the workpieces.

Workpiece	Condition	Weight (g)	Mass Decrease (%)
1	Without material removal	329.75	0.05
	After pass 3	329.60	
2	Without material removal	327.24	0.02
	After pass 3	327.15	
3	Without material removal	331.24	0.33
	After pass 3	330.15	

The volume of material removed is shown in Figure 5. There is an increase in the volume of material removed according to the number of grinding passes and the selected depth of cut, as expected. The volume of material removed is obtained by Equation (7).

$$Q_w(p) = p * a_p * l_w * b_w \tag{7}$$

where Q_w is the volume of material removed in mm^3, p is the number of grinding passes, a_p is the depth of cut in mm, l_w is the workpiece length in mm and b_w is the workpiece width in mm, as presented in Figure 5a. The volume of material removed as a function of the number of passes is shown in Figure 5b.

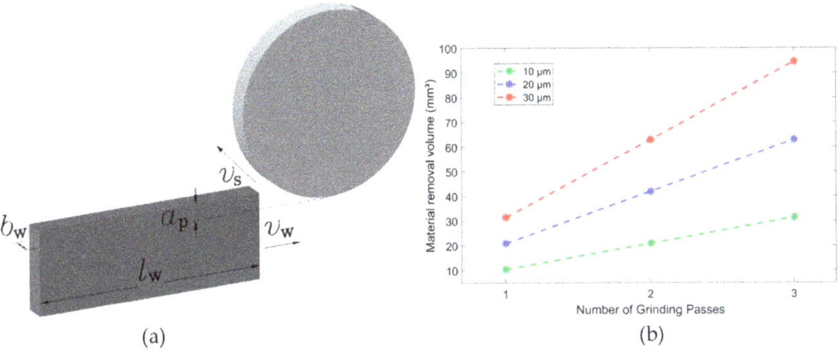

Figure 5. (**a**) Schematic for grinding process with cutting parameters; (**b**) material removal volume as a function of the grinding pass.

6.2. Signal Processing and Selection of Frequency Bands

The emitted and received signals used to evaluate the workpieces are shown in Figure 6. It is observed in Figure 6a,b that the emitted and received signals have five packages (#1, #2, #3, #4, #5), respectively. The amplitude of each emitted package decreases as the frequencies increases; this behavior can be explained by the arrangement of the impedance between the DAQ (chirp signal emitter) and the emitter PZT. The PZT capacitive elements caused a voltage decrease at higher frequencies, as observed in Campeiro et al. [68]. Regarding the received packages in Figure 5b, it is observed that some frequencies propagated with greater effectiveness; this behavior is justified by the material characteristics that play a very important role in the propagation of waves. It is worth mentioning that the emission packages started with amplitudes of 10 V and finished with amplitudes of 3 V, whereas the received packages started with amplitudes of 200 mV, having in certain times amplitudes of 50 mV. This significant attenuation is due to the energy loss through the PZTs' intrinsic characteristics, the workpiece holder medium, interfaces (PZTs/holder; workpiece/holder), coupling medium, workpiece medium, amplifier and cables. However, despite the high energy loss of the ultrasound signal for the setup used in this work, the results will later show that the received signal preserved the process characteristics under study. The procedure for the emitted and received packages shown in Figure 6 was repeated for each grinding pass at each selected depth of cut.

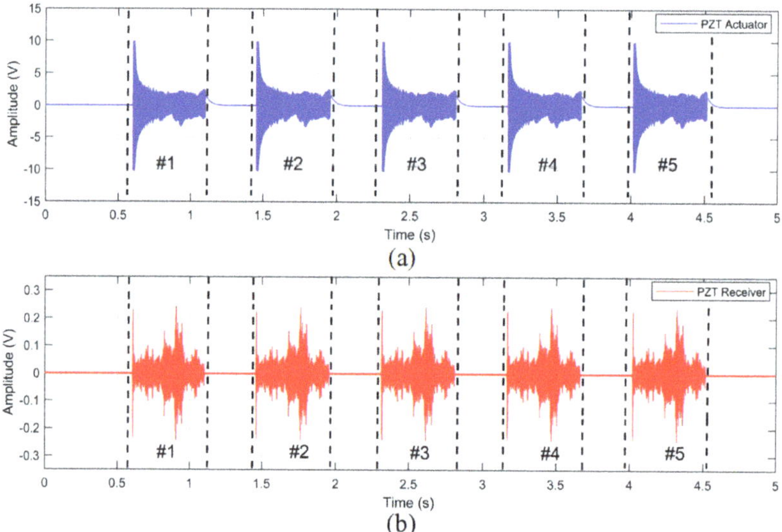

Figure 6. (**a**) Emitted signals packages and (**b**) received signals packages.

The statistics used in the analysis of the received packages are shown in Figure 7. It is worth mentioning that, in order to show the statistics together, it was necessary to normalize the values. However, the analyses, such as the mean and standard deviations, were performed with the raw unfiltered signals without the normalization procedure. The analyses were performed for each package considering the mean values of these statistics (RMS and Counts). Subsequently, a total mean value and standard deviation, with respect to all packages, was calculated.

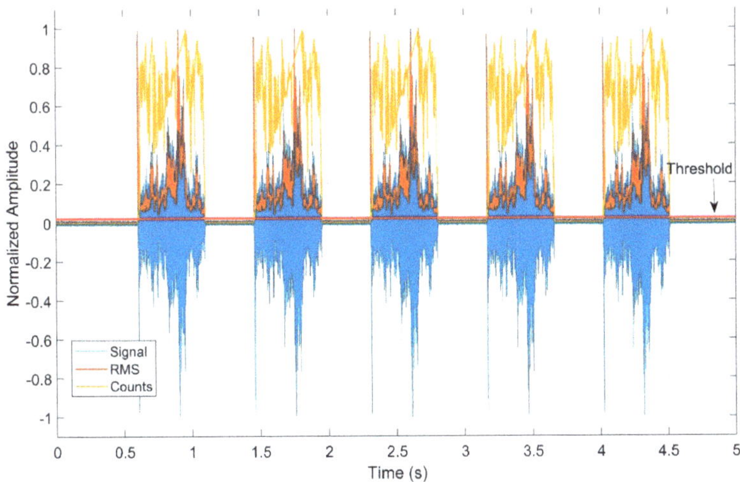

Figure 7. Statistics used to analyze the received packages—root mean square (RMS) and Counts.

The RMS mean and standard deviation values (unfiltered signals) for each grinding pass at each depth of cut are shown in Figure 8. In Figure 8a it can be seen that the result of the RMS mean value without material removal showed a greater amplitude compared to the results of the grinding passes 1 and 2. This non-linear behavior hinders the implementation of this system in the diagnosis of material removal during the grinding process. In relation to Figure 8b,c, a growth tendency is observed as the grinding passes occur; however, the difference of the RMS mean values between the 2nd and 3rd grinding pass for the 20 μm depth of cut is very small, making it difficult to identify the grinding pass and thus, being unattractive for practical implementation. In the same way, this analysis can be applied to the RMS mean values of the 1st and 2nd grinding passes for the 30 μm depth of cut in which there is a small difference, making it difficult to identify this grinding passes. It is worth mentioning that the standard deviations, compared to the RMS mean values, were very small, around 1%, for all the observed conditions. This behavior characterizes the consistency and repeatability of the emission and reception signals used in the proposed technique.

The Counts mean values and the standard deviations (unfiltered signals) for each grinding pass at each depth of cut are shown in Figure 9. As in the behavior shown by the RMS results, the application of the Counts statistic was not effective in identifying a linear growth pattern between the grinding passes. The depth of cuts of 20 μm (Figure 9b) and 30 μm (Figure 9c) showed a non-linear behavior. In Figure 9b it is observed that the Counts mean value of the undamaged workpiece is higher than the Counts mean values for the 1st and 2nd grinding pass. Regarding Figure 9c, the Counts mean value of the workpiece without damage had a very close result in relation to the 1st and 2nd grinding pass, however, it was higher than the Counts mean value of the 3rd grinding pass. Finally, regarding Figure 9a with a 10 μm depth of cut, an increasing pattern was observed, however, the difference between the 1st and 2nd grinding pass was small, similar to the behavior shown by the RMS mean values at the depth of cut of 20 μm (Figure 8b) and 30 μm (Figure 8c). The standard deviations, when compared to the mean values, were very small, around 1%, similar to the values found in the RMS results, shown in Figure 8. Again, the behavior of the presented results emphasizes the consistency and repeatability of the technique used in this work. Both statistics (RMS and Counts) did not show a regular tendency in the diagnosis of material removal, thus a spectral analysis of the received signals for each grinding pass was performed in order to find the frequency bands that are more strongly related to the process conditions.

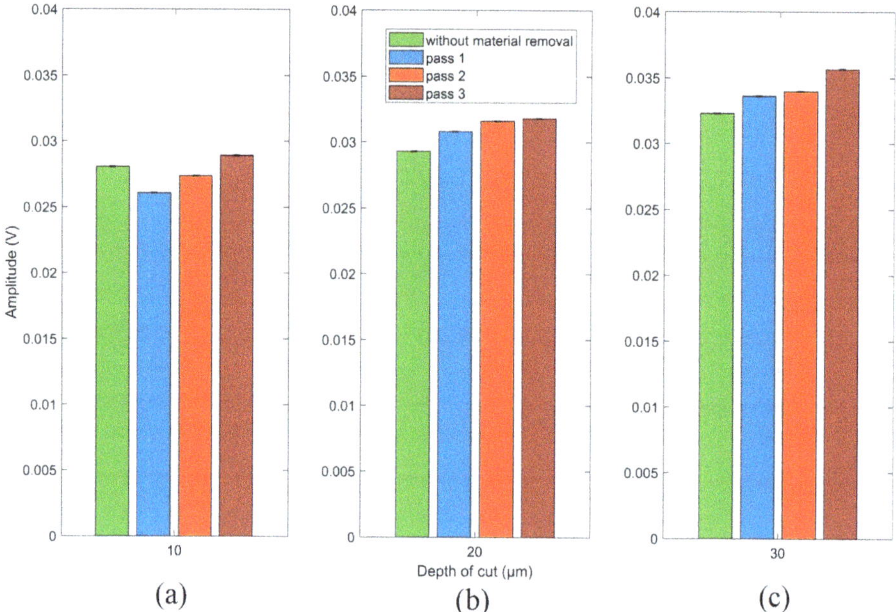

Figure 8. RMS mean and standard deviation values of the raw unfiltered signals at (**a**) 10 µm, (**b**) 20 µm and (**c**) 30 µm.

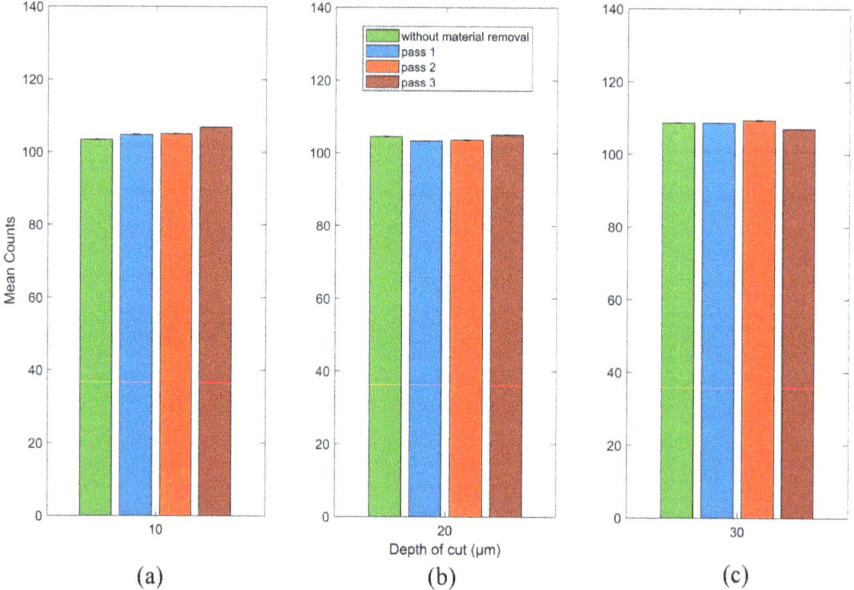

Figure 9. Counts mean and standard deviation values of the raw unfiltered signals at (**a**) 10 µm, (**b**) 20 µm and (**c**) 30 µm.

The mean spectrums for two workpiece conditions, without material removal and after the 3rd grinding pass, at each depth of cut, are shown in Figure 10. It can be seen in Figure 10 that the spectrums, in general, showed similar behavior, with higher spectral activity between 150 and 190 kHz.

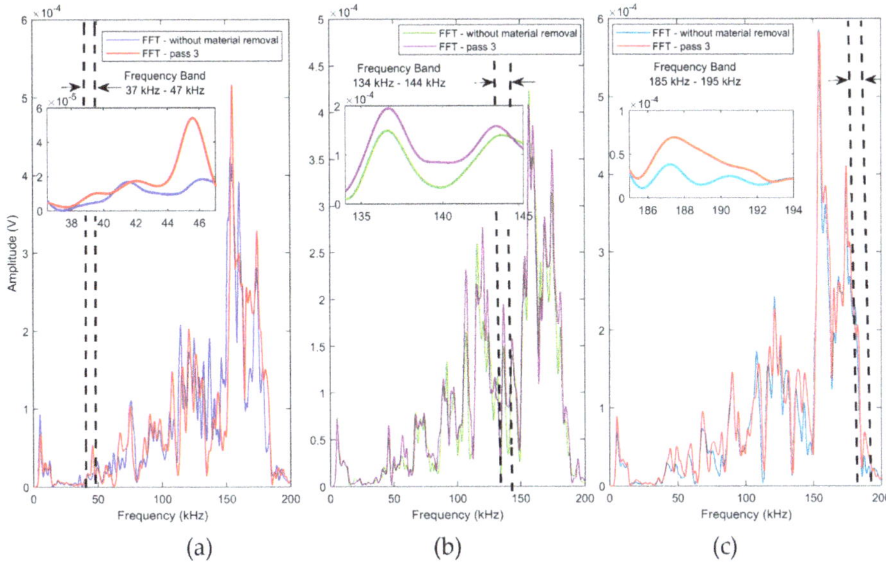

Figure 10. Spectrum of two workpiece conditions at (**a**) 10 µm, (**b**) 20 µm and (**c**) 30 µm.

Furthermore, it is observed that the spectrum of the 3rd grinding pass at the three depth of cuts, 10 µm (Figure 10a), 20 µm (Figure 10b) and 30 µm (Figure 10c), presented the greatest amplitudes along most part of the spectrum when compared with the workpiece without material removal. By means of the frequency band selection criterion, presented in Section 5.2, the 37 to 46 kHz frequency band was selected and subsequently filtered into the received packages of the 10 µm depth of cut, as shown in the magnification of Figure 10a. For the 20 µm depth of cut, a frequency band from 135 to 144 kHz was selected, as shown in Figure 10b. Regarding the 30 µm depth of cut, a frequency band of 186 to 192 kHz was selected. The frequency bands magnifications of both depth of cuts (20 and 30 µm) are shown in Figure 10b,c, respectively. The RMS mean values and standard deviations of the filtered signals for each grinding pass at each depth of cut are shown in Figure 11. In contrast to the results observed in Figure 8, the filtered RMS values showed a uniform increase trend. An increase in the RMS values was identified according to the grinding passes. The increase in the RMS values was due to the behavior of the amplitude of the signals in the selected frequency bands. After the application of digital filters in the selected frequency bands and subsequent computation of the RMS statistic, an increasing trend was observed at all depths of cut (Figure 11a–c). This behavior can be explained by the signal amplitudes in the selected bands, that is, the changes in the workpiece (material removal) presented higher amplitude levels [26]. It is worth mentioning that the sample changed after each grinding pass and, as a consequence, the propagated waves also showed changes that are evidenced in the selected frequency bands. Thus, events produced during the grinding process that indicated changes in the workpiece structure, such as material loss, burning, high roughness and cracks, were evidenced in the selected frequency bands, allowing better diagnosis of the process. Finally, it is worth mentioning that the standard deviation values remained small, again showing the consistency and repeatability of the method. Thus, the appropriate selection of frequency bands is very important for the application of this technique.

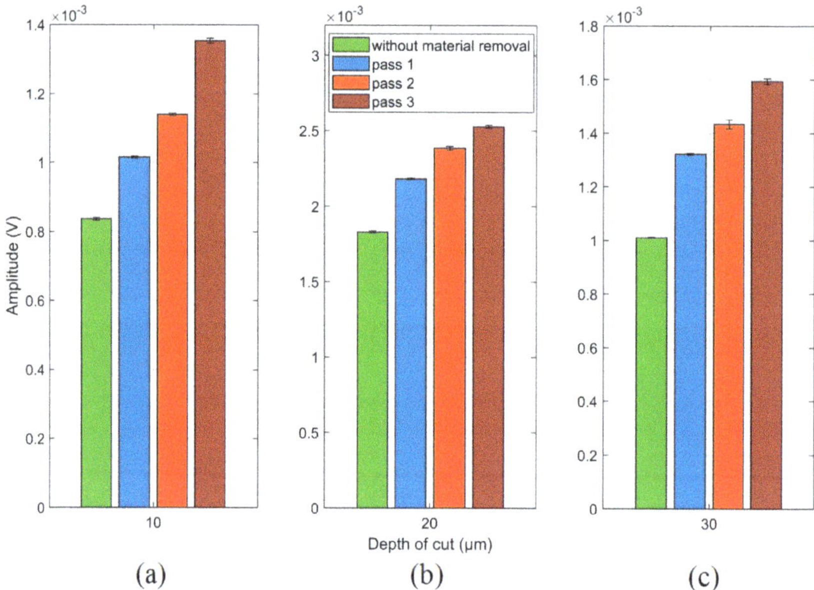

Figure 11. RMS mean and standard deviation values of the raw filtered signals at (**a**) 10 μm; (**b**) 20 μm and (**c**) 30 μm.

The Counts and standard deviation values of the filtered signals for each grinding pass at each depth of cut are shown in Figure 12. As with the filtered RMS values shown in Figure 11, the mean Counts values for the same signals showed an increasing trend in all the conditions observed. However, it is worth mentioning that the depth of cut of 10 μm shown the best result for this statistic, whereas in the depths of cut of 20 and 30 μm the Counts presented a small improvement when compared to the optimal result obtained in the RMS application of Figure 11. Thus, the Counts and RMS statistics can be used together, with priority for the RMS due to the optimal results; the Counts statistic can be used with the purpose of validating the RMS results and strengthening the damage diagnosis system in the grinding process.

The percentages of difference between the mean RMS and Counts values of the unfiltered and filtered packages for two workpiece conditions, without material removal and after the 3rd grinding pass, at each depth of cut are shown in Figure 13. In Figure 13a it can be observed that the highest percentage variation is around 38% for the filtered signal and depth of cut of 10 μm. Similarly, the percentage of difference for the Counts mean values, shown in Figure 13b, presented the greatest difference at the same depth of cut, around 22%. Finally, the differences in the observed conditions are higher in the results of the mean RMS, denoting that the RMS statistic is more adequate for the application of this technique due to the better results.

As demonstrated by Webster et al. [51], the 1 ms interval is best suited for calculating the RMS value in the monitoring of grinding processes. The same interval, corresponding to 4096 points, was applied in the Counts statistic. Thus, it is possible that this interval is not the most appropriate for this statistic, which explains the better result of the RMS statistic when compared to Counts. In addition, the choice of the most appropriate frequency band is crucial for the performance of the statistics. Therefore, other intervals and frequency bands can be studied in order to optimize the Counts statistics.

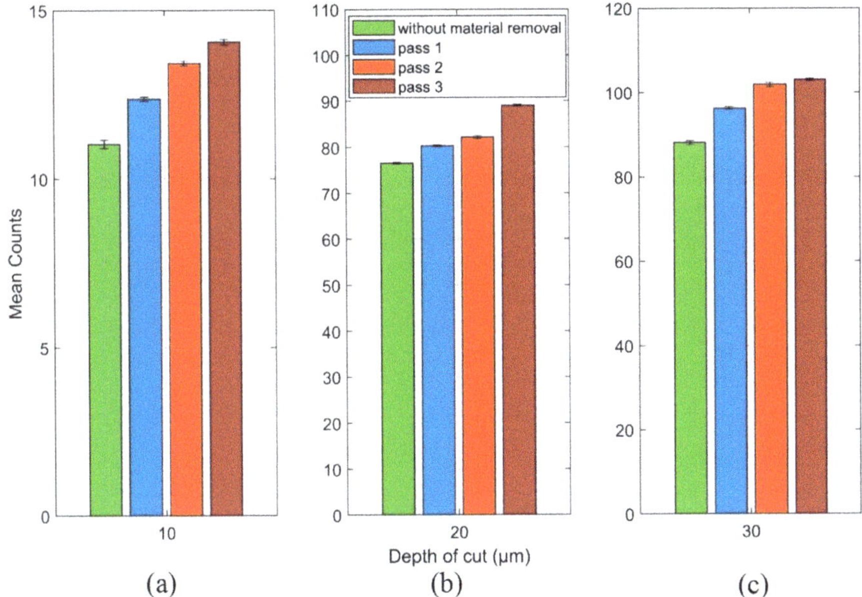

Figure 12. Counts and standard deviation values of the raw filtered signals at (**a**) 10 μm; (**b**) 20 μm and (**c**) 30 μm.

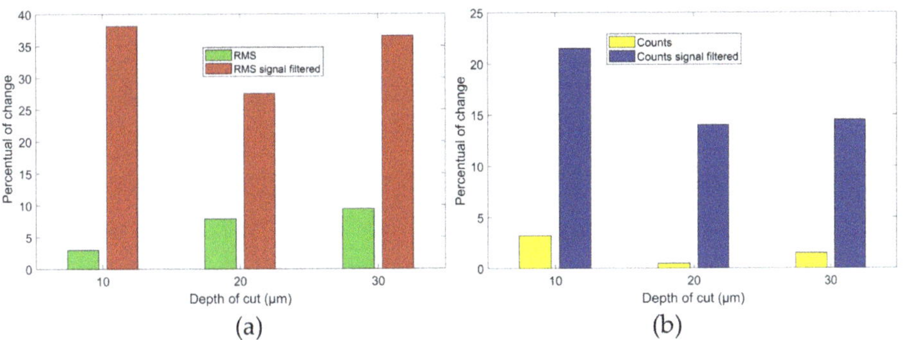

Figure 13. Percentage variation between the unfiltered and filtered signals of the workpiece without removal material and after the 3rd grinding pass—(**a**) RMS and (**b**) Counts values.

The correlations between the mean values of the statistics (RMS and Counts) and the volume of material removed are shown in Figure 14. The correlation analysis is performed by means of the coefficient of determination (R), where $R = 1$ represents a linear fit of 100% and $R = 0$ represents the complete lack of correlation between the values [21]. It can be observed that for the RMS statistic (Figure 14a), the R values were close to 1. Regarding the Counts statistic (Figure 14b), the coefficient of determination showed a high degree of correlation; however, the coefficient was lower than the coefficient found for the RMS statistic. At the 10 μm depth of cut, the RMS statistic presented an R of 0.9883 and the Counts statistic an R of 0.98794. Thus, the RMS was slightly better than the Counts in the estimation of the volume of material removed. At the depth of cuts of 20 and 30 μm, a higher sensitivity of the RMS statistic was observed when compared to the Counts statistic.

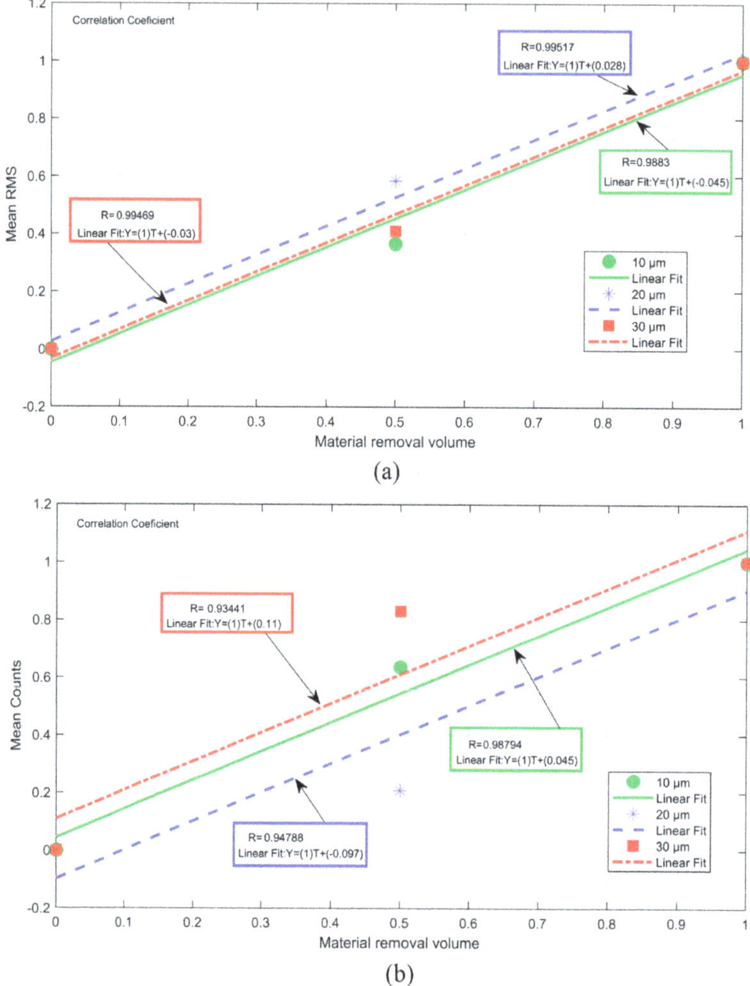

Figure 14. Correlation between the statistics and the volume of material removed (**a**) RMS and (**b**) Counts.

At 20 μm, the RMS had an $R = 0.99517$ while Counts had an $R = 0.94788$. Finally, at 30 μm the largest difference was observed, where the RMS presented an $R = 0.99469$ and Counts presented an $R = 0.93441$. Thus, in the application of this technique, the RMS statistic has a better performance than the Counts statistic, however, the high degree of correlation shows that both statistics can be used to indirectly determine the volume of material removed.

In a system where both statistics can be applied, it is possible to obtain results that confirm the material removal diagnoses, since both statistics showed the same increasing trend. It is worth mentioning that, in both statistics, the filtered signals produced a much greater percentage of change when compared to the unfiltered signals. Thus, the selection of frequency bands that best characterize the events occurred during the grinding process is very important for the application of the technique proposed in this work.

7. Conclusions

The interest of the scientific community in ultrasound techniques has increased in recent years due to its wide range of applications. Thus, a new method of ultrasound wave monitoring is proposed in this study as an alternative to traditional ultrasound methods. The main advantage of this method is the use of low-cost transducers for the emission and reception of ultrasound waves in conjunction with traditional signal processing statistics, thus allowing the non-invasive monitoring of structures. Unlike traditional methods of ultrasound analysis, the method presented in this paper does not depend on reflections and wavelengths to identify damage. Moreover, the proposed method does not require the use of the traditional ultrasound parameters (wavelength, propagation velocity and time of flight) to detect changes in the structures. Finally, this method uses traditional statistics for monitoring the material removal volume in grinding, allowing the study of other statistics, indexes and parameters, such as RMSD and CCDM used in SHM.

A study of the frequency bands that best correlate with the workpiece characteristics during the grinding passes was performed. Thus, it can be concluded that the unfiltered RMS and Counts mean values have low sensitivity to the material removal changes and are not attractive for practical purposes. On the other hand, the RMS and Counts mean values of the filtered signals in the selected frequency bands showed excellent results, obtaining an increasing tendency according to the number of grinding passes that occurred in the tests. It is worth mentioning that the filtered RMS values presented the best results when compared with the Counts values in the same condition. The highest percentage changes between the workpiece without material removal and after the 3rd grinding pass for each depth of cut was observed for the mean RMS values. The frequency band selection and the interval period were determinant for the sensitivity of the statistics. In addition, the correlation analysis between the statistics (RMS and Counts) and the volume of material removed reinforced the results obtained, proving the viability of applying the chirp-through-transmission technique in the monitoring of material removal, as all the coefficients of determination were higher than 90%.

Nevertheless, the results presented in this work are preliminary and new studies must be conducted for improvements of the proposed technique with application in grinding, SHM as well as in other machining processes.

Author Contributions: Conceptualization, F.A.A. and P.R.A.; Formal analysis, T.G.L.; Methodology, F.A.A., R.G., M.A.A.V. and T.G.L.; Supervision, P.R.A. and E.C.B.; Writing—review & editing, F.A.A., P.R.A., M.A.A.V. and E.C.B.

Funding: This research was funded by the Sao Paulo Research Foundation (FAPESP), under grant #2017/18148-5 and National Council for Scientific and Technological Development (CNPQ), under grant 306435/2017-9 for supporting this research work.

Acknowledgments: The authors would like to thank the Norton Company for the grinding wheel donation.

Conflicts of Interest: The authors declare no conflict of interest.

References

1. Black, J.T.; Kohser, R.A. *DeGarmo's Materials and Processes in Manufacturing*, 10th ed.; John Wiley & Sons, Inc.: River Street, Hoboken, NJ, USA, 2008; ISBN 13-978-0470-05512-0.
2. Lizarralde, R.; Montejo, M.; Barrenetxea, D.; Marquinez, J.I.; Gallego, I. Intelligent grinding: Sensorless instabilities detection. *IEEE Instrum. Meas. Mag.* **2006**, *9*, 30–37. [CrossRef]
3. Winter, M.; Li, W.; Kara, S.; Herrmann, C. Determining optimal process parameters to increase the eco-efficiency of grinding processes. *J. Clean. Prod.* **2014**, *66*, 644–654. [CrossRef]
4. Wiederkehr, P.; Siebrecht, T.; Potthoff, N. Stochastic modeling of grain wear in geometric physically-based grinding simulations. *CIRP Ann.* **2018**, *67*, 325–328. [CrossRef]
5. Agarwal, S.; Rao, P.V. Experimental investigation of surface/subsurface damage formation and material removal mechanisms in SiC grinding. *Int. J. Mach. Tools Manuf.* **2008**, *48*, 698–710. [CrossRef]

6. Ding, W.-F.; Xu, J.-H.; Chen, Z.-Z.; Su, H.-H.; Fu, Y.-C. Wear behavior and mechanism of single-layer brazed CBN abrasive wheels during creep-feed grinding cast nickel-based superalloy. *Int. J. Adv. Manuf. Technol.* **2010**, *51*, 541–550. [CrossRef]
7. Teti, R.; Jemielniak, K.; O'Donnell, G.; Dornfeld, D. Advanced monitoring of machining operations. *CIRP Ann.* **2010**, *59*, 717–739. [CrossRef]
8. Stavropoulos, P.; Chantzis, D.; Doukas, C.; Papacharalampopoulos, A.; Chryssolouris, G. Monitoring and Control of Manufacturing Processes: A Review. *Procedia CIRP* **2013**, *8*, 421–425. [CrossRef]
9. Lopes, W.N.; Ferreira, F.I.; Alexandre, F.A.; Ribeiro, D.M.S.; de Oliveira Conceição, P., Jr.; de Aguiar, P.R.; Bianchi, E.C. Digital signal processing of acoustic emission signals using power spectral density and counts statistic applied to single-point dressing operation. *IET Sci. Meas. Technol.* **2017**, *11*, 631–636. [CrossRef]
10. Chen, X.; Li, C.; Tang, Y.; Xiao, Q. An Internet of Things based energy efficiency monitoring and management system for machining workshop. *J. Clean. Prod.* **2018**, *199*, 957–968. [CrossRef]
11. Dimla, D.E. The Correlation of Vibration Signal Features to Cutting Tool Wear in a Metal Turning Operation. *Int. J. Adv. Manuf. Technol.* **2002**, *19*, 705–713. [CrossRef]
12. de Freitas, E.S.; Baptista, F.G. Experimental analysis of the feasibility of low-cost piezoelectric diaphragms in impedance-based SHM applications. *Sens. Actuators A Phys.* **2016**, *238*, 220–228. [CrossRef]
13. Budoya, D.; Baptista, F. Signal Acquisition from Piezoelectric Transducers for Impedance-Based Damage Detection. *Proceedings* **2017**, *2*, 130. [CrossRef]
14. Lucas, G.B.; de Castro, B.A.; Rocha, M.A.; Andreoli, A.L. Study of a Low-Cost Piezoelectric Sensor for Three Phase Induction Motor Load Estimation. *Proceedings* **2019**, *4*, 46. [CrossRef]
15. Marchi, M.; Baptista, F.G.; de Aguiar, P.R.; Bianchi, E.C. Grinding process monitoring based on electromechanical impedance measurements. *Meas. Sci. Technol.* **2015**, *26*, 045601. [CrossRef]
16. Ribeiro, D.M.S.S.; Aguiar, P.R.; Fabiano, L.F.G.G.; D'Addona, D.M.; Baptista, F.G.; Bianchi, E.C. Spectra Measurements Using Piezoelectric Diaphragms to Detect Burn in Grinding Process. *IEEE Trans. Instrum. Meas.* **2017**, *66*, 3052–3063. [CrossRef]
17. Alexandre, F.; de Aguiar, P.; Götz, R.; Aulestia Viera, M.; Lopes, T.; D'addona, D.; Bianchi, E.; Silva, R.B. da Emitter-Receiver Piezoelectric Transducers Applied in Monitoring Material Removal of Workpiece during Grinding Process. *Proceedings* **2018**, *4*, 9. [CrossRef]
18. de Oliveira Conceição, P., Jr.; D'Addona, D.M.; Aguiar, P.R. Dressing Tool Condition Monitoring through Impedance-Based Sensors: Part 1—PZT Diaphragm Transducer Response and EMI Sensing Technique. *Sensors* **2018**, *18*, 4455. [CrossRef]
19. de Freitas, E.S.; Baptista, F.G.; Budoya, D.E.; de Castro, B.A. Equivalent Circuit of Piezoelectric Diaphragms for Impedance-Based Structural Health Monitoring Applications. *IEEE Sens. J.* **2017**, *17*, 5537–5546. [CrossRef]
20. Castro, B.; Clerice, G.; Ramos, C.; Andreoli, A.; Baptista, F.; Campos, F.; Ulson, J. Partial Discharge Monitoring in Power Transformers Using Low-Cost Piezoelectric Sensors. *Sensors* **2016**, *16*, 1266. [CrossRef]
21. Viera, M.A.A.; de Aguiar, P.R.; de Oliveira Conceição, P., Jr.; da Silva, R.B.; Jackson, M.J.; Alexandre, F.A.; Bianchi, E.C. Low-Cost Piezoelectric Transducer for Ceramic Grinding Monitoring. *IEEE Sens. J.* **2019**, *19*, 7605–7612. [CrossRef]
22. Wegener, K.; Hoffmeister, H.-W.; Karpuschewski, B.; Kuster, F.; Hahmann, W.-C.; Rabiey, M. Conditioning and monitoring of grinding wheels. *CIRP Ann.* **2011**, *60*, 757–777. [CrossRef]
23. de Oliveira, J.F.G.; Dornfeld, D.A. Application of AE Contact Sensing in Reliable Grinding Monitoring. *CIRP Ann.* **2001**, *50*, 217–220. [CrossRef]
24. Viera, M.A.; Alexandre, F.; Lopes, W.; de Aguiar, P.; da Silva, R.B.; D'addona, D.; Andreoli, A.; Bianchi, E. A Contribution to the Monitoring of Ceramic Surface Quality Using a Low-Cost Piezoelectric Transducer in the Grinding Operation. *Proceedings* **2018**, *4*, 16. [CrossRef]
25. Yeih, W.; Huang, R. Detection of the corrosion damage in reinforced concrete members by ultrasonic testing. *Cem. Concr. Res.* **1998**, *28*, 1071–1083. [CrossRef]
26. Alexandre, F.; Aguiar, P.R.; Götz, R.; Fernandez, B.O.; Lopes, W.N.; Viera, M.A.A.; Bianchi, E.C.; D'addona, D.; D'addona, D. Damage detection in grinding of steel workpieces through ultrasonic waves. *MATEC Web Conf.* **2018**, *249*, 02002. [CrossRef]
27. Baptista, F.G.; Vieira, J.F. A new impedance measurement system for PZT based structural health monitoring. *IEEE Trans. Instrum. Meas.* **2009**, *58*, 3602–3608. [CrossRef]

28. de Castro, B.A.; Baptista, F.G.; Ciampa, F. New signal processing approach for structural health monitoring in noisy environments based on impedance measurements. *Measurement* **2019**, *137*, 155–167. [CrossRef]
29. Na, S.; Lee, H.K. Steel wire electromechanical impedance method using a piezoelectric material for composite structures with complex surfaces. *Compos. Struct.* **2013**, *98*, 79–84. [CrossRef]
30. Annamdas, V.G.M.; Rizzo, P. Monitoring concrete by means of embedded sensors and electromechanical impedance technique. In *Sensors and Smart Structures Technologies for Civil, Mechanical and Aerospace Systems 2010*; Tomizuka, M., Ed.; SPIE: San Diego, CA, USA, 2010; p. 76471Z.
31. Zhu, H.; Luo, H.; Ai, D.; Wang, C. Mechanical impedance-based technique for steel structural corrosion damage detection. *Measurement* **2016**, *88*, 353–359. [CrossRef]
32. Budoya, D.E.; Baptista, F.G. A Comparative Study of Impedance Measurement Techniques for Structural Health Monitoring Applications. *IEEE Trans. Instrum. Meas.* **2018**, *67*, 912–924. [CrossRef]
33. Farrar, C.R.; Worden, K. *Structural Health Monitoring: A Machine Learning Pespective*, 1st ed.; John Wiley & Sons, Inc.: West Sussex, UK, 2012.
34. Lopes, B.G.; Alexandre, F.A.; Lopes, W.N.; de Aguiar, P.R.; Bianchi, E.C.; Viera, M.A.A. Study on the effect of the temperature in Acoustic Emission Sensor by the Pencil Lead Break Test. In Proceedings of the 2018 13th IEEE International Conference on Industry Applications (INDUSCON), Sao Paulo, Brazil, 11–14 November 2018; pp. 1226–1229.
35. Baptista, F.; Budoya, D.; Almeida, V.; Ulson, J. An Experimental Study on the Effect of Temperature on Piezoelectric Sensors for Impedance-Based Structural Health Monitoring. *Sensors* **2014**, *14*, 1208–1227. [CrossRef]
36. de Oliveira Conceição, P., Jr.; D'Addona, D.; Aguiar, P.; Teti, R. Dressing Tool Condition Monitoring through Impedance-Based Sensors: Part 2—Neural Networks and K-Nearest Neighbor Classifier Approach. *Sensors* **2018**, *18*, 4453. [CrossRef]
37. da Silveira, R.Z.M.; Campeiro, L.M.; Baptista, F.G. Performance of three transducer mounting methods in impedance-based structural health monitoring applications. *J. Intell. Mater. Syst. Struct.* **2017**, *28*, 2349–2362. [CrossRef]
38. Martowicz, A.; Sendecki, A.; Salamon, M.; Rosiek, M.; Uhl, T. Application of electromechanical impedance-based SHM for damage detection in bolted pipeline connection. *Nondestruct. Test. Eval.* **2016**, *31*, 17–44. [CrossRef]
39. Liang, Y.; Feng, Q.; Li, D.; Cai, S. Loosening Monitoring of a Threaded Pipe Connection Using the Electro-Mechanical Impedance Technique-Experimental and Numerical Studies. *Sensors* **2018**, *18*, 3699. [CrossRef]
40. Ihn, J.-B.B.; Chang, F.-K.K. Pitch-catch Active Sensing Methods in Structural Health Monitoring for Aircraft Structures. *Struct. Health Monit. Int. J.* **2008**, *7*, 5–19. [CrossRef]
41. Jata, K.V.; Kundu, T.; Parthasarathy, T.A. An Introduction to Failure Mechanisms and Ultrasonic Inspection. In *Advanced Ultrasonic Methods for Material and Structure Inspection*; Kundu, T., Ed.; ISTE: London, UK, 2010; pp. 1–42, ISBN 9780470612248.
42. Awad, T.S.; Moharram, H.A.; Shaltout, O.E.; Asker, D.; Youssef, M.M. Applications of ultrasound in analysis, processing and quality control of food: A review. *Food Res. Int.* **2012**, *48*, 410–427. [CrossRef]
43. Ihara, I.; Takahashi, M. Non-invasive monitoring of temperature distribution inside materials with ultrasound inversion method. *Int. J. Intell. Syst. Technol. Appl.* **2009**, *7*, 80–91. [CrossRef]
44. Takahashi, M.; Ihara, I. Ultrasonic Monitoring of Internal Temperature Distribution in a Heated Material. *Jpn. J. Appl. Phys.* **2008**, *47*, 3894–3898. [CrossRef]
45. Raišutis, R.; Voleišis, A.; Kažys, R. Application of the through transmission ultrasonic technique for estimation of the phase velocity dispersion in plastic materials. *Ultragarsas "Ultrasound"* **2016**, *63*, 15–18. [CrossRef]
46. Resa, P.; Bolumar, T.; Elvira, L.; Pérez, G.; de Espinosa, F.M. Monitoring of lactic acid fermentation in culture broth using ultrasonic velocity. *J. Food Eng.* **2007**, *78*, 1083–1091. [CrossRef]
47. Aguiar, P.R.; Serni, P.J.A.; Bianchi, E.C.; Dotto, F.R.L. In-process grinding monitoring by acoustic emission. In Proceedings of the 2004 IEEE International Conference on Acoustics, Speech and Signal Processing, Montreal, QC, Canada, 17–21 May 2004; Volume 5, p. V-405-8.
48. Moia, D.F.G.; Thomazella, I.H.; Aguiar, P.R.; Bianchi, E.C.; Martins, C.H.R.; Marchi, M. Tool condition monitoring of aluminum oxide grinding wheel in dressing operation using acoustic emission and neural networks. *J. Braz. Soc. Mech. Sci. Eng.* **2015**, *37*, 627–640. [CrossRef]

49. Aguiar, P.R.; Serni, P.J.A.; Dotto, F.R.L.; Bianchi, E.C. In-process grinding monitoring through acoustic emission. *J. Braz. Soc. Mech. Sci. Eng.* **2006**, *28*, 118–124. [CrossRef]
50. D'Addona, D.M.; Matarazzo, D.; Teti, R.; de Aguiar, P.R.; Bianchi, E.C.; Fornaro, A. Prediction of Dressing in Grinding Operation via Neural Networks. *Procedia CIRP* **2017**, *62*, 305–310. [CrossRef]
51. Webster, J.; Dong, W.P.; Lindsay, R. Raw Acoustic Emission Signal Analysis of Grinding Process. *CIRP Ann. Manuf. Technol.* **1996**, *45*, 335–340. [CrossRef]
52. Reiweger, I.; Mayer, K.; Steiner, K.; Dual, J.; Schweizer, J. Measuring and localizing acoustic emission events in snow prior to fracture. *Cold Reg. Sci. Technol.* **2015**, *110*, 160–169. [CrossRef]
53. Lissek, F.; Kaufeld, M.; Tegas, J.; Hloch, S. Online-monitoring for Abrasive Waterjet Cutting of CFRP via Acoustic Emission: Evaluation of Machining Parameters and Work Piece Quality Due to Burst Analysis. *Procedia Eng.* **2016**, *149*, 67–76. [CrossRef]
54. Alexandre, F.A.; Lopes, W.N.; Ferreira, F.I.; Dotto, F.R.L.; de Aguiar, P.R.; Bianchi, E.C. Chatter Vibration Monitoring in the Surface Grinding Process through Digital Signal Processing of Acceleration Signal. *Proceedings* **2017**, *2*, 126. [CrossRef]
55. Griffin, J.; Chen, X. Classification of the acoustic emission signals of rubbing, ploughing and cutting during single grit scratch tests. *Int. J. Nanomanuf.* **2006**, *1*, 189–209. [CrossRef]
56. Sadegh, H.; Mehdi, A.N.; Mehdi, A. Classification of acoustic emission signals generated from journal bearing at different lubrication conditions based on wavelet analysis in combination with artificial neural network and genetic algorithm. *Tribol. Int.* **2016**, *95*, 426–434. [CrossRef]
57. Ahirrao, N.S.; Bhosle, S.P.; Nehete, D.V. Dynamics and Vibration Measurements in Engines. *Procedia Manuf.* **2018**, *20*, 434–439. [CrossRef]
58. Kang, I.S.; Kim, J.S.; Kang, M.C.; Lee, K.Y. Tool condition and machined surface monitoring for micro-lens array fabrication in mechanical machining. *J. Mater. Process. Technol.* **2008**, *201*, 585–589. [CrossRef]
59. Delrue, S.; Van Den Abeele, K.; Blomme, E.; Deveugele, J.; Lust, P.; Matar, O.B. Two-dimensional simulation of the single-sided air-coupled ultrasonic pitch-catch technique for non-destructive testing. *Ultrasonics* **2010**, *50*, 188–196. [CrossRef] [PubMed]
60. Diamanti, K.; Soutis, C. Structural health monitoring techniques for aircraft composite structures. *Prog. Aerosp. Sci.* **2010**, *46*, 342–352. [CrossRef]
61. Coramik, M.; Ege, Y. Discontinuity inspection in pipelines: A comparison review. *Measurement* **2017**, *111*, 359–373. [CrossRef]
62. Buckin, V.; Altas, M.C. High-resolution ultrasonic spectroscopy. In *Proceedings Sensors 2017*; AMA Conferences, Ed.; AMA Association for Sensors and Measurement: Nuremberg Germany, 2017; Volume 7, pp. 298–303.
63. Giurgiutiu, V.; Rogers, C.A. Recent advancements in the electromechanical (E/M) impedance method for structural health monitoring and NDE. In *Proceedings SPIE 3329, Smart Structures and Materials*; Regelbrugge, M.E., Ed.; SPIE: San Diego, CA, USA, 1998; pp. 536–547.
64. Giurgiutiu, V.; Rogers, C.A. Modeling of the electro-mechanical (E/M) impedance response of a damaged composite beam. In *ASME Aerospace and Materials Divisions, Adaptive Structures and Material Systems Symposium*; ASME Winter Annual Meeting: Nashville, TN, USA, 1999; pp. 39–46.
65. Baptista, F.G.; Filho, J.V.; Inman, D.J. Influence of Excitation Signal on Impedance-based Structural Health Monitoring. *J. Intell. Mater. Syst. Struct.* **2010**, *21*, 1409–1416. [CrossRef]
66. Giurgiutiu, V.; Zagrai, A. Damage Detection in Thin Plates and Aerospace Structures with the Electro-Mechanical Impedance Method. *Struct. Health Monit. Int. J.* **2005**, *4*, 99–118. [CrossRef]
67. de Oliveira Conceição, P., Jr.; Aguiar, P.R.; Foschini, C.R.; França, T.V.; Ribeiro, D.M.S.; Ferreira, F.I.; Lopes, W.N.; Bianchi, E.C. Feature extraction using frequency spectrum and time domain analysis of vibration signals to monitoring advanced ceramic in grinding process. *IET Sci. Meas. Technol.* **2019**, *13*, 1–8. [CrossRef]
68. Campeiro, L.M.; da Silveira, R.Z.M.; Baptista, F.G. Impedance-based damage detection under noise and vibration effects. *Struct. Health Monit. Int. J.* **2017**, *17*, 654–667. [CrossRef]

Article

Analysis of the Influence of Different Settings of Scan Sequence Parameters on Vibration and Noise Generated in the Open-Air MRI Scanning Area †

Jiří Přibil *, Anna Přibilová and Ivan Frollo

Institute of Measurement Science, Slovak Academy of Sciences, 841 04 Bratislava, Slovakia; anna.pribilova@savba.sk (A.P.); umerollo@savba.sk (I.F.)

* Correspondence: Jiri.Pribil@savba.sk; Tel.: +421-2-59104543

† This paper is an extended version of the conference paper: Přibil, J.; Přibilová, A.; Frollo, I. Analysis of Energy Relations between Noise and Vibration Signals in the Scanning Area of an Open-Air MRI Device, In Proceedings of the 5th International Electronic Conference on Sensors and Applications, 15–30 November, 2018.

Received: 14 August 2019; Accepted: 25 September 2019; Published: 27 September 2019

Abstract: A system of gradient coils of the magnetic resonance imaging (MRI) device produces significant vibration and noise. Energetic relations of these phenomena are analyzed depending on MRI scan parameters (sequence type, repetition time (TR), echo time (TE), slice orientation, body weight). This issue should be investigated because of negative physiological and psychological effects on a person exposed to vibration and acoustic noise. We also measured the sound pressure level in the MRI scanning area and its vicinity in order to minimize these negative impacts, depending on intensity and time duration of exposition. From the recorded vibration and noise signals, the energy parameters were determined and statistically analyzed, and the obtained results were visually and numerically compared. Finally, subjective evaluation by a listening test method was used to analyze the influence of the generated MRI noise on the human psyche.

Keywords: magnetic resonance imaging; mechanical vibration; acoustic noise; signal processing

---

1. Introduction

Magnetic resonance imaging (MRI) is an effective method for structure investigation of biological samples [1] or human body parts, such as a head [2,3], a thorax [4], etc. In addition to the standard closed-bore MRI scanner, the open-air one is increasingly used in special cases, e.g., in claustrophobic patients [5]. In this MRI device, two parallel permanent magnets form a static magnetic field between them [6]. A gradient system consisting of two internal planar coils parallel to the magnets is used to select slices in three dimensions. A tested object is placed in the magnetic field, together with an external radio frequency (RF) receiving/transmitting coil.

The coil current changes quickly during gradient switching, resulting in undesirable vibration of the whole structure [7], and subsequent acoustic noise disturbing the speech recorded during articulation and concurrent three-dimensional (3D) MRI scanning for examination of dynamic changes in the shape of the vocal tract and vocal folds [8]. In this case, a speech denoising method must be applied on the recorded signal. Another solution includes the recording of a speech signal by a special fiber optical microphone that can be located in the MRI scanning area [9]. Here, real-time speech processing is enabled during MR scanning at the expense of rather complicated practical realization involving synchronization of both processes, special hardware for an MRI device, etc. A cheaper type of an optical microphone (e.g., the first or the second generation of the Optoacoustics FOMRI microphone) has a limited frequency response in the range between 50 and 4000 Hz that is insufficient for our purpose.

The third generation of the Optoacoustics FOMRI microphone system using fiber-optic technology and active noise reduction [10] can solve this problem; however, this solution is more expensive.

Acoustic noise interference is not only a technical problem, but a physiological and psychological one as well. These negative effects on humans are well known in industrial environments with long-term noise exposure [11]; however, they evolve gradually and can be observed first on a short-term scale. The intensity of the vibration and the resulting acoustic noise and the time duration of its exposure are crucial factors affecting the degree of their physiological, as well as psychological impact.

Motivation of our work was an exploration of energetic relations between the vibration and the noise in the scanning area of the low-field open-air MRI device and its surroundings to find a proper choice of a scan sequence and its parameters—sequence type, repetition time (TR), echo time (TE), slice orientation, etc. In addition, the volume inserted in the scanning area influenced the intensity of the vibration and noise produced in the scanning area of the MRI device. Finally, a tested object with its mechanical properties caused the change in the overall mechanical impedance when it loaded the system of lower gradient coils placed in the patient's bed.

The experimental part has the following structure: In the preliminary phase, the sensitivity and the frequency response of the used vibration sensor were determined. Time-domain vibration and noise signals were recorded and processed, along with the measured sound pressure level (SPL), to find energetic features that were then evaluated statistically. Next, analysis of the influence of scan parameters on the time duration of the executed MR sequence and on the quality factor of the obtained MR images was carried out. Finally, the energetic features determined from the measured data were compared with the results of the subjective evaluation based on the listening test method.

2. Subject and Methods

As the open-air MRI is used primarily in medical practice, planes perpendicular to three axes of the Cartesian coordinate system are called according to medical terminology [12]. Anterior and posterior parts of a human body are divided by a frontal (coronal) plane. Another vertical plane divides a body to its left and right sides and is called a sagittal plane. The third plane dividing a body horizontally into superior and inferior parts is called a cross-sectional (transverse) plane. The scan orientation is selected by activation of the corresponding gradient coils, so the current flowing through them, as well as the resulting vibration and acoustic noise during execution of the scan sequence, depend on this selection. Two of the fundamental pulse sequences are preferably used in this MRI device: A spin echo (SE), being an excitation pulse followed by one or more refocusing pulses, and a gradient echo (GE), produced by conjunction of an excitation pulse with a gradient field reversal [13]. For optimal operation of the MRI unit, different parameters of the used scan sequence (the field of view, the number of slices, the thickness of a slice, etc.) are used for different scanned objects. The intensity of the vibration and noise produced by the MRI system depends not only on the setting of these parameters, but on the volume and the weight of the examined object as well. The tested object (a person, a sample, or a phantom) loads the lower gradient coil structure and thus becomes a part of the mechanical vibration system with its mass, stiffness, and damping.

If the vibration is to be picked up while the MRI sequence is executed, the vibration sensor placed in the MRI scanning area must not contain any ferromagnetic part to prevent its interaction with the static magnetic field, which may decrease the quality of the acquired image. It is essential that the sensor has good sensitivity with as small a dependence on frequency as possible. Using the reference sensitivity B_{a0} at a chosen reference frequency, the frequency-dependent sensitivity (frequency response) of an accelerometer B_a [mV/ms^{-2}] may be expressed in [dB] by a relation:

$$G_{a\log}(f) = 20 \cdot \log_{10}(B_a(f)/B_{a0}). \tag{1}$$

The sensor's frequency range should cover harmonic frequencies of the vibration and noise signals concentrated in the low band due to limited frequency spectrum of the gradient pulse sequence.

These requirements can be fulfilled by sensors constructed for acoustic pickup in musical instruments, or other ones based on the piezoelectric principle. First usage of all these vibration sensors must be preceded by measurement of their sensitivity and frequency response.

For obtaining high-quality MR images without artifacts, the acoustic sensors (the measuring microphone and the sound level meter) containing ferromagnetic parts must be placed beyond the influence of this static magnetic field, although, adequately close to the noise source. The sensitivity of the recording microphone and its directional pattern are selected in regard to effective rejection of the ambient noise. The sound level meter enables choice of the type of frequency weighting to match human perception of silent sounds (A weighting with more suppressed low and high frequencies) and loud sounds (C weighting with much less suppression of low frequencies than A weighting). Due to high measured sound pressure levels, we chose C weighting.

Several approaches can be applied for determination of the signal energy. Three of them represent the basis of our comparisons:

1. The standard root mean square (RMS) is calculated from a signal $x(n)$ in a defined frame (window) with the length of M samples:

$$Signal_{RMS} = \sqrt{\frac{1}{M}\sum_{n=1}^{M}\left|x(n)\right|^2}, \tag{2}$$

2. Another energetic parameter is determined from the Teager–Kaiser energy operator O_{TK} [14]:

$$O_{TK} = x(n)^2 - x(n-1)\cdot x(n+1), \qquad En_{TK} = abs\left(\frac{1}{M-2}\sum_{n=1}^{M-2} O_{TK}(n)\right). \tag{3}$$

3. The third approach uses the short-term N_{FFT}-point fast Fourier transform (FFT) to compute the power spectrum $|S(k)|^2$, and in each frame, the energy is assessed from the first cepstral coefficient c_0 or from the autocorrelation coefficient r_0:

$$En_{c0} = \sqrt{\left[\prod_{k=1}^{N_{FFT}/2}\left|S(k)\right|^2\right]^{\frac{2}{N_{FFT}}}}, En_{r0} = \frac{2}{N_{FFT}}\sum_{k=1}^{N_{FFT}/2}\left|S(k)\right|^2. \tag{4}$$

Next, the basic and supplementary spectral properties are determined from the recorded noise and vibration signals. Methods similar to those used in speech signal analysis can be applied for processing of these signals whose spectral content lies within the standard audio frequency range. The basic spectral properties (basic resonance frequencies F_{V1} and F_{V2} and their ratios, spectral decrease, etc.) are usually determined from the spectral envelope. The supplementary spectral features (spectral centroid (SC), harmonic-to-noise ratio (HNR), spectral entropy, etc.) describe the shape of the power spectrum of the analyzed signal.

3. Performed Experiments and Results

The experiments were carried out on the open-air, low-field (up to 0.2 T) MRI system E-scan Opera manufactured by the company Esaote S.p.A., Genoa, Italy [6]. This MRI system is located at the Institute of Measurement Science (IMS) in Bratislava, in the laboratory of the department of imaging methods. The piezoelectric vibration sensor was located in the scanning area of the MRI device; the microphone and the sound level meter were placed in its vicinity. The time course of the output voltage signals of the vibration sensor and the microphone was recorded, together with the measured noise SPL. The stored data were further processed to analyze the vibration and noise conditions by the energetic signal properties.

3.1. Automatic Measurement of Relative Sensitivity and Frequency Response of Vibration Sensors Suitable for Working in the Low Magnetic Field Environment

The preliminary measurements of the sensitivity and the frequency response of the vibration sensors were done using the vibration exciter ESE 201 on which the sensors were mounted. The Audio Precision System One (AP S1) with two programmable input and output channels was used as a signal generator for the exciter, as well as an input for simultaneous measurement of electrical signals from the sensors. The AP S1 output voltage for the exciter and the AP S1 input signal from the measured sensor were checked in parallel by the digital oscilloscope Rigol DS1102E. Three types of vibration sensors with the main element based on the piezoelectric principle were measured and tested:

- Cejpek SB-1 with a circular 27.5-mm brass disc designed primarily for pickup of a musical sound of a contrabass [15,16];
- RFT heart microphone device HM 692 with a 20-mm disc transducer comprising a piezoelectric element integrated in the aluminum metal cover with a 30-mm diameter; and
- SDT1-028K sensor (by Sensor Solutions—TE Connectivity) consisting of a rectangular piezo film element (with dimensions of 28.6 × 11.2 × 0.13 mm), together with a molded plastic housing, typically used in safety and industrial applications.

The vibration sensors were measured and compared by two types of parameters:

- Relative sensitivity at the reference frequency f_{ref} = 125 Hz; and
- Frequency response in the range of 20 Hz to 2 kHz at a chosen AP S1 output voltage for the vibration exciter (U_{Ba0} = 360 mV_{ef} = 1 V p-p).

Dependence of all four sensors' sensitivities on the excitation voltage can be seen in Figure 1a, and their frequency responses in [dB] in the range of 20 Hz to 2 kHz are depicted in Figure 1b.

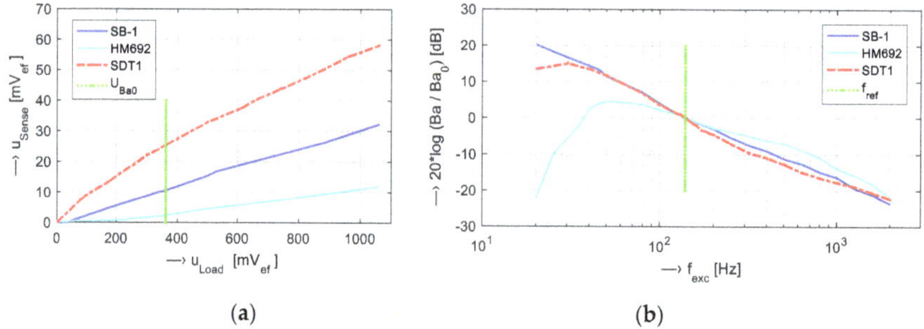

Figure 1. Graphs of (**a**) measured sensors' sensitivities and (**b**) relative frequency responses [dB] in the range of 20 Hz to 2 kHz; f_{ref} =125 Hz; $Uexc_{Ba0}$ =360 mV; B_{a0} = {12.9 (SB-1), 2.45 (HM692), 26.0 (SDT1)} mV/m·s^{-2}.

3.2. Description of Main Measurement and Auxiliary Experiments

For all basic measurements, a plastic sphere-phantom of a 140-mm diameter filled with doped water [6] was placed inside the knee RF coil; see the arrangement photo in Figure 2. For acoustic noise SPL measurement in the MRI device vicinity, the multi-function environment meter Lafayette DT 8820 was used. Its distance from the center of the scanning area was 60 cm, its height from the floor was 75 cm (at the level of the bottom gradient coils), and it was oriented at 30 degrees from the left corner where the temperature stabilizer was placed. First, the noise SPL was mapped in the range of distances from 45 to 90 cm—the obtained values for Hi-Res SE and GE sequences are presented in Figure 3. As a result of the preliminary measurements, the SB-1 sensor was used to record the vibration signal from the solid surface of the plastic holder of the bottom gradient coils in the MRI scanning area—see position P0 in Figure 2. This figure also shows positions of the sound level meter (Lafayette DT 8820) and the dual diaphragm condenser microphone (Behringer B-2 PRO) on a stand with shock mounting for the noise SPL and signal pick up. The signals of the vibration and noise were routed through the Behringer Podcast Studio equipment by USB connection to PC. The signals with duration of about 15 s were sampled at 32 kHz and then processed in the sound editor program Sound Forge 9.0a.

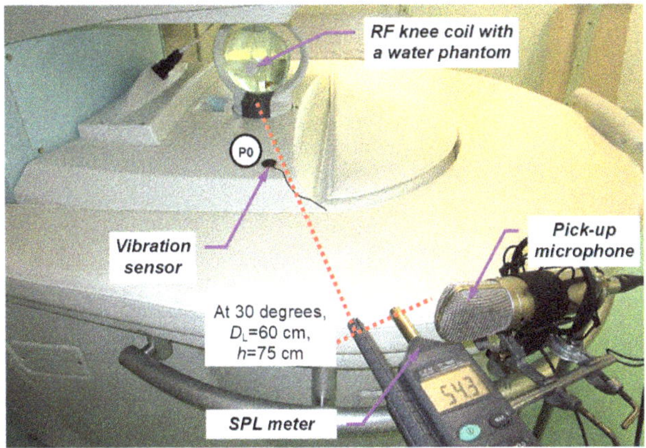

Figure 2. Arrangement photo of SPL noise measurement and parallel recording of noise and vibration signals of the open-air MRI device Opera using the testing phantom.

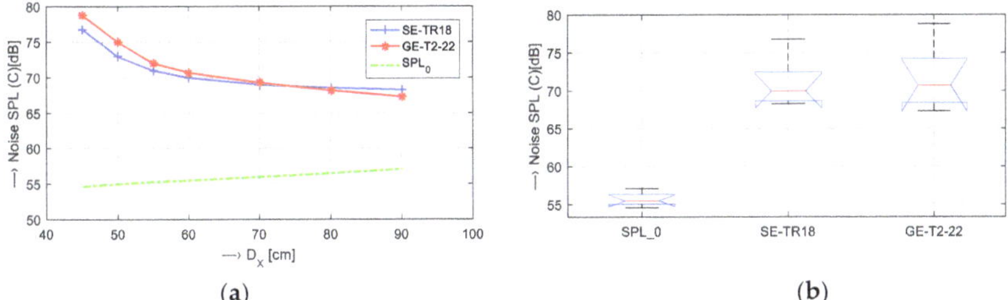

(**a**) (**b**)

Figure 3. Mapping of the acoustic noise SPL at different distances D_X = {45, 50, 55, 60, 70, 80, and 90} cm from the middle of the scanning area of the MRI device for SE/GE sequences: (**a**) SPL values together with the background ones (SPL_0), and (**b**) box-plot of their basic statistical parameters.

The main experimental measurement was aimed at investigation of the impact of MRI scan parameters on the recorded vibration and noise. As presented in our earlier paper [17], only five types of scanning sequences are implemented in the investigated MRI device [6]:

1. SE sequences (11 sub-types);
2. GE sequences (9 sub-types);
3. Turbo (multi echo) sequences (4 sub-types);
4. 3D sequences (5 sub-types); and
5. Hi-Res sequences (8 sub-types).

In practice, two basic types of scan sequences are commonly used for non-invasive examination of human body parts by acquiring high-quality MR images in this type of MRI device:

a. High-resolution SE/GE pulse sequences (Hi-Res); and
b. 3D sequences to create 3D models of various biological objects.

The baseline measurement and recording of the vibration and noise signals were carried out during the execution of MR scan sequences typical for 3D imaging of the human vocal tract. Figure 4 shows how the energetic and spectral parameters of the picked-up vibration and noise signals are determined and processed.

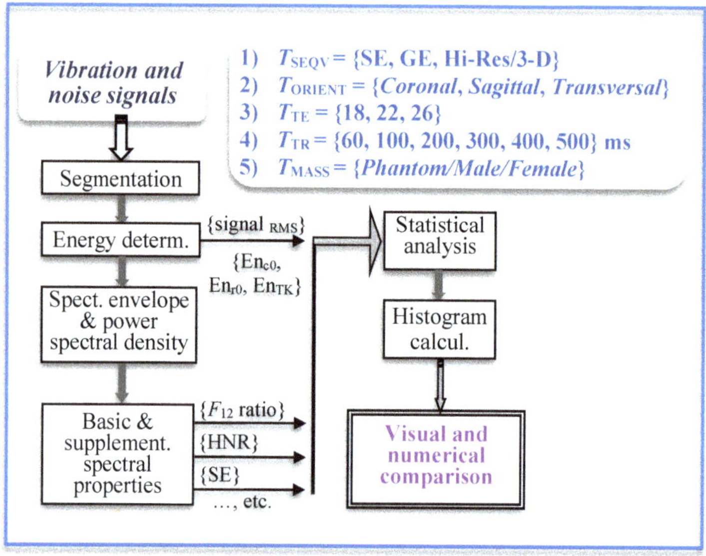

Figure 4. Block diagram of processing and comparison of vibration and noise signals.

Scan parameters of five tested MR sequences (T_{SEQV} = {Hi-Res SE 18 HF, Hi-Res SE 26 HF, Hi-Res GE T2, SS-3Dbalanced, 3D-CE}) were set as shown in Table 1. Graphical comparison of energetic relations between the measured vibration and noise signals can be seen in Figure 6. Then, analysis of the influence of scan parameters on properties of the vibration and noise signals was executed for different parameters:

- Scan slice orientation T_{ORIENT} = {Coronal, Sagittal, Transversal}—results for energetic and basic spectral features are visualized in Figure 5;
- Echo time T_{TE} = {18, 22, 26} ms and repetition time T_{TR}= {60, 100, 200, 300, 400, 500} ms—visualizations shown in Figures 7 and 8; and
- Mass of the object/subject T_{MASS} = {Phantom/Male/Female} placed in the MRI scanning area (the testing phantom with the total weight of 0.75 kg or a head and a neck of a lying male/female person weighing approximately 80/55 kg was placed in the RF scan coil between the upper and lower gradient coils of the MRI device)—numerical comparison can be found in Table 2.

Table 1. Basic scan parameters of used MR sequences.

Type	Name of Sequence	TE (ms)	TR (ms)	FOV	Matrix Size
Hi-Res	SE 18 HF	18	500	250 × 250 × 200	256 × 256
Hi-Res	SE 26 HF	26	500	250 × 250 × 200	256 × 256
Hi-Res	GE T2	22	60	250 × 250 × 200	256 × 256
3D	SS 3D balanced	5	10	200 × 200 × 192	200 × 200
3D	3D-CE	30	40	150 × 150 × 192	192 × 192

Table 2. Comparison of mean energy values of vibration and noise signals for different objects placed in the scanning area of the MRI device.

Subject Type [1]	Vibrations (SB-1)				Noise (B2-Pro)			
	RMS	En_{TK}	En_{c0}	En_{r0}	RMS	En_{TK}	En_{c0}	En_{r0}
Water phantom	34.6	4.69	0.0380	24.0	20.1	4.05	0.0255	8.5
Female	28.7	4.93	0.0402	16.6	23.2	4.19	0.0286	10.6
Male	26.8	4.96	0.0404	14.4	25.5	4.51	0.0328	15.9

[1] Used Hi-Res SE-HF scan sequences with TE = 18 ms, TR = 400 ms, and sagittal orientation.

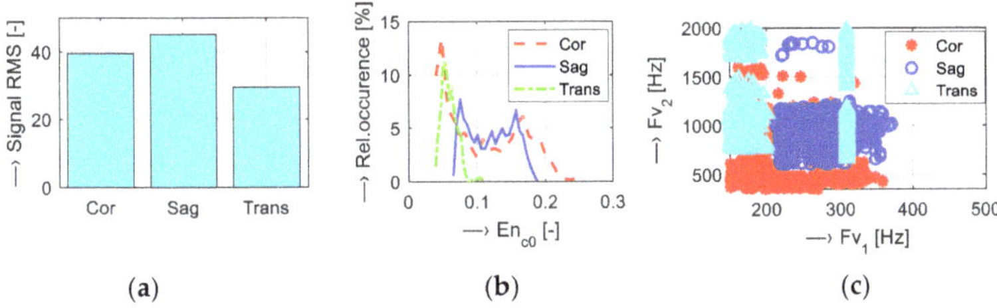

(**a**) (**b**) (**c**)

Figure 5. Visualization of vibration signal features for different slice orientations: {Coronal, sagittal, transversal}; (**a**) bar-graph of signal RMS values; (**b**) histograms of En_{c0}; (**c**) mutual F_{v1}/F_{v2} positions for Hi-Res SE scan sequences with TE = 18 ms and TR = 500 ms.

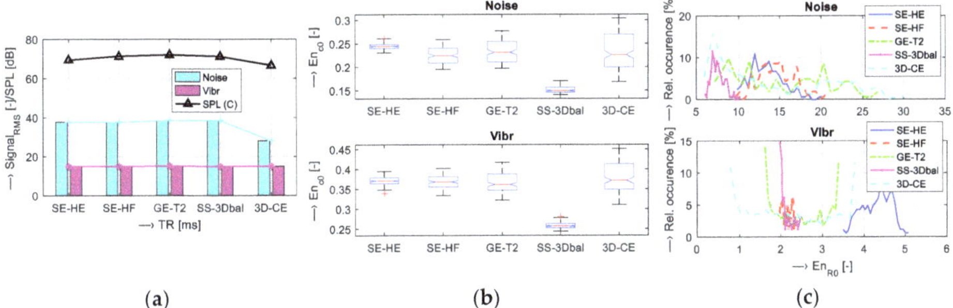

(**a**) (**b**) (**c**)

Figure 6. Comparison of energetic relations of vibration and noise signals for different sequence types—Hi-Res {SE-HE, SE-HF, GE-T2} and 3D {SS-3Dbal, 3D-CE}; (**a**) signal RMS together with SPL values; (**b**) bar-graphs of basic statistical parameters of En_{c0} values; (**c**) corresponding histograms for En_{r0} parameter. In all cases, a sagittal slice orientation was used.

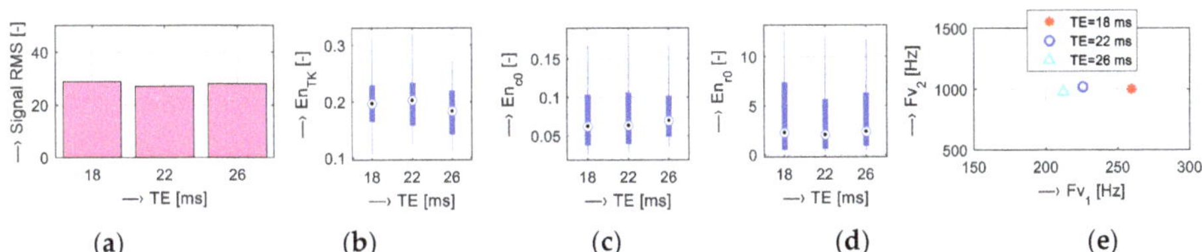

(**a**) (**b**) (**c**) (**d**) (**e**)

Figure 7. Visualization of vibration signal features for different TE times: {18, 22, 26} ms; (**a**) bar-graphs of signal RMS values and basic statistical parameters: (**b**) En_{TK}; (**c**) En_{c0}; (**d**) En_{r0}; (**e**) mean mutual F_{v1}/F_{v2} positions for Hi-Res SE-HF sequences (TR = 500 ms, sagittal orientation).

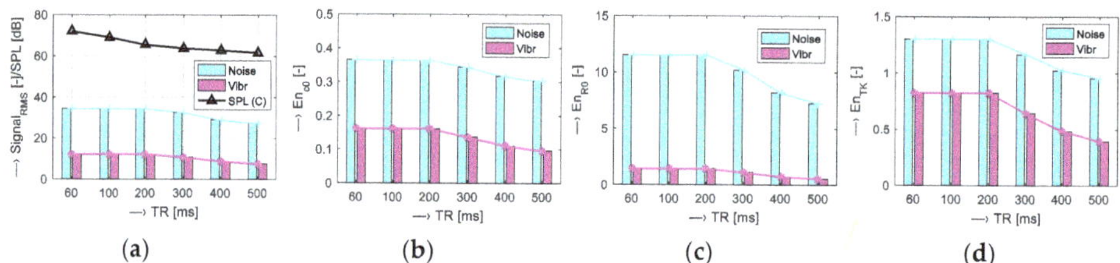

Figure 8. Visualization of energetic relations of vibration and noise signals for different TR times: {60, 100, 200, 300, 400, 500} ms; (**a**) signal RMS together with noise SPL values; (**b**) mean En_{c0}; (**c**) mean En_{r0}; (**d**) mean En_{TK}; used Hi-Res GE-T2 sequences with TE = 22 ms, and sagittal orientation.

3.3. Parameters Determinig the Scan Sequence Duration and the MR Image Quality Factor

The final scanning time is practically defined by a chosen scan sequence and basic scan parameters (TR and TE). The default configuration [6] is often modified manually by changing these two parameters; however, other scan settings (number of slices, slice thickness, number of accumulations N_{ACC} of the free induction decay (FID) signal [13], etc.) also have influence on duration, as well as on the MR image quality. The main aim is to obtain final MR images with maximum quality factor (Q_F) and minimum scan time duration (T_{DUR}).

Using the operating console of the Esaote Opera MRI device [6], we analyzed how the predicted MR image Q_F and the scan-sequence time duration are affected by the following scan parameters:

- Slice thickness {2, 2.5, 3, 4, 4.5, 7, 10} mm—predicted Q_F and T_{DUR} for the Hi-Res scan sequences SE18 HE and GE-T2 22 can be seen in a bar-graph in Figure 9;
- Repetition time T_{TR} = {60, 100, 200, 300, 400, 500} ms and the number of FID signal accumulations N_{ACC} = {1, 8, 16} for SE and GE types of Hi-Res sequences—numerical comparison of the obtained values is shown in Tables 3 and 4; and
- Number of FID signal accumulations on the predicted Q_F and T_{DUR} for scanning sequences of 3D types—see the graphical comparison in Figure 10.

Table 3. Dependence of the MR image quality factor Q_F and the scanning time duration T_{DUR} [min:sec] on TR and N_{ACC} parameters; used Hi-Res SE26 HF sequence, slice thickness = 4.5 mm.

N_{ACC} [-]	Parameters	TR [ms]					
		60	100	200	300	400	500
1	Q_F [-] TDUR [min:sec]	6 0:09	12 0:14	23 0:25	32 0:36	38 0:48	42 0:59
8	Q_F [-] TDUR [min:sec]	16 0:57	33 1:33	66 3:04	90 4:34	107 6:04	500 7:35
16	Q_F [-] TDUR [min:sec]	23 1:51	47 3:03	93 6:04	127 9:05	151 12:06	168 15:07

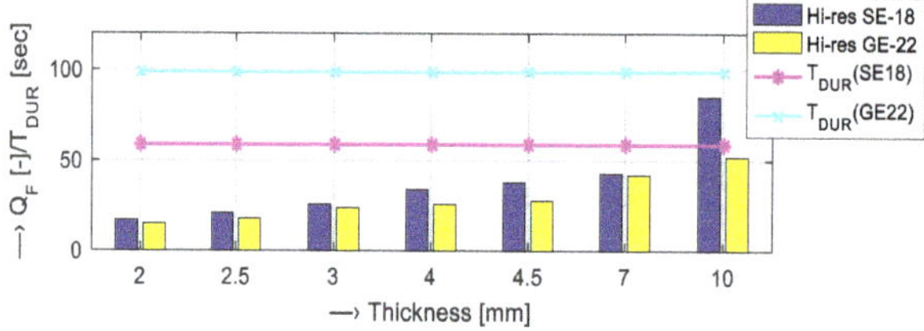

Figure 9. Influence of the thickness of the slice sample on the predicted quality factor and the time duration for the scan sequences Hi-Res SE18 HE and GE T2 22 (TR = 500 ms, N_{ACC} = 1).

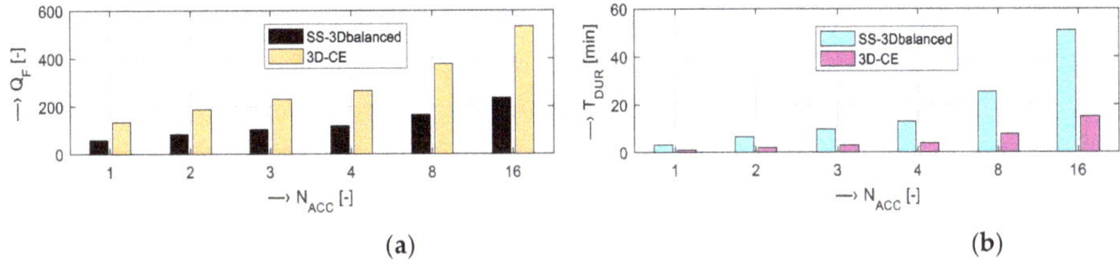

(a) (b)

Figure 10. Influence of the number of FID signal accumulations on (**a**) the predicted image quality factor and (**b**) the time duration; analyzed scan sequences SS-3Dbalanced (TE = 10 ms, TR=20 ms, 3D phases = 24) and 3D-CE (TE = 30 ms, TR = 40 ms, 3D phases = 8).

Table 4. Dependence of the MR image quality factor Q_F and the scanning time duration T_{DUR} [min:sec] on TR and N_{ACC} parameters; used Hi-Res GE-T2 22 sequence, slice thickness = 4.5 mm.

N_{ACC} [–]	Parameters	TR [ms]					
		60	100	200	300	400	500
1	Q_F [–]	15	18	24	26	27	28
	T_{DUR} [min:sec]	0:14	0:24	0:42	1:00	1:20	1:39
8	Q_F [–]	42	52	67	74	77	79
	T_{DUR} [min:sec]	1:35	2:37	5:12	7:46	10:20	12:55
16	Q_F [–]	59	74	95	104	109	112
	T_{DUR} [min:sec]	3:08	5:11	10:20	15:29	20:38	25:47

3.4. Subjective Evaluation by the Listening Test Method

Various emotions can be induced in humans by various auditory and/or visual stimuli. A set of six basic emotions is conventionally used by psychologists, although these discrete emotions can be projected into a two-dimensional plane with continuous values of pleasure and arousal axes [18]. The pleasure can change from negative (anger, sadness, fear, disgust) through neutral to positive (surprise, joy). The arousal can acquire values from low (passive emotions like apathetic sleepiness or boredom) to high (active emotion like frantic excitement) [19].

Our subjective method of evaluation based on the listening test was focused on analysis of influence of vibration and noise generated by the gradient system of the MRI device on the human psyche. The aim of the test was to assess how listeners perceive different noises from emotional point of view. The server realization of the listening test "Evaluation of pleasure and arousal of sound" was operated at the internet site http://www.lcf.um.savba.sk/Scripts/itstposl2.dll. This automatic application runs on the server PC as an MS ISAPI/NSAPI DLL script and communicates with the user within the framework of the HTTP protocol by means of the HTML pages—see an example of a screenshot in Figure 11. This system, designed originally for assessment of speech signal quality [20], serves at present the purpose of the MRI noise and vibration evaluation. Twenty-seven evaluators (8 women and 19 men) within the age range of 20 to 57 years participated in the listening test experiment open from 26 February to 20 March 2019. For compatibility with other emotional databases, like the International Affective Digitized Sounds [21], the listeners were instructed to use two parameters: Pleasantness (pleasure) and intensity (arousal), within the evaluation range of 1 to 9. The entire test consisted of 10 sets with 4 sounds each, so 40 different noises were evaluated. The listeners were allowed to play the audio stimuli as many times as they wished; low acoustic noise conditions and headphones were advised. The obtained evaluation results of the recorded MRI noises were presented by bar-graph comparisons. Figure 12 shows the emotional influence of the SE sequence with different slice orientations and the GE sequence with different TR times. Figure 13 gives the impact of the SE sequence for different masses in the scanning area, and finally, the effect of different scan sequences.

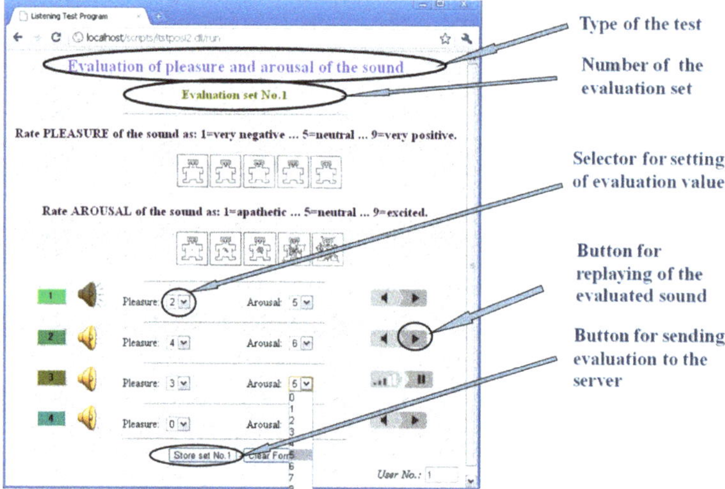

Figure 11. Screen shot of the user communication HTML page used to perform the listening test evaluation.

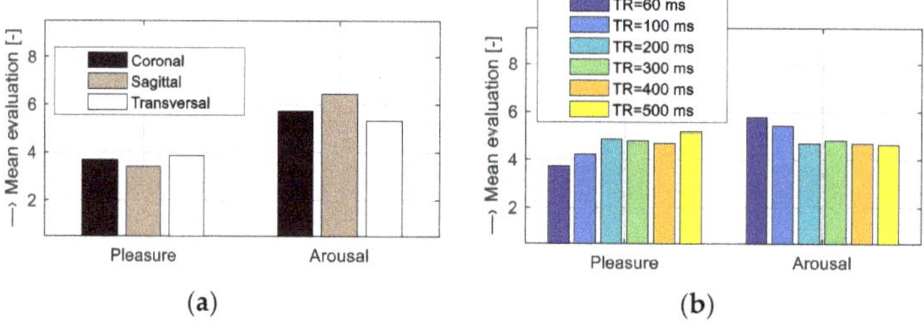

Figure 12. Bar-graph comparisons of evaluated pleasure and arousal parameters for MRI noises of (**a**) the Hi-Res SE sequence with different slice orientations, and (**b**) the Hi-Res GE sequence with different TR times.

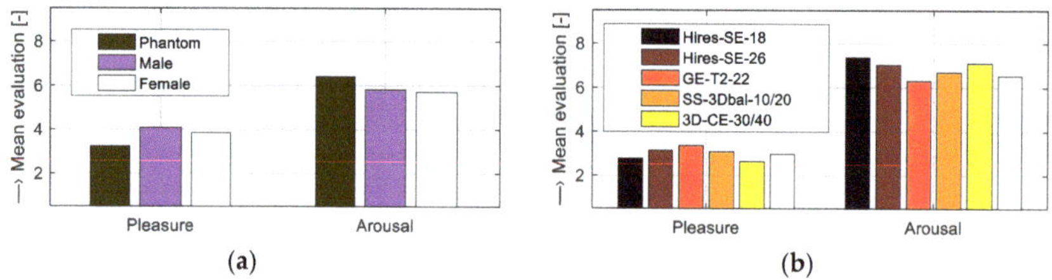

Figure 13. Bar-graph comparisons of evaluated pleasure and arousal parameters for MRI noises of the Hi-Res SE sequence for (**a**) different objects placed in the MRI scanning area and (**b**) different sequence types {Hi-Res SE-HE, Hi-Res SE-HF, Hi-Res GE-T2, SS-3Dbal, 3D-CE} using a water phantom; sagittal slice orientation in all cases.

4. Discussion of Obtained Results

For 3D MR scanning of the vocal tract [1,8] during phonation, the frequency range of 25 Hz to 3.5 kHz covering pitch and formant frequencies must be included in the recordings. A preliminary analysis of the properties of the vibration sensors suitable for measurement in the low magnetic field environment confirmed a general presumption of inverse relationship between the diameter of the used sensor and the minimum frequency of vibration picked up from the measured surface. At the same time, the maximum frequency, as well as the sensitivity, were decreased in the massive aluminum microphone capsule of the phonocardiographic sensor HM692—its frequency response in Figure 1b shows a local maximum at about 50 Hz, which makes it useless for our purpose. The piezo film vibration sensor SDT1 has the best sensitivity but its frequency response shows higher non-linearity in comparison with the sensor SB-1. In addition, the vibration sensor SB-1 has the highest low frequency sensitivity, so it was finally chosen for all next recording and measurement experiments.

The acoustic noise intensity was measured at several distances from the center of the scanning area, beginning with 45 cm because, for shorter distances, the magnetic field inhomogeneity due to interaction with metal parts of the sound level meter causes the display of a warning message on the MRI control console and interruption of further scanning to prevent failure of the MRI device [6]. At distances longer than 90 cm, the measured sound level approached the background noise level of the temperature stabilizer. The results of these measurements for high-resolution SE and GE pulse sequences can be seen in Figure 3. All the following measurements were carried out at the distance of 60 cm.

Results of the first comparison of energetic relations of vibration and noise signals are documented in Figure 6. Small differences were found for all sequence types, but the 3D-CE sequence produced the noise with minimal intensity expressed by the signal RMS parameter and SS-3Dbal calculated using the En_{c0} parameter. Further investigation was aimed at the influence of the choice of slice orientation on the energy of the produced vibration and noise signals. The graphs in Figure 5 show maximum energy in the sagittal plane and minimum energy in the transversal plane. Sagittal orientation was used in the remaining experiments to explore the worst case. In accordance with our previous research [15–17], the current experiments confirm the influence of TR and TE times on the vibration and acoustic noise properties. Prolongation of the TE time was accompanied by a slight lowering of the final signal energy with decreased first dominant frequency F_{V1}, as documented in Figure 7. Higher TR parameter determining the fundamental frequency F_{V0} had greater influence on lowering the signal energy, especially for TR = 400 and 500 ms, as can be seen in Figure 8. The obtained vibration signal energy was higher for the water phantom than for the lying person (see Table 2), due to partial attenuation of vibration pulses by higher effective weight pressing on the bottom plastic holder of the gradient coils. At the same time, the noise signal energy had its maximum for the lying male person because of higher noise induced by vibration of the upper gradient coils that are not loaded by the tested object. From a physical point of view, larger volume of a person in the MRI scanning area needs higher electrical currents in the gradient coils, causing higher Lorentz forces and greater vibration of upper gradient coils for the examined human body in comparison with the 140-mm diameter spherical testing phantom.

Preliminary analysis shows positive influence of increased slice thickness on the predicted MR image Q_F for both SE and GE Hi-Res scanning sequences—compare Figure 9. The effect of TR and N_{ACC} on T_{DUR} and predicted Q_F of the executed scanning sequence is shown in Tables 3 and 4. While the increased repetition time caused only slightly greater overall time duration, the number of accumulations affected the final time duration very much. This applies for both types of Hi-Res sequences: Increasing N_{ACC} from 1 to 16 resulted in about 4 times greater Q_F and about 15 times greater T_{DUR}. The same trend was observed for increased TR, its change from 60 to 500 ms caused about 7 (1.9) times greater Q_F and about 7 (8) times greater T_{DUR} for SE (GE). This course is valid also for two 3D scanning sequences, as seen in Figure 10. Here, the increased time duration was influenced also by the number of 3D phases being an equivalent to the number of slices (selected by the slice thickness) in Hi-Res sequences.

5. Conclusions

The results of the performed measurements help in precise description of the process of mechanical vibration excitation and the acoustic noise radiation in the scanning area and vicinity of the MRI device [22,23]. Comparison with a similar low-field MRI tomograph can be used for optimization of acoustic noise suppression in parallel with speech recording applied in 3D modeling of the human vocal tract [8]. The main usage of the obtained results is expected in experimental practice, when the used scan sequence and its parameters must often be changed depending on the person being tested. So far, our MRI experiments with vocal tract scanning were realized only on healthy people—typically, the trained MRI operators from the IMS and the authors of this paper themselves. In future, we plan cooperation with a medical center certified for work with patients. Then, our MRI equipment can be used for monitoring progress in treatment of diseases of the vocal cords and the vocal tract. For other users of this type of an open-air MRI device (in our country or in neighboring countries), we can only recommend choice of proper MR sequences, as well as modification of scan parameters as a compromise between noise and vibration exposition of an examined person and obtaining high-quality MR images.

We are aware of the limits of this study—at present, there exist a lot of modern MRI devices enabling parallel processing of MR images, thus rapidly shortening the necessary scanning time duration and subsequently minimizing the negative noise and vibration effect on patients. As the gradient coils are the integral part of the whole MRI equipment [6], we as users cannot make any changes to them; only external RF coils can be developed and connected to the system [24]. For the same reason, any system interference (e.g., adding new materials for damping of mechanical vibrations) cannot be recommended or advised as the output of this study.

Combination of the vibration and acoustic analysis can also be used in fault diagnosis of vibrating mechanical systems [25]. In our case, abnormal vibration and acoustic noise could reflect some fault in the MRI sequence generation. The performed measurement can also be reproducible on other types of low-field MRI devices; however, in the case of whole-body tomographs, picking up the vibration, as well as recording and measurement of the acoustic noise, are more difficult. In this case, the results obtained with an open-air device can be applied. A comparison study, including a discussion to this problem, has already been published in our last journal article [26].

The maximum noise SPL of about 78 dB (C) was measured at the closest distance from the central point of the MRI scanning area (45 cm), while the GE scan sequence with short TE and TR was running and the sagittal orientation was set. Although special hearing protection aids are not unconditionally necessary, ear plugs or ear muffs can contribute to the comfort of the examined person. If the scanned part of the human body inserted between the upper and the lower gradient coils is farther from the head, the ears are exposed by much lower noise. However, the absolute values of the vibration and noise signal RMS depend on signal amplification by the recording device, microphone directional pattern, and vibration sensor placement. The measurement of the noise itself is affected by a chosen measuring range (Lo/Hi) and a weighting filter (A/C) of the SPL meter. Therefore, direct comparison with the three relative energetic parameters calculated by Equations (3) and (4) is not possible. The scanning time duration depends on the chosen number of slices and their thickness. For 3D and Hi-Res sequences, it is usually lower than 15 minutes (typically about 3 to 5 minutes), so the vibration and noise exposition of the examined person does not mean a threat to his/her health. If more detailed MR images with higher quality factor Q_F must be done (e.g., scans of particular parts of the human brain, the eye, the middle and inner ear, etc.), the patient will be exposed to the vibration and acoustic noise for a longer time T_{DUR} (exceeding half an hour), which might be followed by a rather great physiological and psychological stress. Only urgent cases justify using these scan parameters in medical practice.

Our next research will be focused on detailed investigation of the negative influence of the generated acoustic noise on the physiological and psychological state of the examined person lying in the scanning area of the low-field MRI device. This negative influence on a human body can be monitored by measuring the blood pressure (BP) and heart rate (HR), as the stress is manifested by changes in the bloodstream. These changes can also be successfully determined directly from MR images of different parts of the human body—typically from the veins in the handbreadth area [27]. In addition, we plan parallel measurement of BP and HR parameters of the currently scanned person during execution of the whole MR scan sequence. For this purpose, we will apply photo-plethysmography (PPG) using optical sensors for non-invasive retrieval of vital information about the cardiovascular system from the skin surface [28,29]. Variations in the photo-detector signal are related to changes in the blood volume inside the tissue. Signal filtering and further processing will be necessary to obtain a clean PPG waveform, which can then be used to derive the instantaneous heart rate. This optical-based approach is fully in compliance with the requirements for sensors working in the magnetic field environment where RF and electromagnetic disturbances are present.

Author Contributions: Conception and design of the study, J.P., A.P., and I.F.; measurement, J.P.; data collection and processing, J.P.; writing, J.P. and A.P.); English correction, A.P.; paper review and advice, I.F.

Funding: This work was funded by the Slovak Scientific Grant Agency project VEGA 2/0001/17 and the Slovak Research and Development Agency, project no. APVV-15-0029.

Acknowledgments: We would like to thank all of our colleagues and other volunteers who participated in the listening test evaluation experiment.

Conflicts of Interest: The authors declare no conflicts of interest.

References

1. Wellard, R.M.; Ravasio, J.P.; Guesne, S.; Bell, C.; Oloyede, A.; Tevelen, G.; Pope, J.M.; Momot, K.I. Simultaneous magnetic resonance imaging and consolidation measurement of articular cartilage. *Sensors* **2014**, *14*, 7940–7958. [CrossRef]
2. Filippi, M.; Preziosa, P.; Rocca, M.A. Brain mapping in multiple sclerosis: Lessons learned about the human brain. *NeuroImage* **2019**, *190*, 32–45. [CrossRef] [PubMed]
3. Masqsood, M.; Nazir, F.; Khan, U.; Aadil, F.; Jamal, H.; Mehmood, I.; Song, O. Transfer learning assisted classification and detection of Alzheimer's disease stages using 3D MRI scans. *Sensors* **2019**, *19*, 2645. [CrossRef] [PubMed]
4. Nedoma, J.; Kepak, S.; Fajkus, M.; Cubik, J.; Siska, P.; Martinek, R.; Krupa, P. Magnetic resonance imaging compatible non-invasive fibre-optic sensors based on the Bragg gratings and interferometers in the application of monitoring heart and respiration rate of the human body: A comparative study. *Sensors* **2018**, *18*, 3713. [CrossRef] [PubMed]
5. Fischbach, K.; Kosiek, O.; Friebe, B.; Wybranski, C.; Schnackenburg, B.; Schmeisser, A.; Smid, J.; Ricke, J.; Pech, M. Cardiac Magnetic Resonance Imaging Using an Open 1.0T MR Platform: A Comparative Study ith a 1.5T Tunnel System. *Pol. J. Radiol.* **2017**, *82*, 498–505. [CrossRef] [PubMed]
6. Esaote SpA. *E-Scan Opera. User's Manual*; Revision A; Esaote SpA: Genoa, Italy, 2008.
7. Panych, L.P.; Madore, B. The physics of MRI safety. *J. Magn. Reson. Imaging* **2018**, *47*, 28–43. [CrossRef] [PubMed]
8. Schickhofer, L.; Malinen, J.; Mihaescu, M. Compressible flow simulations of voiced speech using rigid vocal tract geometries acquired by MRI. *J. Acoust. Soc. Am.* **2019**, *145*, 2049–2061. [CrossRef] [PubMed]
9. Bresch, E.; Nielsen, J.; Nayak, K.; Narayanan, S. Synchronized and noise-robust audio recordings during real-time magnetic resonance imaging scans. *J. Acoust. Soc. Am.* **2006**, *120*, 1791–1794. [CrossRef] [PubMed]
10. Optoacoustics, Ltd. FOMRI-III™ Fiber Optic Microphone for Functional MRI (Version 1.2, January 2010). Available online: http://www.optoacoustics.com/ (accessed on 11 April 2012).
11. Alimohammadi, I.; Kanrash, F.A.; Abolaghasemi, J.; Afrazandeh, H.; Rahmani, K. Effect of Chronic Noise Exposure on Aggressive Behavior of Automotive Industry Workers. *Int. J. Occup. Environ. Med.* **2018**, *9*, 170–175. [CrossRef] [PubMed]
12. Diedrichsen, J.; Balsters, J.H.; Flavell, J.; Cussans, E.; Ramnani, R. A probabilistic MR atlas of the human cerebellum. *NeuroImage* **2009**, *46*, 39–46. [CrossRef] [PubMed]
13. Bernstein, M.A.; King, K.F.; Zhou, X.J. *Handbook of MRI Pulse Sequences*; Elsevier Academic Press: Burlington, MA, USA, 2004.
14. Boudraaa, A.-O.; Fabien Salzenstein, F. Teager–Kaiser energy methods for signal and image analysis: A review. *Digit. Signal Process.* **2018**, *78*, 338–375. [CrossRef]
15. Přibil, J.; Přibilová, A.; Frollo, I. Mapping and spectral analysis of acoustic vibration in the scanning area of the weak field magnetic resonance imager. *J. Vib. Acoust.* **2014**, *136*, 51005. [CrossRef]

16. Přibil, J.; Přibilová, A.; Frollo, I. Comparison of mechanical vibration and acoustic noise in the open-air MRI. *Appl. Acoust.* **2016**, *105*, 13–23. [CrossRef]
17. Přibil, J.; Přibilová, A.; Frollo, I. Analysis of spectral properties of acoustic noise produced during magnetic resonance imaging. *Appl. Acoust.* **2012**, *73*, 687–697. [CrossRef]
18. Trabelsi, I.; Bouhlel, M.S.; Dey, N. Discrete and continuous emotion recognition using sequence kernels. *Int. J. Intell. Eng. Inform.* **2017**, *5*, 194–205. [CrossRef]
19. Nicolau, M.A.; Gunes, H.; Pantic, M. Continuous Prediction of Spontaneous Affect from Multiple Cues and Modalities in Valence-Arousal Space. *IEEE Trans. Affect. Comput.* **2011**, *2*, 92–105. [CrossRef]
20. Přibil, J.; Přibilová, A. Internet Application for Collective Realization of Speech Evaluation by Listening Tests. In Proceedings of the International Conference on Applied Electronics (AE2013), Plzeň, Czech Republic, 10–12 September 2013; pp. 225–228.
21. Bradley, M.M.; Lang, P.J. *The International Affective Digitized Sounds (IADS-2): Affective Ratings of Sounds and Instruction Manual*, 2nd ed.; Technical Report B-3; University of Florida: Gainesville, FL, USA, 2007.
22. Přibil, J.; Přibilová, A.; Frollo, I. Analysis of Energy Relations between Noise and Vibration Signals in the Scanning Area of an Open-Air MRI Device. In Proceedings of the 5th International Electronic Conference on Sensors and Applications, 15–30 November 2018. [CrossRef]
23. Přibil, J.; Přibilová, A.; Frollo, I. Analysis of Energy Relations between Noise and Vibration Produced by a Low-Field MRI Device. (to be published). [CrossRef]
24. Přibil, J.; Gogola, D.; Dermek, T.; Frollo, I. Design, Realization and Experiments with a new RF Head Probe Coil for Human Vocal Tract Imaging in an NMR device. *Meas. Sci. Rev.* **2012**, *12*, 98–103. [CrossRef]
25. Glowacz, A. Fault detection of electric impact drills and coffee grinders using acoustic signals. *Sensors* **2019**, *19*, 269. [CrossRef] [PubMed]
26. Přibil, J.; Přibilová, A.; Frollo, I. Vibration and noise in magnetic resonance imaging of the vocal tract: Differences between whole-body and open-air devices. *Sensors* **2018**, *18*, 1112. [CrossRef] [PubMed]
27. Valkovič, L.; Juráš, V.; Dermek, T.; Vojtíšek, L.; Frollo, I. The effect of stress on MR image contrast in the human hand at low field NMR. In *ELITECH '10, Proceedings of the 12th Conference of Doctoral Students*; Kozáková, A., Ed.; Faculty of Electrical Engineering and Information Technology, Slovak University of Technology: Bratislava, Slovakia, 2010; ISBN 978-80-227-3303-8.
28. Hu, S.; Peris, V.A.; Echiadis, A.; Zheng, J.; Shi, P. Development of effective photoplethysmographic measurement techniques: From contact to non-contact and from point to imaging. In Proceedings of the 2009 Annual International Conference of the IEEE Engineering in Medicine and Biology Society, Minneapolis, MN, USA, 3–6 September 2009. [CrossRef]
29. Daimiwal, N.; Sundhararajan, M.; Shriram, R. Comparative analysis of LDR and OPT 101 detectors in reflectance type PPG sensor. In Proceedings of the International Conference on Communication and Signal Processing, Melmaruvathur, India, 3–5 April 2014. [CrossRef]

Article

Real-Time Human-In-The-Loop Simulation with Mobile Agents, Chat Bots, and Crowd Sensing for Smart Cities

Stefan Bosse [1,*] **and Uwe Engel** [2]

1 Faculty Computer Science, University of Koblenz-Landau, 56070 Koblenz, Germany
2 Department of Social Science, University of Bremen, 28359 Bremen, Germany; uengel@uni-bremen.de
* Correspondence: sbosse@uni-bremen.de

Received: 7 August 2019; Accepted: 1 October 2019; Published: 9 October 2019

Abstract: Modelling and simulation of social interaction and networks are of high interest in multiple disciplines and fields of application ranging from fundamental social sciences to smart city management. Future smart city infrastructures and management are characterised by adaptive and self-organising control using real-world sensor data. In this work, humans are considered as sensors. Virtual worlds, e.g., simulations and games, are commonly closed and rely on artificial social behaviour and synthetic sensor information generated by the simulator program or using data collected off-line by surveys. In contrast, real worlds have a higher diversity. Agent-based modelling relies on parameterised models. The selection of suitable parameter sets is crucial to match real-world behaviour. In this work, a framework combining agent-based simulation with crowd sensing and social data mining using mobile agents is introduced. The crowd sensing via chat bots creates augmented virtuality and reality by augmenting the simulated worlds with real-world interaction and vice versa. The simulated world interacts with real-world environments, humans, machines, and other virtual worlds in real-time. Among the mining of physical sensors (e.g., temperature, motion, position, and light) of mobile devices like smartphones, mobile agents can perform crowd sensing by participating in question–answer dialogues via a chat blog (provided by smartphone Apps or integrated into WEB pages and social media). Additionally, mobile agents can act as virtual sensors (offering data exchanged with other agents) and create a bridge between virtual and real worlds. The ubiquitous usage of digital social media has relevant impact on social interaction, mobility, and opinion-making, which has to be considered. Three different use-cases demonstrate the suitability of augmented agent-based simulation for social network analysis using parameterised behavioural models and mobile agent-based crowd sensing. This paper gives a rigorous overview and introduction of the challenges and methodologies used to study and control large-scale and complex socio-technical systems using agent-based methods.

Keywords: simulation; agent-based modelling; mobile agents; crowd sensing; smart traffic control; social interaction

---

1. Introduction

The key concept of this work is the consideration of humans as sensors and the fusion of real and virtual worlds, in particular, virtual worlds in terms of social simulation, providing a closed-loop simulation. The outcome of simulations performed in real-time can be feedback to real worlds, e.g., to control crowd behaviour and flows, thus considering humans as actors. This human-in-the-loop simulation methodology enables a better understanding of crowd behaviour in real worlds and the opportunity to influence real worlds by simulation. This concept is highly interdisciplinary and is a merit of social and computer science if human sensor data will be coupled with social interaction and networking models. Social interaction has a high impact on the control of complex environmental and technical systems, which should be addressed in this work.

The novelty of this work is the seamless fusion of real and virtual worlds by using mobile agents and a unified agent platform that can be deployed in strong heterogeneous digital network environments and in simulation. The real world is sensed by mobile crowd sensing techniques (MCWS) providing input for an agent-based simulator representing the virtual world. Mobile agents are used to bridge both worlds and to provide the loose coupling required in heterogeneous and mobile environments with ad-hoc connectivity and operation on a wide range of host platforms with varying software versions.

The approach of an augmented agent-based social simulation is a powerful way to study large-scale social and socio-technical interactions. The strength of the framework lies in the potential inclusion of environmental and real-life interactions and providing feedback to real-world environments and humans. This agent platform, when used as the core component of both the MCWS and the simulator enables ubiquitous participation and interaction.

Different use-cases show the strength of the extended simulation framework using self-organising parameterised behaviour and interaction models with parameter sets derived by crowd sensing. The crowd sensing is performed by mobile agents that are used to create individual parameterised digital twins in the simulation from surveys performed by real humans (with respect to the social interaction model and mobility).

The following introduction gives an overview of agent-based modelling, simulation, and crowd sensing in the context of social and computer science, and enlightens the key concepts of agent-based human-in-the-loop simulation. After the Introduction, the role of intelligent systems and the relation to system control are discussed in Section 2, and modelling of socio-technical systems is discussed in Section 3. In Sections 4 and 5, the proposed simulation architecture and the workflow are described. Section 6 describes the agent-based real-world sensing using crowd sensing techniques. In Section 7, a parameterised social interaction model is discussed, and Section 8 introduces a crowd mobility and traffic model. Finally, in Section 9 three use-cases demonstrate the suitability and utilisation of the augmented simulation approach in different fields of application.

1.1. Preliminary Notes on the Involved Computational Social Science (CSS) Perspective

Even though the present article essentially describes the computer-science basis for a new digital-twin modelling and simulation approach, this work aims to contribute to the broader framework of computational social science (CSS). This designation is certainly at risk of being misunderstood as a branch of social science only. Quite the contrary, however, CSS is an emerging interdisciplinary field of research of growing importance at the intersection of computer science, statistics, and social science. On the one hand, CSS includes a well-known strong formal modelling/simulation branch on artificial societies [1]. On the other hand, it includes the trend towards a dynamically increasing field of automated collection of digital-trace and text data on the Internet that goes hand in hand with an equally important rise of widespread applications of machine learning techniques in society [2]. Though both CSS branches undoubtedly offer valuable scientific insights on their own right, the link between the modelling/simulation branch, on the one hand, and the social-research and social-media research branch [3] respectively, on the other hand, currently remains a quite challenging task still.

1.2. Why Agent-Based Models?

This work focuses on agent-based modelling (ABM) of social interaction as well as socio-technical systems. The main advantage of ABM over analytical or machine learning methods is its implementation of self-organisation using simple behaviour and interaction models. Typical analytical methods that can be combined with agent-based modelling are pattern recognition and cluster graph analysis [4], but the required functional social models on a global interaction scope are not always available. Common social models oversimplify the real world, a disadvantage that can be avoided by agent-based simulation only using neighbouring interaction. Agent-based models are especially relevant to simulating social phenomena that are inherently complex and dynamic [5]. Dealing with complexity is a challenge in social modelling, and the decomposition of complex systems in many simple interacting systems (divide-and-conquer approach) is well established and reflected by agents. The simplest agent model only consists of condition–action rules, but is powerful enough to model traffic and crowd flows.

1.3. Why Not Analytical Models?

ABM is in principle the counterpart to analytical modelling and analysis. ABM addresses local interaction between agents posing emergence (the global behaviour), whereas analytical models often cover the mapping of individual behaviour on global behaviour directly (the hard-to-predict aggregate outcome) [6]. Analytical models are well established in social science and address both empirical and formal methods that pose their strength with respect to some of the fundamental analysis challenges in social science [7]. The main issue of analytical modelling, e.g., spatial analysis, is their limitation when it comes to covering a broad diversity in social models, i.e., addressing minor variances that occur in real-world systems [8]. This work demonstrates the introduction of the variance of real humans by crowd sensing integrated into the agent-based simulation. It could be shown that small local disturbances have a significant effect on global structures and that interaction is not covered by current analytical models. But ABM can rely on analytical micro-scale models, shown, e.g., by the Sakoda segregation model and the social expectation function (see Section 7).

1.4. Explaining Simulation-Based Aggregate Effects

This Section concerns, for instance, the ways and possible improvements of empirically testing assumptions and predictions of such simulation models by experiments and survey research, as detailed e.g., in [9,10]. As outlined there, the validation of empirically grounded agent-based models certainly has to consider the targeted degree of realism and the related importance attached, particularly to mechanism-based explanations. This means, notably, the validation of the mechanisms assumed to produce the expected aggregate effects in the micro-to-macro transition, and in doing so, the rejection of the popular as-if attitude in model-based explanations [6]. Adoption of this approach essentially implies the view that validation should not be restricted to empirical tests of the observable model implications alone, simply because different models can imply the same implications; validation should also extend to the very model assumptions from which the observable implications were deduced.

1.5. Why Humans-In-The-Loop?

Such assumptions are essentially assumptions about the behaviour of individual agents and why they act as they act. From a sociological point of view, this brings the humans in the loop and the factors underlying their behaviour (e.g., their preferences, expectations, values, attitudes, personality traits, habits, and resources). It also introduces the social relations that agents maintain in dyads, networks, and larger groups. This, in fact, is important for an understanding of the intended and unintended, even paradoxical emergent effects which result from the behaviour of individual agents and their interactions in society. This, moreover, is important for intervention and steerage purposes too. Cases in point of relevant effects certainly include shapes of structural differentiation (segregation vs. intersection) and opinion formation, currently with strong scientific attention to opinion polarisation (e.g., [11,12]) and extremism due to propaganda in digital social networks [13].

Among considering humans as sensors feeding the simulation at run-time, the loop considers humans as actors, too. i.e., the loop provides output data for crowd and flow control, human decision making via social media, or at least influencing the real world with data from simulations.

1.6. Why Simulations in Real-Time?

Social simulation using mathematical, statistical, and agent-based models is well established to investigate and predict social and socio-technical interaction. Often there is a gap between simplified modelling and real-world observations. The factors underlying human behaviour can change over time. Some tend to volatility, some to persistence instead, in any case these factors represent genuine sources of variation. Given that a relevant factor tends to create changes in shorter rather than longer periods of time, and given that such a factor has a share in producing an aggregate effect, models which are capable of both sensing such changes and simulating the resulting effects in real-time appear most suitable for the testing of relevant assumptions of simulation models and the possibly wanted derivation of policy recommendations. Most notably, especially in simulations that are capable of dynamically adapting to changing preconditions at the agent level, afford the opportunity to develop realistic scenarios in use-fields where the level of behavioural change per unit of time is rather high than low.

Crowd simulation can be utilised to feedback data from the simulation to the real-world to control crowd flows, e.g., in cities or domestic services. But this feedback requires the real-time and time-lapse capability of the simulation, discussed in Section 4. The de-facto standard in traditional social simulation is the *Netlogo* simulator [14]. Integrating data mining in agent-based modelling and simulation was first introduced in [15], but data mining uses real-world data collected prior to simulation (delayed coupling of real and virtual worlds).

1.7. Why Sensing?

A real-time human-in-the-loop simulation needs continuously updated empirical information on the relevant model parameters. The sensing and collecting of continuous measurements represent the first hub, and the integration of related subjective and objective measurement a second one. It then simply depends on the angle of view which type of measurement is regarded as the primary source and which one is the possible enrichment. In the same way as one can enrich survey data with objective sensor measurements [16], one can enrich sensed physical data with information usually obtained through survey research. One can even discard the weighting inherent to highlighting one source as primary, and just talk of empirical (objective and subjective) information obtained by sensing. In this spirit a key concept of this work is the consideration of humans as sensors and mobile devices such as smartphones are the instruments for sensing the required empirical information. Data fusion approaches can further enhance analytical as well simulative methods [4].

1.8. Why Crowd Sensing?

In order to investigate complex and dynamical societies, appropriate data are required. Unfortunately, acquiring such data is a challenge. The traditional methods of analysis in sociology gather qualitative data from interviews, observation, or from documents and records, and carry out surveys of samples of people [5].

The crowd comes in for three related reasons. First, the nearly ubiquitous use of mobile devices makes it simply possible to sense the required information on a large scale, at least in principle. Secondly, a crowd is necessary: without the information obtained from the set of persons that constitutes a crowd, no simulation would be possible at all. Thirdly, the focus on crowds goes along with two ongoing paradigm shifts in social research. The first of these paradigm shifts concerns the changing survey landscape (trend towards the use of non-probability samples) and the second one concerns the tendency from survey research towards social media research. Moreover, just recently, social research encountered the suggestion towards the creation of mass collaboration for data collecting purposes [17]—a trend in line with similar suggestions towards citizen sciences.

The analysis of crowd sensing data and crowd behaviour can be performed with different methods: 1. analytical methods [4]; 2. machine learning (ML); 3. simulation. Simulation and ML can be seen as counterparts to analytical methods or can be used in conjunction.

One major drawback of classical surveys is that they come from measurements made at one moment in time [5]. Crowd sensing combined with real-time simulation can extend the data time scale significantly.

1.9. The Role of Incentive Mechanisms

Crowd sensing can be participatory or opportunistic. Furthermore, crowd sensing can be classified in platform- or source- and user-centric architectures [18]. In all cases incentive mechanisms have a high impact on crowd user selection and participation, directly affecting the quality of sensed data [19]. Although the usage of mobile phones introduces new opportunities in social data sensing, adapted incentive mechanisms are required. Distributed and self-organising crowd sensing commonly requires the installation of software, a high barrier for most users, with an impact on participation and data bias (young users vs. older users). Mobile chat-bot agents, e.g., embedded as (*JavaScript*) code in a WEB page (but executed on the user device) as considered in this work can overcome this limitation.

Among participation, the quality of the sensed data contributed by individual users varies significantly by crowd sensing [19]. Sensor fusion and localised pre-processing (filtering) are required. Correlating the quality of information provided by user to the incentive rewarded and as part of the auction model can improve overall data quality of mobile crowd sensing significantly.

1.10. Smart City Management

Today administration and management of public services and infrastructure relies more and more on the user and ubiquitous data collected by many domestic and private devices including smartphones and Internet services. People use social digital media extensively and provide private data, enabling profiling and tracing. User data and user decision making have a large impact on public decision-making processes, for example, plan-based traffic flow control. Furthermore, intelligent behaviour, i.e., cognitive, knowledge-based, adaptive, and self-organising behaviour based on learning, emerges rapidly in today's machines and environments. Social science itself exerts influence on public opinion and decision formation.

Smart city infrastructures and management are characterised by adaptive and self-organising control algorithms using real-world sensor data. In [20], crowds are considered a collective intelligence that can be employed in smart cities, although such self-organising and self-adapting intelligence driven by a broad range of individual goals is inaccurate and can compromise society's goals.

Traffic is a classic example of group decision making with (social) neighbouring interaction, but commonly, traffic control relies on physical sensor processing only [21]. In [21], adaptive and partly self-organising traffic management was achieved by using agents with multi-levels of decision making and a hierarchical organisational structure. Car traffic control in smart cities can be achieved by global traffic sign synchronisation. But pedestrian, bicycle, and domestic traffic usage cannot be controlled this way. A future vision to control domestic traffic flows is the deployment of digital and social media with chat bots to influence people's decision making regarding goal-driven mobility, sketched with the methods introduced in this paper.

Traffic control can be performed by perception and analysis of vehicle and/or crowd flows. Furthermore, vehicle-flows can be classified, e.g., introducing weights for individual and public vehicles.

Surveys play another important role in the modelling and understanding of interaction patterns. Artificial Intelligence (AI) and chat bots shift classical survey methods towards computational methods introducing possible implications, i.e., usage of different data, different methods, exposition to different threats to data quality (concerning sampling, selection effects, measurement effects, and data analysis), and different rules of inference are likely to result in comparably different conclusions and public reports, and hence in different input to public opinion formation. Among field studies, simulations can contribute to the investigation and understanding of such interactions.

1.11. The Underlying Computer-Science Perspective: A New Augmented Simulation Paradigm

In real worlds, hardware and software robots can be considered close together in a generalised way by an association with the agent model. One prominent example of a software robot is a chat or social bot. Additionally, real and artificial humans can be represented by the agent model, too.

Multi-agent systems enable modelling of agent interaction and emergence behaviour. Agent-based modelling and simulation is a suitable methodology to study interaction and mobility behaviour of large groups of relevant human actors, hardware, and software robots, using methods with inherent elements of AI (e.g., learning algorithms) or which use data which may be influenced partly by bots.

Agent-based methods are established for the modelling and studying of complex dynamic systems and for implementing distributed intelligent systems, e.g., in traffic and transportation control (see [21,22]). Therefore, agent-based methods can be described by the following taxonomy [15]:

1. Agent-based Modelling (ABM) - Modelling of complex dynamic systems by using the agent behaviour and interaction model ⇒ **Physical agents**
2. Agent-based Computing (ABC) - Distributed and parallel computing using mobile agents related to mobile software processes ⇒ **Computational agents**
3. Agent-based Simulation (ABS) - Simulation of agents or using agents for simulation
4. Agent-based Modelling and Simulation (ABMS)
5. Agent-based Computation, Modelling, and Simulation (ABX) - **Combining physical and computational agents**

The fifth paradigm is the novelty introduced in this work with the application to social mobility and crowd interaction simulation.

Two promising fields of current studies in computer science are data mining (DM) and Agent-based modelling and simulation (ABMS) [15]. An agent-based simulation is suitable for modelling complex social systems with respect to interaction between individual entities, manipulation of the world, spatial movement, and emergence effects of groups of entities. The main advantage is the bottom-up modelling approach composing large-scale complex systems by simple entity models. The main disadvantage of ABM is the (over-) simplified entity behaviour and simplification of the world the entities are acting in. Commonly, simulations are based on synthetic data or data retrieved by field studies. Many simulations and models are lacking in diversity that exists in the real world. Commonly, sensor and model data (parameters) used in simulations (virtual world) are retrieved from experiments or field studies (real world), shown in Figure 1.

But there is neither feedback from the virtual to the real world nor an interaction of the real world with the virtual world. For example, in [21] self-organised traffic control was simulated with agents by using the JACK software for intelligent agents, which is an agent-oriented development environment built on top of and integrated with the Java programming language and Matlab software. The simulation was based on a parameterised traffic model. Sensor data were created synthetically within the simulation without real-world coupling.

Parameterised models (modelling behaviour and interaction), discussed in Section 3, play an important role in ABM and ABS. The achieved results and conclusions rely on the parameter settings, and the selection of appropriate and representative parameter sets is crucial for modelling real-world scenarios accurately. Furthermore, parameter sets derived from real-world sensing enables variations and evaluation for different real-world situations.

To overcome the limitations of closed and synthetic simulation worlds, we investigated a new augmented agent-based simulation paradigm coupling real and virtual worlds in real-time, although this is an optional feature.

Controlling ABS by human participation appears in early work in stock market analysis and prediction [23].

ABS and ABMS are commonly constructed using collections of condition–action rules to be able to percept and react to their situation, to pursue the goals they are given, and to interact with other agents, for example, by sending them messages [5]. The *NetLogo* simulator is an example of an established ABS tool used in social and natural sciences [14], but is limited to behavioural simulation only (ABM domain). In this work, a different simulation approach combining ABM, ABC, and ABS methodologies, is used by deploying the widely used programming language JavaScript. JavaScript is increasingly used as a generic programming language for ABC and ABS [24,25].

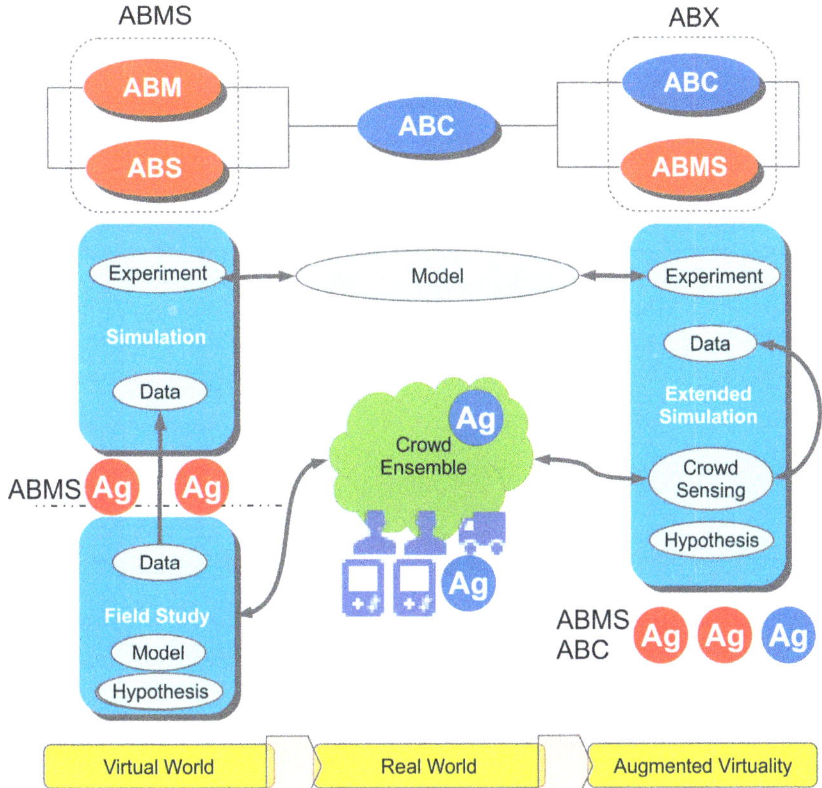

Figure 1. (**Left**) Traditional field studies and agent-based simulation of social systems. (**Right**) New combined agent-based simulation and agent-based crowd sensing enabling bidirectional data exchange.

1.12. Sensing the World

In crowd sensing, users of mobile devices are sensors and actuators [20]. Mobile devices like smartphones are valuable sources for social data [26], either by participatory crowd sensing with explicit participation of users providing first-class data (e.g., performing surveys or polls) or implicitly by opportunistic crowd sensing collecting secondary class data, i.e., traces of device sensor data delivering, e.g., actual position, ambient conditions, network connectivity, and digital media interaction. Crowd sensing and social data mining as a data source contribute more and more to investigations of digital traces in large-scale machine–human environments characterised by complex interactions and causalities between perception and action (decision making). This is relevant for the algorithmic and architectural design of smart cities [20]. Agent-based software is already used in crowd sensing applications [27], which has advantages in adaptivity over traditional static client–server architectures.

But mobile devices are not limited to being sensors. They can be used as actuators, too, by interacting with the user via the chat dialogue or social media that can affect the user behaviour (e.g., mobility decisions).

It is difficult to study such large-scale data collection, data mining, and their effect on societies, domestic services, and social interaction in field studies due to a lack of reliable data and complexity. Controlling all aspects of the virtual world allows explaining the source of structural variation in human interactions, i.e., tackle the hard-to-observe mechanisms of micro–macro interactions. For instance, it is known that the creation processes of social networks rely heavily on the (organisational) environment [28]. Utilising real-world interactions together with ABMs in carefully selected, systematically diverging environments creates a framework in which we can experimentally distinguish between ideal typical situations and understand how different structures result in similar/diverging behaviour. First applications might involve hierarchical/unstratified environments represented by different types of organisations. In so doing, we would add to the question of how the same micro-mechanisms (e.g., homophily, balance) are created by and reproduce different network structures and, hence, extend emerging approaches on network ecological theory.

Agent-based modelling of socio-technical systems is well established [29] and can be applied for smart city management, however commonly applied in an artificial world, i.e., a simulation is performed in virtual reality worlds only to derive and proof models under hard limitations. In this work, a new concept and framework for augmented virtual reality simulation are introduced, suitably, but not limited to, investigate large-scale socio-technical systems. Mobile agents are used already successfully in field crowd sensing [30]. In this work, mobile agents are used to combine in-field ubiquitous crowd sensing, e.g., performed by mobile devices with simulation.

A chat bot agent is capable of performing dynamic dialogues with humans to get empirical data into the simulation and to propagate synthetic data from the simulation world into the real world. The chat bot can act as an avatar providing information for users, e.g., for optimised and dynamic traffic control based on real (covering actual and history data) and simulated data (addressing future predictions).

Chat bot agents (as computational agents) can operate both in real and virtual worlds including games and providing a fusion of both worlds by seamless migration. Mobile computational agents, i.e., mobile software, have advantages in strong heterogeneous systems, existing in opportunistic and ad-hoc crowd sensing environments. Examples are traffic control by smartphone data. Agents are loosely coupled to their environment and platform and interact with each other, e.g., via tuple spaces (generative data-driven communication) and via unicast or broadcast signals (addressed messages) [31].

The novel MAS crowd sensing and simulation framework, introduced in the next sections, is suitable to combine social and computational simulations with real-world interaction at run-time and in real-time, e.g., by integrating crowd sensing, using mobile agents. There are two classes of mobility that can be modelled in the simulation: (1) Social mobility on short- and long-term time scales, e.g., creating group formation and segregation, and (2) goal-driven mobility, e.g., creating traffic.

Agent-based traffic management simulation still neglects social interactions and the influence of social media, e.g., MATISSE, the Multi-Agent based Traffic Safety Simulation System [32], which is a large-scale multi-agent-based simulation platform designed to specify and execute simulation models for agent-based intelligent transportation systems.

Among the capability of real-time simulation, the simulation framework is capable of creating simulation snapshot copies that can be simulated in parallel in non-real-time to get future predictions (simulation branches) from actual simulation states that can be backpropagated into the current real-time simulation world.

2. Intelligent Machines and Socio-Technical Systems

There are basically three classes of entities existing in real and virtual worlds: 1. Humans, 2. machines, and 3. robots. Automatic and autonomous vehicles belong to class 3; semi-automatic driving is, e.g., relevant for smart traffic flow management.

Human–machine interaction is an emerging field. The interaction with intelligent machines is much more sophisticated and created two classes of interaction: 1. human–machine (HM), and 2. machine–machine (MM). Today, robots can be considered as intelligent machines (i.e., AI bots) posing some kind of autonomous behaviour, learning, and interacting with environments and humans. The main advantage of ABMS and the proposed agent-based framework is the unified handling of totally different entities using the same agent model. Autonomous vehicles are examples of both HM and MM interaction.

There is a micro-context of everyday life and the macro-context of public decision and opinion formation related to human–robot interaction, affecting, e.g., future city management like adaptive and dynamic traffic control considering a broad range of participants and vehicles (cars, buses, trains, pedestrians, and bicycles). Automatic and probably autonomous driving systems can be considered as intelligent machines (i.e., robots) interacting with humans and environments, too. Social behaviour of humans and interaction with intelligent machines will depend on the way intelligent machines will be controlled and the way they interact with humans and other intelligent machines. For example, chat bots influence human opinions, but humans and other chat bots can influence the behaviour and knowledge base of chat bots, too. There are different involved ways of mutual affection that are difficult to investigate in classical field studies. The proposed simulation method integrating real-world environments by using mobile agents can boost investigations of such complex interacting environments. For example, decisions of traffic participants (e.g., car drivers) cause wide-ranging effects on global behaviour (emergence), and centralised or decentralised traffic control will influence individual behaviour.

Since both sides (humans and bots) may learn from each other in this interaction, the studies shall explore possibilities of targeted responses of the human side. That way we expect to reveal options of understanding the influence on the possible ways that AI will shape future society. In addition, we expect to find out the factors in robot–human interaction that let AI gain acceptance in society. This list of factors shall, amongst others, include the visibility of AI during a communication (e.g., can a human discern whether a comment comes from a bot or a human and what difference this makes to his/her response?) and different programmable behavioural and ethical principles in robot–human interaction (by systematically varying the reward function). Furthermore, since we expect in the medium- and long-term, respectively, a trend towards the use of robots that not only assists but replace humans in daily communication, we project the development and exploration of the first occurrence of this in a home environment, and the evaluation of its applicability in practice.

At society's macro-level, robot–human interaction takes place too, though less visible and less direct. The investigation of public decision and opinion formation is of high relevance for the design of public management services. We project an exploration of how this process is shaped in the interaction of groups of relevant human actors who interact with intelligent machines through its reports of findings of projects which either uses methods with inherent elements of AI (e.g., learning algorithms) or which use data which may be influenced partly by bots. In this regard, the study is targeted on possible emergent effects of AI at the system level of society.

To summarise, the agent model used in this work captures different entity classes, mapping activities (performing actions like environmental interaction) A, transitions T between activities (that can be conditional based on V and S), data variables V. and sensors S, on actions X, discussed in Section 4:

$$Ag(A,T,V,S) : S \times A \times T \times V \rightarrow X. \tag{1}$$

3. Parameterised Modelling and Emergence

Modelling of social and socio-technical systems can be classified in macro-scale and micro-scale models addressing a macro- and micro-context discussed in the previous section.

Commonly, social behaviour is studied on a macro-scale level with a unified model assuming a common mean of individual behaviour or on a micro-scale level with a unified behaviour model. Social behaviour and interaction models with parameters are used to study more heterogeneous and individual behaviour beyond a unified mean.

On a micro-scale level, it is assumed that there is an individual parameterised behaviour model M for individual entities or groups of entities, i.e., a model of social interaction, which is used to model physical entities and digital twins representing individual real humans (with P: Parameter set, S: Sensor set, X: Action set):

$$M(S, P) : S \times P \rightarrow X. \tag{2}$$

A set of individual entities (agents) modelled by $M(P)$ interacting with each other within a bound space will create a specific observation on system level, i.e., emergence effects. One example is a traffic jam as a result of individual behaviour and constraints.

4. Simulation Architecture and Agent Platform

The proposed framework couples virtual and real worlds by integrating simulations with human interactions by using two classes of agents: (1) Computational agents (chat bots), and (2) physical agents, both executed inside the simulation world by a unified agent processing platform. The agent-based simulation assumes a spatial context and agent interaction within a spatial domain (or range), similar to cellular automata.

The entire crowd sensing and simulation architecture consists of the following components:

1. Unified **Agent Processing Platform** programmed in *JavaScript* for high portability and deployment in strong heterogeneous environments: JavaScript agent machine (JAM) [31];
2. **Crowd Sensing Software** (Mobile App and WEB Browser using JAM);
3. Agent-based simulation with Internet connectivity supporting three different agent types:
   - **Physical behavioural agents** representing physical entities, e.g., individual artificial humans;
   - **Computational agents** representing mobile software, i.e., used for distributed data processing and digital communication, and implementing chat bots;
   - **Simulation agents** controlling the simulation and performing simulation analysis (e.g., creation of physical and computational agents, reading and writing sensor data, accessing databases)
4. Chat dialogues, chat bots, and mobile agents collecting user and device sensor data;
5. **Knowledge-Based Question-Answer Systems, Natural Language Processing**.

All agents are programmed in *JavaScript* and executed by the JAM platform in a sandbox environment. Computational agents can migrate between platform nodes (as mobile process snapshots). This widely-used programming language offers a steep learning curve and *JavaScript* can be executed on a wide range of host platforms and software. This feature includes the execution of mobile agents with a WEB browser by embedding the agent platform in WEB pages. *JavaScript* programs and agents offer high portability which is essential for the deployment in strong heterogeneous environments and to reach a broad range of crowdsensing data providers.

Simulation is performed by the Simulation Environment for *JAM* (*SEJAM*, details are discussed in [31,33]). In this work, the *SEJAM* simulator is used to create a simulation world (consisting of entities represented by agents) that is attached to the Internet enabling remote crowd sensing with mobile computational agents. The *SEJAM* simulator is basically a graphical user interface (GUI) on the top of *JAM*, shown in Figure 2.

SEJAM extends *JAM* with a simulation layer providing virtualisation, visualisation of 2D- and 3D-worlds, chat dialogues and avatars, and an advanced simulation control with a wide range of integrated analysis and inspection tools to investigate the run-time behaviour of distributed systems. A *SEJAM* world consists of one physical *JAM* node and an arbitrary number of virtual/logical *JAM* nodes associated with a visual object (shape), which can be connected via arbitrary connection graphs to establish communication and migration of computational agents. In the graphical 2D world virtual *JAM* nodes can be either placed at any coordinate or aligned on a grid world.

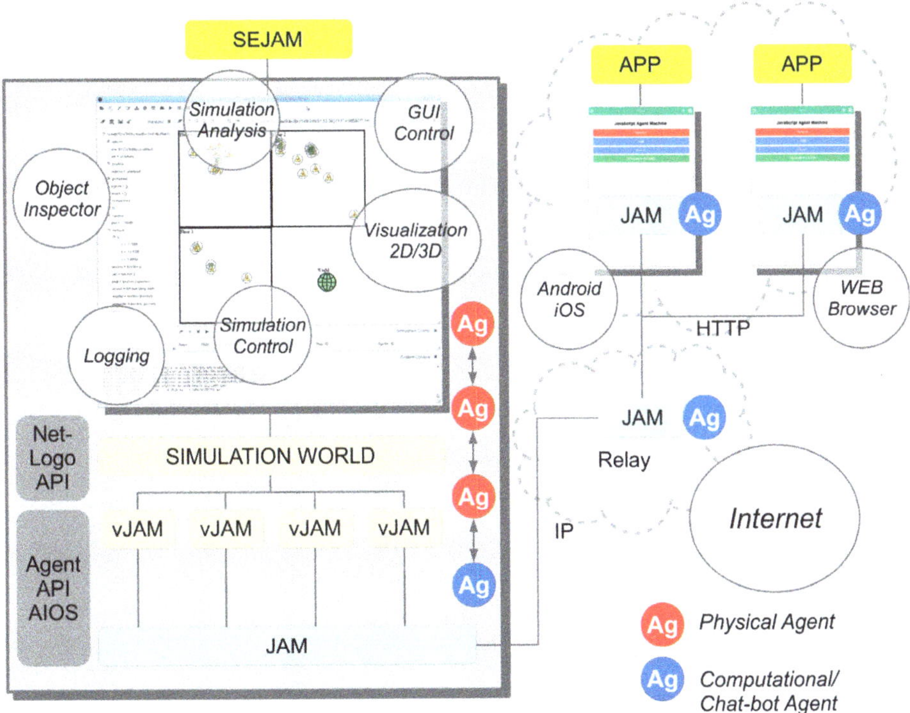

Figure 2. The principle concept of closed-loop simulation for augmented virtuality: (**Left**) Simulation framework based on the JavaScript agent machine (JAM) platform (**Right**) Mobile and non-mobile devices executing the JAM platform connected with the virtual simulation world (via the Internet) [24].

Each *JAM* node is capable of processing thousands of agents concurrently. *JAM* and *SEJAM* nodes can be connected in clusters and on the Internet (of Things) enabling large-scale "real-world"-in-the-loop simulations with millions of agents. Agents executed in the simulation have access to an extended simulation API. Physical agents can use an extended *NetLogo* compatible simulation API extension, too, enabling simulation world analysis and control, based on object iterators (*ask* and *create* statements, see [14], with a detailed discussion in Appendix A).

The mixed-model simulation world consists of physical and computational agents bound to logical (virtual) platforms (host of the agent) that are arranged or located on a lattice (patch grid world) to provide world discretisation for the sake of simplicity. The patch grid world was derived from the *NetLogo* simulator, although the *SEJAM* simulator is not limited to this world model and can handle arbitrary two-dimensional non-discretised world coordinate systems (e.g., GPS). The agents are mobile. Computational agents, as mobile software processes, can migrate between platforms (both in virtual and real digital worlds), whereas physical agents are fixed to their platform and only the platform is mobile (in the virtual world only).

The *JAM* agent behaviour model is based on activity-transition graphs (ATG, details in [34]). An ATG decomposes the agent behaviour into activities performing actions (computation, mobility, and interaction with other agents and the world). An activity is related to a sub-goal of a set of goals of the agent. There are transitions between activities based on the internal state of the agent, basically related to reasoning behaviour under both pro-active (goal-directed) and reactive (event-driven) stimuli. Modification of agents (visual, shape, position, body variables, and state) can be performed globally by the world agent via a simulation interface providing a *NetLogo* API compatibility layer, or by the individual agents.

Both *JAM* and *SEJAM* are programmed entirely in *JavaScript*, enabling the deployment on a wide range of host platforms (mobile devices, servers, IoT devices, and WEB browser, details described in [35]). *JAM* and *SEJAM* can be connected via IP-based communication links. *JAM* provides virtualisation and security (encapsulation) by the agent input–output system (AIOS), tuple spaces for generative or signal-driven inter-agent communication, and virtual (logical) nodes bound to a world contained in one physical *JAM* node. Each physical or logical *JAM* node can be connected with an unlimited number of remote *JAM* nodes by physical links (UDP/TCP/HTTP using the AMP protocol), shown in Figure 2. Logical nodes can be connected by virtual communication links. Links provide agent process migration, signal (message) and tuple propagation.

The agent behaviour defines:

- a set of sensors and beliefs about the world (its data set), basically stored in the agent's body variables;
- a set of events that it will respond to (signals, tuples, sensor changes);
- a set of goals that it may desire to achieve, basically implemented by the set of activities, and;
- a set of plans that describe how it can handle the goals or events that may arise activating activities.

A simulation world consists of multiple virtual *JAM* nodes (*vJAM*), which can be connected by virtual links. In contrast to pure ABM simulators like *Netlogo*, the *SEJAM* simulator simulates computational and physical agents and the agent processing platform itself and is primarily an ABC simulator, but can be used for ABM like simulations, too. There is a global simulation model defining agent classes (behaviour and visuals), nodes (visuals), resources, simulation parameters, and the construction of the simulation world.

Computational agents are always bound to a virtual *JAM* node. A node can be created dynamically at simulation run-time or during the initialisation of the simulation world. A node is related to a virtual position in the two-dimensional world (similar to a patch in the *Netlogo* world model). Virtual nodes can be grouped in tree structures and the movement of a parent node (in the virtual world) moves all children, too.

In contrast to a *Netlogo* simulation model using one main script describing the entire world, agents, and resources, a dedicated world agent (operating on a dedicated world node) controls the simulation in *SEJAM*. A node position can be changed by agents (related to the movement of turtles in the *Netlogo* model. To summarise, *JAM* agents play two different roles in the simulation: (1) ABC: mobile crowd sensing, digital interaction between physical agents and the environment, and simulation control and; (2) ABM: Representation of humans, bots, and digital twins of humans.

There is a significant difference between traditional closed-world simulations and simulations coupled with the real world and real-time environments (human-in-the-loop simulation). Closed simulations are performed on a short time-scale with pre-selected use-cases and input data. The simulation can be processed step-wise without a relation to a physical clock. In contrast, open-loop simulation requires continuous simulation on a large time-scale creating big data volumes. A relation to a physical clock is required, too.

The mapping of the physical onto virtual worlds and vice versa is another issue to be handled. There are basically three possible scenarios and world mapping models (illustrated in Figure 3):

1. The real and virtual worlds are isolated (no agent/human of one world knows from the existence of the other);
2. The real world is mapped on the virtual simulation world (simulating the real world with real and artificial humans or entities);
3. The virtual world extends the real world, e.g., by a game world, and real and virtual entities known from each other.

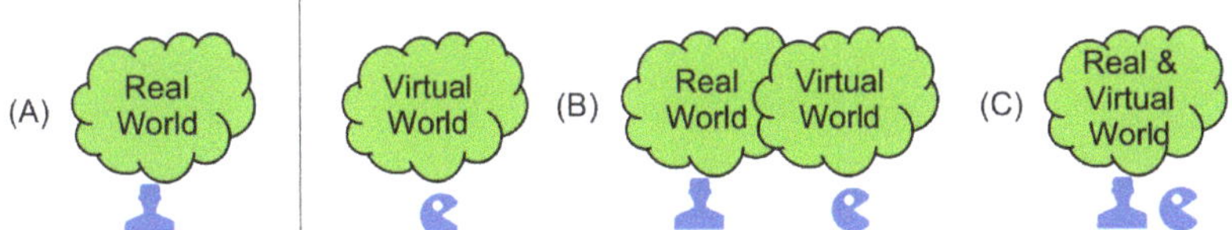

Figure 3. (**A**) Real-world only deployed with humans. (**B**) non-overlapping real and virtual world, and (**C**) overlapping real and virtual world.

Simulation Snapshots

One key capability of the simulator and the agent platform is the creation of agent, world, and entire simulation snapshots. This feature enables the forking of simulation runs either continuing on a different time scale (speed-up) or with different parameters and sets of agents. A time-lapsed simulation can be used to get future time predictions, e.g., of crowd or traffic flows, clusters, or social networks.

5. Workflow

The following Figure 4 shows the principle work and data flow of the proposed architecture. The ABMS relies on generated data stored in a database provided by synthetic data, Monte Carlo simulation, and data mining (DM) of sensor data from mobile crowd sensing (MCWS). The MCWS is performed by mobile chat bot agents, discussed in Section 6.3, which establishes the connection between virtual and real worlds. But the data flow is bidirectional, and agents can carry data generated by the simulation to mobile devices and users in the real world, e.g., by posting messages in chat blogs or other social media.

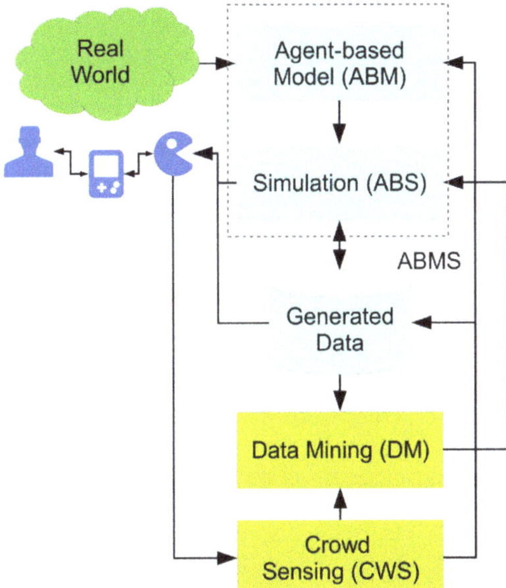

Figure 4. Principle work and data flow integrating agent-based crowd sensing in agent-based modelling and simulation.

Simulation data is stored in the database, too. The simulation data, e.g., monitoring data or artificial sensor data, can be analysed numerically or statistically by DM. Sensor data can be collected off-line (classical surveys) before the simulation or on-line during the simulation creating incremental simulation runs.

6. Crowd Sensing and Simulation

6.1. Crowd Sensing, Surveys, and Cloning of Digital Twins

The basic crowd sensing and social survey architectures can be classified in (see Figure 5):

1. Centralised crowd sensing architecture with a single master instance (crowd sourcer);
2. Decentralised crowd sensing architecture with multiple master crowd sourcer instances;
3. Self-organising and ad-hoc crowd sensing architecture without any dedicated crowd sourcer master instances.

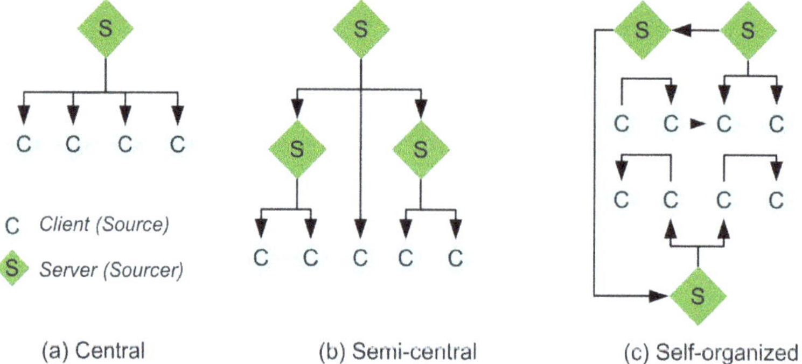

Figure 5. Different crowd sensing architectures and strategies with data sourcer and data sources.

The survey used in this work relates to the classical central approach, although it can be extended with a decentralised and self-organised approach, too, handled by the same framework and mobile chat bot agents.

Among the crowd sensing architecture, there are two different crowd sensing models:

1. Participatory crowd sensing negotiated between a crowd sourcer and user based on utility and reward; there is strong coupling between crowd sourcer and crowd sensing participants;
2. Opportunistic or ad-hoc crowd sensing with weak coupling between crowd sourcer and participant based on immediate reward.

Classical surveys are participatory crowd sensing and the selection of data provider (user) is done carefully to avoid biased data. Ubiquitous computing enables opportunistic crowd sensing using mobile devices, e.g., smartphones, from a broader range of participants (usually not known) in real-time.

Initially, an AB simulation is performed with artificial agents relying on model parameters derived from theoretical considerations, prior sampled experimental data, and survey data, which can be combined with Monte Carlo methods. Augmented virtuality enables dynamic simulations with agents representing real humans (or crowds). By using crowd sensing it is possible to create digital twins of real humans based on a parameterised behaviour and interaction model, including spatial context. The parameters of artificial humans in the simulation represented by agents are collected by sensor data, i.e., surveys optionally fused with physical sensors like GPS. One simple example is shown in Section 7.

The crowd sensing via mobile chat bot agents enables the interaction of real humans with agents and digital twins in the simulation world in real-time and vice versa. The digital twins, as well as the artificial physical agents in the simulation, can interact by dynamically creating (influenced) dialogues reflecting the state of the simulation world.

The survey performed by chat bot agents (computational agents) aims to create digital twins in the virtual world from survey participants in the real world by deriving twin behaviour model parameters P from the survey feedback answers F retrieved by a user dialogue D. The survey is performed by chat bot agents that can execute dynamic dialogues via a chat dialogue platform based on previous answers and context.

$$survey(D) : D \rightarrow F, analyse(F) : R \rightarrow P, twin(P) : P \rightarrow AG_{twin} \quad (3)$$

Crowd sensing combined with ABM/ABS introduces randomisation and variance in the simulation and model verification, shown in Figure 6. The addition of digital twins with behaviour parameters derived from real humans and CWS can lead to different outcomes of the simulation:

1. The expected structures (emergence) are observed as given by the model (same case as the pure synthetic world simulation);
2. No structures are observed caused by the disturbance of the real-world interaction;
3. Expected structures and new structures are observed normally not occurring in pure synthetic and closed worlds.

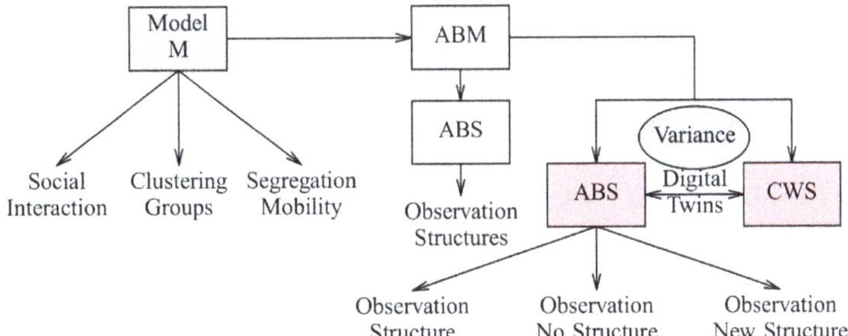

Figure 6. Crowd sensing combined with agent-based simulation (ABS) adds variance to the synthetic simulation world by digital twins.

The ABS/CWS coupling via digital twins enables iterative surveys in multiple rounds, i.e., results from one round combined with new simulation results can be used for the next survey round.

One major issue in simulation and experimental field studies is the reproducibility of results. To ensure reproducibility and to avoid data bias, participants of surveys are commonly chosen carefully and not randomly (participators crowd sensing). Opportunistic crowd sensing selects users randomly and requires further data analysis and user classification to ensure simulation stability.

6.2. Security

Participatory crowd sensing has to ensure data privacy and protection against data manipulation. This crowd sensing strategy can be established via secure channels and by using client–server architectures. Opportunistic crowd sensing, on the other hand, poses a loose coupling of data producers and consumers, raising security issues.

Security is provided on different levels, either by the *JAM* platform, e.g., capability protected agent roles, or by system-level agents, e.g., the chat mediator agent, which authorises and filters survey requests, or detects spam or malware agents. A crowd sensing platform can be open (like any other WEB page), or protected via a capability.

6.2.1. Agent Roles

Physical agents are caught in the simulation world. They can access and modify any other agent in the virtual world and the simulation world itself. They are executed on a privileged level. But computational agents can leave the virtual world and enter the real world and vice versa. The loose coupling of the simulation with the crowd sensing network and to enable opportunistic sensing requires different agent privilege roles. Four different roles are supported by the platform:

1. Guest (not trustful, non-mobile);
2. Normal (maybe trustful, mobile);
3. Privileged (trustful, mobile);
4. System (highly trustful, local processing only, non-mobile, creation by platform only).

The agent platform assigns a security level to new received agents (commonly level 0 or 1). The agents have to negotiate a higher level using the capability-based approach discussed in the next sub-section. The privilege level determines the set of API functions that an agent can access (e.g., migration, forking, and communication).

6.2.2. Capabilities

Capabilities are keys to enable specific agent API functions like migration. A capability consists of a service port, a rights bit field, and an encrypted protection field generated with a random port known by the server (node) only and the rights field. The rights field enables specific rights. e.g., the right to negotiate a higher privilege level. A capability is created by the respective service (e.g., the platform itself) by encoding the rights field in a secure port using a private port and a one-way function, shown in Figure 7. Major issues are the secret handling of capabilities on agent migration (requiring secured channels) and passing the right capability to an agent. A broker server is commonly required to distribute and pass capabilities to agents based on classical authorisation and authentication.

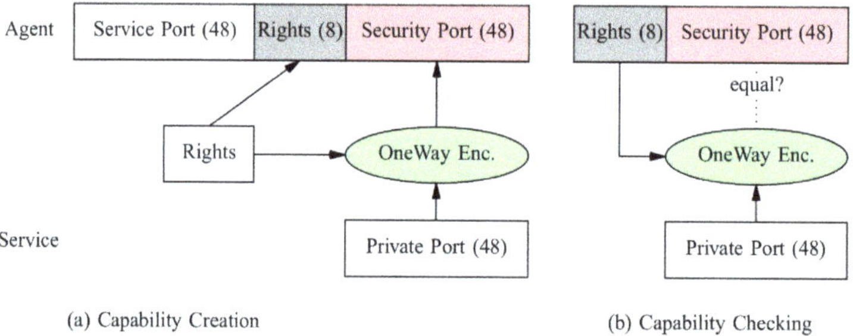

Figure 7. (**Top**) Capability format. (**a**) Capability creation by encoding private security port and new rights field. (**b**) Capability checking by encoding requested rights field and private security port and comparison with provided security port.

Capabilities enable fine-grained control of operations that can be performed by agents and allow loosely coupled self-organising systems, e.g., participation in chats. In self-organising systems, the identity of agents is not of primary interest. Instead the actions they can perform are of primary interest.

6.3. Chat Bots as Mobile Agents in Both Worlds

Mobile agents are used in this work for distributed data processing and data mining in the real and virtual simulation world seamlessly. It is assumed that the real-world environments consist of communication access points (i.e., beacons, e.g., using WLAN and WWAN communication technologies) and mobile devices (smartphones, IoT devices). Commonly there are no globally visible organisational and network structure, i.e., agents cannot rely on a network world model to reach and identify specific devices. Each *JAM* node provides connectivity information, i.e., a list of all connected *JAM* nodes. Commonly, *JAM* nodes are connected peer-to-peer in mesh-like networks via TCP/HTTP or UDP connections.

Agents can communicate with each other either by exchanging tuples via a tuple space database or by using signals (lightweight messages) that can be propagated remotely along agent migration paths. Tuple space access is generative communication, i.e., the lifetime of tuples can exceed the lifetime of the generating agent. Additionally, tuple space communication is data-driven and anonymously (sender and receiver need no knowledge about each other).

Mobile agents, e.g., chat bots, can be used to carry information from one location to another. Moreover, mobile devices carrying mobile agents can be used to collect, carry, and distribute data within large regions via tuple spaces, shown in principle in Figure 8. The virtual simulation world is just another region in the mobility regions of agents, and chat bots performing crowd sensing can migrate between real and virtual worlds seamlessly.

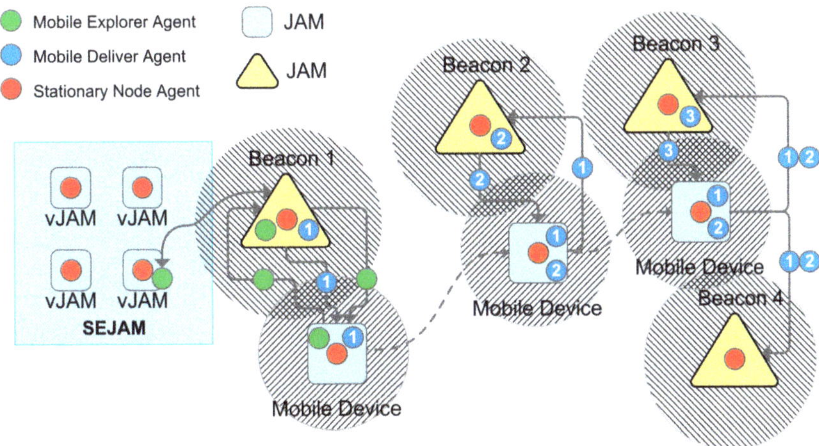

Figure 8. Virtual simulator world connected to spatially distributed non-mobile beacons, mobile devices connected temporarily to beacons (cellular or local WIFI networks), and mobile agents (crowd sensing and chat bots) used for wide-range interaction.

6.4. Chat Bots and Human–Agent Interaction

Mobile agents can migrate between different devices. The main goal of explorer agents is to collect crowd and social data. This data can be retrieved by device sensors (including aggregated virtual sensors) and information provided by users. Privacy and security have to be addressed. Among device sensor data like position, velocity, and ambient light, which can be collected by mobile explorer agents automatically, user data can be gained by a human–agent dialogue chat via a WEB Browser or Android/iOS App (see Figure 9). Therefore, an explorer agent is capable of interacting with humans by performing textual dialogues (question–answer surveys). Due to the distributed and parallel processing of loosely coupled agents filtering and scheduling of agent dialogue request has to be performed. On each device participating in crowd sensing there is a mediator agent that manages and assesses interaction requests from incoming (explorer) agents. Passed requests are forwarded to a user–bot chat dialogue. Responses to questions (requests) are passed back to the requesting explorer agents.

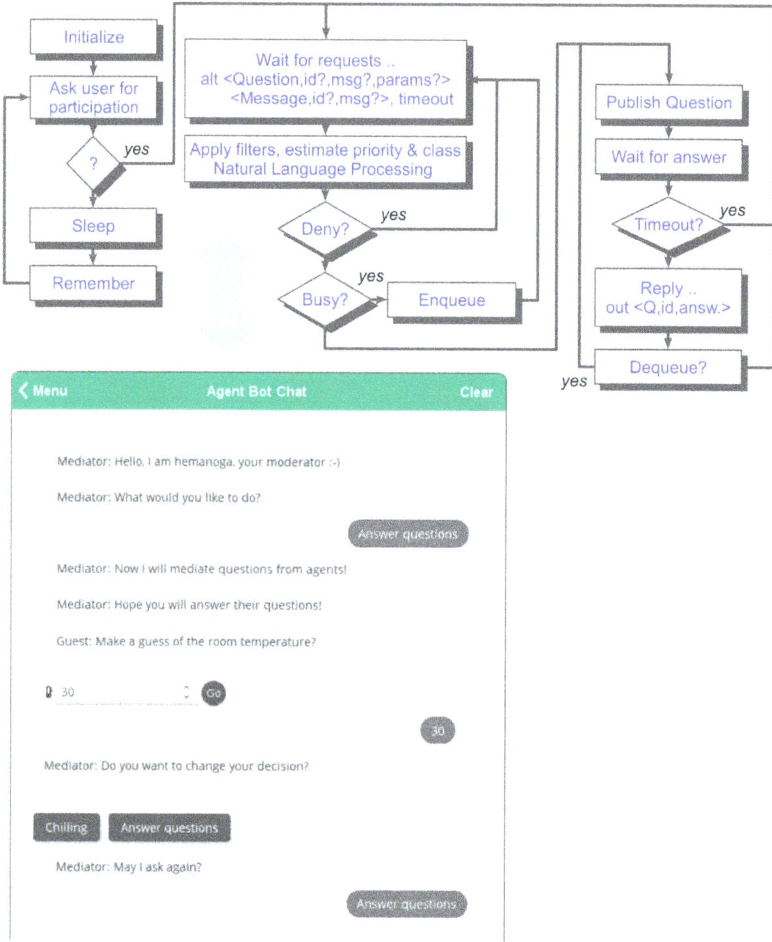

Figure 9. (**Bottom**) Chat dialogue of the JAM App showing messages and questions from chat bot agents and principle mediator agent behaviour scheduling the user chat (**Top**).

Each agent migrating to a device operated by humans (i.e., smartphones, navigation computer, or fixed terminals) can ask questions via the APP tuple space by inserting a question tuple, which is handled by the node mediator agent. The tuple space enables data-driven communication between agents that have different roots, a situation arising in opportunistic crowd sensing. See [35] for details on tuple-space communication.

A question request tuple is evaluated by the mediator agent performing security and privacy checks, finally passing the question to the chat blog waiting for human participation. The answer (if any within a defined timeout) is passed back to the original requesting agent by an answer tuple, again previously checking privacy and security concerns.

The principle communication format via the tuple space is shown in Algorithm 1. The dialogues can be composed of dialogue trees, i.e., questions and messages that can be dynamic and can depend on previously answered questions or environmental perception (sensors).

Algorithm 1: Communication of remote chat bot agents with user dialogue mediator agents via an anonymous tuple space

```
out(['Message',<id>,<message>])
out(['Question',<id>,<question>,
{value: <default>,
          type:'number'|'string',
          icon:<icon-name>}]
     inp.try(timeout,
          ['Answer',<id>,<question>,_],
          function (tuple) {
              if (tuple) this.result = tuple[3]})
```

In doubt, the question or answer is discarded or modified (e.g., filtered or annotated with additional warnings). The mediator agent performs question request scheduling to satisfy agent requests and human interaction capabilities (high question rates will decrease the user interest and motivation to answer questions). Additionally, the mediator agent has to monitor chat bot dialogues by using natural language processing (NLP).

A typical chat dialogue is shown in Figure 9. It consists of question–answer snippets processed and controlled by the mediator agent operating on each device with user interaction. This chat dialogue is embedded in a *JAM* platform application software (*JAMapp*). *JAMapp* is available for smartphones or as a WEB-page that can be processed by any *JavaScript* capable WEB browser.

The mediator agent can pass questions (or just stand-alone messages) to the user chat via a simple platform API providing a question text, possible answers or input fields, and a callback function handling the result, shown in Algorithm 2.

Algorithm 2. Chat dialogue API accessed by the mediator agent

```
chat.message(<id>,<message>);
chat.question(<id>,<question>,{
        size: <test-size>,
          icon: <icon-name>,
          value: <default value>
          sub_type: 'number'|'string',
          placeholder: <placeholder-value>
     },function (result) {
        this.answer = result;
     },timeout);
```

The following Definition 1 describes the dialogue script passed to a chat bot agent that is used to perform the crowd sensing on mobile devices by the chat bot agents. The question dialogue is dynamic: (1) Some questions are only presented based on answers of previous questions → *cond* function, and (2) some question texts are updated based on previous answers → *eval* function replacing text variables $ in the question text with values returned by an array. The optional *param* function can be used to change a default value or a choices list (returning an array of alternatives). Among user dialogues, device sensors (e.g., light intensity or position) can be requested within the dialogue script, too.

Definition 1. *Survey job type definition.*

```
type dialog = message|question|sensor
type message = { message:string }
type question = {
  question:string,
  choices?:string [] | number [],
  value?:number|string,

  eval? : function (script) -> string [],
  param? : function (script) -> string|string [],
  cond? : function (script) -> boolean,
  tag : string,
  answer?: string
}
type sensor = { sensor:string, default?:*, answer?:* }
type script = dialog []
type job = {script:dialog [], action: function}
```

7. Social Interaction and Mobility by the Sakoda Model

To demonstrate the augmented virtuality approach combining agent-based simulation with agent-based crowd sensing, social interaction modelled by the Sakoda model [36] was chosen as a simple social interaction and behaviour model between groups of individual humans posing self-organising behaviour (emergence) and structures of social groups by segregation.

The more well-known Shelling model used to study segregation effects (long-range mobility) bases on the less known but more general Sakoda social interaction model. The Sakoda model can be used for long- and short-range mobility, required, e.g., to study interaction and group aggregation effects, e.g., in cities and traffic.

Modelling social interaction is an example of an individual parameterised behaviour model explained below. The parameters can be set a priori or derived by surveys, i.e., using crowd sensing. In this case, the crowd sensing is only used to provide input data for the simulation. There is no immediate feedback to the crowd.

The Sakoda behaviour model can be applied to two different spatial and time scales resulting in different self-organising behaviour:

1. Long-time and long-range scale of mobility addressing classical segregation, e.g., in cities and countries, and
2. Short-time and short-range scale of mobility addressing, e.g., city mobility.

Both time scales are relevant for future city management including traffic management.

The original Sakoda behaviour model [36] consists of social interactions among two groups of individuals evolving in a network according to specific attitudes of attraction, repulsion, and neutrality. An individual evaluates its social expectative at all possible available locations (starting at its current location), preferring originally those near individuals associated with attractive (positive) attitudes and avoiding locations near individuals associated with repulsive (negative) ones. This procedure is repeated randomly among all possible individuals; henceforth Sakoda's algorithm is iterated repeatedly developing a spatially distributed social system to an organised spatial pattern, although this depends on the parameter set of the model, crowd densities, and individualisation, introduced below.

7.1. Interaction Model

For the sake of simplicity, there is a two-dimensional grid world that consists of places at discrete locations (x,y). An artificial agent occupies one place of the grid. A maximum of one agent can occupy a place. The agents can move on the grid and can change their living position. It is assumed that there are two groups related to the classes a and b of individuals. The social interaction is characterised by different attitudes [31] of an individual between different and among same groups given by four parameters:

$$S = (s_{aa}, s_{ab}, s_{ba}, s_{bb}). \tag{4}$$

The model is not limited to two groups of individuals. The S vector can be extended to four groups (or generalised) by the matrix:

$$S_{abcd} = \begin{pmatrix} s_{aa} & s_{ab} & s_{ac} & s_{ad} \\ s_{ba} & s_{bb} & s_{bc} & s_{bd} \\ s_{ca} & s_{cb} & s_{cc} & s_{cd} \\ s_{da} & s_{db} & s_{dc} & s_{dd} \end{pmatrix}. \tag{5}$$

The world model consists of N places x_i. Each place can be occupied by none or one agent either of group α or β, expressed by the variable $x_i = \{0,-1,1\}$, or generalised $x_i = \{0,1,2,3,4,..,n\}$ with n groups. The social expectation of an individual i at place x_i is given by:

$$f_i(x_i) = \sum_{k=1}^{N} J_{ik}\delta_s(x_i, x_k). \tag{6}$$

The parameter J_{ik} is a measure of the social distance (equal to one for Moore neighbourhood with a distance of one), decreasing for longer distances. The parameter δ expresses the attitude to a neighbouring place, given by (for the general case of n different groups):

$$\delta_s(x_i, x_k) = \begin{cases} s_{\alpha\beta}, & \text{if } x_i \neq 0 \text{ and } x_k \neq 0 \text{ with } \alpha = x_i, \beta = x_k \\ 0, & \text{otherwise} \end{cases}. \tag{7}$$

Self-evaluation is prevented by omitting the current place (i.e., $k \neq i$). There is a mobility factor m giving the probability for a movement.

An individual agent ag_i of any group α (class from the set of groups) is able to change its position by migrating from an actual place x_i to another place x_m if this place is not occupied ($x_m = 0$) and if $f_i(x_m) > f_i(x_i)$ and the current mobility factor m_i is greater than 0.5. Among social expectation (resulting in segregation), transport and traffic mobility have to be considered by a second goal-driven function $g(x_m)$, commonly consisting of a destination potential functions with constraints (e.g., streets). If $g_i(x_m) > f_i(x_m) > f_i(x_i)$ than the goal-driven mobility is chosen, otherwise the social-driven is chosen.

$$g : (x_i, x_m, x_k, v, t) \rightarrow \mathbb{R} \tag{8}$$

The mobility function g returns a real value [0,1] that gives the probability (utility measure) to move from the current place x_i to a neighbouring place x_m to reach the destination place x_k with a given velocity v. The g function records the history of movement. Far distances from the destination increase g values. Longer stays at the same place will increase the g level with time t. Social binding (i.e., group formation) will be preferred over goal-driven mobility.

The computation of the neighbouring social expectation values f is opportunistic, i.e., if f is computed for a neighbouring node assuming the occupation of this neighbouring place by the agent if the place is free, and the current original place is omitted x_i for this computation. Any other already occupied places are kept unchanged for the computation of a particular f value. From the set of neighbouring places and their particular social and mobility expectations for the specific agent the best place is chosen for migration (if there is a better place than the current with the above condition). In this work, spatial social distances in the range of 1–30 place units are considered.

Originally, the entire world consists of individual agents interacting in the world based on one specific set of attitude parameters S. In this work, the model is generalised by assigning individual entities its own set S retrieved from real humans by crowd sensing, or at least different configurations of the S vector classifying social behaviour among the groups. Furthermore, the set of entities can be extended by humans and bots (intelligent machines) belonging to a group class, too.

Segregation effects inhibit individual movement until a different social situation enables a movement. Transportation mobility triggers movement even if there is no social enabler. This is reflected in the extended Sakoda model by the mobility factor m and the goal-driven expectation function g that control mobility and overlays social and transportation and traffic mobility. The mobility function g includes random walk and diffusion behaviour, too. Constrained mobility is one major extension of the original Sakoda model presented in this work.

7.2. Model Parameters and Crowd Sensing

Creation of virtual digital twins is the aim of the crowd sensing. The crowd sensing is performed with chat bot agents. One stationary agent is operating on a user device, e.g., a smartphone, and another mobile agent is responsible for performing a survey (either participatory with a former negotiation or opportunistic ad-hoc). The results of the survey, a set of questions, are used to derive the following simulation model parameters, shown in Definition 2.

Definition 2. *Sakoda model parameters.*

```
type parameters = {
   group : string "a"|"b"|..,
   social-distance: number [1-100],
   mobility-distance: number [1-100],
   social-attitudes : [saa,sab,sba,sbb] | number [][],
   mobility : number [0-1],
   position : {x:number,y:number},
   destination : {x:number,y:number}
}
```

The *group* parameter sorts the user in one of two classes *a*/*b*, the *social-distance* parameter is an estimation of the social interaction distance, the *social-attitudes* parameter is the S vector, but limited to a subset of all possible S vector combinations (discussed below), and the *mobility* parameter is a probability to migrate from one place to another. The position (in Cartesian coordinates) is derived from the living centre of the user (global position data, GPS) and mapped on the simulation world (x,y). The S vector parameter determines the spatial social organisation structure. Typical examples of the S vectors with relation to social behaviour are [36]:

- (1,−1,−1,1): Typical segregation with strong and isolated group clusters;
- (0,−1,−1,0): Mutual suspicion;
- (1,−1,1,−1): Social climbers;
- (1,1,−1,−1): Social workers;
- (1,1,1,1): Inclusion;
- (1,0,−1,1): Traffic (a:cars, b:bicycles).

The crowd sensing extends the simulation with the following dynamics and changes:

1. Enhancement of the synthetic simulation with real-world data by digital twins not conforming to an initial parametrisation of the artificial individuals (affecting S vector, spatial social interaction distance, mobility);
2. Adaptation and change of the fraction of different groups (commonly an equally distributed fraction from each group is assumed);
3. Convergence and divergence of the emergent behaviour of group formation (spatial organisation structures);
4. Influence of real-world users by chat bots with data from the virtual simulation world.

The dynamics in the simulation world can be backpropagated to the real world by chat bot agents, too, delivering information and opinions formed by the artificial entities. Among classical surveys, chat bots can perform manipulation by distributing biased information or opinions via the chat dialog interface.

8. Traffic Prediction and Control

The previously introduced functional social interaction model is suitable to model group formation, but not group control. Traffic control and prediction of traffic emergence effects, e.g., jams, require a different algorithmic and iterative agent-based model introducing constraints, i.e., traffic signs and signals, security distance, and crash avoidance. In this model, there are vehicles and humans.

Traffic control can be performed by the perception and analysis of vehicle and/or crowd flows. Furthermore, vehicle-flows can be classified, e.g., introducing weights for individual and public vehicles, and crowd-flows can be classified the same way, e.g., distinguishing individuals from public mobility people or leisure from working mobility. In this work, both analysis approaches should be considered to optimize vehicle- and crowd-flows.

Among driver and passenger behaviour estimation by crowd sensing, accurate and robust localisation can be a challenge, especially if it should be performed in a distributed way. Distributed sensor fusion is a key method for deriving data with high quality and strength. Mathematical and statistical methods are well established, e.g., Gaussian processes. In [37], mobile sensors networks and Gaussian Markov random fields are proposed for accurate spatial prediction. The authors could proof the distribution of such method and the deployment in low-resource networks. Clustering effects (e.g., of users and their mobile devices) can have relevant impact on simulation and traffic prediction, too. Gauss–Poisson models as a class of clustered point processes are able to capture such clustering effects [38].

8.1. Traffic Model

The traffic model consists of a city map (streets, places, and so on), vehicles, and humans. There are four sub-classes of vehicle agents:

1. An individual vehicle with one driver agent and zero passengers and dynamic route planning;
2. A shared vehicle with one driver and [1,n] passengers and semi-dynamic route planning;
3. A public vehicle with one driver and [1,m] and m >> n passengers and static route planning;
4. Bicycles with one driver.

Each vehicle requires at least one twin agent controlling the vehicle (the driver), although a vehicle (except bicycles) can be driven automatically or autonomously. In the latter case a driver sets the destination position only.

Similar to vehicle agents, there are three different sub-classes of twin agents:

1. A driver agent (can be a passenger, too);
2. A passenger agent using a shared or public vehicle;
3. Pedestrians or currently non-mobile humans.

The crowd sensing provides input data for the simulation by creating parameterised digital twins of real humans, and it provides crowd feedback (output of the simulation), e.g., by suggesting alternative mobility routes or using incentive mechanism to control crowd flows.

8.2. Crowd Sensing and Traffic Control

The crowd sensing data can be used to analyse current traffic situations (driver class) and to predict crowd flows resulting in traffic in the near future (passenger and pedestrian classes). The flow, behaviour, and clustering prediction can be fed back to the real world, e.g., by controlling traffic signal switching or by influencing people's decision making via social or domestic media.

Traffic control can address optimised signal switching as well as dynamic street sign (e.g., velocity), and control of vehicles by influencing drivers (or passengers). Signal control can be static (with respect to switching times) or dynamic, including signal group switching. Commonly signals are controlled by evaluating the current traffic flow situation without evaluating the drivers and passengers (e.g., with respect to destination, attitudes, or goals). If the goal is to optimise the crowd flow rather than the vehicle flow, then public or shared vehicles have to be preferred by priorities.

9. Use-Cases and Evaluation

The next sub-sections show three different use-cases showing the deployment of the augmented simulation approach.

The first use-case is related to social science and addresses segregation as well as short-term and short-range mobility arising in city traffic (clustering driven by social interaction and behaviour). It is used primarily to study social networks and interactions. Real and virtual worlds can be different.

The second use-case demonstrates the suitability of the framework for smart light management in cities by using crowd sensing to deliver real-time data from physical sensors and humans. The simulation is used to derive optimal street illumination under different constraints, e.g., low-energy constraints. The simulation is primarily used to control real-world city infrastructure and the real world is mapped on the virtual world.

Finally, the third use-case demonstrates the study of smart traffic control using crowd sensing delivering real-time data from physical sensors and humans. Depending on the goal simulation (control or modelling), the real world can be mapped on the virtual world, but can be different, too, using artificial city maps only.

The first use-case relies on a pure mathematical interaction model, whereas the second and third use-cases rely on pure algorithmic interaction models.

9.1. Segregation and Social Interaction

This use-case should demonstrate the impact of real-world humans on the simulation of social interaction and social networks (clusters). Crowd sensing was used to create digital twins in the simulation with model parameters derived by surveys performed with a WEB-page or suing mobile App software. For the sake of simplification, the world consists of a regular grid (patch grid). The Sakoda interaction model introduced in Section 7 is basically a local interaction model, in contrast to algorithmic models shown in the use-cases in Sections 9.2 and 9.3.

9.1.1. Long-Term Mobility

The initial simulation was performed with 200 class a and 200 class b agents and a unified $S = (1,-1,-1,1)$ setting for all agents (blue and red squares) of both classes and unconstrained mobility leading to classical segregation structures (strong isolated clusters), shown in Figure 10.

During the simulation run, crowd sensing was performed using chat bot agents. Up to 200 digital twins (triangles) retrieved form crowd sensing surveys (chat dialogues, see Example 1) were added to the simulation dynamically. The S vector and the social distance r of digital twins now depend on the answers given by the (real) humans, which can differ from the initial S setting. These agents (if their S differs from the basic model) create a disturbance in the segregation patterns. Agents with $s_{\alpha\beta} = 0/1$ and $s_{\alpha\alpha} = 0/1$ can be integrated into both groups and are able to bind different groups close together (see the development and movement of the blue and red clusters inside the red circle in Figure 10). The figure shows different simulation snapshots at 500, 1000, and 1500 simulation steps. Each simulation step corresponds to a real-time interval of 10 s (to enable crowd sensing in real-time). The survey participants were chosen randomly by the chat bot agents.

Using an initial unified stable parameter set for all agents, digital twins with varying and different parameter sets based on survey data can be used to test the stability and convergence criteria of structure formation or mobility patterns like in traffic management on a fine-grain scale.

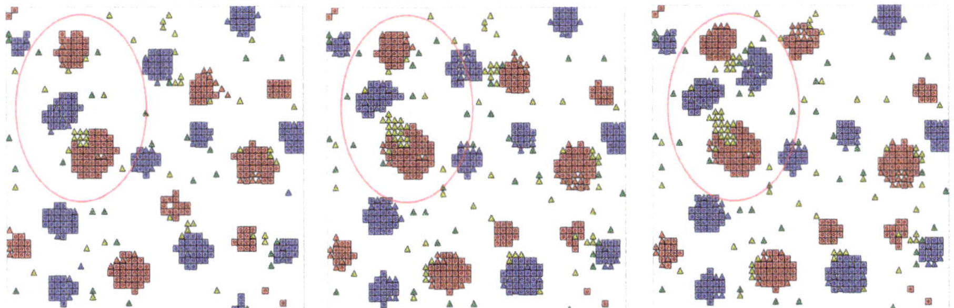

Figure 10. Simulation world with long-term mobility, without spatial and context constraints at different simulation times (500/1000/1500 steps) consisting of 200/200 a/b class agents (blue/red squares) all with S = (1,−1,−1,1) and r = 3 parameter settings and additionally up to 200 digital twins (triangles with colour based on individual S/r parameters).

Example 1. *A typical survey job script for the Sakoda model.*

```
job = [
 script : [
  {question:'Who are you? Do you think you are poor or rich?',
   choices:['Poor','Rich','Do not know'],
   tag:'class'},
  {question:'How do you rate your relation to people of your own group $1?'+
          'Do you like (1) or dislike (-1) them?',
     choices:['-1','0','1'],
   eval:function (dialog) return [dialog[0].answer] end,
   cond:function (dialog) return dialog[0].answer!='Do not know' end,
   tag:'attitudeP'},
  {question:
  'How do you rate your relation to people of the other group $1?'+
          'Do you like (1) or dislike (-1) them?',
   choices:['-1','0','1'],
   eval:function (dialog) return
[dialog[0].answer=='Poor'?'Rich':'Poor'] end,
   cond:function (dialog) return dialog[0].answer!='Do not know' end,
   tag:'attitudeN'},
  {question:'Do you have $1 friends?',choices:['Yes','No'],
   eval:function (dialog) return
[dialog[0].answer=='Poor'?'rich':'poor'] end,
   cond:function (dialog) return dialog[0].answer!='Do not know' end,
   tag:'friendsN'},
  {question:'How old are you?',tag:'age'},
  {question:'What is your monthly salary?',
   choices :['5000'],tag:'salary'},
  {question:'Where do you live [Enter ZIP Code]?',
   value:function () return myIpLocation().city end, tag:'location'},
  {question:'How many social contacts do you have?',tag:'contacts'},
  {question:'Do you like to move?',
   choices:['Yes','Maybe','No'],tag:'mobility'},
  {message:'Thank you!'}
 ],
 action : function (script) {
   // create digital-twin from script data;
   ..
 }
]
```

9.1.2. Short-Term Mobility

The experiment aims to study crowd formation in cities (places, streets, and buildings) based on physical and virtual social interaction via social media and opinions and were performed with an artificial map of a city (simplified real world). Distributed crowd sensing via chat bots introduce updates and disturbance into the social formation structures. The mobility of physical agents is modelled by the extended Sakoda model introduced in Section 7 and depends on local social interaction (temporary crowd formation) and a potential field attraction regarding the destination in the city that has to be reached. Mobility is spatially constrained (streets).

The crowd sensing tries to estimate the local people's movement and compound crowd formation constrained by urban structures (like streets) based on an estimated individual *S* matrix. For the sake of simplicity, two groups are assumed (*a*,*b*), but members of each group differ in attitudes and behaviour.

The two groups, for instance, could be car and bicycle users acting as public traffic participants, differing in social, mobility, and group formation behaviour. The survey can be opportunistic, e.g., ad-hoc and occasional with a situation-aware dialogue for specific traffic situations, or participatory with a more general survey character.

The position of the digital twins added to the two-dimensional simulation world is estimated by GPS and IP localisation collected by the chat bot agent or estimated by user answers.

Mobility constraints by streets were added, shown in Figure 11. The constraints are synthetic. The mobility of agents was driven by reaching a destination in the simulation world (either chosen randomly or by user information in the case of digital twins) and damped or delayed by social interaction, depending on individual *S* vector, social, and mobile interaction distances. The simulation was carried out with 400/400 class *a*/*b* agents.

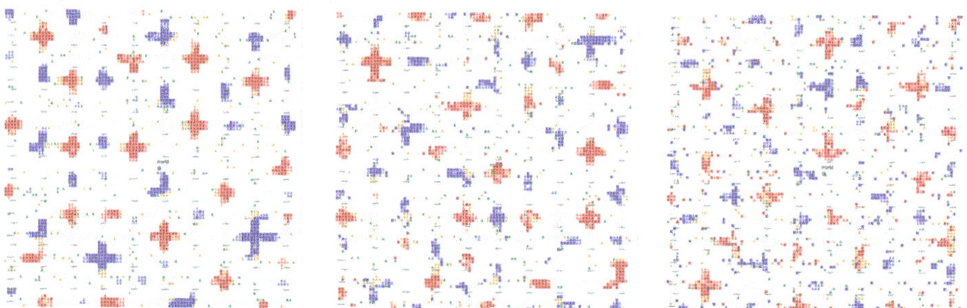

Figure 11. World consisting of mobility constraints (streets), 400/400 class a/b agents (blue and red squares) and digital twins (triangles) with different model parameters. (**Left**) Only social-driven mobility and clustering. (**Middle**) Social and traffic driven mobility with low attraction of the destination. (**Right**) Social and traffic driven mobility with high attraction of the destination.

The results (snapshots after 3500 simulation steps) show an increasing disturbance of social aggregation patterns resulting from social interaction (mobility due to social attraction) by goal driven traffic mobility (attraction by destination). Again, $s_{\alpha\beta} = 0/1$ and $s_{\alpha\alpha} = 0/1$ twins act as compound glue and bring homogenous groups closer together. With the increasing influence of goal-driven traffic mobility over social attraction and mobility, the social clustering gets fuzzier and more diffuse, shown on the right simulation snapshot of Figure 11.

9.1.3. Evaluation

In both examples (with respect to long- and short-term mobility), digital twins with varying parameter sets have a significant impact on crowd structures and group behaviour. The digital twins pose individual behaviour not initially considered in the synthetic simulation. The behaviour is based on crowd sensing surveys performed by real humans. Short-term mobility and interaction based on the Sakoda model can be used in smart city crowd behaviour prediction.

9.2. Smart City Light Management

The goal of this use-case is to provide a simple demonstrator that uses simulation to investigate crowd interaction with smart city infrastructures by implementing a self-organised control of ambient light conditions (e.g., in streets or buildings). Crowd sensing provides real-world feedback as input for the augmented simulation to evaluate and predict real-world changes, finally used for decision-making processes (light management) [24]. The crowd gets feedback from the simulation by the environmental light control (output of the simulation).

9.2.1. World

This simple demonstrator consists of a simulation containing an artificial city area with mobile physical agents representing humans interacting with mobile devices or machines), beacons (access points) for sensor aggregation and distribution, and some external beacons connected to the Internet enabling the connection to mobile devices via the Internet, and finally smart light devices illuminating streets and buildings, shown in Figure 12.

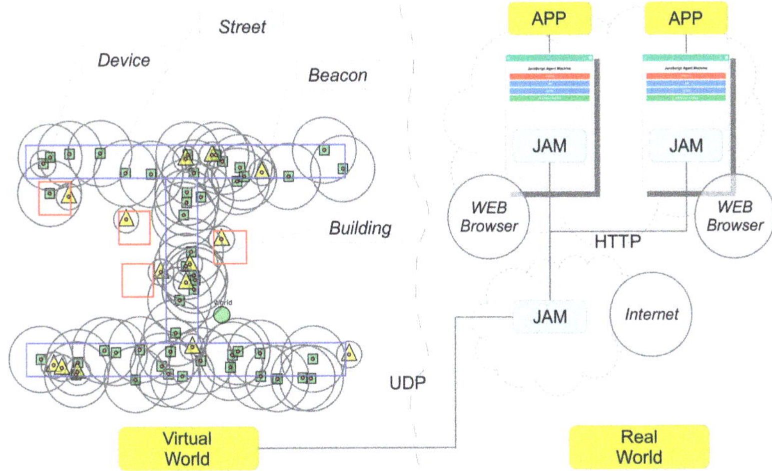

Figure 12. A demonstrator: An artificial street area consisting of street segments (blue rectangle), buildings (red rectangles), and beacons (yellow triangles), populated with stationary and mobile devices (green squares). The grey circles around beacons and nodes show the wireless communication range. Only overlapping circles connect.

9.2.2. Interaction and Control Model

The interaction model is rather simple and purely algorithmic in this use-case. The goal of the simulation is to control and predict light illumination in city areas based on activity, requirements (constraints), and people's feedback retrieved by crowd sensing.

9.2.3. Crowd Sensing

The goal of explorer agents (sent out by mobile or light devices) is information mining in the outside world via Internet deployed agent processing platforms that can be accessed by simulation nodes and WEB browser Apps. The collected information is passed back to the root node (e.g., a mobile device of a passenger) to assist decision making and navigation.

The explorer agents have to estimate the position of the root node by performing sensor mining from surrounding devices they are visiting and from the outside world (far away) by asking questions answered by humans via the WEB App chat dialogue. An example dialogue of the explorer agent is shown below in Example 2. The explorer asks a user for its current place and location within a region of interest and an assessment of the current light situation. Depending on the answer a specific action is suggested.

Example 2. *Survey job defining a dialogue for an explorer agent that migrates to mobile devices (smart phones) to participate in a chat dialogue with the user of the mobile device. The aim is to get environmental perception. Finally, an action function evaluating the survey is executed to deliver data.*

```
job = {
 script:[
  {question:'Where are you?',
   choices:['Street 1','Street 2','Street 3',
          'Anywhere','Other place']},
  {question:'How do you rate ambient light?',
   choices:['Dark','Good','Bright'],
   cond: function (s) {
     return s[0].answer &&
          s[0].answer.indexOf('Street')==0},
  {message:'Thank you!'
 ],
 action: function (script) {
  var place;
  switch (script[0].answer) {
     case 'Street 1':
     case 'Street 2':
     case 'Street 3':

       place=script[0].answer; break;
  }
  if (place) switch (script[1].answer) {
    case 'Dark': return {light:150,place:place};
    case 'Bright': return {light:50,place:place};
  }
 }
}
```

Based on answered questions regarding the current user location, the satisfaction of ambient light condition, and an optional fusion with device sensor data (light, position, etc.), actions are directed to smart light control devices to change the light condition in streets and buildings by using mobile notification agents. The action planning is based on crowd demands and energy-saving constraints. If action is required, mobile notification agents are sent out to neighbouring nodes to change light intensity based on directed diffusion, random walk, and divide-and-conquer approaches.

9.2.4. Evaluation

The simulation with the *SEJAM* simulator was able to be performed with more than 1000 nodes, hundreds of beacons, and more than 10000 explorer agents. Real-time values for one simulation step depend on the number of active agents to be processed, node and agent mobility (graphics and communication), and ranges from 1 ms to 1 s. For this use-case, a real-time simulation step of 1 s is more than sufficient to control lights in streets and on places (or within buildings). The minimal response time from sensor input to actuator control is about 5 s.

The user feedback from crowd sensing surveys performed by mobile agents is used to optimise illumination in public areas like streets with respect to user and energy management constraints. Energy saving up to 30% can be achieved, light pollution can be decreased, and citizen satisfaction can be optimised. Combined with machine learning, the simulation (mapping a real-world city region) can be used to predict future light demands based on crowd sensing and using other physical sensors (e.g., crowd and traffic flow sensors).

9.3. *Smart City Traffic Management*

The goal of the traffic simulation is the recognition of flows, jams, and structures (clusters) on streets and places. In [21], adaptive and partly self-organising traffic management was achieved by using a Multi-agent System (MAS) with multi-levels of decision making and a hierarchical organisational structure.

The simulation results can be finally used to manage real-world traffic by agents in real-time. Real-world traffic flows can be influenced by technical traffic infrastructure, by traffic information systems, by vehicle driver behaviour (which can be diverse), or by influencing people's decision making (on a temporally and spatially short-range scale)

9.3.1. World

In general, the simulation world consists of a graph $G = (P,S)$ with nodes P representing places or spatial regions and edges S representing streets. Here, the world consists of a simplified street map arranging streets orthogonally in a regular grid, although the traffic simulation is not limited to any specific map or graph geometry. An example street map is shown in Figure 13. Streets have a major orientation (horizontal or vertical) and consist of three tracks (ride, left, and middle). A vehicle can move in directions north (N), south (S), east (E), and west (W).

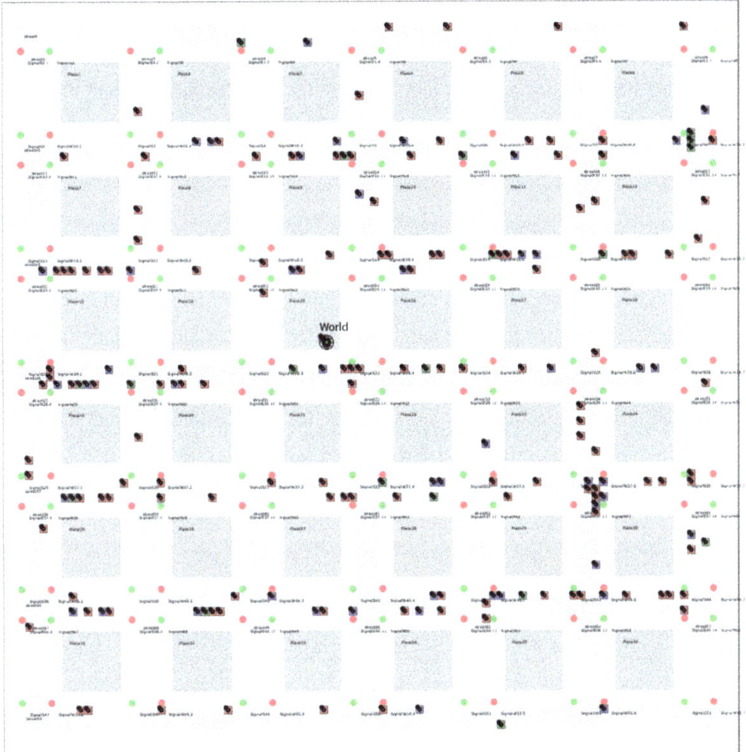

Figure 13. A typical simulation world consisting of a street map, traffic signals, places, and different vehicle classes (red: individual, blue: shared, and green: public).

9.3.2. Interaction and Mobility Model

The model parameters of vehicles and their driver agent or passengers (digital twins) are defined in Definition 3. They can be set with default values, by Monte Carlo simulation, or they can be selected and derived from crowd sensing surveys from real humans (in real-time).

Definition 3. *Typical model parameters of drivers and automatic vehicles.*

```
parameters = {
     class : string "driver"|"passenger"|"co-driver"|..,
     sub-class : string,
     security-distance δ : string "correct"|"lower"|"higher"  | [0,2],
     driving-style ε : string "normal"|"risk"|"defense",
     sharing Ω : number [0,1],
     group γ : number [1,100],
     straight-routing σ : number [0,1],
     ignore-signal χ : number [0,1],
     position : {x : number,y : number},
     from : {x : number,y : number},
     destination : {x : number,y : number},
     duration-expectation τ : number [1,100],
   }
```

The *sub-class* parameter set can address specific classes like gender, age, or more general attributes like citizen status, or living conditions. The *security-distance* behaviour (between vehicles), the *driving-style* parameter, and the *straight-routing* parameter (for route planning) are only relevant for the *driver* class. A *straight-routing* = 0 parameter results in a zig-zag routing, otherwise a longer straight-line routing is used following the current street as long as possible (range depends on parameter value). The *ignore-signal* parameter is a probability to ignore stop signals (red signal), e.g., if the signal state changes.

A typical survey job script is shown in Example 3, showing a job being passed to a crowd sensing agent. It is used to derive the actual position, e.g., of a mobile device, motion, and to get information about the user (e.g., passenger or driver). Finally, a parameter set for the behaviour model is derived and digital twins are created.

Example 3. *A typical crowd sensing script job (JavaScript) for traffic flow exploration passed to a crowd sensing agent.*

```
job = {
 script:[
  {sensor:'GPS' },
  {sensor:'ACCEL' },
  {question:'Where are you?',
   choices:['Building','Street','Walkway',
           'Anywhere','Other place']},
  {question:'Is this position $1 correct?',
   choices:['Yes','No'],
   cond : function (s) {return script[0].answer},
   eval : function (s) {return gps2loc(script[0].answer)}
  }
  {question:'How do you rate traffic density?',
   choices:['Full','Medium','Low'],
   cond : function (s) { return s[2].answer=='Street' &&
                             s[3].answer=='Yes'},
  },
  {question:'Do you like fast driving?',
   choices:['Yes','No'],
   cond : function (s) { return s[2].answer=='Street'},
  },
  {question:'Do you like risks?',
```

```
    choices:['Yes','Maybe','No'],
   },
   {question:'How old are you?'},
   {question:'Do you use a navigation device?',
    choices:['Yes','No'],
    cond : function (s) { return s[2].answer=='Street'},
   ...
   {message:'Thank you!'}
  ],
  action: function (script) {
    // create digital-twin from script data;
    ..
  }
}
```

9.3.3. Multi-agent System

The MAS in the simulation consists of a world agent that controls the simulation and establishes the connection to the outside world (Internet), vehicle, and twin agents. The twin agent can either be the active controller of the vehicle agent (influencer) or a passive passenger. The vehicle agents implement different overlaid behaviour:

1. Constraint-based traffic mobility following specific rules (generalised behaviour)
2. Social interaction (individual behaviour)
3. Route-planning and optimisation (goal-based individual behaviour)

The vehicle agent consists of a set of activities performing specific actions and represent different micro-goals, basically: {*percept*, *plan*, *move*, *collect*, *wait*}. A vehicle agent represents a mobile entity and at least one human (a twin agent). The *collect* activity is able to bind more vehicle or twin agents (humans) to this vehicle entity, i.e., implementing vehicle grouping or sharing. Twin agents can be coupled to vehicle and communicate with each other via the tuple space of the mobile platform both sharing.

A vehicle has a direction *dir* and uses a set of sensors {*f*,*l*,*r*,*b*} providing information about the neighbourhood in front, left, right, and backward relative to current direction. Each sensor is an array with distance increasing with the array index. New sensor values are collected in the *percept* activity, shown in Algorithm 3.

Algorithm 3. Perception activity of the vehicle agent. Sensors k0, v0, and i can depend on twin driver agents, too.

```
k0 ← update keep going probability (twin)
do update sensors {p,s,d,v,i,f,l,r,b}:
p ← get current position (x,y)
    s ← get street at this position
    d ← compute distance to destination
    v ← compute speed
    v0 ← update highest speed
    i ← compute default security distance (s,v)
    do in current direction:
    f ← get front resources {street,vehicle}
    l ← get left resources {street,vehicle}
    r ← get right resources {street,vehicle}
    b ← get back resources {street,vehicle}
```

The vehicle mobility and its planning are immediately influenced by its velocity v, the sensors d and $i(s,v)$ limiting the distance to the next vehicle, and the force to keep the current direction (not turning) k. There is a current long-range direction *dir* and a short-range position correction Δ, an (x,y) delta vector, e.g., required for turns or street and direction changes.

The current street is given by the *s* sensor. A street *s* consists of coordinates $\{x,y,w,h\}$, a unique identifier *id* and a set of possible directions *dirs*. It is assumed that a street has three tracks (left, right, middle). A street has either a horizontal (*N,S*) or vertical alignment (*W,E*). $P(x,y)$ is place in the world at position (x,y) with attributes *street, vehicle, signal*.

In automatic driving mode, the distance to a vehicle in front is limited to $d = v^{\delta}/dk$ and depends on the security distance parameter δ and the current velocity of the vehicle.

The vehicle plans the next movement based on perception in activity *plan*, shown in Algorithm 4 (simplified pseudo code), finally executing the movement in activity *move*, shown in Algorithm 5.

A twin agent is divided in two sub-class behaviour: (1) Driver (2) Passenger. The twin agents implement basically the goal-driven behaviour of mobility and provides higher planning and routing levels influencing the vehicle. A driver twin agent directly controls traffic routing, whereas passenger twin agents influence the temporal flow of vehicles (by entering or leaving the vehicle).

Algorithm 4. Simplified planning activity of the vehicle agent, which can be depend on twin driver and partly on passenger agents, too

```
do find new direction and position correction:
if Δ = 0 then
  //0. Check correct direction
  if k = 0 ∧ (dir∉s ∨ p∉{(x,y):(x,y)∈s ∧ on right side of dir}) then
    (dir,Δ)←new dir(s,p)
  //1. Check traffic signal (red?)
  stop ← r[1].signal = red ∧ r[1].signal.dir = dir
  //2. Navigate to destination
  if k<k0 then
    case dir of
      N => if p.x+ε < dst.x ∧ f[1].street≠s then next←E,Δ←(0,−1)
           if p.x−ε > dst.x ∧ f[1].street≠s then next←W,Δ←(−3,−3)
      S => if p.x+ε < dst.x ∧ f[1].street≠s then next←E,Δ←(3,3)
           if p.x−ε > dst.x ∧ f[1].street≠s then next←W,Δ←(−1,1)
      W => if p.y+ε < dst.y ∧ l[3].street≠s then next←S,Δ←(−2,3)
           if p.y−ε > dst.y ∧ r[1].street≠s then next←N,Δ←(0,−1)
      E => if p.y+ε < dst.y ∧ r[1].street≠s then next←S,Δ←(0,1)
           if p.y−ε > dst.y ∧ l[13].street≠s then next←N,Δ←(2,−3)
    if next then dir←next,k←k0+3
  //3. Check road crossing; keep dir.
  if ¬next ∧ f[1].street≠s then k←k+4
 else if blocked ∧ random(1) then
  //4. Escape blocking situation
  do ∀ dir ∈ {N,S,W,E}:
    find one neighbour place q of p
        with |q−p] = 1 ∧ street∈q = s ∧ vehicle∉q ∧ Δ.x/y = 0:
        Δ←Δ+(q−p)
```

Algorithm 5. Simplified move activity of the vehicle agent. The gotoXY function moves the vehicle one step to a new free place minimizing minΔ→0

```
go ← true
//1. Check free distance
    do ∀ n in {1,2,,..i}:
      if f[n].vehicle then go ← false
    if Δ = 0 ∧ go then
      case dir of
        N => next←{x:p.x,y:p.y−1}
        S => next←{x:p.x,y:p.y+1}
        W => next←{x:p.x−1,y:p.y}
        E => next←{x:p.x+1,y:p.y}
      if vehicle∉P(next) then blocked←0,goto(next)
      else if v = 0 then incr(blocked)
    else Δ←gotoXY(Δ,p)
```

9.3.4. Evaluation

The traffic simulation was initially performed with a closed simulation world using a default behaviour model and traffic control parameter set. The outcome of this simulation was compared with an open simulation using agents with the same default behaviour extended with digital twins created from user surveys via crowd sensing agents. The major goal was to show the influence of variation by real-word digital twins on the simulation outcome and to identify relevant model parameters with significant impact. The traffic flow was controlled by light signals (red and green signals) positioned at each crossing corner. The default traffic signal switching is periodic with a state sequence {rr,gr,rr,rg}, with the first character assigned to all North-South signals, and the second character assigned to all West-East signals (r:red/g:green). Transition times were t_1 for gr|rg → rr transitions, and $t_0 < t_1$ for rr → gr|rg transitions.

In Figure 14, a closed simulation with vehicles (individual class) and different traffic signal switching times were investigated. The simulation started with an initial population of 200 vehicles at random starting positions. The plots show the time-resolved decrease of driving passengers and increase of arriving passengers (in this case vehicles with on driver as a passenger) due to arrival at a destination position. The first plot (a) uses a signal switch time (red ↔ green signal state) of 500, the second plot of 200 simulation time units. Either North-South (*NS*) streets get green lights, or the West-East (*WE*) streets. The switching time has a significant impact on the mean arrival time of vehicles, although vehicles uses *NS* and *WE* streets, but the arrival time do not depend on individual parameters, i.e., security distance and routing strategy (zig-zag versa straight-line). The vehicle flow depends on the routing behaviour (straight-line shows lower dependence on signal switching times).

In Figure 15, the simulation was extended with MCWS and digital twins posing individual behaviour. The MCWS users were chosen randomly. The goal of this simulation was to identify jams and the dependence on driver behaviour parameter. Two main parameters were identified: The security distance and routing (straight-line vs. zig-zag shortest path) behaviour. Again, about 200 vehicles were created, a fraction with a default behaviour, and another fraction with individual behaviour. The value of the security distance parameter δ has significant impact on the jam probability (defined by the number vehicles involved in jams relative to the overall number of vehicles). A value δ = 1.0 relates to a law-conforming driving style, a lower value means a riskier driving style with lower distances to other vehicles, and a higher number a more defence driving style. Lower security distances (relative to current vehicle velocity) increases the jam probability due to a modified vehicle flow (accidents are not considered here). The is more important if a straight-line routing style (parameter greater than zero) is chosen by the driver. The last three bars in the plot show a distribution of the security-distance parameter (25%, 33%, 50%). Even a low fraction of 25% risky drivers increases the jam probability significantly.

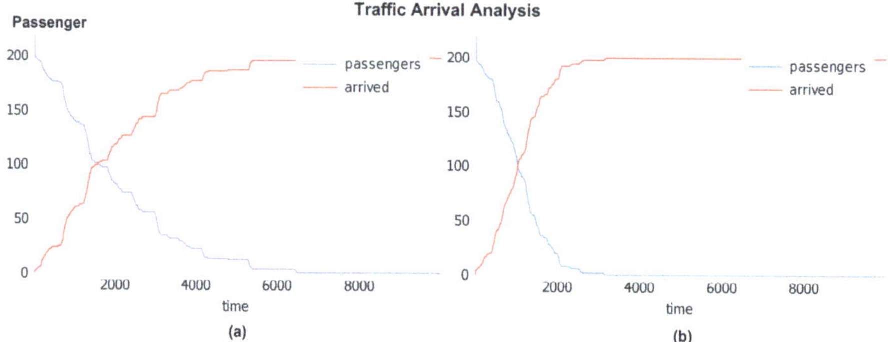

Figure 14. Simulation with 200 vehicles with static traffic control showing the accumulated number of moving vehicles/passengers and the number of arrived vehicles/passengers (**Left**) t_1/t_0 = 500/60 signal change times (**Right**) t_1/t_0 = 200/60.

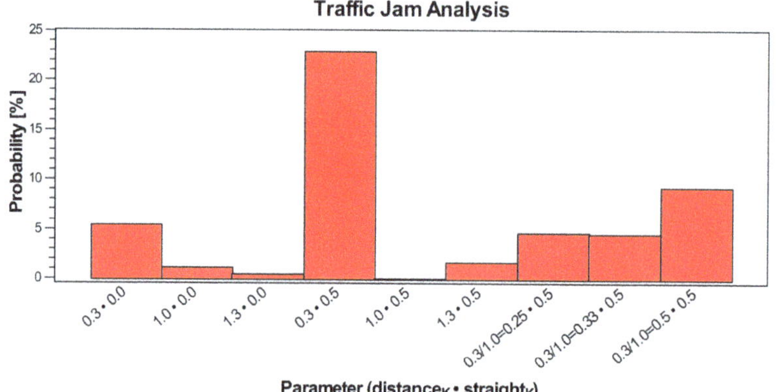

Figure 15. Simulation with 200 vehicles and different mobility parameters showing the probability of a jam.

Finally, the traffic control algorithm was evaluated in Figure 16. Again, the traffic flow of 200 vehicles with zig-zag routing was evaluated (δ = 1.0, σ = 0). Two traffic signal control algorithms were compared: On the left side the static global switching algorithm (with t_1/t_0 = 500/60), and on the right side a dynamic local switching algorithm with dynamic switching times based on vehicle flow densities before each signal of a crossing. The global mode switches all signals of crossings globally and periodically, the local mode switches signals of each crossing individually. The dynamic flow-based switching increases arrival times by 2 times, and the velocity of vehicles shows low modulations and is near by the default velocity v = 50 (arb. units).

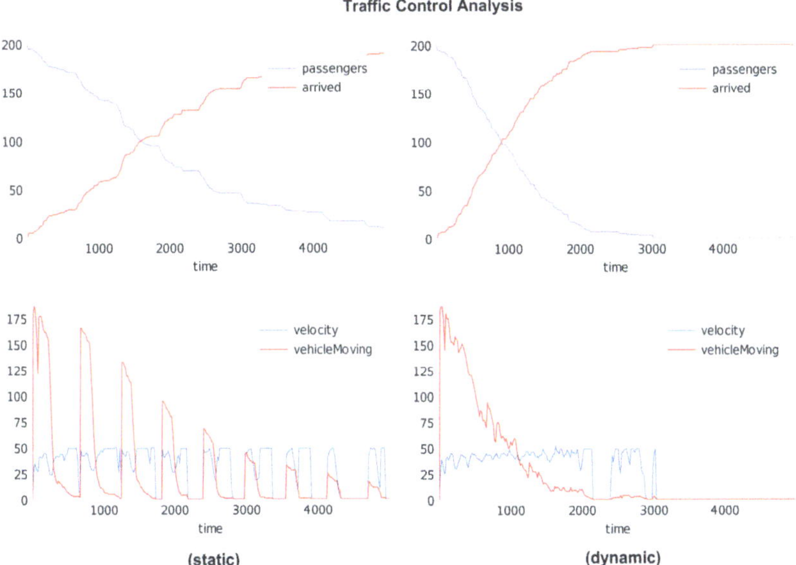

Figure 16. Simulation of 200 vehicles ($\sigma = 0.0, \delta = 1.0$): (**Top**) Temporal development of number of mobile vehicles/passengers and number of vehicles/passengers reached destination (**Bottom**) Mean velocity of moving vehicles and number of moving vehicles (**Left**) Without traffic control with t_1/t_0 = 500/60 signal change times (**Right**) With dynamic flow traffic management.

To summarise, individualism of behaviour model parameters derived by real-world sensing (eventually in real-time) like driving style (security distance, routing) can influence crowd and vehicle flows significantly and hence has to be considered by traffic control to optimise traffic and crowd flows. It is important to get a representative model parameter set from an actual traffic situation to be able to predict flows and to control traffic.

10. Conclusions

The MAS simulation framework presented in this work provides the bidirectional integration of real-world data via agent-based crowd sensing, agent-based modelling, and agent-based simulation via data mining, i.e., applying DM in ABMS by creating generated simulation data and applying ABMS in DM by backpropagating generated simulation data. The framework is suitable to combine social and computational simulations with real-world interaction at run-time and in real-time by using mobile computational agents. The simulator supports two classes of agents. Physical agents only exist in the virtual world and correspond to *NetLogo* agents, whereas computational agents can exist in real and virtual worlds and create the bridge between the two worlds.

The *JAM* agent processing platform is the core component in the simulator and crowd sensing system that can be deployed in heterogeneous networks on a broad range of devices including simulation software. This is enabled by programming *JAM* and agents in *JavaScript*. Chat bots with dynamic (situation and context-dependent) dialogues can be implemented with *JAM* agents directly using a unified dialogue structure and API.

Two key concepts in the augmented simulation approach are parametrizable behaviour models and digital twins created from surveys via crowd sensing.

Different use-cases demonstrate the augmented simulation approach with social or socio-technical interaction models. First, a classical social interaction study was performed based on a parametrizable Sakoda model, second, a reactive and interactive city light management was sketched, and third, a crowd-based city traffic control was investigated. Digital twins retrieved by agent-based crowd sensing extended agent-based simulation with variance and a broader range of model parameter settings. The crowd sensing performed by mobile agents is used to create digital twins of real humans (with respect to the social interaction model and mobility) based on individual surveys via a chat bot dialogue. Chat bots are the link between virtual and physical worlds.

It could be shown that social interaction and individualism (variance) has a relevant impact on traffic control and crowd or vehicle flows, hence social models overlay technical models. It is important to get a representative model parameter set and distributions from real-world situations by crowd sensing for accurate system prediction and control.

In the future, we aim to use the framework in order to investigate different environments in quasi-experimental set-ups with broader data and user range.

Finally, the system behaviour and cross interaction effects should be studied with hybrid models, i.e., the combination of the social interaction model describing group formations, e.g., Sakoda model, with mobility models, e.g., mobile vehicles, or by combining agent-based with analytical models to strength the simulation output.

Author Contributions: S.B. provided content and methodologies related to the computer science parts, social models, and performed the experiments, U.E. provided content for the general concepts and methodologies related to the social science parts of the paper.

Funding: There was no founding.

Conflicts of Interest: There are no conflicts of interest.

Appendix A Comparison of NetLogo and SEJAM Simulations

The well-known and established *NetLogo* simulator is (commonly) used for the ABM domain only. In contrast, the *SEJAM* simulator is primarily used for the ABC domain, although any physical simulation can be transformed into a computational task implementing the behaviour of a physical entity. The fusion of real worlds deploying computational agents (that interact with machines and humans) with virtual simulation models requires a mapping methodology to be able to simulate physical agents (i.e., artefacts of real entities) using computational agents.

Computational agents as mobile software processes require a connected communication network of host platforms (computers) to migrate along a path *AB*. A human being, in contrast, does not require such a digital transport network. In a simulation world, a physical (purely behavioural) agent can overcome arbitrary distances in one step (in principle).

To combine social interaction and computational simulation models, the *SEJAM* simulation model was extended by two super classes of agents: (1) computational and (2) physical agents. A physical agent can only exist in the virtual simulation world and consists of a *<logical node, physical agent>* tuple. The physical agent is bound to his node and can change its position only by moving the node carrying the agent. In contrast, computational agents are mobile and can hop from one platform to another if there is a communication link between both platforms by performing process serialisation and deserialisation. The link can be virtual (inside the simulation world) or physical connecting a logical node of the simulation world with another *JAM* platform in the real world (e.g., a smartphone) via the Internet.

The mapping of *NetLogo* model constructs and statements on the *SEJAM* simulation model basically consists of a simulation API extension. Physical agents represent turtles (active agents) and resources representing patches. The patch grid world is discretised, although *SEJAM* is not limited to and can model any two-dimensional world. In *NetLogo*, the agent behaviour is entirely controlled and executed by a global observer (centralised macro control), whereas in *SEJAM* the agent behaviour is controlled and executed by each individual agent (decentralised micro-control).

References

1. Cioffi-Revilla, C. *Introduction to Computational Social Science*; Springer Science and Business Media LLC: Berlin/Heidelberg, Germany, 2017.
2. The Royal Society, Machine Learning: The Power and Promise of Computers That Learn by Example. Available online: https://royalsociety.org/~{}/media/policy/projects/machine-learning/publications/machine-learning-report.pdf (accessed on 2 August 2019).
3. Quan-Haase, A.; Sloan, L. Introduction to the Handbook of Social Media Research Methods: Goals, Challenges and Innovations. In *The Sage Handbook of Social Media Research Methods*; Sloan, L., Quan-Haase, A., Eds.; SAGE: Los Angeles, CA, USA, 2017; pp. 606–859.
4. Roggen, D.; Wirz, M.; Tröster, G.; Helbing, D. Recognition of crowd behavior from mobile sensors with pattern analysis and graph clustering methods. *Netw. Heterog. Media* **2011**, *6*, 521–544.
5. Gilbert, N. *Agent-Based Social Simulation: Dealing with Complexity*; University of Surrey: Guildford, UK, 2004.
6. Keuschnigg, M.; Lovsjö, N.; Hedström, P. Analytical sociology and computational social science. *J. Comput. Soc. Sci.* **2018**, *1*, 3–14. [CrossRef]
7. De Marchi, S. *Computational and Mathematical Modeling in the Social Sciences*; Cambridge University Press: Cambridge, UK, 2005.
8. Carrington, P.J.; Scott, J.; Wasserman, S. *Models and Methods in Social Network Analysis*; Cambridge Press: Cambridge, UK, 2005.
9. Flache, A.; Mäs, M.; Feliciani, T.; Chattoe-Brown, E.; Deffuant, G.; Huet, S.; Lorenz, J. Models of Social Influence: Towards the Next Frontiers. *J. Artif. Soc. Soc. Simul.* **2017**, *20*, 2. [CrossRef]
10. Bruch, E.; Atwell, J. Agent-Based Models in Empirical Social Research. *Sociol. Methods Res.* **2015**, *44*, 186–221. [CrossRef] [PubMed]
11. DellaPosta, D.J.; Macy, M.W. The Center Cannot Hold. Networks, Echo Chambers, and Polarization. In *Order on the Edge of Chaos, Social Psychology and the Problem of Social Order*; Lawler, E.J., Thye, S.R., Yoon, J., Eds.; Cambridge University Press: Cambridge, UK; pp. 86–104.
12. Bail, C.A.; Argyle, L.P.; Brown, T.W.; Bumpus, J.P.; Chen, H.; Hunzaker, M.F.; Lee, J.; Mann, M.; Merhout, F.; Volfovsky, A. Exposure to opposing views on social media can increase political polarization. *Proc. Natl. Acad. Sci. USA* **2018**, *115*, 9216–9221. [CrossRef] [PubMed]
13. Ferrara, E. Contagion dynamics of extremist propaganda in social networks. *Inf. Sci.* **2017**, *418*, 1–12. [CrossRef]
14. Wilensky, U.; Rand, W. *An Introduction to Agent-Based Modeling Modeling Natural, Social, and Engineered Complex Systems with NetLogo*; MIT Press: Cambridge, MA, USA, 2015.
15. Baqueiro, O.; Wang, Y.J.; McBurney, P.; Coenen, F. Integrating Data Mining and Agent Based Modeling and Simulation. In *Computer Vision – ECCV 2012*; Springer Science and Business Media. LLC: Berlin/Heidelberg, Germany, 2009; Volume 5633, pp. 220–231.
16. Schnell, R. Enhancing Surveys with Objective Measurements and Observer Ratings. In *Improving Survey Methods. Lessons from Recent Research*; Engel, U., Jann, B., Lynn, P., Scherpenzeel, A., Sturgis, P., Eds.; Routledge: London, UK, 2015; pp. 288–302.
17. Salganik, M. *Bit by Bit: Social Research in the Digital Age*; Princeton University Press: Princeton, NJ, USA, 2019.
18. Yang, D.; Xue, G.; Fang, X.; Tang, J. Crowdsourcing to Smartphones: Incentive Mechanism Design for Mobile Phone Sensingin. In Proceedings of the ACM International Conference on Mobile Computing and Networking, Istanbul, Turkey, 22–26 August 2012.
19. Jin, H.; Su, L.; Chen, D.; Nahrstedt, K.; Xu, J. Quality of Information Aware Incentive Mechanisms for Mobile Crowd Sensing Systems. In Proceedings of the 16th ACM Symposium on Mobile Ad Hoc Networking and Computing (MobiHoc 2015), Hangzhou, China, 22–25 June 2015.
20. Cardone, G.; Foschini, L.; Bellavista, P.; Corradi, A.; Borcea, C.; Talasila, M.; Curtmola, R. Fostering ParticipAction in Smart Cities: A Geo-Social Crowdsensing Platform. *IEEE Commun. Mag.* **2013**, *6*, 112–119. [CrossRef]
21. Hamidi, H.; Kamankesh, A. An Approach to Intelligent Traffic Management System Using a Multi-agent System. *Int. J. ITS Res.* **2018**, *16*, 112–124. [CrossRef]
22. Wang, F.-Y. Agent-Based Control for Networked Traffic Management Systems. *IEEE Intell. Syst.* **2005**, *20*, 92–96. [CrossRef]

23. Gulyás, L.; Adamcsek, B. Charting the Market: Fundamental and Chartist Strategies in a Participatory Stock Market Experiment. In *EES 2004: Experiments in Economic Sciences - New Approaches to Solving Real-world Problems*; Springer: Berlin/Heidelberg, Germany, 2007.
24. Bosse, S.; Engel, U. Augmented Virtual Reality: Combining Crowd Sensing and Social Data Mining with Large-Scale Simulation Using Mobile Agents for Future Smart Cities. *Proc.* **2019**, *4*, 1. [CrossRef]
25. Calenda, T.; De Benedetti, M.; Messina, F.; Pappalardo, G.; Santoro, C. Mobile Crowd Sensing: Current State and Future Challenges. *IEEE Commun. Mag.* **2011**, *49*, 11.
26. Ganti, R.K.; Ye, F.; Lei, H. Mobile Crowd Sensing: Current State and Future Challenges. *IEEE Commun. Mag.* **2011**, *49*, 32–39. [CrossRef]
27. Leppänen, T.; Lacasia, J.Á.; Tobe, Y.; Sezaki, K.; Riekki, J. Mobile crowdsensing with mobile agents. *Auton. Agent Multi-Agent Syst.* **2017**, *31*, 1–35. [CrossRef]
28. McFarland, D.A.; Moody, J.; Diehl, D.; Smith, J.A.; Thomas, R.J. Network Ecology and Adolescent Social Structure. *Am. Soc. Rev.* **2014**, *6*, 1088–1121. [CrossRef] [PubMed]
29. Van Dam, K.H.; Lukszo, I.; Zofia, N. *Agent-Based Modelling of Socio-Technical Systems*; Springer: Berlin/Heidelberg, Germany, 2013.
30. Raykar, V.C.; Yu, S.; Zhao, L.H.; Valadez, G.H.; Florin, C.; Bogoni, L.; Moy, L. Learning From Crowds. *J. Mach. Learning Res.* **2010**, *11*, 1297–1322.
31. Bosse, S.; Pournaras, E. An Ubiquitous Multi-Agent Mobile Platform for Distributed Crowd Sensing and Social Mining, FiCloud. In Proceedings of the 5th International Conference on Future Internet of Things and Cloud, Prague, Czech Republic, 21–23 August 2017.
32. Al-Zinati, M.; Wenkstern, R. An agent-based self-organizing traffic environment for urban evacuations. In Proceedings of the The Sixteenth International Conference on Autonomous Agent and Multiagent Systems, Sao Paulo, Brazil, 8–12 May 2017.
33. Bosse, S. Smart Micro-scale Energy Management and Energy Distribution in Decentralized Self-Powered Networks Using Multi-Agent Systems. *Proc. 2018 Fed. Conf. on Comput. Sci.Inf. Syst.* **2018**, *15*, 203–213.
34. Bosse, S.; Lechleiter, A. A hybrid approach for Structural Monitoring with self-organizing multi-agent systems and inverse numerical methods in material-embedded sensor networks. *Mechatronics* **2016**, *34*, 12–37. [CrossRef]
35. Bosse, S. Mobile Multi-Agent Systems for the Internet-of-Things and Clouds using the JavaScript Agent Machine Platform and Machine Learning as a Service. In Proceedings of the IEEE 4th International Conference on Future Internet of Things and Cloud, Vienna, Austria, 22–24 August 2016. [CrossRef]
36. Medina, P.; Goles, E.; Zarama, R.; Rica, S. Self-Organized Societies: On the Sakoda Model of Social Interactions. *Complexity* **2017**, *2017*, 1–16. [CrossRef]
37. Xua, Y.; Choi, J. Spatial prediction with mobile sensor networks using Gaussian processes with built-in Gaussian Markov random fields. *Automatica* **2012**, *48*, 1735–1740. [CrossRef]
38. Guo, A.; Zhong, Y.; Zhang, W.; Haenggi, M. The Gauss–Poisson Process for Wireless Networks and the Benefits of Cooperation. *IEEE Trans. Commun.* **2016**, *64*, 1916–1929. [CrossRef]

Article

A Radio Channel Model for D2D Communications Blocked by Single Trees in Forest Environments

Imanol Picallo [1], Hicham Klaina [2], Peio Lopez-Iturri [1,3], Erik Aguirre [1,3], Mikel Celaya-Echarri [4], Leyre Azpilicueta [4], Alejandro Eguizábal [1], Francisco Falcone [1,3] and Ana Alejos [2,*]

1 Department of Electric, Electronic and Communication Engineering, Public University of Navarre, 31006 Pamplona, Navarra, Spain; imanol.picallo@unavarra.es (I.P.); peio.lopez@unavarra.es (P.L.-I.); erik.aguirre@unavarra.es (E.A.); alex.egui.2@gmail.com (A.E.); francisco.falcone@unavarra.es (F.F.)

2 Department of Signal theory and Communications, University of Vigo, 36310 Vigo, Spain; hklaina@uvigo.es

3 Institute for Smart Cities, Public University of Navarre, 31006 Pamplona, Navarra, Spain

4 School of Engineering and Sciences, Tecnologico de Monterrey, Monterrey 64849, Mexico; mikelcelaya@gmail.com (M.C.-E.); leyre.azpilicueta@tec.mx (L.A.)

* Correspondence: analejos@uvigo.es; Tel.: +34-986-811-949

Received: 29 July 2019; Accepted: 18 October 2019; Published: 23 October 2019

Abstract: In this paper we consider the D2D (Device-to-Device) communication taking place between Wireless Sensor Networks (WSN) elements operating in vegetation environments in order to achieve the radio channel characterization at 2.4 GHz, focusing on the radio links blocked by oak and pine trees modelled from specimens found in a real recreation area located within forest environments. In order to fit and validate a radio channel model for this type of scenarios, both measurements and simulations by means of an in-house developed 3D Ray Launching algorithm have been performed, offering as outcomes the path loss and multipath information of the scenarios under study for forest immersed isolated trees and non-isolated trees. The specific forests, composed of thick in-leaf trees, are called Orgi Forest and Chandebrito, located respectively in Navarre and Galicia, Spain. A geometrical and dielectric model of the trees were created and introduced in the simulation software. We concluded that the scattering produced by the tree can be divided into two zones with different dominant propagation mechanisms: an obstructed line of sight (OLoS) zone far from the tree fitting a log-distance model, and a diffraction zone around the edge of the tree. 2D planes of delay spread value are also presented which similarly reflects the proposed two-zone model.

Keywords: device-to-device; internet of things; wireless sensor networks; vegetation; ray launching; 5G; radio channel model; scattering; log-distance

---

1. Introduction

In the last few decades, Wireless Sensor Networks have become indubitably one of the most important technologies for both development and investigation in various fields as health care [1], agriculture [2], domestic monitoring [3], and means of land transport [4], among others. However, although in most of applications, numerous sensor nodes are distributed around a wide geographical area while the reliability of the wireless communication in terms of network coverage is a crucial requirement, some environments can be tough and make it hard to achieve the desired communication between nodes; thus resulting in a non-responsive application. Inhomogeneous vegetation environment is one of the most common cases that cause wireless communication complexities. Although WSN are expected to automatically monitor the ecological evolution and wildfire habits in forests [5,6], as over 400,000 wildfires have occurred in Spain over the last 30 years [7] also many rescue interventions are carried out to retrieve people lost in the woods using mobile wireless communication, inhomogeneous vegetation environments have the special feature of acting as scatterers of electromagnetic waves.

The signal scattering is translated into an excess of attenuation which can limit the performance of the Device-to-Device (D2D) communications envisaged at low power, high data rates and low latency expected for the upcoming 5G mobile wireless communications. D2D is expected to be used in almost every environment and scenario, and very especially in smart cells. The reduced coverage area and need for high data rate performance of this type of cell deployment can be limited by the attenuation due to vegetation.

In fact, the appearance of the trees in the path of the communication link between the transmitter and the receiver has significant effects on the quality of the received signal. This is because a forest is characterized by vegetation of different canopies and components with different physical natures in terms of trees height, leaves pattern and thickness, trunk sizes and number of branches, which determine the rate of attenuation of radio waves that propagate through it via scattering, absorption, refraction and diffraction of the waves. The signal scattering is translated into an excess of attenuation which can limit the performance of the Internet of Things (IoT) envisaged at high data rates and low latency expected for the upcoming 5G mobile wireless communications. In this context, radio planning tasks become necessary in order to assess the validity of future D2D communications operating in vegetation environments.

For that purpose, path loss models for scenarios with vegetation play a key role since they provide RF power estimations that allow an optimized design and performance of the wireless network [8]. Moreover, a path loss model may contribute to the evaluation of the maximum effective distance between adjacent terminals and hence to the estimation of the number of sensors needed to cover a certain area. Finally, the signal strength loss is related to the quality of service (QoS), causing unreliable communication between nodes that will increase both the number of data packet retransmissions and the power consumption of the nodes, causing radio link failure in last term. Therefore, there is a need for reliable through-vegetation radio channel modelling for vegetation environments, which will assess the propagation behavior in terms of both path loss and multipath propagation.

Since the 1960s, a significant amount of work has been done to investigate the characterization of the radio channel blocked by vegetation elements [9,10], which proposed different analytical and empirical models to estimate the power attenuation or level excess loss introduced by a signal blockage due to vegetation obstacles, mainly trees [9–23]. Among the conventional forest models, we can mention the COST 235 [11] and the Weissberger [12]. These models were developed to model the excess attenuation encountered in a forest beyond that predicted by either free-space or two-ray propagation. Moreover, the path loss was empirically modeled in [11,13–15]. In [16–18], path loss models for a specific 1.9 GHz radio frequency are presented. More characteristics such as the shadowing loss [19] and small-scale fading [19,20] for trees as well as tall food grass fields [21] were empirically modeled. In addition, a model for ultra-wideband (UWB) channels is presented in [22].

However, although a significant amount of research has been performed on the empirical propagation loss modeling, it is still a challenge to describe the radio wave propagation within the forest environment accurately. Especially in inhomogeneous vegetation environments, since a forest can be made up of mixed or homogeneous tree types resulting in different effects on radio waves even at the same frequency by the same group of trees, depending on the geometry of the link. A model combining path loss and multipath is rarely considered.

In this contribution, which is an extension of a conference paper [24], we present a simple model to characterize the attenuation due to the isolated trees in an air-to-air communication channel occurring between a static transmitter and a mobile user which moves linearly toward the tree. The developed channel model also considers parameters due to the multipath presence obtaining the value of the delay spread parameter.

Since an accurate modeling of the propagation of radio waves through tree foliage, generally requires accurate electromagnetic description of the tree geometry, including its branches and leaves, the radio characterization was performed by means of simulations based on 3D Ray Launching software, where the specific material parameters of the vegetation elements are considered, such as dielectric constant and conductivity. A geometrical and dielectric model of the trees were created and introduced in the simulation software for two species: pine and oak. The scenario simulated and measured at 2.4 GHz to determine the effects of signal level blockage consisted of a medium-size single oak tree immersed in a forest. The path loss was estimated as dependent of the radio link range. Simulation and measurement results corroborate the existence of a double zone within the path loss curve due to different dominant propagation mechanisms: scattering (near the tree) and line of sight obstruction (far from the tree). Furthermore, 2D planes of delay spread values obtained by simulation similarly reflect this two-zone propagation model.

The measurements were carried out using devices operating at the band of 2.4 GHz. For the isolated tree case, a specific forest (called Chandebrito, situated in Galicia, Spain) composed of different type of trees was the chosen scenario, since it suffered a fire and some isolated trees can be found there nowadays. On the other hand, the Orgi Forest, located in Navarra, Spain, was the selected scenario for the analysis of non-isolated tree case.

The paper is organized as follows. In Section 2, the radio channel characterization for a D2D communications blocked by an isolated single tree is presented. Both simulations by the in-house developed 3D Ray Launching algorithm and measurements in a real forest environment are shown. In Section 3, the same approach is used for the radio channel characterization when the blocking tree is a non-isolated single tree, i.e., when dense vegetation is present in the surroundings. Section 4, Discussion, closes the paper.

2. Isolated Single Tree Radio Channel Characterization

In this section, the radio channel characterization at 2.4 GHz when an isolated single tree is blocking the communication path is presented. First, wireless channel estimation methodology based on deterministic 3D Ray Launching algorithm is described and employed. For that purpose, a simulation scenario is implemented, including dimensions and electromagnetic frequency dispersive properties of the employed materials. Then, measurements in a real scenario have been performed in order to complete the obtained simulation-based radio channel model.

2.1. Simulation Software

In order to perform wireless channel estimation, in which multiple elements such as vegetation are considered, different approaches can be employed. These can be based on analytical models, usually employing first degree approximations to simplify them, or empirical/semi-empirical models, which rely on measurement-based regressions.

Theoretical or analytical approaches are based on statistical theory and usually provide accurate propagation phenomena predictions [25–28]. Two different types of theoretical models can be distinguished in the literature, namely the radiative energy transfer (RET) model [29] and the analytical theory approach by Foldy [30]. These two models have successfully been used to simulate radio wave propagation in vegetation environments but, due to the complexity of the mathematical equations on which they are based, and the difficulty to extract the input parameters of the models, such as density, area or thickness of leaves and branches, they tend to be unaffordable for real-sized scenarios, such as forests [31].

On the other hand, empirical techniques are based on extensive measurement campaigns in the considered environment [32–36]. They have successfully characterized path loss attenuation impact in diverse scenarios with inhomogeneous vegetation within them, and at different frequency bands. However, their main drawback is their lack of accuracy when different site-specific environments are analyzed. As an example, the work in [37] verifies that the well-known propagation empirical models can lead to an error percentage of 30% in propagation predicted values for a classic tomato greenhouse.

To overcome these limitations, ray-tracing based methods have been proposed in the literature. In [38], a propagation model based on uniform theory of diffraction (UTD) has been presented for urban environments, considering the impact of propagation over buildings and the vegetation attenuation and scattering. In [39], another propagation model based on geometrical optics (GO) is presented to assess the scattered field caused by vegetation elements in the radio path. Methods based on GO such as ray tracing or ray launching achieve a good trade-off between simulations accuracy and computational cost [40]. Due to this fact, in this work, an in-house developed 3D Ray Launching (3D-RL) algorithm has been used to characterize inhomogeneous vegetation environments. The 3D-RL model is a deterministic technique based on GO and UTD and it is divided in three main steps:

1. The first step consists in the design and creation of a realistic scenario, considering all the obstacles and scatterers within it. All the geometries and dimensions of all the objects within the scenario are taken into account, which lead to a realistic scenario which consider every detail of the environment.
2. The second step is the simulation procedure, in which the physical wavefront is approximated by a set of rays, which travel from the emitter in the same direction as the corresponding wave vector. The rays are launched following a solid angle, with angular resolution in both horizontal and vertical planes. These rays interact with the surrounding media following Fresnel equations, considering the frequency dispersive characteristics of the dielectric constant and conductivity of the corresponding materials of all the elements within the implemented scenario.
3. The last step deals with data processing from the complete 3D simulation volume, in order to extract power characterization as well as time domain characterization of the corresponding system within the simulation scenario.

The ad-hoc 3D ray launching algorithm has been described in detail in [40] and validation in complex environments including vegetation has been described in [41]. In the algorithm, the consideration that the vegetation medium is homogeneous has been assumed, thus it has been treated as an isotropic dielectric material with constant permittivity and conductivity. The 3D RL algorithm considers ray/object interaction in terms of reflection, refraction and diffraction. In particular, diffraction phenomena can be optionally activated in the simulation parameters, which is the case of the simulation results presented in this work.

2.2. Model of Tree

For the presented analysis, the radio propagation through both pines (*Pinus pinaster*) and oaks (*Quercus robur*) has been assessed and specific models have been created in order for them to be considered in the 3D RL simulation code. The structure of oak tree is complex to make a 3D model with RL. Therefore, an approach has been made to respect the main shape of the tree, modelling the mass of leaves as a cube. Usually, three fifths of the height of the tree is mass of leaves and two fifths trunk, although this will depend on the particular oak tree. The oak tree simulation model implemented is composed by a solid trunk and a homogeneous leaf canopy, as depicted in Figure 1a. In the case of the pine simulation model, homogeneous branches with air gaps have been considered. The pine tree branches have a cone-shaped structure with superposition of greater to less length layers starting from the branches closer to the ground (see Figure 1b). The size of the branches depends on the particular pine tree. It is worth noting that both the widths and the heights of the tree models have been parameterized in order to create a computational tree model as closer to the real trees. Moreover, the forest could have a strong influence on the propagation of radio waves. The substrates generally are basic, and rich in acids. In addition, they are usually wet and with accumulation of little leaves and flora remains. Therefore, an appropriate ground has been modelled and included for simulations, since it is an inherent part of a real scenario, which also generates reflections on the propagating electromagnetic wave. The implemented oak and pine tree models are depicted in Figure 1, whereas the corresponding simulation parameters are detailed in Table 1, obtained from [42].

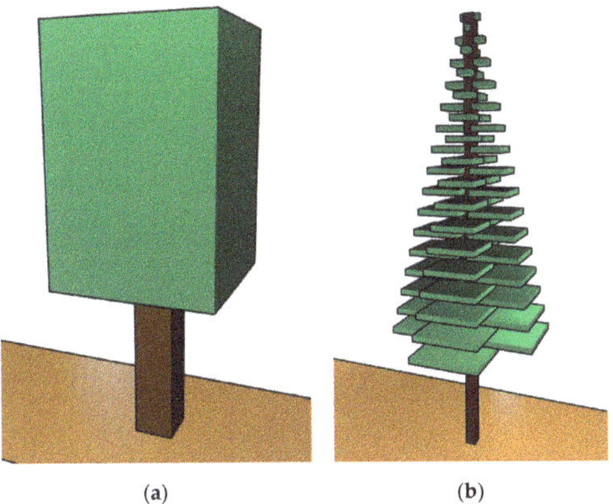

(a) (b)

Figure 1. Created tree models for the 3D Ray Launching simulation tool: (**a**) Oak tree; (**b**) Pine tree.

Table 1. Material properties for Ray Launching simulations (at 2.4 GHz).

Material	ε_r	Conductivity (S/m)
Foliage	4.48	0.02
Pine trunk	1.97	0.052
Oak trunk	2.51	0.148
Forest Ground	4.8	0.98

2.3. Simulation Results

In order to perform wireless channel characterization considering blockage, the implemented simulation scenario considers the location of a static transmitter and a mobile terminal with a thick tree in the propagation path, as depicted in Figure 2. The transmitter source (TX), operating at 2.4 GHz is located at a height of 2 m and is highlighted by a red circle. Material properties and simulation parameters are detailed in Tables 1 and 2, respectively. The simulation scenario has been implemented with boundary conditions defined by air in order to avoid unwanted reflection components. Note that the parameters regarding the radio signal communications (transmitted power level, antenna type and operation frequency) have been chosen in order to fit the parameters used by the real devices. It is important to mention that the 3D-RL simulation tool has the option to include the effect of diffraction phenomenon. For the current analysis, the diffraction has been activated since it is a relevant phenomenon in the proposed case under study. Simulation results have been obtained for the complete scenario volume. For the sake of clarity, as a specific example, propagation losses for the linear TX-RX radials defined by the yellow dashed lines depicted in Figure 2b,d have been considered for height cut planes of 1, 2 and 3 m.

Table 2. Ray Launching simulation parameters.

Parameters	Values
Transmitted Power	10 dBm
Operation Frequency	2.4 GHz
Antenna Type/Gain	Monopole/0 dB
Launched rays resolution	1 degree
Permitted maximum rebounds	6
Cuboids size (Mesh resolution)	10 cm × 10 cm × 10 cm
Diffraction phenomenon	Activated

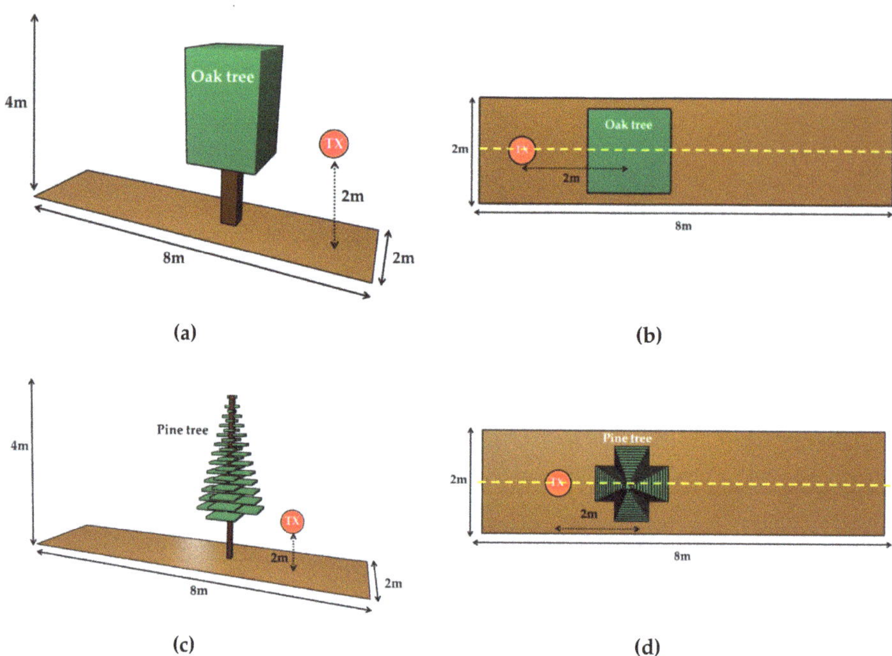

Figure 2. Created scenario for the 3D Ray Launching simulations. (**a**) General view with the oak tree; (**b**) Upper view with the oak tree; (**c**) General view with the pine tree; (**d**) Upper view with the pine tree.

2.3.1. Path Loss Model

Once the simulation results have been obtained for the scenario, path loss was analyzed considering relative mobile receiver displacement. Results have been obtained considering both oak and pine tree models, for an operating frequency of 2.4 GHz and with increasing distance from the tree location. From the results it can be observed that field scattering owing to the tree location can be divided in two zones: a diffraction dominant zone within the tree vicinity and a free-space zone, depicted in Figure 3. Consequently, at larger distances relative to the tree location, power decay follows free space conditions, owing to dominant line-of-sight (LOS) condition. When the distance from the receiver to the tree location is decreased, the observed response in terms of power decay corresponds to dominant multipath or scattering conditions. In this case, a linear variation of opposite trend to the free-space in received signal is observed, corresponding to a diffraction dominant zone in which relevant tree blockage attenuation recovers, subsequently following a free-space component trend.

The results obtained follow a similar trend as those described in [14], in which a scenario with low elevation is presented. In that case, the propagation path corresponds to an air-to-ground radio link with blockage owing to an isolated tree. The transmitter is located over both the tree and a ground mobile receiver, considering several tree species at frequencies within the X band (8–12 GHz) and Ku band (12–18 GHz). The results presented in [14] identified an OLoS region as well as two different scattering zones: a diffuse scattering-dominant region within the tree trunk vicinity, in which only a statistical distribution function model was followed by the signal level; and a colliding region in which prevalence of tree crown diffraction is observed, modelled considering knife-edge diffraction loss with a correction of the tree height. Beyond this second scattering zone, the signal recovers the power decay corresponding to the OLoS model.

In the presented model, the region within the vicinity of the tree canopy corresponding to diffuse scattering phenomena has not been clearly identified. For the oak tree, this may be due to the absence of air gaps in the simulation model adopted for the mass of leaves which would turn the propagation media into a multi dispersive material. In the case of the pine tree, the simulation model already includes a sufficient volume of air gaps offering a more realistic approach. An experimentally derived model based on anechoic chamber measurements is presented in [14]. For actual trees the mass of leaves is not homogeneous as in the simulation. A diffuse scattering is created, given by random nature of fading owing to interaction with the leaves. It is worth noting that signal attenuation as observed in Figure 3a, corresponding to the oak tree case, strongly decreases, which is in principle given by the plausible existence of a diffuse scattering zone.

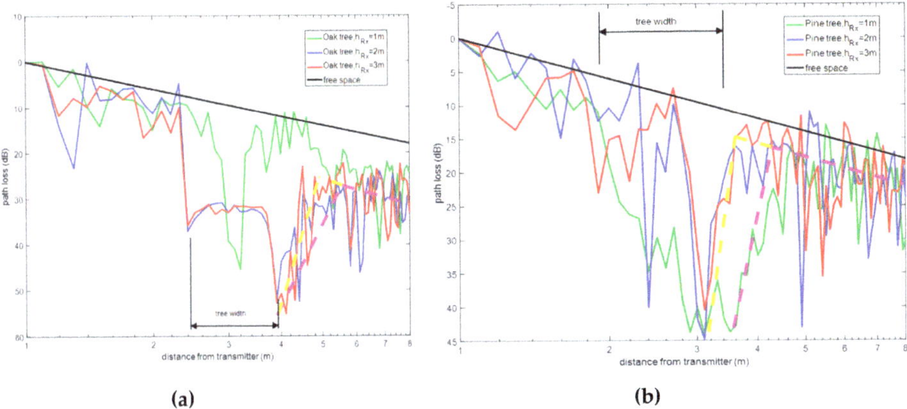

Figure 3. Simulation path loss for a single tree, (**a**) Oak tree; (**b**) Pine tree.

Experimental measurements later shown in Section 2.4 have corroborated this fact. Figure 3b depicts the results considering the pine, which are similar with lower definition, given mainly by lower attenuation considering air gaps in the pine model vs. the homogeneous model employed in the case of the oak tree.

2.3.2. Delay Spread

One of the characteristics of the deterministic 3D RL algorithm is the capability to provide multipath characterization, a feature seldom considered in previous models. In Figure 4 it is presented the simulation results of the Delay Spread (DS) as seen at different 2D height planes, for the single tree scenario simulated in Section 3 with the pine and oak tree models. A 2D plane of DS indicates the values of DS detected at each point of a plane located at a specific height, assuming a static transmitter positioned as in the simulations of Section 3, at a height of 2m. DS values depend on the tree species, the distance of the observation point to the transmitter, the percentage of signal blockage, and the observation 2D height plane. A range between 2 and 12 ns was obtained for the simulated scenarios. It is important to note that in the presented results, the RF power level estimations within obstacles (trunk and mainly the vegetation part of the oak) have not been represented in order to show clearer the radio propagation zones around the trees.

For instance, at height 1 m, the points of the observation plane located after the tree show low-medium values of DS: 6–8 ns for the pine and 6–10 ns for the oak. This situation is due to the tree trunk, which is the most influent part in signal blockage.

However, for the pine case at 3 m height plane, the signal blockage is weaker due to the many air gaps of the pine tree top. Therefore, while pine tree behaves like a low density medium, the oak tree shows the behavior of a homogeneous medium. Generally, the obtained results are in accordance with [43,44]. Specifically, [43] presents UHF radio propagation estimations through a trunk-dominated area, for distances greater than 100 m. The work presented in [44] is closer to the case analyzed in this paper: the authors compute the RMS delay spread of the measured power delay profiles at 5 GHz band along different trails within different forests. They obtained delay spread mean values between 60 ns and 90 ns depending on the trail (and therefore, the distance). These values agree with our simulation results, which correspond to shorter distances (less than 8 m in any case).

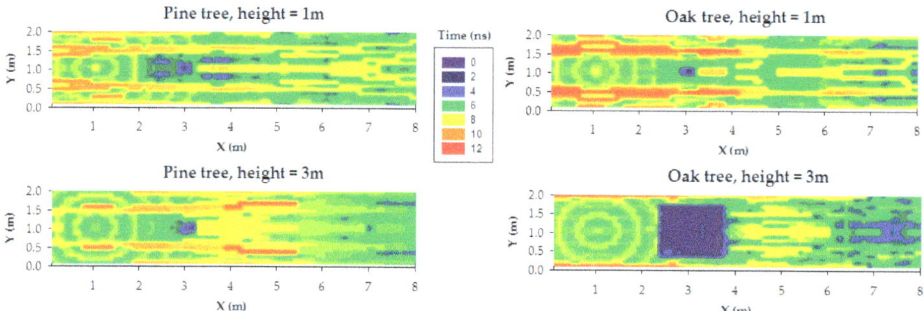

Figure 4. Simulation results of the Delay Spread, for both pine tree (**left**) and oak tree (**right**).

Similarly to the results in [45], it is noticed that, for any 2D height plane, the presence of vegetation contributed to the received signal level enhancement, predominantly in the side- and back-scattering regions, contrasted with the significant attenuation caused by absorption and scattering in the forward-scattering region produced after the tree [45].

Furthermore, delay spread values achieved along a linear path reflect a two-zone model similarly to path loss in Section 2.1: in the diffraction predominant zone, the number of signal components coming from scattering inside the tree is larger, and once mostly-coherently added produce larger values of DS; in the OLoS zone, away from the tree, the number of multipath components decrease considerably.

2.4. Experimental Measurements

In order to complete the simulation outcomes for path loss, measurements were carried out in a forest and one medium size specimen of oak was chosen. The measurements were carried out using devices operating at the band of 2.4 GHz. The specific forest, composed of different type of trees, is called Chandebrito and is situated in Galicia, Spain. As shown in Figure 5a, the chosen tree's dimensions are larger in height and width than the model used for simulations: 5.5 m tall and 4.25 m wide. The larger width compensates the lesser homogeneity of the canopy that presents considerably more air gaps than in the simulated model becoming a polidispersive medium [46]. Figure 5b shows the surroundings of the chosen isolated tree, which is at least around 10 m away from any other tree.

The transmitter consisted of a programmable signal generator (WindFreak SynthHD) connected to a directional log-periodic antenna with vertical polarization, model Electro-Metrics EM6952, via a low-loss coaxial cable. The antenna radiation pattern shows 70° of azimuth (E-plane) and 125° of elevation (H-plane). The frequency of the transmitted tone was 2.4 GHz, and its power was 0 dBm. The antenna gain at this frequency is 5.5 dBi.

For acquiring the received signal we used a spectrum analyzer (SA) Rohde Schwarz FSH-6 that registered the RF signal via an antenna identical to the one used in transmission. No amplifiers or filters were used in transmission or reception.

The transmitter antenna was placed on a tripod at a height of 2m, in a fix location at a distance of 3.5 m from the tree trunk base. The receiver antenna was also fixed on the top of a tripod at 2 and 3 meters high, moving along a radial from the tree trunk base to 4.6m apart. Both the transmitter and receiver antennas pointed to the tree canopy.

(a) (b)

Figure 5. (**a**) Oak tree specimen chosen for experimental measurements. (**b**) Surroundings of the isolated oak tree.

The measurement was done moving the tripod with the receiver antenna along a linear path from near (0.25 m) to far (4.6 m) the tree. The initial distance to the tree is 0.25 m. This initial distance was increased in steps of 0.10 m (>λ/4 increments) until the final position at 4.6 m from the tree was reached. At each position of the receiver antenna along the linear path the SA was configured to acquire a data trace of 301 power samples at 2.4 GHz, with a resolution bandwidth of 1 KHz and zero span, in order to average temporal power variations. Measurements were completed for receiver antenna heights of 2 m and 3 m given that in these cases the influence of the tree canopy is what creates the situation of a double propagation zone. A picture of the employed equipment is shown in Figure 6.

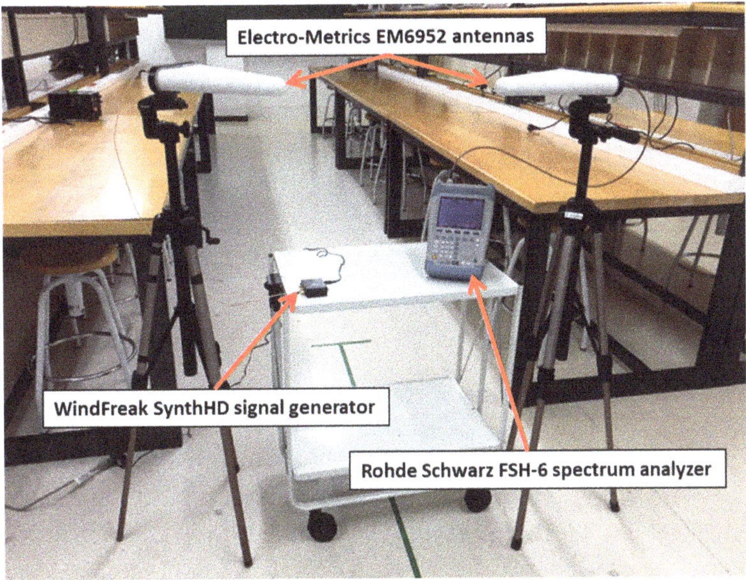

Figure 6. Employed equipment for the measurement campaigns.

2.4.1. Path Loss

The SA measured values were processed. First, an averaging is applied to the 301 power samples composing each trace. Later all the data traces were normalized with respect to the power received at a reference distance of 1m from the transmitter, for each receiver height. In the case of the experimental data, the antenna gains are then compensated. The result is a set of 59 points that determines the signal level loss as a function of the distance between the tree and the receiver.

The measured path loss is compared to simulated data in Figure 7 (corresponding datasets are provided as supplementary material). Only the portion of the distance axis corresponding to the forwarding zone (after the tree) matches for both datasets. The experimental outcomes corroborate the two-zone propagation model. The difference in dimensions between the virtual tree and the actual tree are primarily reflected in the lower power level received. Moreover, the radiation diagrams, and the power gains of the antennas used differ: omnidirectional in the virtual case and directive in the measurements. These and other factors are responsible for dissimilarities observed between the simulated and the experimental datasets. The not averaged multipath contributions observed in the simulation curves in Figure 7 are likely due to the lack of temporal averaging in the simulated data. The differences in the antenna radiation diagrams of the virtual and measurement cases can also explain the fit discrepancy. The following values of RMSE are observed between the experimental and simulated curves according to the antenna height and the propagation zone:

- For diffuse zone: RMSE = 8.6370 dB at 2 m, and RMSE = 5.2804 dB at 3 m.
- For OLoS zone: RMSE = 5.8596 dB at 2 m, and RMSE = 6.2672 dB at 3 m.

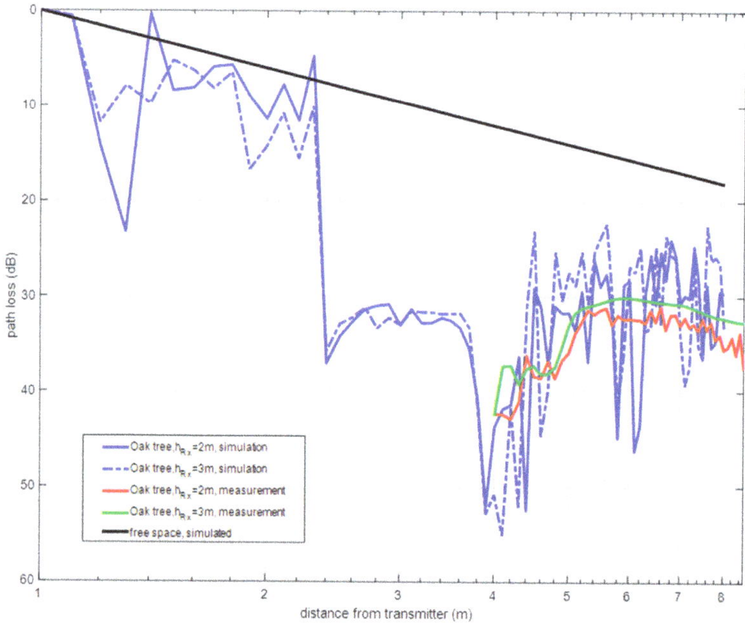

Figure 7. Comparison between simulation and measurement path loss for an isolated oak tree.

The diffuse scattering zone, located between 4m and around 5.5 m of the virtual distance axis, presents a less pronounced slope for both receiver heights. The slope of the OLoS zone decays faster than in the simulated scenario case.

Finally, the averaging introduced in the SA acquisition and processing provided lower signal variability than in the simulation results. It is not possible to introduce averaging in the electric field results estimated by the simulation; however, in a simulator, the scenario is static and averaging does not seem necessary. In the measurement scenario, e.g., the wind blowing produces non trivial signal level variations that the averaging filters. The result is that the experimental curves are smoother than the virtual ones. Spatial averaging was not applied; however, it is recommended for windy scenarios or irregularly shaped trees.

Typically, the path loss variation as a function of distance is modeled by means of a log-distance or a linear regression. In Figure 8a the results of the model fitting are shown for the measured values, and in Figure 8b for the simulation case. It is noticed that depending upon the radio propagation zone a fit model is more suitable than any other, even despite the receiver height.

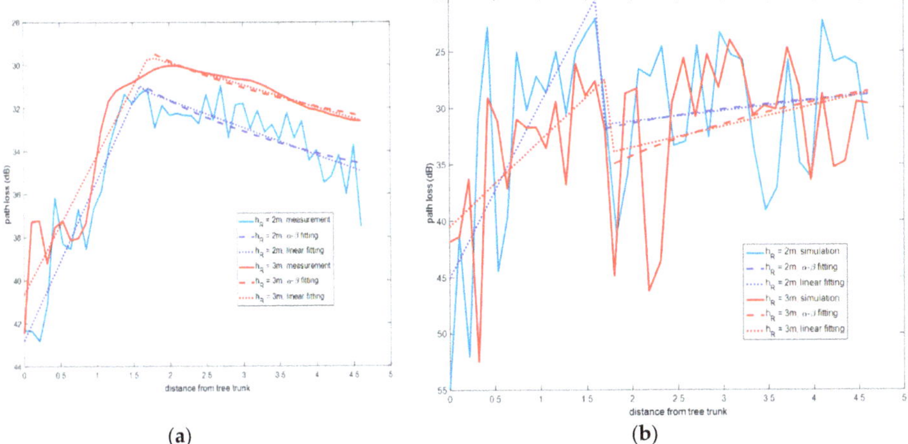

Figure 8. Comparison of path loss curves fitting for an isolated oak tree: (**a**) experimental results and (**b**) simulation-based results.

The experimental and simulated path loss curves resulted in the following fitting:

1. Diffuse scattering zone: it occurs near the tree, between 0.25 m and 1.6 m from the tree for 2 m height, and up to 1.8 m for 3 m case. In this zone only the fitting could be obtained by linear regression that meets the expression given in Equation (1):

$$PL(d) = P_0 + n{\cdot}d \tag{1}$$

where P_0 is the reference power loss in dB at the start distance, 0.25 meter from the tree; d is the distance in meters between the start distance and receiver; n is the factor that determines the power decay rate with the distance. Parameters P_0 and n in Equation (1) have been obtained for each one of the two receiver heights.

2. OLoS zone: it occurs for a distance from the tree beyond 1.6 m for 2 m height, and beyond 1.8 m for 3 m case. In this zone the path loss admits a linear regression according to the floating intercept (FI) model Equation (2) applied in [15,20]:

$$PL(d) = \alpha + 10{\cdot}\beta{\cdot}\log_{10}(d) \tag{2}$$

where d is the distance, β the line slope, and α the floating-intercept in dB. The slopes β are −0.818 for 2 m and −0.7047 for 3 m, and the fit errors are 1.053 and 0.2681 dB, respectively. A linear regression was tried with smaller fit errors than the FI model.

As noticed in Figure 8a,b, the division into two propagation zones is more remarkable for the experimental data. The values of the parameters α, β, P_0 and n obtained for each zone and height are summarized in Table 3. The RMS error was used as an estimation of the error of the models.

Table 3. Path loss fitting parameters for measurement and simulation results.

	Height	Zone	Linear	FI model	RMSE
Measured	2 m	Diffuse	$P_0 = -42.81, n = 7.412$	—	1.491
		OLoS	$P_0 = -29.07, n = -1.275$	$\alpha = -29.18, \beta = -0.818$	0.9687/1.053
	3 m	Diffuse	$P_0 = -40.64, n = 6.413$	—	1.329
		OLoS	$P_0 = -27.81, n = -1.037$	$\alpha = -27.68, \beta = -0.7047$	0.188/0.2681
Simulation	2 m	Diffuse	$P_0 = -45.02, n = 15.45$	—	7.256
		OLoS	$P_0 = -33, n = 0.9304$	$\alpha = -33.41, \beta = 0.6963$	5.565/5.552
	3 m	Diffuse	$P_0 = -40.50, n = 7.67$	—	5.263
		OLoS	$P_0 = -37.18, n = 1.863$	$\alpha = -38.95, \beta = 1.582$	6.201/6.096

2.4.2. Signal to Noise Ratio

Following the procedure described in [15], it is possible to estimate the coverage distance for an IEEE 802.15.4 connection if the Signal to Noise Ratio (SNR) is obtained as a function of distance. As described in [15], for a 22 byte frame length and a PER of 2%, then a BER < 1.14×10^{-4} would be needed. According to [15] and Figure 9 the SNR required is 0 dB approximately. The noise power is measured using the SA, at a central frequency of 2.45 GHz and for a span of 5 MHz that simulates a 5 MHz bandwidth ZigBee channel. The value measured was −82 dBm.

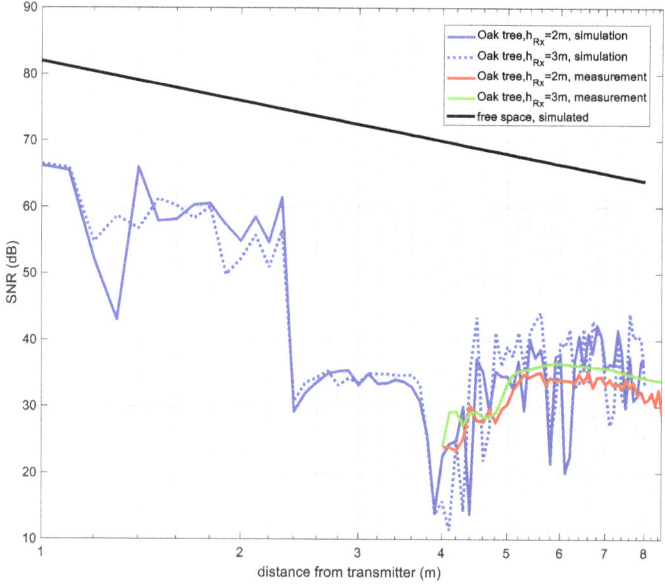

Figure 9. Signal to Noise Ratio (SNR) vs. distance to transmitter for a ZigBee device.

The resulting SNR is shown in Figure 9. It can be deduced that for a ZigBee system with a threshold of SNR = 0 dB connectivity is possible in the vicinity of the tree. The connectivity is kept for both simulated and experimental cases, even if 2-5.5 dB are subtracted to compensate the antenna gain used in the measurements. This result agrees with [15,20].

This result reinforces the conclusion given in Section 2.3.2 in which it was indicated that the presence of the tree, thanks to the diffraction phenomenon, reinforced the signal reached in certain points around the tree, as also stated in [45].

3. Non-Isolated Single Tree Radio Channel Characterization

Once the radio channel characterization of an isolated single tree has been performed by means of measurements and simulations, in this section a step towards a more realistic scenario has been taken. Specifically, simulations and measurements of a single tree nearby a dense forest zone with thick in-leaf trees have been performed. This scenario is located in the Orgi Forest. This forest is a 77-hectares millennial forest with a high ecological value which extends into the Ultzama Valley (Navarra). The oak groves populated the Navarra valleys 4000 years ago. Orgi Forest was declared a Natural Recreational Area in 1996, after the intense forest exploitation of wood, pasture for livestock and hunting. This was to encourage the natural regeneration of forest´s flora and fauna, while regulating the use of people.

Orgi has been included in the European Natura 2000 network. Natura 2000 is the largest network of protected areas in the world and it offers protection to most valuable and threatened species in Europe. In the case of Orgi, the oak (Quercus robur) is the main species protected. There are oak trees of two types where many of them are centenarians: the common oak and the American oak.

The forest is organized into three zones (shown in Figure 10). The first called Arigartzeta, is the welcome zone that gives access for the oak grove. It has parking and picnic areas. The second zone is called Tomaszelaieta, and has a large walk area, where the visitor can discover other tree species such as holly and elms as well as some animal species such as birds or amphibians. The third zone is called Muñagorri, which is a conservation area. The public cannot access this last zone due to restrictions in order to facilitate the process of natural regeneration. The measurements have been performed within the accessible Tomaszelaieta zone (see zoomed zone in Figure 10), which is prepared for visitors (see Figure 11).

3.1. D Ray Launching Simulations

The specific scenario under analysis is shown in Figure 12a, and it corresponds to the zoomed area shown in Figure 10 (i.e., measurements zone). The scenario has been created for simulations by the 3D Ray Launching tool. All the materials as well as the real distances and sizes of the surrounding trees have been taken into account in order to obtain accurate estimations of the RF power distribution and therefore, the path loss. Both the material properties and simulation parameters are the same as those used for the isolated tree case, and can be seen in Tables 1 and 2, except the transmitted power, which in this case was 8.35 dBm.

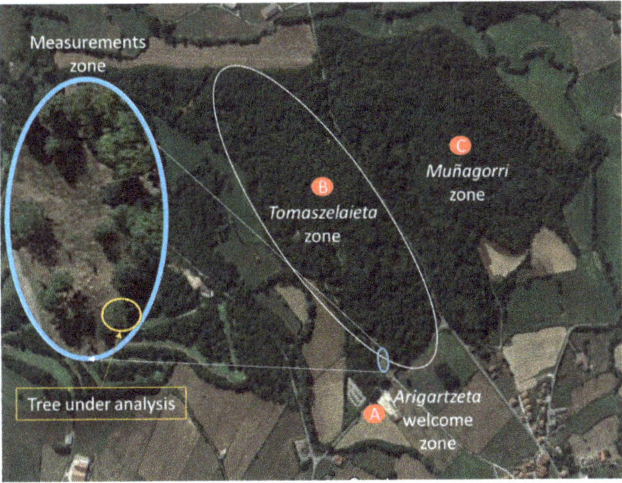

Figure 10. Location of the Orgi Forest and its different zones, including the measurement zone (zoomed). Extracted from Google maps.

Figure 11. Picture of *Tomaszelaieta* zone within the Orgi Forest.

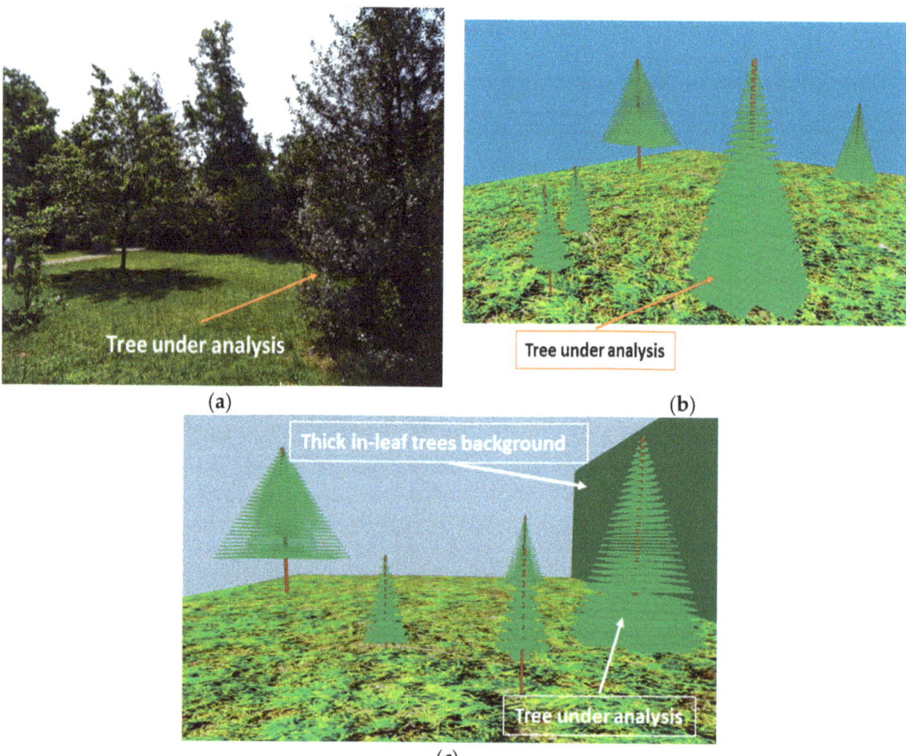

Figure 12. (**a**) Picture of the measurements zone. (**b**) The scenario created for the simulations. (**c**) Detail of the position of the dense vegetation wall.

Figure 12b shows the schematic view of the created scenario for simulation purposes. As can be seen, the morphology and size of each of the single trees surrounding the tree under analysis have been taken into account. In the same way, a specific wall has been created in the scenario in order to model the thick in-leaf trees background that exists behind the tree under analysis (the wall in green in Figure 12c). This wall has been defined with the material Foliage (see Table 1), while the rest have been defined as air.

3.2. Experimental Analysis

In this case of non-isolated tree, measurements have been also performed in the real scenario. The transmitter has been placed at 2.5 m distance from the tree under analysis at a height of 1 m, supported by a plastic structure. The tree is 6 m height and 3.7 m of maximum width (see Figure 13). Measurements have been taken in a linear path (represented by a white dashed line in Figure 14) from the transmitter (red dot in Figure 14) to a distance of 15 m. The measured path loss values are depicted in Figure 15, where the simulation results have also been included in order to comparing them. As can be seen, in this non-isolated tree case, the simulator provides very accurate estimations. This good agreement is due to the multipath propagation, which starts gaining relevance because of the dense vegetation present in the nearer area of the tree under analysis.

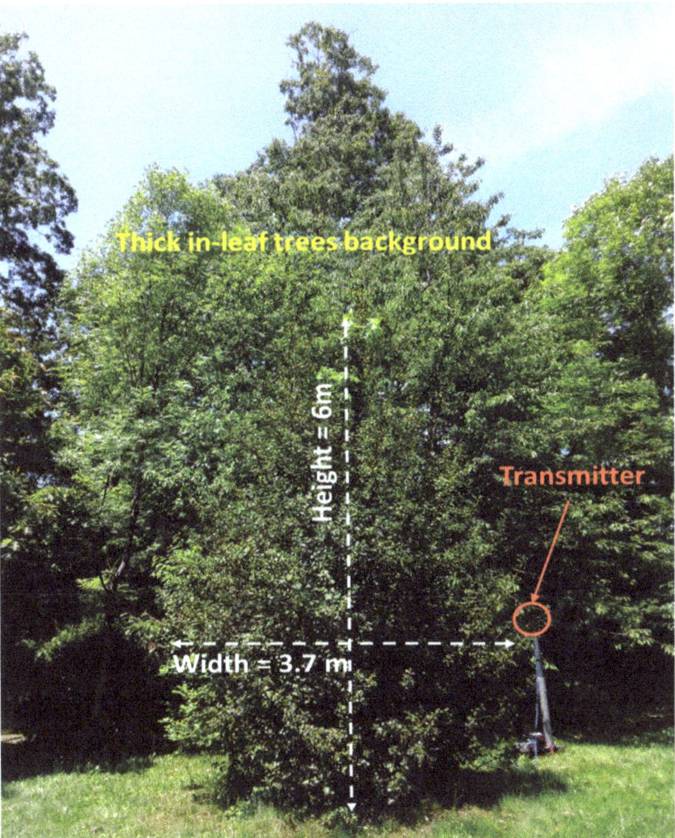

Figure 13. Detail of the Non-Isolated tree under analysis with the dense forest background.

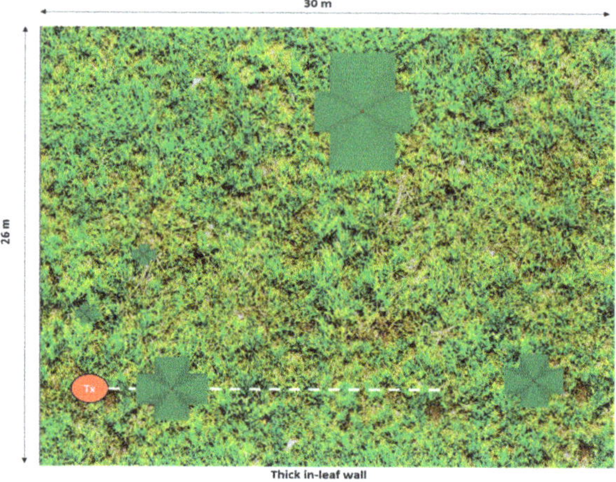

Figure 14. Upper view of the created scenario. The white dashed line represents the linear path where the measurements have been taken.

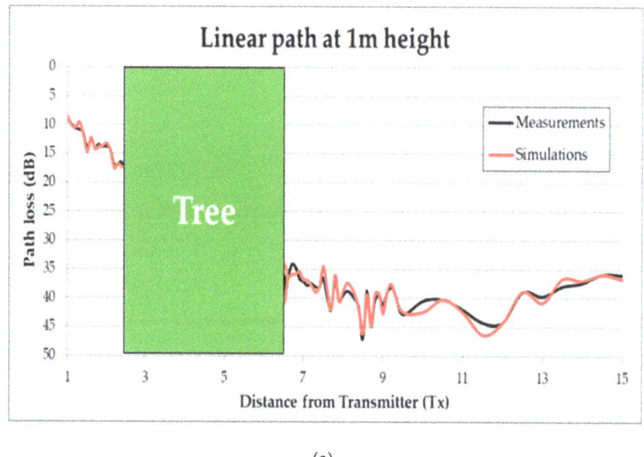

(**a**)

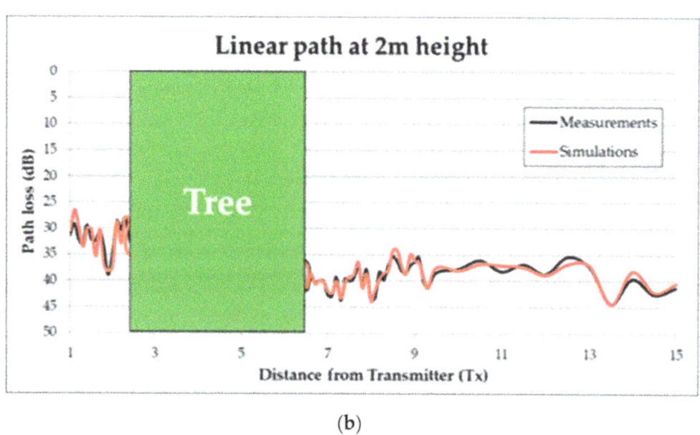

(**b**)

Figure 15. Measured vs. Simulated Path loss for the linear path corresponding to the white dashed line of Figure 14. (**a**) at 1m height; (**b**) at 2 m height.

4. Discussion

Observing the values in Table 3, the rate of decay with distance, n, appears to be smaller than the value for open scenarios under free-space conditions that is around 2. This indicates the re-radiation effect probably due to the tree. The fitting of the simulation data for the 3 m case exhibits the decay rate value closest to the free-space condition, 1.863.

The parameters values summarized in Table 3, for linear or log-distance models are not in agreement with values reported in [14,15,20]. Reasons for this dissimilarity are found, among others, in the frequency, the scenario, and the radio link setup. Largest RMSE values are observed for the simulation results likely due to the lack of temporal averaging in the simulated data.

In [14], the transmitter antenna points to the top of the tree, and the receiver antenna is pointed upwards, so that the set simulates a transmitter – satellite or helicopter – located over the tree and the receiver. The frequency band analyzed was 8–12 GHz, so that the signal loss is stronger than for the 2.4 GHz: As explained in Section 3, this setup led to three-zone propagation.

In [15,20], the peer-to-peer transmission is analyzed, without blocking the radio link by a tree, only locating the transmitter on the tree trunk. Using a log-distance model, reported values of the decay rate n are larger than 2. In [15,20], a log-distance model is also applied for the radio link blocked by an undefined number of trunks, but not canopies.

In summary, with the exception of [14], most of the literature related to propagation in vegetation media does not reflect a scenario as described herein. Only the RET model [29] could be used to try to replicate the results presented here. However, the difficulty of programming this model, given the high number of parameters to be evaluated, has made this task unbearable.

For our study of an isolated tree, it is reasonable to consider that any signal fading analysis should not depend on the distribution of the foliage in the species or specimen of the tree considered as would occur for the case of a thin tree. It is for these cases that up-to-date prediction models may not fit the experimental results. For a case of an isolated tree as a vegetation obstacle, some models as ITU-R P.833-7 [13] consider the signal excess loss produced by the diffraction occurring on the obstacle top as well as on the left and right sides of the tree, and these diffractions are modeled according to a knife-edge case.

Our objective is not to offer a model for the components due to the diffraction or scattering phenomena. For modeling those components, a more exhaustive measurement campaign should be carried out, and a wideband channel sounder would be needed to provide power and delay metrics. In our present contribution we provide a path loss model that results sufficient to determine the quality of a D2D link in terms of probability error.

Regarding the results presented in this paper, a diffraction dominant zone in the vicinity of the tree has been identified. Near the tree the channel model corresponds to a multipath response or scattering situation. This region would be dominated by diffraction that is responsible for producing that the large attenuation introduced by the tree blockage starts recovering and continue as an open-space component.

In summary, compared to previous works, in this work experimental results achieved under anechoic chamber have been now demonstrated for outdoors. In addition, it has been also demonstrated that the neighborhood of trees affects the diffuse zone's positive effect on the received power level. This helps to understand the effect of the single trees on the SNR and the achieved results could find a practical applicability when deploying a network in forest areas.

As previously said, further research is required to deepening in the phenomenon of the propagation in vegetation media, and specifically, more experimental measurements should be carried out at different locations to complete the statistical analysis as well as the results presented in this work. Possible future work would be to analyze the effect of line of trees blocking the wireless propagation.

It is worth noting that the use of the presented 3D Ray Launching simulation tool gains importance when the scenario under analysis becomes more complex in terms of morphology and the number of elements/obstacles/vegetation present within it, as it can be seen in presented results. Thus, obtaining accurate path loss estimations will lead to optimized radio planning decisions in scenarios with dense vegetation such as parks and forests.

Supplementary Materials: The following are available online at http://www.mdpi.com/1424-8220/19/21/4606/s1, Data set DS_sim and DS_meas corresponding to Figure 7: Comparison between simulation and measurement path loss for an isolated oak tree.

Author Contributions: Conceptualization, A.A.; Methodology, A.A., P.L.-I. and F.F.; Software and Simulations, P.L.-I., I.P., A.E., M.C.-E. and L.A.; writing—original draft preparation, A.A., L.A., I.P. and P.L.-I.; writing—review and editing, A.A., P.L.-I. and F.F.; Visualization, E.A., I.P., H.K. and A.A.; measurements, A.A., H.K.; Supervision, A.A.

Funding: This research was funded by Xunta de Galicia under grant ED431C-2019/26, Spanish Government under grant TEC2017-85529-C03-3R, AtlantTIC Research Center and project RTI2018-095499-B-C31, Funded by Ministerio de Ciencia, Innovación y Universidades, Gobierno de España (MCIU/AEI/FEDER, UE).

Acknowledgments: We would like to recognize the work, courage and sacrifice of the professionals and volunteers who worked hard against the terrifying fire that in October 2017 devastated the forest to which the tree of our study belonged. In the present, wonderful actions struggle to bring back the forest.

Conflicts of Interest: The authors declare no conflict of interest. The funders had no role in the design of the study; in the collection, analyses, or interpretation of data; in the writing of the manuscript, or in the decision to publish the results.

References

1. Lopez-Iturri, P.; Aguirre, E.; Trigo, J.D.; Astrain, J.J.; Azpilicueta, L.; Serrano, L.; Villadangos, J.; Falcone, F. Implementation and operational analysis of an interactive intensive care unit within a smart health context. *Sensors* **2018**, *18*, 389. [CrossRef] [PubMed]
2. Kamienski, C.; Soininen, J.-P.; Taumberger, M.; Dantas, R.; Toscano, A.; Salmon Cinotti, T.; Filev Maia, R.; Torre Neto, A. Smart Water Management Platform: IoT-Based Precision Irrigation for Agriculture. *Sensors* **2019**, *19*, 276. [CrossRef] [PubMed]
3. Monteriù, A.; Prist, M.R.; Frontoni, E.; Longhi, S.; Pietroni, F.; Casaccia, S.; Scalise, L.; Cenci, A.; Romeo, L.; Berta, R.; et al. A Smart Sensing Architecture for Domestic Monitoring: Methodological Approach and Experimental Validation. *Sensors* **2018**, *18*, 2310. [CrossRef] [PubMed]
4. Granda, F.; Azpilicueta, L.; Vargas-Rosales, C.; Lopez-Iturri, P.; Aguirre, E.; Astrain, J.J.; Villandangos, J.; Falcone, F. Spatial characterization of radio propagation channel in urban vehicle-to-infrastructure environments to support WSNS deployment. *Sensors* **2017**, *17*, 1313. [CrossRef]
5. Grassini, S.; Ishtaiwi, M.; Parvis, M.; Vallan, A. Design and deployment of low-cost plastic optical fiber sensors for gas monitoring. *Sensors* **2015**, *15*, 485–498. [CrossRef]
6. Sun, G.; Hu, T.; Yang, G. Real-time and clock-shared rainfall monitoring with a wireless sensor network. *Comput. Electron. Agric.* **2015**, *119*, 1–11. [CrossRef]
7. Moreno, M.V.; Malamud, B.D.; Chuvieco, E.A. Wildfire frequency-area statistics in Spain. *Procedia Environ. Sci.* **2011**, *7*, 182–187. [CrossRef]
8. Klaina, H.; Vazquez Alejos, A.; Aghzout, O.; Falcone, F. Narrowband Characterization of Near-Ground Radio Channel for Wireless Sensors Networks at 5G-IoT Bands. *Sensors* **2018**, *18*, 2428. [CrossRef]
9. Tamir, T. On radio-wave propagation in forest environments. *IEEE Trans. Antennas Propag.* **1967**, *15*, 806–817. [CrossRef]
10. Burrows, C.R. Ultra-short-wave propagation in the jungle. *IEEE Trans. Antennas Propag.* **1966**, *14*, 386–388. [CrossRef]
11. Hall, M.P.M. *Radiowave Propagation Effects on Next Generation Fixed Services Terrestrial Telecommunications Systems*; Final Report; COST 235; Commission European Union: Brussels, Belgium, 1996.
12. Weissberger, M.A. *An Initial Critical Summary of Models for Predicting the Attenuation of Radio Waves by Foliage*; ECAC-TR-81-101; Electromagnetic Compatibility Analysis Center: Annapolis, MD, USA, 1981.
13. International Telecommunication Union. *Recommendation ITU-R P.833-8: Attenuation in Vegetation*; United Nations: New York, NY, USA, 2013.
14. Cid, E.L.; Alejos, A.V.; Sanchez, M.G. Signaling Through Scattered Vegetation: Empirical Loss Modeling for Low Elevation Angle Satellite Paths Obstructed by Isolated Thin Trees. *IEEE Veh. Technol. Mag.* **2016**, *11*, 22–28.

15. Gay-Fernandez, J.A.; Garcia Sanchez, M.; Cuiñas, I.; Alejos, A.V. Propagation analysis and deployment of a wireless sensor network in a forest. *Prog. Electromagn. Res.* **2010**, *106*, 121–145. [CrossRef]
16. Blaunstein, N.; Kovacs, I.Z.; Ben-Shimol, Y.; Andersen, J.B.; Katz, D.; Eggers, P.C.F.; Giladi, R.; Olesen, K. Prediction of UHF path loss for forest environments. *Radio Sci.* **2003**, *38*, 1059. [CrossRef]
17. Adewumi, A.S.; Olabisi, O. Characterization and Modeling of Vegetation Effects on UHF Propagation through a Long Forested Channel. *Prog. Electromagn. Res. Lett.* **2018**, *73*, 9–16. [CrossRef]
18. Tokunou, T.; Yamane, R.; Hamasaki, T. Near earth propagation loss model in forest for low power wireless sensor network. In Proceedings of the 2017 USNC-URSI Radio Science Meeting (Joint with AP-S Symposium), San Diego, CA, USA, 9–14 July 2017; pp. 19–20.
19. Oestges, C.; Villacieros, B.M.; Vanhoenacker-Janvier, D. Radio channel characterization for moderate antenna heights in forest areas. *IEEE Trans. Veh. Technol.* **2009**, *58*, 4031–4035. [CrossRef]
20. Cuiñas, I.; Gay-Fernandez, J.A.; Alejos, A.V.; Sánchez, M.G. A comparison of radioelectric propagation in mature forests at wireless network frequency bands. In Proceedings of the Fourth European Conference on Antennas and Propagation, Barcelona, Spain, 12–16 April 2010; pp. 1–5.
21. Srisooksai, T.; Kaemarungsi, K.; Takada, J.; Saito, K. Radio Propagation Measurement and Characterization in Outdoor Tall Food Grass Agriculture Field for Wireless Sensor Network at 2.4 GHz Band. *Prog. Electromagn. Res. C* **2018**, *88*, 43–58.
22. Anderson, C.R.; Volos, H.I.; Buehrer, R.M. Characterization of low antenna ultrawideband propagation in a forest environment. *IEEE Trans. Veh. Technol.* **2013**, *62*, 2878–2895. [CrossRef]
23. Dal Bello, J.C.R.; Siqueira, G.L.; Bertoni, H.L. Theoretical analysis and measurement results of vegetation effects on path loss for mobile cellular communication systems. *IEEE Trans. Veh. Technol.* **2000**, *49*, 1285–1293. [CrossRef]
24. Lopez-Iturri, P.; Aguirre, E.; Celaya-Echarri, M.; Azpilicueta, L.; Eguizábal, A.; Falcone, F.; Alejos, A. Radio Channel Characterization in Dense Forest Environments for IoT-5G. In Proceedings of the 5th International Electronic Conference on Sensors and Applications, 15–30 November 2018.
25. Torrico, S.A.; Bertoni, H.L.; Lang, R.H. Modeling tree effects on path loss in a residential environment. *IEEE Trans. Antennas Propag.* **1998**, *46*, 872–880. [CrossRef]
26. Chee, K.L.; Torrico, S.A.; Kürner, T. Radio wave propagation prediction in vegetated residential environments. *IEEE Trans. Veh. Technol.* **2013**, *62*, 486–499. [CrossRef]
27. Didascalou, D.; Younis, M.; Wiesbeck, W. Millimeter-wave scattering and penetration in isolated vegetation structures. *IEEE Trans. Geosci. Remote Sens.* **2000**, *38*, 2106–2113. [CrossRef]
28. Morgadinho, S.; Caldeirinha, R.F.S.; Al-Nuaimi, M.O.; Cuinas, I.; Sanchez, M.G.; Fernandes, T.R.; Richter, J. Time-variant radio channel characterization and modelling of vegetation media at millimeter-wave frequency. *IEEE Trans. Antennas Propag.* **2012**, *60*, 1557–1568. [CrossRef]
29. Johnson, R.; Schwering, F. *A Transport Theory of Millimeter Wave Propagation in Woods and Forest*; Technical Report; CECOM-TR-85-1; Center for Communications Systems: Fort Monmouth, NJ, USA, 1985.
30. Foldy, L.L. The Multiple Scattering of Waves. I. General Theory of Isotropic Scattering by Randomly Distributed Scatterers. *Phys. Rev.* **1945**, *67*, 107–119. [CrossRef]
31. Wang, F. Physics-based Modeling of Wave Propagation for Terrestrial and Space Communications. Ph.D. Thesis, Dept. Electrical and Computer Engineering, University of Michigan, Ann Arbor, MI, USA, 2006.
32. Al-Nuaimiand, M.O.; Hammoudeh, A.M. Measurements and predictions of attenuation and scatter of microwave signals by trees. *IEEE Proc. Microw. Antennas Propag.* **1994**, *141*, 70–76. [CrossRef]
33. COST 235. *Radio Propagation Effects on Next-Generation Fixed-Service Terrestrial Telecommunications Systems—Final Report*; European Commission: Luxembourg, 1995.
34. Meng, Y.; Lee, Y.; Ng, B.N. Empirical near ground path loss modeling in a forest at VHF and UHF bands. *IEEE Trans. Antennas Propag.* **2009**, *57*, 1461–1468. [CrossRef]
35. Seville, A.; Craig, K. Semi-empirical model for millimeter wave vegetation attenuation rates. *Electron. Lett.* **1995**, *31*, 1507–1508. [CrossRef]
36. Seville, A. Vegetation attenuation modeling and measurements at milli- metric frequencies. In Proceedings of the 10th International Conference Antennas Propagation, Edinburgh, UK, 14–17 April 1997; Volume 2, pp. 5–8.

37. Cama-Pinto, D.; Damas, M.; Holgado-Terriza, J.A.; Gómez-Mula, F.; Cama-Pinto, A. Path Loss Determination Using Linear and Cubic Regression Inside a Classic Tomato Greenhouse. *Int. J. Environ. Res. Public Health* **2019**, *16*, 1744. [CrossRef]
38. Goncalves, N.; Correia, L. A propagation model for urban micro- cellular systems at the UHF band. *IEEE Trans. Veh. Technol.* **2000**, *49*, 1294–1302. [CrossRef]
39. Mani, F.; Oestges, C. A ray-based method to evaluate scattering by vegetation elements. *IEEE Trans. Antennas Propag.* **2012**, *60*, 4006–4009. [CrossRef]
40. Azpilicueta, L.; Rawat, M.; Rawat, K.; Ghannouchi, F.; Falcone, F. Convergence Analysis in Deterministic 3D Ray Launching Radio Channel Estimation in Complex Environments. *ACES J.* **2014**, *29*, 256–271.
41. Azpilicueta, L.; Lopez-Iturri, P.; Aguirre, E.; Mateo, I.; Astrain, J.J.; Villadangos, J.; Falcone, F. Analysis of Radio Wave Propagation for ISM 2.4 GHz Wireless Sensor Networks in Inhomogeneous Vegetation Environment. *Sensors* **2014**, *14*, 23650–23672. [CrossRef] [PubMed]
42. Komarov, V.V. *Handbook of Dielectric and Thermal Properties of Materials at Microwave Frequencies*; Artech House: Boston, MA, USA; London, UK, 2012.
43. Lang, R.H.; Schneider, A. Radiowave Propagation Within Trunk-dominated Forests: Coherence Bandwidth and Delay-spread. In Proceedings of the 12th Canadian Symposium on Remote Sensing Geoscience and Remote Sensing Symposium, Vancouver, BC, Canada, 10–14 July 1989; Volume 5, p. 2847.
44. Yang, F.-C. Forest Channel Characterization in the 5 GHz Band. Master's Thesis, Russ College of Engineering and Technology of Ohio University, Athens, OH, USA, 2008. Available online: https://etd.ohiolink.edu/rws_etd/document/get/ohiou1226076756/inline (accessed on 15 October 2018).
45. Gómez, P.; Cuiñas, I.; Alejos, A.V.; Sánchez, M.G.; Caldeirinha, R.F.S. Shrub-blown time variability in attenuation and scattering at cellular frequencies. *IET Microw. Antennas Propag.* **2010**, *4*, 526–542. [CrossRef]
46. Alejos, A.V.; Dawood, M.; Medina, L. Experimental dynamical evolution of the Brillouin precursor for broadband wireless communication through vegetation. *Prog. Electromagn. Res.* **2011**, *111*, 291–309. [CrossRef]

Article

Underwater Acoustic Impulsive Noise Monitoring in Port Facilities: Case Study of the Port of Cartagena[†]

Ivan Felis Enguix [1,*], Marta Sánchez Egea [1], Antonio Guerrero González [2] and David Arenas Serrano [1]

1 Naval Technology Center (CTN), 30320 Fuente Álamo, Murcia, Spain; martasanchez@ctnaval.com (M.S.E.); davidarenas@ctnaval.com (D.A.S.)

2 Department of Automation and Systems Engineering, Technical University of Cartagena, 30202 Cartagena, Murcia, Spain; antonio.guerrero@upct.es

* Correspondence: ivanfelis@ctnaval.com; Tel.: +34-968-197-521

† This paper is an extension version of the conference paper: Enguix, I.F.; Sánchez Egea, M.; Guerrero González, A. and Arenas Serrano, D. Acoustic Characterization of Impulsive Underwater Noise Present in Port Facilities: Practical Case in the Port of Cartagena, In Proceedings of the 5th International Electronic Conference on Sensors and Applications, 15–30 November 2018.

Received: 11 August 2019; Accepted: 26 October 2019; Published: 28 October 2019

Abstract: Recording underwater impulsive noise data is an important aspect of mitigating its environmental impact and improving maritime environmental management systems. This paper describes the method used and results of the spatial monitoring of both the baseline noise level and the impulsive noise sources in the Port of Cartagena. An autonomous vessel was equipped with a smart digital hydrophone with a working frequency range between 10 and 200 kHz and a received voltage response (RVR) of, approximately, −170 dB re 1V/μPa. A GIS map was drawn up with the spatiotemporal distribution of the basal sound pressure levels by coupling the acoustic data with the vessel's GPS positions to identify the sources of the impulsive noise of interest and their temporal characteristics. The loading of cargo containers was identified as the main source of impulse noise. This study is the first of a series designed to obtain accurate information on underwater noise pollution and its potential impact on biodiversity in the Port of Cartagena.

Keywords: underwater noise monitoring; impulsive underwater noise; continuous underwater noise; marine contamination; hydrophone; marine strategy framework directive; ASV

---

1. Introduction

Intensive port activity is sending an ever-growing amount of artificial underwater sound into the sea, mainly due to maritime traffic. This situation has been well documented for many years, especially by the world's navies [1]. Continuous underwater maritime transport noise has increased in recent decades [~3 dB per decade] [2]. In [3], recordings made in the 1960s were compared with others from the 1990s on the continental slope off Point South, California, showing how underwater noise related to maritime traffic has increased by up to 10 dB at frequencies of 20 to 300 Hz.

Underwater noise affects marine organisms in many ways [4], disrupts acoustic communications and auditory systems, alters behavior, and changes population distribution and abundance. Underwater noise generates physiological stress in marine organisms, stimulating nervous stress, increasing the metabolism, and reducing their immunity. These sounds affect not only aquatic mammals but also fish, amphibians, reptiles, and even invertebrates [5–7]. A study conducted in the Sado estuary in Setúbal, Portugal, one of the country's largest ports, revealed changes in the behavior of bottlenose dolphins. During daylight hours, they spent less time in areas with the highest underwater noise levels [8].

Given this information on the adverse effects of noise pollution, regulatory bodies increasingly require an assessment of the acoustic underwater noise in a marine habitat before a construction project. In a review of noise modeling processes in [9], the factors that affect underwater noise generation and basic aspects of environmental impact assessment (EIAs) are highlighted.

Human activities at sea generate different types of noise that vary in frequency and intensity and can be stationary and low frequency or impulsive and high intensity. The latter are produced in pile driving, underwater blasting, seismic exploration, or sounding activities, among others. Ports can also be subjected to impulsive noise due to activities such as pumping systems, air conditioning, civil engineering works in ports, and loading containers. In [10], recordings of underwater noises were made in the port of Fremantle Inner Harbor for several months. Subsequent analyses of these data detected underwater noise from various sources, these being vessel traffic, vehicles passing over a nearby bridge, machinery noise from regular operations in the port, and vibratory or impact pile driving recorded during during wharf construction.

Impulsive noises have adverse effects on marine life. In [11], which measured and modeled pile driving noise from a large number of points around Australia, an acoustic propagation model was proposed that allowed these results to be extrapolated to other locations. Another example, [12], on pile driving in the Ligurian Sea, found that fin whales (*Balaenoptera physalus*) avoided areas with impulsive noise. Another study, [13]. also analyzed whales' reactions to underwater impulsive noise; even 72 h after an airgun shooting period in the Alboran basin, fewer whale and received song levels were detected.

The aim of this study was to perform an acoustic characterization of impulsive underwater noise at the Port of Cartagena. This characterization was carried out according to the methodological criteria for the implementation of the Marine Strategy Framework Directive [14], including:

- Location and identification of impulsive noise emission sources.
- Quantification of background levels and continuous noise.
- Quantification of impulsive noise levels.
- Propagation of the values obtained beyond the domain of the port.

For this, a series of underwater acoustic measurements were taken in the bay, as described in Section 2.

2. Identification of Activities Potentially Generating Impulsive Noise in Port Facilities

In accordance with the methodological guidelines published by the European Commission [15–17], the main sources of impulsive sound are

- Pile driving [18]—installing piles requires the use of a large mechanical hammer to drill the seabed. Sound levels of mechanical driving can reach up to 260 dB re 1 μPa re 1 m in the absence of noise reduction measures. In [18], the impact of sound measurements had a duration of 0.2 s, the signals showed peak energy at 160 Hz, and also had significant energy up to and beyond 100 kHz.
- Explosions [19]—explosives are used as an alternative to heavy machinery to break up hard substrates, for example, in the construction of docks or deepening ports. In [19], the noise measured from the explosions had an sound exposure level (SEL) of ≥ 190 dB re 1 μPa^2·s.
- Compressed air guns [20]—compressed air cannons are commonly used in seismic surveys to determine the geological structures of the seabed. These guns send shots of compressed air towards the seabed. The sound bounces off the bed and returns to the surface, where it is recorded by hydrophones. In [20], a 0.33 L Bolt PAR 600B air gun was used at 1500 psi (5 m depth). In these operating conditions it had a source level of 192 dB re 1 μPa^2·s at 1 m.
- Sonar [21]—used especially by naval vessels to detect underwater objects. Simple sonar systems direct sound (short pulses) in one direction, although there are more complex systems that can emit beams of sound in multiple directions. Examples of acoustic characteristic of sound sources

of this type include: multi-beam sonar with center frequency 15.5 kHz, sound pressure level (SPL): 237 dB re 1 μPa at 1 m (RMS) with omnidirectional beam; sub-bottom profiler with center frequency 3. 5 kHz, SPL: 204dB re: 1uPa at 1 m (RMS); or AN/SQS 56 sonar with center frequency of 6.8 kHz; 7.5 kHz; 8.2 kHz, SPL: 223 dB re 1 μP at 1 m (RMS) horizontal beam.

- Echosounders [21]—use sound production to locate the depth of the marine environment or schools of fish. They are used on most vessels, on fishing boats, and also on a large number of pleasure craft. For example, the SimradEK 500 Scientific echo sounder has a frequency of 38 kHz, 3 dB beam width 6.9 deg, peak transmit power 2000 W, pulse duration 1 ms and a 3.8 kHz bandwidth.
- Acoustic deterrent devices [22]—are used to alert or frighten away marine animals with sounds to avoid their interfering with fishing gear or aquaculture cages. An example of this type of device is the Ferranti-Thomson MK2 4X, which has a source level (re 1 μPa) of about 200 dB at 25 kHz. Other commercial devices have similar characteristics.

Some of these sources are regularly present in the port area, such as bathymetries and construction projects. In addition to the sources mentioned above, this study also considered the possibility of analyzing other routine activities, such as the loading and unloading of containers to analyze their potential impulsive noise production.

3. Materials and Methods

3.1. Characterization of the Submarine Bottom

The characterization of the properties of the water column and the seabed is fundamental for the implementation of acoustic propagation models. For this purpose, two types of deployment were made:

1. Visual inspections were made with underwater ROVs (remote operated vehicles) in order to determine the type of seabed in Cartagena Bay. For this, the following vehicles were used: the Deep Trecker DTG2 and Seabotix VLBV950. Figure 1 shows the location of the deployment point inside the port (left) and one of the robots used (right). Photographic and video reports were obtained.
2. A bathymetric survey was carried out to profile the seabed by means of an echo sounder using a Seaking SBP (parametric sub-bottom profiler). This sonar emits two frequencies: low frequency 20 kHz (parametric) with a beam width of 4.5°, and high frequency at 200 kHz with a beam width of 4°. The pulse width is 100 μs. When producing a 20 kHz pulse, the device is able to penetrate the seabed and highlight structural differences that conventional echo sounders cannot detect.

Figure 1. Deployment point inside the port (**left**) and one of the robotic equipment used (**right**).

For the measurement campaign, a special fastening element was designed and built for the SBP sonar. This clamping element allowed the sonar to be kept at the required depth without movement and vibrations. Figure 2 shows the coupling of the support to the sonar on the boat.

Figure 2. Tooling for joining the echosounder to the side of the boat.

The bathymetry survey was performed by an autonomous surface vehicle ASV (see Figure 3). The ASV hull was 5.10 m long and 1.97 m wide. The displacement of the vessel under normal operating conditions was approximately 1000 kg, with an average draft of 0.325 m. The ASV was steered by two independent outboard propellers anchored to the transom. Further information on the ASV can be found in [23].

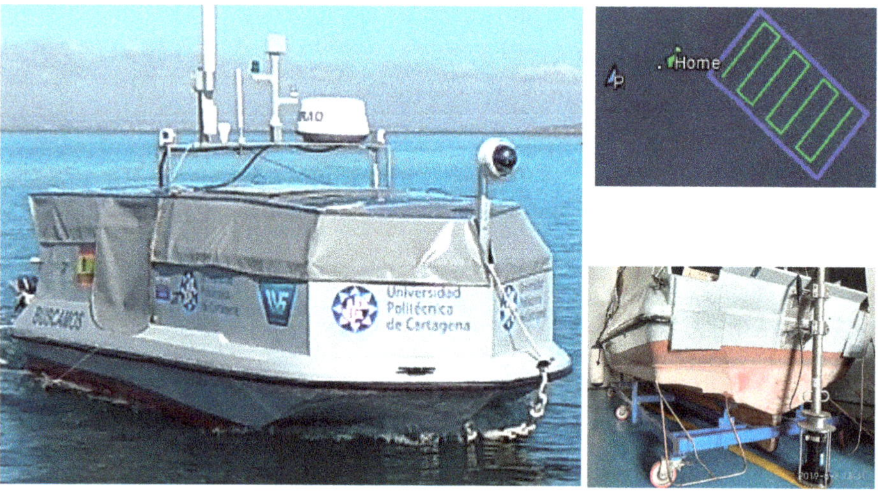

Figure 3. ASV with SBP (parametric sub-bottom profiler) (**left**), coupling of support to the boat (**bottom right**), ASV course definition (**top right**).

The SBP was coupled to the stern of the ASV, as shown in Figure 3 (bottom right). The sub-bottom system received the GPS positions used to reference all the sub-bottom information obtained. The sub-bottom system was connected to the hardware and software architecture of the ASV control system and synchronized the data with the position and course of the vessel. This system was capable of obtaining accurate and high-resolution sub-bottom information of the study area.

3.2. Characterization of Underwater Noise

Two types of measurements were carried out for the monitoring of underwater sound in the Port of Cartagena:

- Type 1: Recordings at different points around the port perimeter by manual deployment of a hydrophone.
- Type 2: Recordings in the interior of the port as well as in the access zone outside the port infrastructure by submerging a hydrophone from a boat.

The GPS position of each measuring point was also recorded for the subsequent spatial processing of the noise data. Figure 4 shows the position of the points.

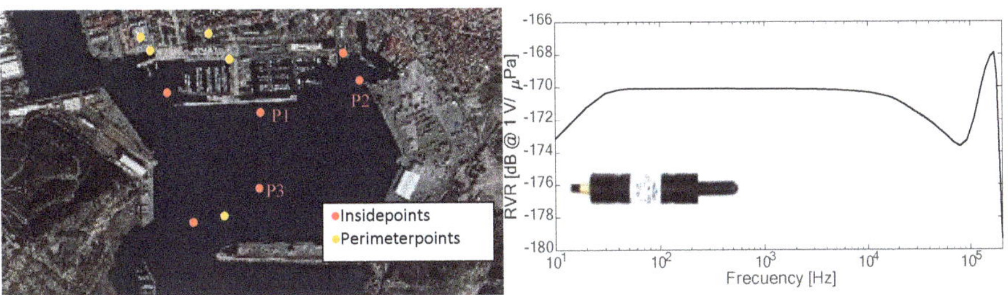

Figure 4. (**left**) Location of all the measurements points. (**right**) Received voltage response (RVR) of the hydrophone used.

All the measurements were made by an IcListen HF smart hydrophone. This transducer has a sensitivity of approximately −170 dB re 1V/µPa with a bandwidth between 10 Hz and 98 kHz. The device acquires signal with a sampling frequency of 192 kS/s and its bit-depth is 24 bits. Figure 4 (right) shows its received voltage response (RVR) sensibility according to the frequency band.

In both types of measurements, the hydrophone was submerged 6 meters in a series of 3 min recordings of events due to:

- Continuous sound—different types of boats passing close to the measurement points to be contrasted with the background noise. These recordings helped us to characterize the sources of continuous noise with levels above that of the background noise.
- Impulsive sound—due to boat sonars and especially to loading containers onto cargo ships. An initial estimate of potential impulsive noise inherent to the activity in the port was made from the spectrum, duration, and periodicity of these events. The recorded levels were also the basis for the further development of the numerical propagation model.

3.3. Processing of Undewater Noise

The signals were processed to obtain information on the frequency domain as well as the time-frequency domain.

For the frequency information, each 5 min recording was filtered in each third octave band with center frequencies from 10 Hz to 100 kHz. The average and standard deviation of the sound pressure level (SPL) was calculated from the signals obtained by windowing the filtered signals in 10 s sections. The average was calculated from $p_{rms} = \frac{1}{N}\sum_{i=1}^{N} p_{i,rms}$ and the standard deviation from $\delta p_{rms} = \sqrt{\frac{1}{N}\sum_{i=1}^{N}(p_{i,rms} - p_{rms})^2}$, where $p_{i,rms}$ is the root mean square pressure of each windowed signal. From the resulting values, the SPL was calculated according to its usual definition [24], $SPL = 20\log\left(p_{rms}/p_{ref}\right)$ and $\delta SPL = 20\log\left(\delta p_{rms}/p_{ref}\right)$, where $p_{ref} = 1\ \mu Pa$. Although the variability of the SPL may change according to the window used, this calculation provided an approximation of the error involved in quantifying the SPL.

A spectrogram with a rectangular window able to identify the variations of SPL with respect to the background was used for the time-frequency domain.

3.4. Underwater Sound Propagation Model

Several mathematical models can be used to study the underwater acoustic propagation (parabolic equation model, the based on normal modes, the spectral integration model, among others). In each case, the implementation of these mathematical models required several parameters whose exact value was not always known, which could lead to unreliable results. In order to simplify this problem, several semi-empirical models can be found that distinguish different types of analytical propagation phenomena [25].

The depth of water in the Port of Cartagena (~12 m) is such that there are multiple rebounds of the signal between the surface and the seabed, which leads to a considerable level of interaction between the propagated and reflected acoustic signals. This interaction is quite complex, as it is necessary to take into account the type of seabed, type of sediment, how it is distributed, and possible depth variations, among others. This behavior means that there is a shallow-water propagation channel.

The transmission loss of sound propagation in shallow water depends upon many natural variables of the sea surface, water medium, and bottom. Because of its sensitivity to these variables, the transmission loss in shallow water is only approximately predictable in the absence of precise values for the variables. The semi-empirical Marsh–Schulkin expressions based on the Colossus models are useful for rough prediction purposes [24,26]. This model can be used for simulations between 0.1 and 10 kHz. From this model, the transmission losses are obtained according to the equations given in:

$$TL = \begin{cases} 20\log(R) + \alpha R + 60 - k_L & R < H \\ 15\log(R) + \alpha R + \alpha_T\left(\frac{R}{H} - 1\right) + 5\log(H) + 60 - k_L & H < R < 8H \\ 10\log(R) + \alpha R + \alpha_T\cdot\left(\frac{R}{H} - 1\right) + 10\log(H) + 64.5 - k_L & R > 8H \end{cases} \quad (1)$$

where R is the distance from the source; α is the absorption coefficient of the water; k_L is a parameter called *near-field1 anomaly*, which measures the gain due to rebounds between the background and the surface; α_T is the so-called effective attenuation coefficient, which takes into account the losses due to the energy coupling between the surface and the bottom [26]; and H is the *distance of jump* or *transmission*, defined as that maximum distance at which a ray contacts the surface or the background, whose shape depends on the depth of the water column, D, and is given by the following equation:

$$H = \sqrt{\frac{L + D}{3}} \quad (2)$$

In the study, the model was applied to the port considering 1 kHz, a calm sea, and a sandy bottom: k_L = 6 dB/bounce and α_T = 1.8 dB/bounce.

It should be noted that this model was simplified and did not account for a large number of effects that should be taken into account in a more detailed propagation study. Some of these effects consider a reflection coefficient at the port perimeter and variations in depth, among others. Therefore, the numerical results of the calculations should be taken with caution. However, our intention of using a simplified propagation model here was to determine the approximate acoustic behavior of the complex propagation processes in a port environment.

4. Results and Discussion

4.1. The Submarine Seabed

The fundamental properties of the water column and the seabed were characterized by two different methods:

1. Visual inspections by an underwater ROV to obtain photographic and video reports. Figure 5 shows the robot deployed from the boat (left) and an image of the bottom at a depth of 9 m (right).

The bottom was found to be soft (composed of sand, mud, small stones, and gravel) and shelved slightly with a distance from the jetty.

2. A bathymetric survey, which obtained information on the depth and types of material below the seabed, plus an updated isopach map, was carried out by a sub-bottom profiler (SBP) coupled to the stern of an ASV. The autonomous vessel was programmed to cover Cartagena Bay following parallel paths. Figure 6 shows the ASV's control, communication, and command software, which included an initial isopach layer of the bay provided by the Cartagena Port Authority.

Figure 5. Robot on deployment (**left**) and an image of the seabed (**right**).

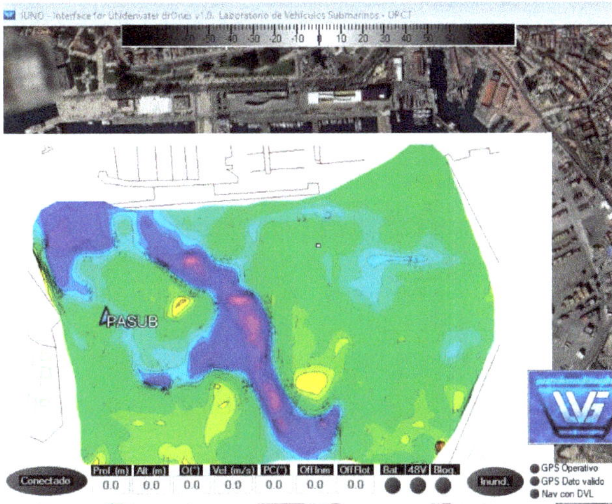

Figure 6. ASV control software.

A screenshot of the sub-bottom profile can be seen in Figure 7. The SBP emits a 20 kHz low frequency pulse into the seabed and highlights seismic structures and layers. The operator can view the raw high frequency profile on the screen.

The survey found a structural change at a depth of around 5 m under the seabed at certain points due to rock formations and possible sunken objects.

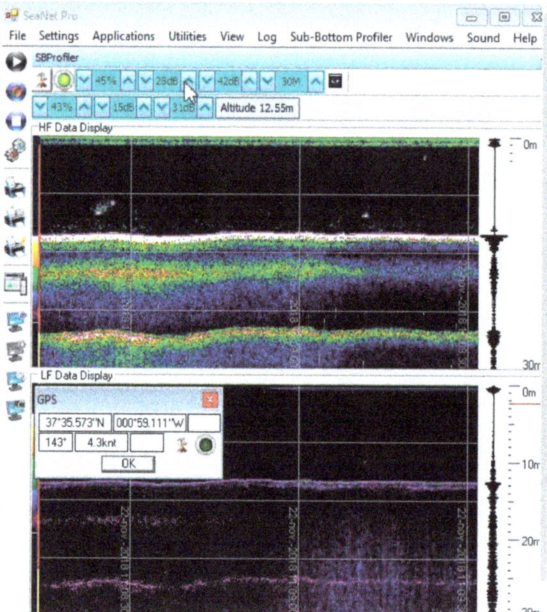

Figure 7. Sub-bottom profile image obtained.

4.2. Continuous Acoustic Sound

Two types of continuous sound were differentiated to characterize the basal sound in the measurement area: (i) background sounds, recorded without the presence of marine traffic from normal port activities, waves, and other environmental factors; and (ii) continuous anthropogenic sounds from maritime traffic entering and leaving the port.

Figure 8 gives the levels of background sound obtained and how they affect the vessels that typically operate in the port, which helped us to establish the reference level of non-impulsive sound. It shows the average SPL obtained in the octave band thirds between 1 and 100 kHz and the standard deviation calculated at all the measurement points. The average recorded SPL was 62 dB with variations of ± 18 dB. On the right is shown the changes of SPL with distance from the measurement point. Some residual sonar signals were also present. Figure 8 shows a source of the 50 kHz echosounder noise from military ships moored at a port of the bay.

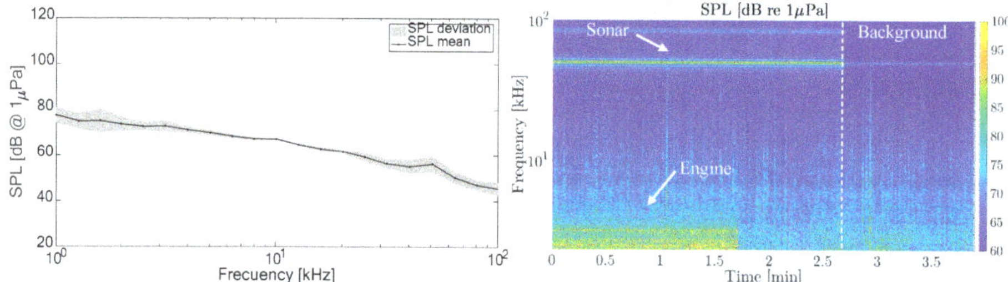

Figure 8. (**left**) Spectrum of background noise. (**right**) Its spectrogram at Point P1. SPL: sound pressure level.

These results gave an initial estimate of the basal sound in the port used as the basis for the study of the continuous and impulsive sound sources. However, some of the continuous sound sources, although not basal, were considered as part of the background, and not specifically impulsive sound, mainly due to passing boats. Figure 9 shows the spectrum of the influence on the background of a pilot boat passing approximately 40 m away from the measurement point.

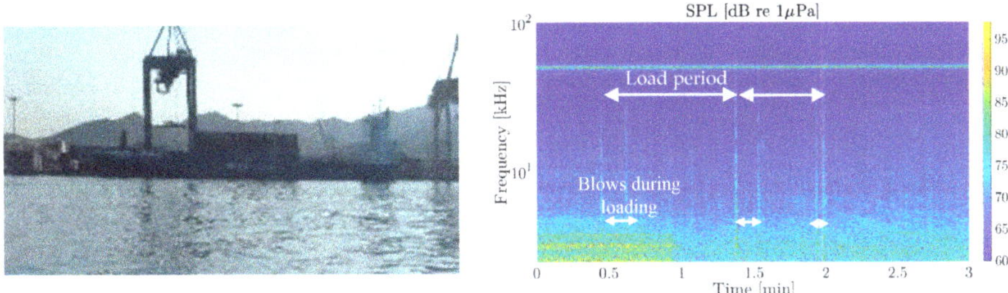

Figure 9. (**left**) Spectrum of continuous noise source. (**right**) Its spectrogram (Point P3).

Some vertical traces could be identified, which, as shown below, were from the nearby container loading operations.

4.3. Impulsive Sounds

During the measurement, only one source of impulsive sound was found from the process of loading sea containers onto a freighter in the eastern part of the port (Figure 10, left). A study of the spectrogram showed the levels and periodicity of these impulses (Figure 10, right). These measurements were made at approximately 50 m from the source. In the spectrogram, a 50 kHz continuous peak also appeared from naval ships docked in the port.

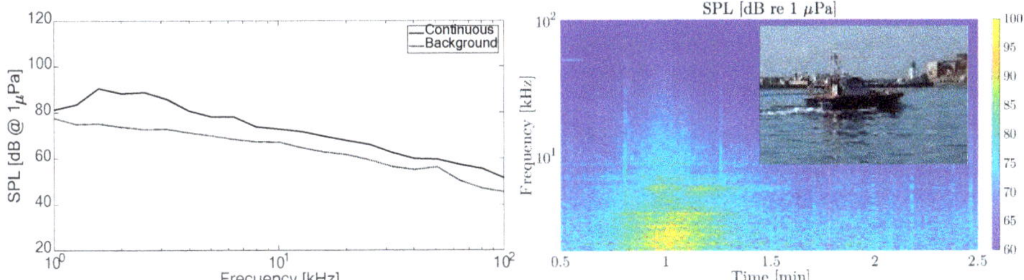

Figure 10. Loading a cargo ship (**left**), and spectrogram of the noise generated (**right**) at P2 in Figure 4.

During the recordings, it was found that the appearance of the vertical lines in the spectrogram could be attributed to the container loading operations (especially banging noises). Some clear vertical lines can be seen in the spectrogram from this loading process. They had a considerable bandwidth (as they were pulses) with peaks of up to 85 dB at frequencies below 20 kHz. It was estimated that this impulsive source emitted around 100 dB re 1μPa re 1m.

Although this source is currently not considered as a relevant sound source [14], it may be desirable to move forward with this analysis in a context of high intensity loading and unloading operations in different ports. The possibility of including this activity as a potential impulsive noise source could also be assessed, being taken into account in noise monitoring guidelines.

4.4. Propagation of Underwater Noise

In these impulse noise source monitoring studies, we were interested in estimating the noise levels at various distances outside the port, as well as applying acoustic propagation models taking into account the characteristics of the water column and the soil.

Here, we applied the simplified model described in Section 3.3 to estimate the propagation range of a source with characteristics similar to the one described in Section 4.3. In addition, a ray tracing algorithm, specifically created by the Centro Tecnológico Naval y del Mar (CTN), was implemented that considered the topography of the Bay of Cartagena.

The ray trace was generated from the position of the detected impulsive noise source within the geometry of the port. Figure 11 shows, on the left, two examples of ray trajectories up to 5 km from the source. It should be noted that one ray leaves the port while the other remains inside the limits, which affected the final propagation levels. Figure 10 shows ray traces for different emission angles.

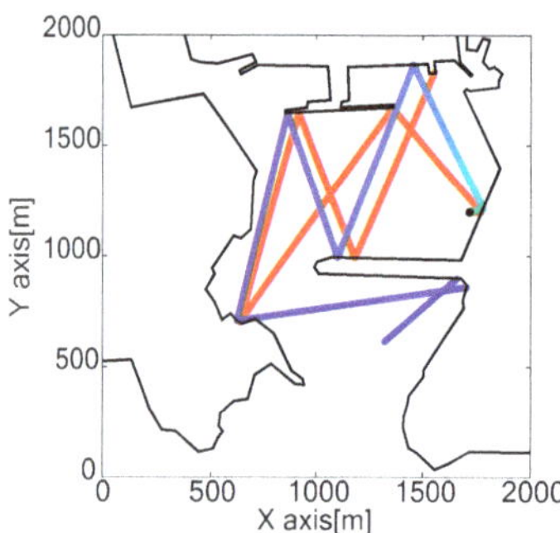

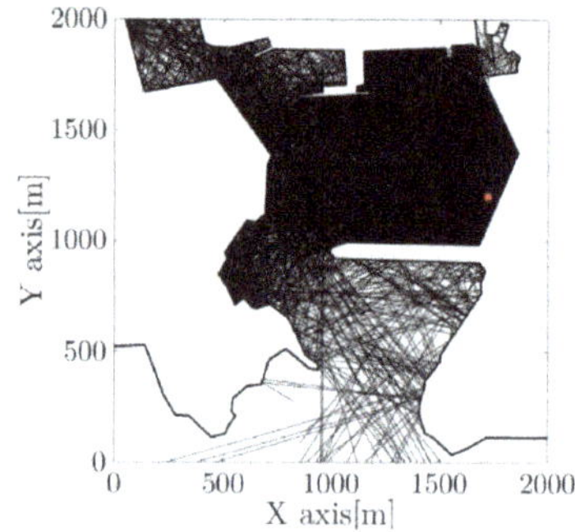

Figure 11. (**left**) Example of two ray traces. (**right**) All the rays traced. The red dot indicates the source.

Figure 12 shows, on the left, the percentage of rays that left the port, according to their emission point. It can be seen that this position is quite independent of the number of rays that left the port, that is, the port geometry influenced whether or not they exited. Only 10% of the rays travelled up to 5 km and 20% travelled up to 7.5 km. As explained below, these distances were enough to reduce the generated impulsive noise level.

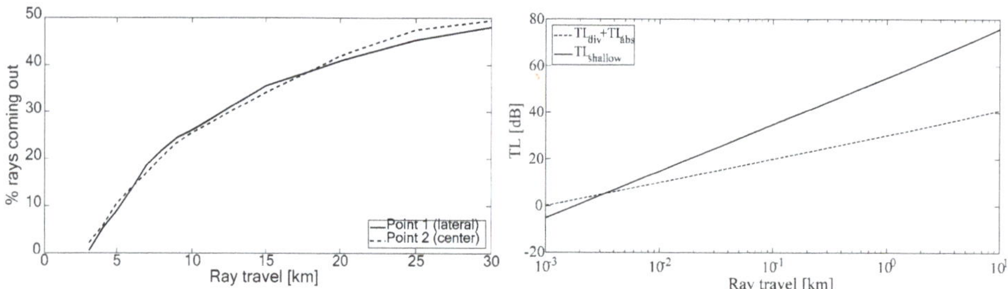

Figure 12. (**left**) Percentage of rays that left the port. (**right**) Comparison of transmission losses with different propagation models.

In order to detect the effect of the resulting underwater acoustic propagation along these rays, we first compared the transmission losses due only to cylindrical geometrical divergence and absorption with the transmission losses of the semi-empirical model given in Section 3.3. Figure 12 right shows the results. It can be seen that both models had the same losses within a few meters of the emission. At closer distances, the semi-empirical model presented fewer losses because, in this region, the sound propagated under a spherical divergence. For longer distances, the large number of reflections led to a sharp increase in losses that limited propagation. Specifically, for 1 kHz, there were reductions of 35 dB in 100 m and 55 dB in 1000 m, much greater than those foreseen by divergence and absorption only.

From the above results we can estimate that a source of impulsive noise with an SPL of, typically, 120 dB re 1 µPaat 1 kHz inside the port, can generate a sound wave that could travel approximately

5 km to reach the background level shown above (70 dB re 1 μPa, see Figure 12). This means the sound could travel outside the port with a level below the background noise. Nevertheless, the real losses would be smaller, as the propagation model does not consider the lost reflections, which could be of about 4 to 7 dB per collision, so that this range would decrease.

5. Geomaritime Representation of the Results

To represent the results, the CTN designed the underwater acoustic propagation model on the basis of a geographic information system (GIS) located in the study area with recorded submarine noise data. This tool can select the points where noise measurements are made and observe their temporal profile and frequency spectrum. It also allows the user to introduce a noise source and apply the simplified models of underwater acoustic propagation described above. With this, the acoustic rays propagated from the source to a previously indicated distance are obtained, plus the final sound pressure levels. Figure 13 shows the result of locating a fictitious 180 dB re μPa at a 1 kHz source next to the San Pedro dock.

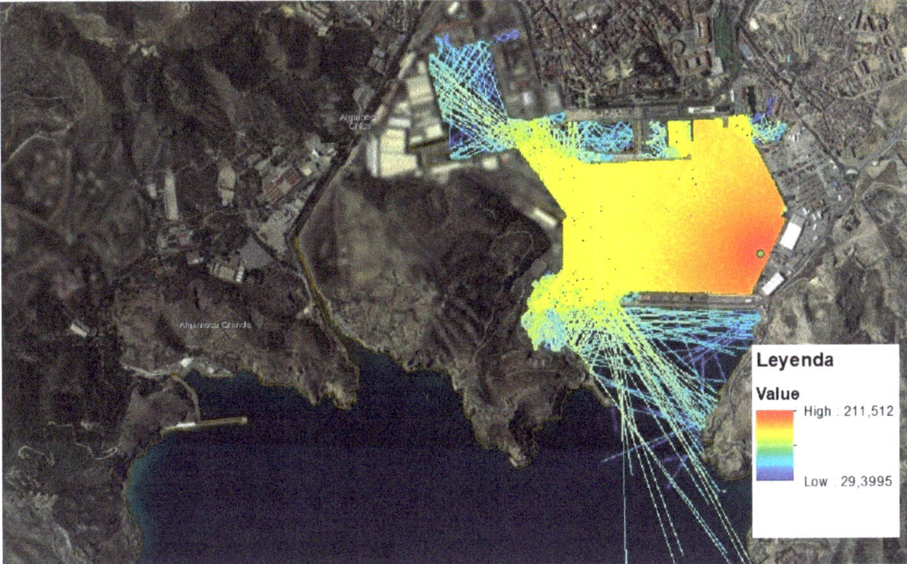

Figure 13. Example of submarine acoustic propagation by the web tool.

6. Conclusions

Through this study it has been proven that

- After analyzing the background and continuous sound levels in the Port of Cartagena, it was found that impulsive sound must have a level above approximately 70 dB re 1 μPa to be detected, depending on the frequency (between 80 and 45 dB re 1 μPa in the frequency range from 10 to 100 kHz, respectively). Only one impulsive sound source was detected, and this was due to container loading operations on the east side of the port. Although this source is not usually indicated in the search for impulsive sources, these operations can be considered as a potentially impulsive noise generating activity, according to the requirements of the Marine Strategy Framework Directive.
- The analysis of the bathymetric survey of the study area was fundamental in the application of acoustic propagation models, as the type of seabed influenced the intensity of the sound, due to the reflections of the waves. Also, having reliable bathymetric information contributed to the

construction of a consistent acoustic propagation model, as the accuracy of the data will also affect the accuracy of the model.

- The application of the acoustic propagation model implemented in this study, despite being approximate, provided information on the potential impact produced by the generation of impulsive noise in the vicinity of the Port of Cartagena.

In addition, this study has provided

- Measurements of impulsive sound, in accordance with the requirements of the Marine Strategy Framework Directive, which will serve to form part of the port's underwater sound record.
- Updated and accurate data on the bathymetry of the port that can be used in subsequent studies and projects.

The results of this study can serve as a starting point for a larger scale strategy against the potential impact of underwater noise (both impulsive and continuous) in the Port of Cartagena.

To continue advancing in this regard, the following recommendations are proposed:

- Continuous monitoring of the levels of underwater sound in the Port of Cartagena, especially during activities that could potentially generate underwater noise during the periods of high sensitivity of any marine species that could be affected.
- Collection of information on marine biodiversity present in the area where activities are carried out and their surroundings. Identifying species with special sensitivity to the potential impact of submarine noise in the vicinity of the Port of Cartagena.
- Incorporating location systems for sensitive species in the port area, through, for example, passive acoustic systems.
- Adopting a set of preventive and corrective measures for activities potentially generating submarine noise, firstly related to the selection and exclusion of areas to carry out activities according to their potential impact and to minimize their environmental impact. Secondly, to control activities in seasons of special sensitivity (migrations, breeding, etc.), including the regulation of activities (navigation, works, etc.) to minimize their possible adverse effects.

Author Contributions: This work was a collaborative development by all authors. Conceptualization, I.F.E. and A.G.G.; Methodology, I.F.E. and A.G.G.; Validation, I.F.E., M.S., A.G.G. and D.A.S.; Formal Analysis, I.F.E.; Investigation, I.F.E., A.G.G. and M.S.; Resources, I.F.E., A.G.G. and M.S.; Data Curation, I.F.E., Writing-Original Draft Preparation, I.F.E., M.S., A.G.G. and D.A.S.; Writing-Review & Editing, I.F.E. and A.G.G.; Visualization, I.F.E. and D.A.S.; Supervision, I.F.E. and A.G.G.; Project Administration, I.F.E., M.S.E. and A.G.G.; Funding Acquisition, A.G.G. and M.S.E.

Funding: This research was funded by the *Cátedra de Medio Ambiente Autoridad Portuaria de Cartagena-Campus Mare Nostrum*.

Acknowledgments: We would like to thank the Cartagena Port Authority for their general support in the operational campaign.

Conflicts of Interest: The authors declare no conflict of interest.

References

1. Pricop, M.; Chitac, V.; Pazara, T.; Popa, A.; Oncica, V. Assessment of underwater noise produced by ships at the entrance of port Constanta. In Proceedings of the 2nd International Conference on Manufacturing Engineering, Quality and Production Systems, Constantza, Romania, 3–5 September 2010.
2. Etter, P.C. *Underwater Acoustic Modeling and Simulation*, 5th ed.; CRC Press, Taylor and Francis: Boca Raton, FL, USA, 2018; pp. 293–313.
3. Andrew, R.K.; Howe, B.M.; Mercer, J.A. Ocean ambient sound: Comparing the 1960s with the 1990s for a receiver off the California coast Noise in the Sea and Its Impacts on Marine Organisms. *Acoust. Res. Lett. Online* **2002**, *3*, 65–70. [CrossRef]

4. Peng, C.; Zhao, X.; Liu, G. Noise in the Sea and Its Impacts on Marine Organisms. *Int. J. Environ. Res. Public Health* **2015**, *12*, 12304–12323. [CrossRef] [PubMed]
5. Weilgart, L. *The Impact of Ocean Noise Pollution on Fish and Invertebrates*. Report for OceanCare. Switzerland. 2018. Available online: https://www.oceancare.org/wp-content/uploads/2017/10/OceanNoise_FishInvertebrates_May2018.pdf (accessed on 27 October 2019).
6. Popper, A.N.; Hastings, M.C. The effects of anthropogenic sources of sound on fishes. *J. Fish Biol.* **2009**, *75*, 455–489. [CrossRef] [PubMed]
7. Williams, R.; Wright, A.J.; Ashe, E.; Blight, L.K.; Bruintjes, R.; Canessa, R.; Clark, C.W.; Cullis-Suzuki, S.; Dakin, D.T.; Erbe, C.; et al. Impacts of anthropogenic noise on marine life: Publication patterns, new discoveries, and future directions in research and management. *Ocean Coast. Manag.* **2015**, *115*, 17–24. [CrossRef]
8. Cruz, E.; Luis, A.R.; dos Santos, M.E. Underwater noise in the Sado Estuary, Portugal, and its potential impact on the resident bottlenose dolphins. In Proceedings of the 26th European Cetacean Society Conference, At Galway, Ireland, 26–28 March 2012.
9. Farcas, A.; Thompson, P.M.; Merchant, N.D. Underwater noise modelling for environmental impact assessment. *Environ. Impact Assess. Rev.* **2016**, *57*, 114–122. [CrossRef]
10. Salgado, C.P.; McCauley, R.D.; Parnum, I.M.; Gavrilov, A.N. Underwater noise sources in Fremantle inner harbour: Dolphins, pile driving and traffic. In Proceedings of the Acoustic 2012, Fremantle, Australia, 21–23 November 2012.
11. Duncan, A.J.; McCauley, R.D.; Parnum, I.; Salgado-Kent, C. Measurement and Modelling of Underwater Noise from Pile Driving. In Proceedings of the 20th International Congress on Acoustics (ICA), Sydney, Australia, 23–27 August 2010.
12. Borsani, J.F.; Clark, C.W.; Nani, B.; Scarpiniti, M. Fin whales avoid loud rhythmic low-frequency sounds in the Ligurian Sea. *Bioacoustics* **2008**, *17*, 161–163. [CrossRef]
13. Castellote, M.; Clark, C.W.; Lammers, M.O. Acoustic and behavioural changes by fin whales (Balaenoptera physalus) in response to shipping and air-gun noise. *Biol. Conserv.* **2012**, *147*, 115–122. [CrossRef]
14. Dekeling, R.; Tasker, M.; Van Der Graaf, S.; Ainslie, M.; Andersson, M.; André, M.; Drensing, K.; Castellote, M.; Cronin, D.; Dalen, J.; et al. Monitoring Guidance for Underwater Noise in European Seas, Part II: Monitoring Guidance Specifications. JRC Scientific and Policy Report EUR 26555. [CrossRef]
15. MSFD. Calibration Guidelines. Available online: www.quietmed-project.eu/wp-content/uploads/2019/01/QUIETMED_D3.1.-Best_practices_guidelines_on_sensor_calibration_final.pdf. (accessed on 27 October 2019).
16. MSFD. Signal Processing Guidelines. Available online: http://www.quietmed-project.eu/wp-content/uploads/2019/01/QUIETMED_D3.2_Best_practices-guidelines_on_signal_processing_final.pdf. (accessed on 27 October 2019).
17. MSFD. Acoustic Modelling Processing Guidelines. Available online: http://www.quietmed-project.eu/wp-content/uploads/2019/01/QUIETMED_D3.3_Best-practices-guidelines-on-acoustic-modelling-and-mapping_final.pdf. (accessed on 27 October 2019).
18. Tougaard, J.; Carstensen, J.; Teilmann, H.; Skov, P. Rasmussen. Pile driving zone of responsiveness extends beyond 20 km for harbor porpoises. *J. Acoust Soc. Am.* **2009**, *126*, 11. [CrossRef] [PubMed]
19. Aarts, G.; von Benda-Beckmann, A.M.; Lucke, K.; Sertlek, H.Ö.; van Bemmelen, R.; Geelhoed, S.C.V.; Brasseur, S.; Scheidat, M.; Lam, F.P.A.; Slabbekoorn, H.; et al. Harbour porpoise movement strategy affects cumulative number of animals acoustically exposed to underwater explosions. *Mar. Ecol. Prog. Ser.* **2016**, *557*, 261–275. [CrossRef]
20. Fewtrell, J.L.; McCauley, R.D. Impact of air gun noise on the behaviour of marine fish and squid. *Mar. Pollut. Bull.* **2012**, *64*, 984–993. [CrossRef] [PubMed]
21. Dolman, S.J.; Evans, P.G.H.; Notarbartolo-di-Sciara, G.; Frisch, H. Active sonar, beaked whales and European regional policy. *Mar. Pollut. Bull.* **2011**, *63*, 27–34. [CrossRef] [PubMed]
22. Götz, T.; Janik, V.M. Acoustic deterrent devices to prevent pinniped depredation: Efficiency, conservation concerns and possible solutions. *Mar. Ecol. Prog. Ser.* **2013**, *492*, 285–302. [CrossRef]
23. González-Reolid, I.; Molina Molina, J.C.; Guerrero-González, A.; Ortiz, F.J.; Alonso, D. An Autonomous Solar-Powered Marine Robotic Observatory for Permanent Monitoring of Large Areas of Shallow Water. *Sensors* **2018**, *18*, 3497. [CrossRef] [PubMed]
24. Urick, R.J. *Principles of Underwater Sound*, 3rd ed.; McGraw-Hill Inc.: New York, NY, USA, 1996.

25. Leighton, T.G. Fundamentals of Underwater Acoustics. In *Fundamentals of Noise and Vibrations*, 1st ed.; Fahy, F.J., Walker, J.G., Eds.; Taylor and Francis: London, UK, 1998.
26. Schulkin, M.; Mercer, J.A. Colossus Revisited: A Review and Extension of the Marsh-Schulkin Shallow Water Transmission Loss Model. Applied Physics Laboratory, University of Washington, APL-UW 8508. 1985.

Article

A Sensor for Spirometric Feedback in Ventilation Maneuvers during Cardiopulmonary Resuscitation Training

Rodolfo Rocha Vieira Leocádio [1,2,*], Alan Kardek Rêgo Segundo [1] and Cibelle Ferreira Louzada [2]

1 Department of Control and Automation Engineering (DECAT), Escola de Minas, Universidade Federal de Ouro Preto (UFOP), Morro do Cruzeiro, 35400-000 Ouro Preto, MG, Brazil; alankardek@ufop.edu.br

2 Department of Pediatric and Adult Clinic (DECPA), Escola de Medicina, Universidade Federal de Ouro Preto (UFOP), Morro do Cruzeiro, 35400-000 Ouro Preto, MG, Brazil; cibelle.louzada@hotmail.com

* Correspondence: rodolfo.leocadio@ufop.edu.br; Tel.: +55-31-98807-3747

Received: 30 September 2019; Accepted: 3 November 2019; Published: 21 November 2019

Abstract: This work proposes adapting an existing sensor and embedding it on mannequins used in cardiopulmonary resuscitation (CPR) training to accurately measure the amount of air supplied to the lungs during ventilation. Mathematical modeling, calibration, and validation of the sensor along with metrology, statistical inference, and spirometry techniques were used as a base for aquiring scientific knowledge of the system. The system directly measures the variable of interest (air volume) and refers to spirometric techniques in the elaboration of its model. This improves the realism of the dummies during the CPR training, because it estimates, in real-time, not only the volume of air entering in the lungs but also the Forced Vital Capacity (FVC), Forced Expiratory Volume (FEVt) and Medium Forced Expiratory Flow ($FEF_{20–75\%}$). The validation of the sensor achieved results that address the requirements for this application, that is, the error below 3.4% of full scale. During the spirometric tests, the system presented the measurement results of (305 ± 22, 450 ± 23, 603 ± 24, 751 ± 26, 922 ± 27, 1021 ± 30, 1182 ± 33, 1326 ± 36, 1476 ± 37, 1618 ± 45 and 1786 ± 56) × 10^{-6} m^3 for reference values of (300, 450, 600, 750, 900, 1050, 1200, 1350, 1500, 1650 and 1800) × 10^{-6} m^3, respectively. Therefore, considering the spirometry and pressure boundary conditions of the manikin lungs, the system achieves the objective of simulating valid spirometric data for debriefings, that is, there is an agreement between the measurement results when compared to the signal generated by a commercial spirometer (Koko brand). The main advantages that this work presents in relation to the sensors commonly used for this purpose are: (i) the reduced cost, which makes it possible, for the first time, to use a respiratory volume sensor in medical simulators or training dummies; (ii) the direct measurement of air entering the lung using a noninvasive method, which makes it possible to use spirometry parameters to characterize simulated human respiration during the CPR training; and (iii) the measurement of spirometric parameters (FVC, FEVt, and $FEF_{20–75\%}$), in real-time, during the CPR training, to achieve optimal ventilation performance. Therefore, the system developed in this work addresses the minimum requirements for the practice of ventilation in the CPR maneuvers and has great potential in several future applications.

Keywords: cardiopulmonary resuscitation; sensor; medical mannequins; spirometric techniques

1. Introduction

Noninvasive methods used to characterize human and animal respiration require advanced techniques and are costly, such as: computed tomography and densitometry [1–6], electrical impedance tomography [7,8], magnetic resonance imaging [9–11], contrast radiology [12], image ultrasound [13,14], ultrasonic sensors [15–17], pulse oscillometry [18,19], electrostatic methods [20], closed circuits with inert gases [21], impedance pneumography and plethysmography [22,23] and electronic noses [24].

One of the procedures used in plethysmography is spirometry, which uses physical concepts to study the air going in and out of the lungs, characterizing human breathing [25,26]. The technique is used to evaluate pulmonary function [1,18], chronic obstructive pulmonary disease [1,27], cystic fibrosis [2], smokers [4], air pollution [28–30], hyperinflation [31,32], exposure of particulates such as nanotubes and nanofibers [33] and airway resistance [34], among others.

To perform spirometry, a spirometer is used, which can be: (i) of volume (sealed in water, piston, and bellows [35]); (ii) of flow (differential pressure or pneumotacometers [36], thermistors, Pitot and turbinometers [35]); or (iii) portable [37] of volume or flow [25]. The respiratory volume sensors used in these types of equipment have a high cost and some of them perform indirect measurements, discarding their application in medical simulators or dummies, as in the case of Hamiltonian sensors coupled to differential pressure sensors [38], flow mass sensors [39] and airflow meter [40] rotary or vibratory beam and shell flow meters [41].

Cardiopulmonary resuscitation (CPR) is a recurring practice in medical urgencies and emergencies. CPR is characterized by a set of maneuvers performed in an attempt to reanimate the victim of cardiac and/or respiratory arrest, to restore the heart and lung to normal functions while maintaining the oxygenation of the brain. These procedures provide continuous improvement in the quality of healthcare professional skills by using automated dummies for teaching.

The pioneers in automating CPR dummies [42] used the Resusci Anne® manikin. The authors developed compression and ventilation sensors using digital logic and normally open contacts. The monitoring of the system was done using a display panel or indicator lamps, which showed indications of "insufficient", "acceptable", and "above acceptable". Currently, identical models are still widely used [43].

In the present study, we propose measuring the volume of air supplied to the lungs in rescue ventilation during CPR training using a rotor-type flow sensor with propellers. Also, we develop a theoretical model to make it equivalent to spirometric models. This brings more realism to the dummies and introduces advantages to possible debriefings after various simulations [44].

This work is an extension of [44], (doi:10.3390/ecsa-5-05724), where only the idea was put forward to assess the feasibility of the application. Additionally, in the present article, we perform the validation of the adapted water flow sensor to measure airflow considering concepts of fluid mechanics. Also, we apply spirometric concepts to the results, defining a theoretical model for the curves obtained.

2. Mathematical Modeling of Propeller Type Flow Sensors

The analysis of the behavior of any material contained in a finite region of space, control volume, solves many problems involving fluid mechanics [45]. The Reynolds transport theorem ensures that the time rate of change of mass within a system is equal to the sum of the time rate of mass within the control volume (CV) and the net flux of mass through the control surface (CS), that is,

$$\frac{DM_{sys}}{Dt} = \frac{\partial}{\partial t}\int_{CV} \rho dV + \int_{CS} \rho v \cdot \hat{n} dS, \tag{1}$$

where M_{sys} is the mass of the system (kg), ρ is the specific mass of the fluid (kg/m^3), V is the control volume (m^3), and v is the velocity vector perpendicular to the differential area dS (m/s).

Using the principle of mass conservation for a system, the material derivative of the mass of the system is

$$M_{sys} = \int_{sys} \rho dV, \tag{2}$$

therefore,

$$\frac{DM_{sys}}{Dt} = 0. \tag{3}$$

At permanent regime, the properties at any point in the system remain constant over time, so

$$\frac{\partial}{\partial t}\int_{CV}\rho dV = 0. \tag{4}$$

Applying (3) and (4) to (1) and adding up all the differential contributions that exist on the control surface, we obtain the net flux of mass in the control volume, that is,

$$\int_{CS}\rho v\cdot\hat{n}dS = \sum \dot{m}_O - \sum \dot{m}_I = 0. \tag{5}$$

Considering that the input of the sensor in question has the same characteristics of the output, that is, $S_I = S_O$, and applying Equation (3) to its control volume, we conclude that:

$$\dot{m}_I = \dot{m}_O. \tag{6}$$

A widely used expression for mass flow assessment $\dot{m}$ (kg/s), in a section of the control surface with area S (m^2), is

$$\dot{m} = \rho Q = \rho S v, \tag{7}$$

where Q is the volume flow (m^3/s), and v is the velocity vector perpendicular to area S (m/s).

We can adequately analyze many mechanical fluid problems considering a fixed and undeformable control volume. In addition, considering a uniform distribution of the specific mass of the fluid in each flow section (of the compressible flows) allows specific mass variations to occur only from one section to another.

Substituting (7) into (6), we obtain

$$v_I = \frac{\rho_O}{\rho_I}v_O. \tag{8}$$

An ideal gas can be characterized by having a large number of molecules, considered as spherical beads with a mass greater than zero and negligible individual volume when compared to the volume containing them [46]. Thus, the evident macroscopic properties of an ideal gas are consequences mainly of the independent movement of the molecules as a whole.

In various conditions the gases deviate from ideality, being characterized as a real gas, which is constituted by particles endowed with chaotic movement, and subjected to the forces of attraction of long-distance and forces of repulsion at a short distance. It is important to know the specific mass range in which an ideal gas equation describes the behavior of one real gas with adequate accuracy. It is important also to know how much the behavior of a real gas can deviate from the ideal gas at a given pressure and temperature. This information originates from the compressibility factor Z. When it is an ideal gas ($Z = 1$), the distance from Z of the unit is a measure of the behavior deviation of the actual gas from that predicted by the ideal gas equation [47]

$$PV = nRT, \tag{9}$$

where P is the absolute pressure of the gas (Pa), V is the volume occupied by the gas (m^3), n is the number of moles of the gas (mols), R is the ideal gas constant (8.31 J/(mol·K)) and T is the temperature (K).

If the temperature ranges from 250 K up to 400 K, and at a pressure of 101,325 Pa, atmospheric air (compressible fluid) approaches to an ideal gas with acceptable accuracy for the system of this work [48]. If the pressure and temperature differences are small, generally less than 10%, air can be considered incompressible. Therefore, we can write Equation (8) as:

$$v_I = v_O. \tag{10}$$

3. Materials and Methods

3.1. YF-S201 Flow Sensor

The flow sensor YF-S201 (Sea brand) has been widely used to measure water flow in pipes. It consists of a valve, containing inside it a propeller rotor and a Hall effect sensor, which is commonly used by supply companies to monitor water consumption [43–47]. The rotor has a toroidal magnet that produces an alternating magnetic field as the rotor rotates [49]. The magnetic field interacts with the Hall effect sensor, which in turn produces digital pulses that correspond to the rotor speed. The rotor speed corresponds to the average speed water flowing through the valve [50,51].

Unlike other applications involving YF-S201 sensor, this work is the first one which uses it for measuring air volume and performs spirometric feedback in ventilation maneuvers during CPR using medical simulators or training manikins, in real-time.

Figure 1 shows the different views of the sensor. According to Figure 1a–c, there is throttling in the diameter of the sensor inlet channel (the area I relative to 1). Also, there is no difference in the diameter of the output channel (the area 2 relative to O) of the YF-S201 sensor. Therefore, considering the model presented in Section (2) and starting from Equation (10), we can write

$$v_I = v_O = v_2. \tag{11}$$

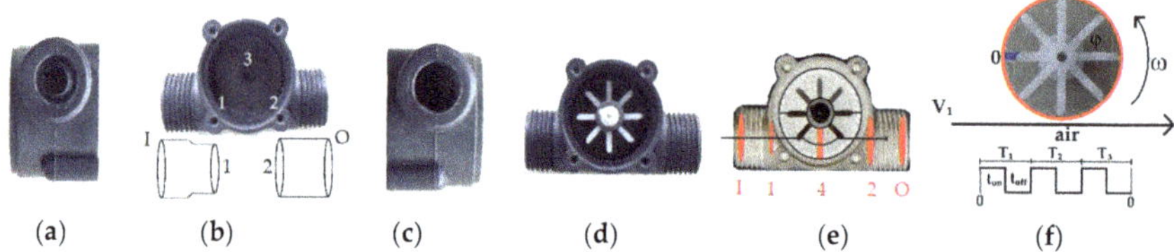

(a) (b) (c) (d) (e) (f)

Figure 1. YF-S201 flow sensor: (**a**) Input profile; (**b**) Control volume; (**c**) Output profile; (**d**) Propeller compartment; (**e**) Airflow profile; (**f**) Detail of the propeller.

According to the details of the control volume, shown in Figure 1e, the mass flow in section I is a function of mass flows in Sections 1, 4 and 2. Considering Equation (5) and the flow in the permanent regime, the volume flow of I, 1, 4, and 2 are constants, that is, $v_1 = v_4 = v_2$ and the rate of temporal variation of the mass contained in the control volume results in:

$$\dot{m}_I = \dot{m}_1 + \dot{m}_4 + \dot{m}_2. \tag{12}$$

Substituting Equation (7) into Equation (12), we obtain

$$v_I = (S_1 + S_4 + S_2)\frac{v_1}{S_I}. \tag{13}$$

In Figure 1e we can see that

$$S_1 = \frac{\pi d_1^2}{4}, \tag{14}$$

$$S_2 = \frac{\pi d_2^2}{4} \tag{15}$$

and

$$S_4 = bh; \tag{16}$$

where b is the base (m), and h is the height (m) of area 4 oriented entering the plane of the paper in Figure 1e. Thus, the input flow is

$$Q_{IO} = Q_I = S_I v_I. \quad (17)$$

Substituting Equations (13)–(16) into Equation (17) we found

$$Q_{IO} = k_1 v_1, \quad (18)$$

where k_1 is a constant which depends on the areas, that is, the geometry of the sensor.

Figure 1d shows the propeller of the YF-S201 sensor, which rotates according to the flow of air passing through it. Figure 1f shows the detail of the propeller, which has a Hall sensor for providing digital pulses proportional to its angular velocity. An ATmega328 microcontroller measures the digital pulses using external interrupt, along with a real-time scheduling and multitasking software [52].

Besides of the Hall effect sensors being widely used in fluids flow measurements [53–55], they are also used as magnetic sensors [56–58] in numerous applications such in water pump flow measurement [41], infiltrometers [49], energy monitoring [59], electromagnetic flowmeters used in industrial and physiological techniques [60], hydrometers [61], induction-frequency converters [62], among others.

To make the sensor suitable for measuring airflow, we use the relation of the linear velocity v_1 (m/s) with the angular velocity ω (rad/s) [47], this is

$$v_1 = \frac{d_h f}{2}, \quad (19)$$

where d_h (m) is the diameter of the helix and f (H_Z) is the rotation frequency of the helix.

Applying Equation (19) in Equation (18), we have

$$Q_{IO} = kf \quad (20)$$

where

$$k = d_h\left[\frac{\pi}{8}\left(d_1^2 + d_2^2\right) + \frac{bh}{2}\right] \quad (21)$$

is a constant equal to $(261 \pm 3) \times 10^{-8}$ m^3, calculated according to the sensor dimensions. The calculation of the geometric constant k, according to the mathematical modeling presented in this work (Equation (21)), is the first step to adjust the sensor output signal to the unit of measurement: flow (m^3/s).

From (20), the volumetric flow is

$$V = Q_{IO} \cdot t, \quad (22)$$

where V is the volume of air (m^3) flowing inside the lung (reservoir) of the dummy during the time interval Δt (s).

3.2. Calibration

After calculating the geometric constant k, according to Equation (21), a calibration procedure was performed to verify the degree of agreement between the measurements made by the YF-S201 sensor and the reference values of (300 ± 2, 450 ± 3, 600 ± 3, 750 ± 4, 900 ± 5, 1050 ± 6, 1200 ± 6, 1350 ± 7, 1500 ± 8, 1650 ± 9 and 1800 ± 9) × 10^{-6} m^3, provided by a syringe especially used in spirometer calibration procedures. As stated in its manual, the syringe was marked at the points corresponding to volumes of interest, according to Table 1, to perform the calibration. Therefore, it was possible to estimate the systematic error and the range of the random error is expected with 95% of probability, achieving application of the bias correction and collecting information about the uncertainty of the instrument along with its measurement range in future measurements, respectively.

Table 1. The relationship between the length of stem and volume provided by the syringe.

Reference volume ($\times10^{-6}$ m^3) ±0.5%	0	300	450	600	750	900	1050	1200	1350	1500	1650	1800
Length of the stem ($\times10^{-3}$ m) $\pm1\times10^{-3}$ m	0	42	64	85	106	127	148	169	191	212	233	254

The acceptable limit of error in spirometry for Forced Vital Capacity (FVC) and Forced Expiratory Volume (FEV) is 3.5% of full scale [25]. Therefore, in this work, we considered the maximum error ε equal to 60×10^{-6} m^3, since it represents the limit of 3.4%, satisfying the spirometric conditions.

Considering a small number of repetitions ($10 \leq n \leq 25$), and assuming that the mean of the indications follows an approximately normal distribution, the t-Student distribution is used to determine the confidence interval. Due to statistical inference, we have:

$$\varepsilon = t_{\alpha/2}\frac{s_0}{\sqrt{n}}. \tag{23}$$

Ten random measurements were taken to estimate the standard deviation s_0, which is approximately equal to 98×10^{-6} m^3. For a 95% confidence interval, the significance level α is 0.05, so $t_{\alpha/2}$= 2.2. Thus, a total of 13 measurements (n) should be performed to ensure the statistical significance of the data according to the sampling rules. According to Student's distribution, we obtained $t = 2.17$, which will be used in subsequent tests [63,64]. The correction is added to the measurements to compensate for the effect of the systematic error. The estimated systematic error corresponds to the average value of the measurement error, i.e., the average of n sensor measurements of the same measurand carried out under repeatability conditions minus the conventional true value of the measurand, provided by a standard or a reference instrument. The correction is equal to the negative of the estimated systematic error [63,64].

The uncertainty of measurement (U) defines an interval about the result of a measurement that may be expected. In this work, it represents the symmetric range of values around the average error where the random error is expected with 95% of probability. As the probability distribution of the sensor measurements follows the Normal distribution (according to the Normal Probability Plot with R-square equal to 0.99927), we considered Student's distribution to take into account the difference between the standard deviation of the mean and the experimental standard deviation of the mean [63,64]. Thus, the uncertainty of measurement is:

$$U = t_{0.95}\, u, \tag{24}$$

where $t_{0.95}$ is the t-factor from Student's distribution considering 95% of confidence level; and u is the Type A standard uncertainty, calculated as the experimental standard deviation of the mean.

In this work, the error curve represents the calibration results. It is formed by the center line, which represents the estimated systematic error; and by the upper and lower limits of the range containing the random errors, i.e., the uncertainty of measurement.

3.3. Spirometric Tests

The procedures required to perform rescue ventilation in the practice of CPR must follow the parameters of the American Heart Association [65], which establishes a breath every five or six seconds, that is, 10 to 12 breaths during each 60 s. Approximately 500×10^{-6} m^3 of air enters and leaves the lungs of a healthy young adult in a resting state at each respiratory cycle [66]. Therefore, efficient ventilation should provide such a volume of air to the lungs by mouth-to-mouth or using devices for this purpose.

As said before, spirometry is the measure of the air that enters (inspired) and exits (expired) from the lungs. It can be performed during slow breathing or forced expiratory maneuvers. One of the results generated by this technique is an inspired/expired volume versus time graph [25].

Figure 2 shows, as an example, a result of a real spirometry test performed at the Collective Health Laboratory of the Federal University of Ouro Preto, in a 72-year-old male, 58 kg and 1.68 m; following the stress protocol [25], generated by a commercial instrument (Koko brand). The Koko spirometer utilizes a differential pressure sensor, also known as a pneumotachometers, which measures a small (but measurable) pressure difference around a low-value resistance. As the variations in pressure to be detected are small, the material that constitutes the resistance has a high cost. Furthermore, like other commercial spirometers, it cannot be installed on the CPR training dummy because it takes up a lot of space inside it and the response time does not meet the real-time prerequisites for performing CPR training spirometric feedback. It is also worth noting that the spirometer response automatically correlates the measurement range with the patient's breathing conditions, what not desirable during CPR, as the goal is to test for optimal ventilatory maneuvers on a cardiorespiratory arrest victim. For volumes between 300 to 600.10^{-6} m^3, the typical CPR range, the spirometer has difficulty to perform measurements, as this is not the spirometric assessment range, which is usually around 3 to 6×10^{-3} m^3.

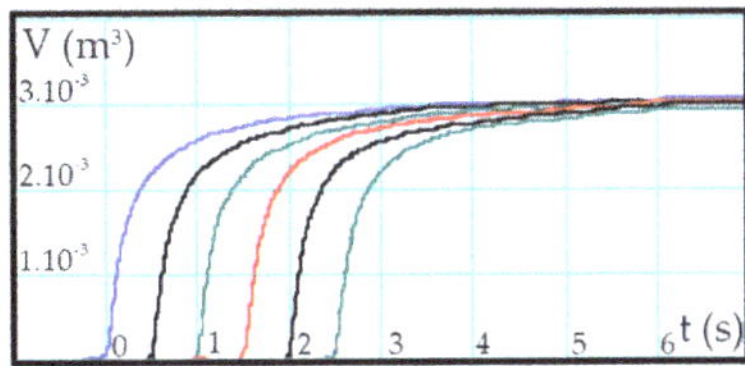

Figure 2. Volume versus Time chart generated by Koko spirometer.

Two parameters obtained from these curves are Forced Vital Capacity (FVC) and Forced Expiratory Volume (FEVt). FVC is measured by asking the individual to breathe out until the total lung capacity and expires as rapidly and intensely as possible in a spirometer (Figure 2, FVC = 3×10^{-3} m^3). FEVt can be measured in the FVC maneuver at predefined intervals. In the blue line of Figure 2, FEV is approximately 2.5×10^{-3} m^3 for 1 s, 2.8×10^{-3} m^3 for 2 s, 2.9×10^{-3} m^3 for 3 s, and practically equal to FVC for 4 s. Besides, the FEV value for 1 s should be approximately 80% of the FVC value [25].

From the blue curve of Figure 2, another parameter is obtained: The Medium Forced Expiratory Flow ($FEF_{25–75\%}$). To calculate $FEF_{25–75\%}$, we mark the points at which 25% and 75% of the FVC were expired on the volume-time curve. A straight line connecting these points is drawn with a duration of 1 s. The vertical distance between the intersection points is $FEF_{25–75\%}$ [25].

After calibration, both the YF-S201 sensor and the Koko spirometer were used in a spirometric test, which consisted of applying known air volumes using the syringe: (300 ± 2, 450 ± 3, 600 ± 3, 750 ± 4, 900 ± 5, 1050 ± 6, 1200 ± 6, 1350 ± 7, 1500 ± 8, 1650 ± 9 and 1800 ± 9) $\times 10^{-6}$ m^3. The total volume of air inside the syringe was passed through the spirometer, lifting the curves from the test. Such curves correspond to the volume of the syringe, considering the measurement error.

Beyond the YF-S201 sensor, Figure 3a,b shows the other components used to perform this test: the syringe outlet and the Koko spirometer, respectively. The spirometer shows the uncertainty of 3% or 100×10^{-6} m^3, reproducibility of 0.5% or 150×10^{-6} m^3, volume range of 16×10^{-3} m^3, flow rate 16×10^{-3} m^3/s, and resistance less than 147.1×10^3 $Pa/(m^3s)$ with the filter.

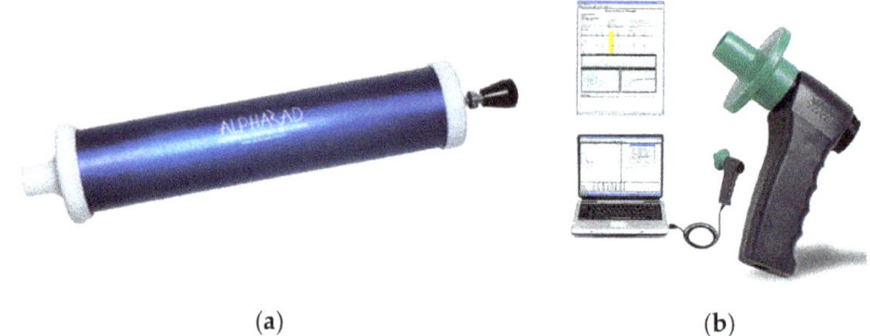

(a) (b)

Figure 3. Components for calibrating the spirometer: (**a**) Calibration syringe; (**b**) Koko flow spirometer model 313105.

The measurement results presented by the Koko spirometer were used to verify the quality of the measurements obtained by the system developed in this work, under the same experimental conditions.

3.4. Spirometric Feedback in Ventilation Maneuvers during Cardiopulmonary Resuscitation Training

We performed a test using both the sensor developed in this work and the system of the automated dummy manufactured by Laerdal®, which uses a linear optical encoder sensor to measure the volume of air entering the lungs. The encoder sensor measures the chest expansion that occurs during the ventilatory maneuver and relates it to the amount of air that has entered in the lung, so it is an indirect measurement. Thus, the range of measured volumes is limited, and it is also impossible to apply spirometric concepts from such indirect measurements.

The sensor installed on the dummy has the configuration of Figure 4, characterizing the system. Air is considered an ideal gas in the temperature range that the sensor works. The one-way valve A is placed in the mouth of the manikin to prevent that contaminated air from returning to the person who is performing the maneuver, due to hygiene. The sensor is coupled between the lung and the one-way valve A, and it has the function of measuring the volume of air entering the lung. The manikin has a single lung with a volume of 3500×10^{-6} m^3 that has the function of storing ventilation air and causing thoracic expansion. The one-way valve B ensures that the amount of air exiting the lung into the atmosphere is less than the amount of air entering the lung, so it is responsible for thoracic expansion.

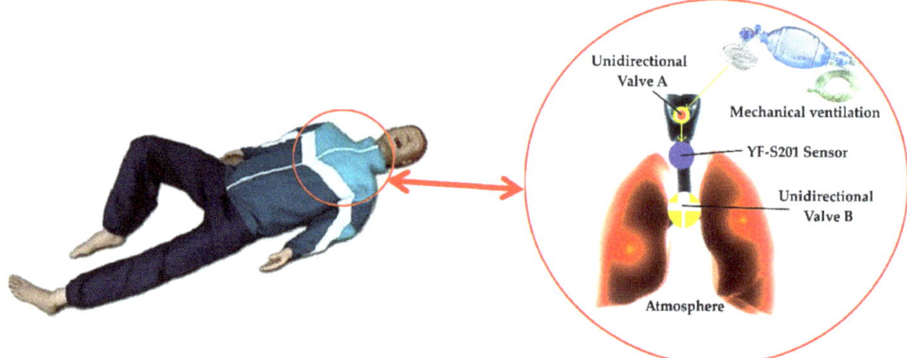

Figure 4. Air route inside the manikin. Adapted from CanStock and SimulaCare.

In 100% of the vital capacity, the inspiratory muscle pressure can reach a maximum of 2942 Pa, and the expiratory muscle pressure can reach at least −2942 Pa [67]. Thus, the maximum pressure difference in the lungs, both expiratory and inspiratory, is 5884 Pa. Most mechanical ventilation devices have a safety valve that operates at a pressure of 4903 to 5884 Pa. Therefore, the maximum pressure in the mechanical ventilation can reach 5884 Pa. Since at the end of the process there is an open tube, the pressure at this point is atmospheric, so the pressure difference at the inlet and outlet of the device does not exceed 6%. In this way, the device installed inside the mannequin addresses the boundary conditions imposed by the theoretical model.

The test procedure involved performing ventilatory maneuvers on the dummy, simulating CPR training, containing inside the arrangement of Figure 4, as well the encoder. The volume of air that enters in the lungs of the dummy and causes chest expansion was measured simultaneously by both the YF-S201 sensor and the encoder.

4. Results and Discussion

4.1. Calibration and Validation

Figure 5 shows the error curve for the YF-S201 sensor, considering $n = 13$. There are repeatability and agreement between the results of the measurements performed. Therefore, after applying the bias corrections in the results, the sensor model presents a minimum uncertainty of 22×10^{-6} m^3 for volumes up to 300×10^{-6} m^3, and a maximum uncertainty of 56×10^{-6} m^3 for volumes up to 1800×10^{-6} m^3. Thus, the systematic and random errors were characterized, with a maximum error of 65×10^{-6} m^3 or 3.6%.

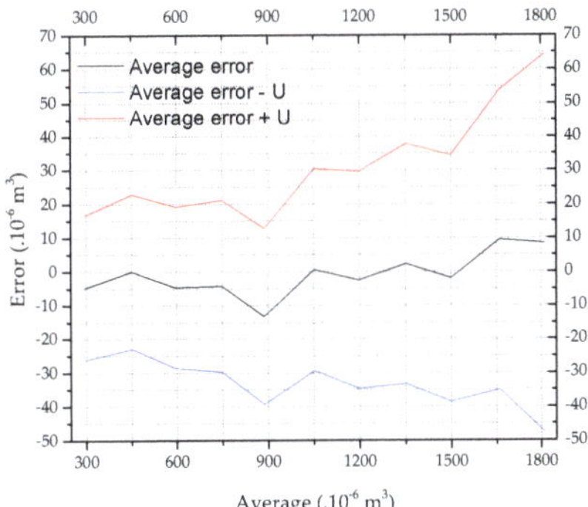

Figure 5. Error curve of the YF-S201 sensor.

After the calibration, the sensor performed the measurements shown in Table 2, using the syringe. The results are according to the spirometric model, and as expected for the performance of the sensor, i.e., the uncertainty is less than 3.4% of the full scale, satisfying the spirometric conditions [25].

Table 2. Measurements after calibration.

Reference Volume ($\times 10^{-6}$ m^3) ±0.5%	300	450	600	750	900	1050	1200	1350	1500	1650	1800
Average Indication ($\times 10^{-6}$ m^3)	301	450	601	751	901	1050	1201	1356	1500	1650	1800
Uncertainty ($\times 10^{-6}$ m^3)	22	23	24	26	27	30	33	36	37	45	56

4.2. Spirometric Tests

Figure 6 shows the measurement results of the YF-S201 sensor (experimental data) and the spirometric model curve obtained from these measurements. We found proximity among the dataset of each graph of Figure 2 and the nonlinear models of Boltzmann's (BTZ), Logistic (LG), Modified Langevin (ML), Doseresp, Gompertz, Slogistic, and Langmuir EXT 1 (LA). Aiming to fit the dataset to these models, we performed the algorithms of Levenberg Marquardt (LM) and Orthogonal Distance Regression (ODR) for each model. We chose the Langevin model along with the Orthogonal Distance Regression algorithm because it reaches the highest coefficient of determination (R-squared), according to Table 3.

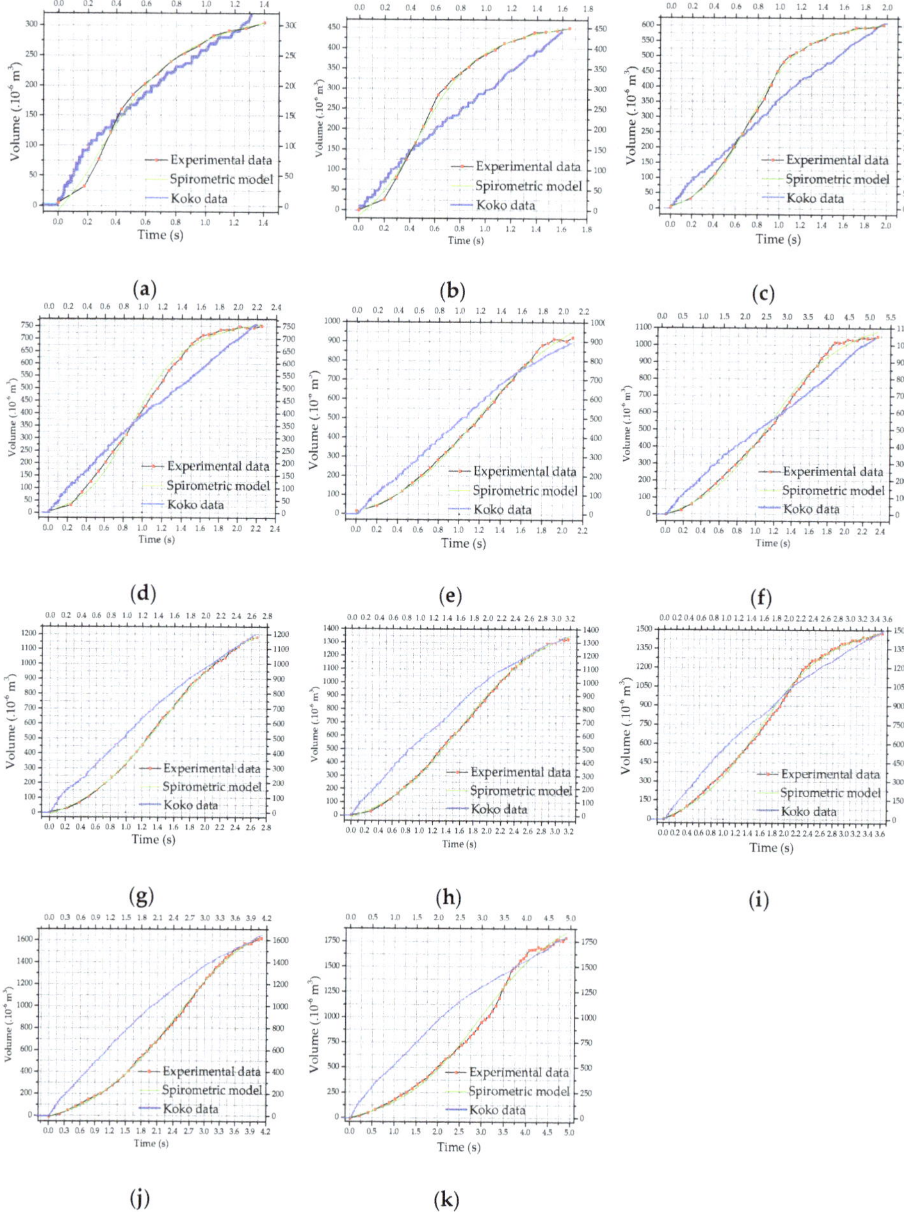

Figure 6. Result of validation of YF-S201 sensor with air: (**a**) 300×10^{-6} m^3; (**b**) 450×10^{-6} m^3; (**c**) 600×10^{-6} m^3; (**d**) 750×10^{-6} m^3; (**e**) 900×10^{-6} m^3; (**f**) 1050×10^{-6} m^3; (**g**) 1200×10^{-6} m^3; (**h**) 1350×10^{-6} m^3; (**i**) 1500×10^{-6} m^3; (**j**) 1650×10^{-6} m^3 and (**k**) 1800×10^{-6} m^3.

Table 3. R-Square of the non-linear adjustments.

Volume ($\times 10^{-6}$ m^3)	R-Square													
	BTZ		LG		ML		Doseresp		Gompertz		Slogistic		LA	
	LM	*ODR*	*LM*	*ODR*	*LM*	*ODR*	*LM*	*ODR*	*LM*	*ODR*	*LM*	*ODR*	*LM*	*ODR*
300						0.99999997354852		0.99999996735480				0.99999968697100		
450						0.99999998774221		0.99999998313179				0.99999956213635		
600						0.99999999387587		0.99999998738478				0.99999994155838		
750						0.99999997787916		0.99999998623968				0.99999996494737		
900						0.99999999363261		0.99999999373928				0.99999998003168		
1050	IA	NC	IA	NC	IA	0.99999998596560	IA	0.99999998676343	IA	NC	IA	0.99999977894465	IA	NC
1200						0.99999999843628		0.99999999853441				0.99999994501474		
1350						0.99999999976504		0.99999999797433				0.99999994523958		
1500						0.99999999614389		0.99999999677671				0.99999993443600		
1650						0.99999999808238		0.99999999767503				0.99999680496200		
1800						0.99999998519947		0.99999998367641				0.99999792109656		

IA—Inadequate accuracy. NC—Not converged.

The Langevin function—a simplified version of Brillouin function—is used for classic cases of solid-state physics in quantum treatments. It has applications in paramagnetism [68–73] and dielectric properties (permittivity) [74–76]. When performing a nonlinear adjustment of experimental data, there may be a need to consider errors in both independent and dependent variables (as in the case of this work). The Orthogonal Distance Regression algorithm [77–79] has applications in metrology [80] because it adjusts data with implicit or explicit functions.

Langevin's function is scale modified to address this application, and the mathematical equation that describes the model is:

$$Y = Y_0 + C\left[coth\left(\frac{x - x_c}{s}\right) - \frac{s}{x - x_c}\right], \tag{25}$$

where Y_0 is the linear coefficient of equation (m^3), x_c is the central coordinate of the curve (s), C is the amplitude of the curve (m^3), and s is the scale. To obtain Langevin's equation, we set the initial guess $Y_0 = 0, C = 1, x_c = 0$ and $s = 1$.

Table 4 shows the convergence parameters of the non-linear adjustment. It is noteworthy that the R-squared of the calculated model is close to unity, so the modified Langevin mathematical model can be used to describe the spirometric curve and, consequently, the results obtained in this work.

Table 4. Parameters of convergence of non-linear adjustment applied to the results.

Volume ($\times 10^{-6}$ m^3)	Y_0	x_c	C	s
300	98 ± 21	0.30 ± 0.07	258 ± 29	0.22 ± 0.02
450	182 ± 10	0.47 ± 0.02	320 ± 14	0.19 ± 0.01
600	284 ± 3	0.72 ± 0.01	372 ± 5	0.18 ± 0.01
750	364 ± 6	0.89 ± 0.02	467 ± 12	0.21 ± 0.02
900	515 ± 12	1.20 ± 0.03	735 ± 39	0.38 ± 0.03
1050	535 ± 9	1.15 ± 0.03	745 ± 33	0.33 ± 0.03
1200	608 ± 4	1.43 ± 0.01	891 ± 15	0.45 ± 0.01
1350	672 ± 5	1.62 ± 0.01	999 ± 19	0.51 ± 0.02
1500	709 ± 6	1.56 ± 0.02	997 ± 14	0.47 ± 0.01
1650	881 ± 6	2.40 ± 0.02	1218 ± 17	0.66 ± 0.02
1800	1006 ± 24	2.98 ± 0.06	1354 ± 54	0.77 ± 0.05

The results of the measurements agree with the conventional values of the measured volume, considering the experimental error. As shown in Table 2, for the reference values of (300 ± 2, 450 ± 3,

600 ± 3, 750 ± 4, 900 ± 5, 1050 ± 6, 1200 ± 6, 1350 ± 7, 1500 ± 8, 1650 ± 9 and 1800 ± 9) × 10^{-6} m^3, the developed system measured (305 ± 22, 450 ± 23, 603 ± 24, 751 ± 26, 922 ± 27, 1021 ± 30, 1182 ± 33, 1326 ± 36, 1476 ± 37, 1618 ± 45 and 1786 ± 56) × 10^{-6} m^3 (Figure 6a-k, respectively).

Comparing the graphs in Figure 6 (Experimental Data and Calculated Spirometric Model) with the spirometric model, from zero to the maximum experimental volume, the behavior follows the spirometer models (Figure 2), characterizing the inspiration. During CPR, there is no muscle activity in the victim's chest, so the victim's expiration occurs due to the chest's weight or due to the resumption of the cardiac massage. Due to these conditions, it is not possible to apply spirometry concepts to the expiration step.

Although the profile of the YF-S201 curves is slightly different from the Koko spirometer results, Table 5 shows that the major part of spirometric results is equivalent. The difference occurs because the Koko spirometer has limitations in use in CPR training, as its measurement range is related to physical breathing parameters such as completely obstructed airway (0-300 × 10^{-6} m^3), partially obstructed airway (300-1000 × 10^{-6} m^3) or severe disease (200-2000 × 10^{-6} m^3). The results of this study refer to cardiorespiratory arrest victims, i.e., a person in conditions of severe disease. However, to perform optimal ventilatory maneuvers during CPR, the result must contain spirometric characteristics such as those obtained by the YF-S201 sensor (Figure 6), whose response does not depend on the measurement range.

Table 5. Comparison between spirometric results of Koko and the sensor developed in this work.

Reference (×10^{-6} m^3)	Measured Volume (×10^{-6} m^3)		FVC (×10^{-6} m^3)		t_{FVC} (s)		$FEV_{t=1\,s}$ (×10^{-6} m^3)		$FEF_{25-75\%}$ (×10^{-6} m^3/s)	
	YF-S201	Koko	YF-S201	Koko	YF-S201	Koko	YF-S201	Koko	YF-S201	Koko
300	305 ± 22	320 ± 23	305 ± 22	320 ± 23	1.4 ± 0.1	1.3 ± 0.1	274 ± 22	260 ± 23	355 ± 26	230 ± 40
450	450 ± 23	450 ± 45	450 ± 23	450 ± 45	1.7 ± 0.1	1.6 ± 0.2	384 ± 23	310 ± 45	500 ± 77	270 ± 81
600	603 ± 24	610 ± 80	603 ± 24	610 ± 80	2.0 ± 0.1	2.0 ± 0.3	463 ± 24	360 ± 80	617 ± 87	330 ± 143
750	751 ± 26	760 ± 100	751 ± 26	760 ± 100	2.1 ± 0.1	2.2 ± 0.3	414 ± 26	390 ± 100	565 ± 78	330 ± 143
900	922 ± 27	890 ± 100	922 ± 27	890 ± 100	2.1 ± 0.1	2.1 ± 0.2	395 ± 27	490 ± 100	585 ± 67	480 ± 150
1050	1051 ± 30	1050 ± 100	1051 ± 30	1050 ± 100	2.4 ± 0.1	2.4 ± 0.2	419 ± 30	490 ± 100	635 ± 78	420 ± 124
1200	1182 ± 33	1200 ± 100	1182 ± 33	1200 ± 100	2.7 ± 0.1	2.7 ± 0.2	334 ± 33	530 ± 100	565 ± 75	490 ± 139
1350	1326 ± 36	1350 ± 100	1326 ± 36	1350 ± 100	3.2 ± 0.1	3.2 ± 0.2	316 ± 36	560 ± 100	597 ± 89	470 ± 129
1500	1476 ± 37	1500 ± 100	1476 ± 37	1500 ± 100	3.6 ± 0.1	3.7 ± 0.2	372 ± 37	570 ± 100	610 ± 96	440 ± 118
1650	1618 ± 45	1650 ± 100	1618 ± 45	1650 ± 100	4.1 ± 0.1	4.2 ± 0.3	197 ± 45	530 ± 100	565 ± 107	430 ± 156
1800	1786 ± 56	1800 ± 100	1786 ± 56	1800 ± 100	4.9 ± 0.1	5.0 ± 0.3	179 ± 56	530 ± 100	525 ± 127	380 ± 136

From these curves, it was possible to obtain information about FVC, $FEV_{t=1\,s}$ and $FEF_{25-75\%}$, shown in Table 5. In spirometry, the FEV value in 1 s time is approximately 80% of the FVC value. One ventilation should be done every 6 s in a forced ventilation maneuver. On average, there are 3 s for expiration and 3 s for inspiration. Therefore, for a time of 3 s, FEV always has to be less than FVC, i.e., ventilation should provide FVC within 3 s. It is noteworthy that it happens in the graphs of (300, 450, 600, 750, 900, 1050 and 1200) × 10^{-6} m^3; around times of 1.40, 1.65, 1.96, 2.25, 2.09, 2.40 and 2.67 s; respectively (Figure 6). In the (1350, 1500, 1650 and 1800) × 10^{-6} m^3 charts; it occurs around 3.1, 3.61, 4.05 and 4.89 s; respectively (Figure 6). It is also worth noting that the values highlighted in blue in Table 5, measured by the YF-S201 sensor, are different from those obtained by the Koko spirometer. This happens because the higher volumes have a capacity which is not supported by the dynamics of ventilation, because air volumes applied at short intervals cause stomach insufflation to occur, differently from the dynamics of Koko spirometry. If the dynamics of ventilation fail, it is still possible to address the requirements of spirometry applying a faster ventilatory maneuver without stomach insufflation.

It is also worth mentioning that FVC provides the instantaneous maximum expired volume. In mechanical ventilation, it represents the amount of air that was introduced into the lung, and therefore air volume provided in the ventilation. When calculating $FEF_{25-75\%}$, note that the values highlighted in green in Table 5 are different from the values measured by the YF-S201 sensor. Meeting the spirometric parameters in this range is difficult, and converges to results from serious diseases, with airway obstruction or dead volumes. To characterize ventilation, performed in humans under the mentioned conditions, the sensor of this work obtains results with adequate spirometric standards [25], unlike the Koko spirometer.

4.3. Spirometric Feedback in Ventilation Maneuvers during Cardiopulmonary Resuscitation Training

Comparing measurements performed with the Laerdal® manikin simultaneously with the system of this study, the latter presents a superior performance when compared to the first one, besides a smaller experimental error, according to Table 6. The Laerdal® model is limited to measures below 1000×10^{-6} m^3, and this work is limited to measures below 1800×10^{-6} m^3, therefore, it caters to all devices used in rescue ventilations. Moreover, only when the Laerdal® indicates $\leq 400 \times 10^{-6}$ m^3, the values are experimentally equal, but below or above this value, there are divergences between the measurements. The Laerdal® model performs indirect measurements of air volume entering the lung based on chest position, which causes errors when the volume is far from 400×10^{-6} m^3. On the other hand, the sensor of this work performs the direct measurement of the air volume, which is much more accurate compared to this kind of indirect measurement.

Table 6. Simultaneous measurements of the Laerdal® and YF-S201 sensors.

Laerdal® ($\times 10^{-6}$ m^3)	Indicators	This Work ($\times 10^{-6}$ m^3)
0	Off	196 ± 2
	Orange	215 ± 2
	Orange	282 ± 2
	Orange	328 ± 3
	Orange	373 ± 3
≤400 ± 60	Orange	419 ± 3
>400 ± 60	Green	557 ± 4
	Green	663 ± 5
≤600 ± 90	Green	851 ± 6
>600 ± 90	Red	1096 ± 2

Comparing the measurements provided by the YF-S201 sensor and the Koko spirometer (Table 5), we observed that the results are experimentally equivalent. Therefore, the YF-S201 achieves the objective of measuring air volume entering the lung of CPR dummies in respiratory maneuvers providing spirometric results. As stated before, the incorporation of sensors such as those presented in [1,2,4,19,27–41] is not feasible for this purpose due to, mainly, its high cost.

Another advantage is the simplicity with which measurements are performed, functioning as a noninvasive method that characterizes the ventilation maneuver. The fact that techniques and sensors presented in [1–24] require advanced techniques also make their application on dummies unfeasible, due to their complexity and, again, because they have a high cost. Therefore, the alternative presented in this manuscript is attractive for the proposed application because it adds spirometric feedback to ventilation practices in medical simulators using a low-cost sensor that is accord to the application requirements.

The main advantage of the prepared mechanism lies in its cost-effectiveness, the direct measurement of the air entering the lung, and the measurements of spirometric parameters during CPR training. Furthermore, we expect to generate feedback to the users, in future works, as expiration charts based on spirometric models, to bring more realism to the simulations, and innumerable debriefing possibilities.

The spirometric parameters, especially the FVC, along with the graphs generated for debriefing, will allow the student to perform an ideal ventilation maneuver during CPR because the system shows the amount of air that entered the lung and its spirometric input profile from the graphical analysis of the smoothness of the curve. For a more rigid control of the parameters, it is still possible to require time intervals considering the FEVt and to make indirect inference of the airflow using the mean FEV parameter.

5. Conclusions

In this work, a sensor was adapted to measure the amount of air supplied to the lungs during ventilation in cardiopulmonary resuscitation (CPR) maneuvers. The calibration and validation of the sensor achieved results that address the CPR requirements. In addition, during the spirometric tests, the system presented the measurement results of (305 ± 22, 450 ± 23, 603 ± 24, 751 ± 26, 922 ± 27, 1021 ± 30, 1182 ± 33, 1326 ± 36, 1476 ± 37, 1618 ± 45 and 1786 ± 56) × 10^{-6} m^3 for reference values of (300 ± 2, 450 ± 3, 600 ± 3, 750 ± 4, 900 ± 5, 1050 ± 6, 1200 ± 6, 1350 ± 7, 1500 ± 8, 1650 ± 9 and 1800 ± 9) × 10^{-6} m^3, respectively. Furthermore, we considered both the spirometry and pressure boundary conditions during the experiments using the mannequin lung, according to the results.

The performance of the proposed sensor was compared with a commercial spirometer, and the experimental results were equivalent. The profile of the curves and some measured parameters by the YF-S201 sensor and Koko spirometer are different. The YF-S201 characterizes normal breathing during ventilatory maneuvers while the Koko characterizes breathing from a person with a completely obstructed airway, partially obstructed airway or severe disease during the same maneuvers. After calibration, the YF-S201 sensor showed a minimum uncertainty of 22 × 10^{-6} m^3 for volumes up to 300 × 10^{-6} m^3, and a maximum uncertainty of 56 × 10^{-6} m^3 for volumes greater than 1800 × 10^{-6} m^3. Thus, the systematic and random errors were characterized, with a maximum error of 65 × 10^{-6} m^3 or 3.6%.

The experiment confirmed that the measurements can be performed in various simulations using the dummies in conjunction with the sensor. It is a cost-effective alternative, and relatively easy to adapt to different mannequins. The results were based on spirometric models, bringing more realism to the simulations, and bringing numerous possibilities of debriefing. Thus, the sensor has great potential in various future applications.

In future work, we intend to use this sensor on mannequin babies and children. In addition, a supervisory software is being developed for training purposes, and to use in conjunction with the sensor on the manikin. It is also intended to perform the instrumentation of manikins dedicated to the teaching of pulmonary intubation maneuvers and tracheostomy, which the authors believe to be a novelty.

Author Contributions: Conceptualization, R.R.V.L., A.K.R.S. and C.F.L.; methodology, R.R.V.L., and A.K.R.S.; software, R.R.V.L. and A.K.R.S.; validation, R.R.V.L., A.K.R.S. and C.F.L.; formal analysis, A.K.R.S.; investigation, R.R.V.L. and A.K.R.S.; resources, R.R.V.L.; data curation, R.R.V.L.; writing—original draft preparation, R.R.V.L.; writing—review and editing, A.K.R.S. and C.F.L.; visualization, R.R.V.L.; supervision, A.K.R.S.; project administration, A.K.R.S.; funding acquisition, R.R.V.L.

Funding: This research received financial support from the Universidade Federal de Ouro Preto (UFOP) and the Coordenação de Aperfeiçoamento de Pessoal de Nível Superior—Brazil (CAPES)—Financing Code 001.

Acknowledgments: The authors acknowledge the Collective Health Laboratory EMED/UFOP for lending the spirometer.

Conflicts of Interest: The authors declare no conflict of interest.

References

1. Gu, S.; Leader, J.; Zheng, B.; Chen, Q.; Sciurba, F.; Kminski, N.; Gur, D.; Pu, J. Direct assessment of lung function in COPD using CT densitometric measures Direct assessment of lung function in COPD using CT densitometric measures. *Physiol. Meas.* **2014**, *35*, 833–845. [CrossRef] [PubMed]
2. Kongstad, T.; Buchvald, F.F.; Green, K.; Lindblad, A.; Robinson, T.E.; Nielsen, K.G. Improved air trapping evaluation in chest computed tomography in children with cystic fi brosis using real-time spirometric monitoring and biofeedback. *J. Cyst. Fibros.* **2013**, *12*, 559–566. [CrossRef] [PubMed]
3. Monfraix, S.; Bayat, S.; Porra, L.; Berruyer, G. Physics in Medicine & Biology Quantitative measurement of regional lung gas volume by synchrotron radiation computed tomography Quantitative measurement of regional lung gas volume by synchrotron radiation computed tomography. *Phys. Med. Biol.* **2005**, *50*, 1–11. [CrossRef] [PubMed]

4. Karimi, R.; Tornling, G.; Forsslund, H.; Mikko, M.; Wheelock, A.M.; Nyrén, S.; Sköld, C.M. Differences in regional air trapping in current smokers with normal spirometry. *Eur. Respir. J.* **2017**, *49*, 1–10. [CrossRef] [PubMed]
5. Lee, D.; Chun, E.; Suh, S.; Yang, J.; Shim, S. Evaluation of postoperative change in lung volume in adolescent idiopathic scoliosis Measured by computed tomography. *Indian J. Orthop.* **2014**, *48*, 360–365. [PubMed]
6. Nyeng, T.B.; Kallehauge, J.F.; Høyer, M.; Petersen, J.B.B.; Poulsen, P.R.; Muren, L.P. Clinical validation of a 4D-CT based method for lung ventilation measurement in phantoms and patients. *Acta Oncol.* **2011**, *50*, 897–907. [CrossRef]
7. Nebuya, S.; Mills, G.H.; Milnes, P.; Brown, B.H. Indirect measurement of lung density and air volume from electrical impedance tomography (EIT) data. *Physiol. Meas.* **2011**, *32*, 1953–1967. [CrossRef]
8. Riera, J.; Corte's, J.; Masclans, J.R.; Rello, J. Effect of High-Flow Nasal Cannula and Body Position on End-Expiratory Lung Volume: A Cohort Study Using Electrical Impedance Tomography. *Respir. Care* **2013**, *58*, 589–596. [CrossRef]
9. Kyriazis, A.; De Alejo, R.P.; Rodriguez, I.; Olsson, L.E.; Ruiz-cabello, J. A MRI and Polarized Gases Compatible Respirator and Gas Administrator for the Study of the Small Animal Lung: Volume Measurement and Control. *IEEE Trans. Biomed. Eng.* **2010**, *57*, 1745–1749. [CrossRef]
10. Sonigo, P.; Mahieu-Caputo, D.; Dommergues, M.; Fournet, J.C.; Thalabard, J.C.; Abarca, C.; Benachi, A.; Brunelle, F.; Dumez, Y. Fetal lung volume measurement by magnetic resonance imaging in congenital diaphragmatic hernia. *Br. J. Obstet. Gynaecol.* **2001**, *108*, 863–868, PII: S0306-5456(00)00184-4.
11. Li, W.; Davlouros, P.A.; Kilner, P.J.; Pennell, D.J.; Gibson, D.; Henein, M.Y.; Gatzoulis, M.A. Doppler-echocardiographic assessment of pulmonary regurgitation in adults with repaired tetralogy of Fallot: Comparison with cardiovascular magnetic resonance imaging. *Am. Heart J.* **2004**, *147*, 165–172. [CrossRef]
12. Kitchen, M.J.; Lewis, R.A.; Morgan, M.J.; Wallace, M.J.; Siew, M.L.; Siu, K.K.W.; Habib, A.; Fouras, A.; Yagi, N.; Uesugi, K.; et al. Dynamic measures of regional lung air volume using phase contrast x-ray imaging. *Phys. Med. Biol.* **2008**, *53*, 6065–6077. [CrossRef] [PubMed]
13. Liu, M.; Jiang, H.; Chen, J.; Huang, M. Ultrasound Imaging System. *IEEE Sens. J.* **2016**, *16*, 9014–9020. [CrossRef]
14. Prina, E.; Torres, A.; Roberto, C.; Carvalho, R. Lung ultrasound in the evaluation of pleural effusion. *J. Bras. Pneumol.* **2014**, *40*, 1–5. [CrossRef]
15. Lynnworth, L.C.; Korba, J.M.; Wallace, D.R. Fast Response Ultrasonic Flowmeter Measures Breathing Dynamics. *IEEE Trans. Biomed. Eng.* **1985**, *BME-32*, 530–535. [CrossRef]
16. Buess, C.; Pietsch, P.; Guggenbuhl, W.; Koller, E.A. Design and Construction of a Pulsed Ultrasonic Air Flowmeter. *IEEE Trans. Biomed. Eng.* **1986**, *BME-33*, 768–774. [CrossRef]
17. Hitomi, J.; Murai, Y.; Park, H.J.I.N.; Tasaka, Y. Ultrasound Flow-Monitoring and Flow-Metering of Air – Oil – Water Three-Layer Pipe Flows. *IEEE Acess* **2017**, *5*, 15021–15029. [CrossRef]
18. Rundell, K.W.; Evans, T.M.; Baumann, J.M.; Kertesz, M.F. Lung function measured by impulse oscillometry and spirometry following eucapnic voluntary hyperventilation. *Can. Respir. J.* **2005**, *12*, 257–264. [CrossRef]
19. Marotta, A.; Klinnert, M.D.; Price, M.R.; Larsen, G.L.; Liu, A.H. Impulse oscillometry provides an effective measure of lung dysfunction in 4-year-old children at risk for persistent asthma. *J. Allergy Clin. Immunol.* **2003**, *112*, 4–9. [CrossRef]
20. Gajewski, J.B. Electrostatic Nonintrusive Method for Measuring the Electric Charge, Mass Flow Rate, and Velocity of Particulates in the Two-Phase Gas—Solid Pipe Flows—Its Only or as Many as 50 Years of Historical Evolution. *IEEE Trans. Ind. Appl.* **2008**, *44*, 1418–1430. [CrossRef]
21. Schwartz, J.G.; Fox, W.W.; Shaffer, T.H. A Method for Measuring Functional Residual Capacity in Neonates with Endotracheal Tubes. *IEEE Trans. Biomed. Eng.* **1978**, *BME-25*, 304–307. [CrossRef] [PubMed]
22. Cohen, K.P.; Ladd, W.M.; Beams, D.M.; Sheers, W.S.; Radwin, R.G.; Tompkins, W.J.; Webster, J.G. Comparison of Impedance and Inductance Ventilation Sensors on Adults During Breathing, Motion, and Simulated Airway Obstruction. *IEEE Trans. Biomed. Eng.* **1997**, *44*, 555–566. [CrossRef] [PubMed]
23. Seppa, V.; Viik, J.; Hyttinen, J. Assessment of Pulmonary Flow Using Impedance Pneumography. *IEEE Trans. Biomed. Eng.* **2010**, *57*, 2277–2285. [CrossRef] [PubMed]

24. Incalzi, R.A.; Pennazza, G.; Scarlata, S.; Santonico, M.; Petriaggi, M.; Chiurco, D.; Pedone, C.; D'Amico, A. Reproducibility and Respiratory Function Correlates of Exhaled Breath Fingerprint in Chronic Obstructive Pulmonary Disease. *PLoS ONE* **2012**, *7*, e45396. [CrossRef] [PubMed]
25. Pereira, C.A.D.C. Spirometry. *J. Bras. Pneumol.* **2002**, *28*, S1–S82.
26. Sim, Y.S.; Lee, J.; Lee, W.; Suh, D.I.; Oh, Y.; Yoon, J.; Lee, J.H.; Cho, J.H.; Kwon, C.S.; Chang, J.H. Spirometry and Bronchodilator Test. *Tuberc. Respir. Dis.* **2017**, *3536*, 105–112. [CrossRef]
27. Ansarin, K.; Chatkin, J.M.; Ferreira, I.M.; Gutierrez, C.A.; Zamel, N.; Chapman, K.R. Exhaled nitric oxide in chronic obstructive pulmonary disease: Relationship to pulmonary function. *Eur. Respir. J.* **2001**, *17*, 934–938. [CrossRef]
28. Santos, U.P.; Garcia, M.L.S.B.; Braga, A.L.F.; Pereira, L.A.A.; Lin, C.A.; André, P.A.; André, C.D.S.; Singer, J.M.; Saldiva, P.H.N. Association between Traffic Air Pollution and Reduced Forced Vital Capacity: A Study Using Personal Monitors for Outdoor Workers. *PLoS ONE* **2016**, *11*, e0163225. [CrossRef]
29. Panis, L.I.; Provost, E.B.; Cox, B.; Louwies, T.; Laeremans, M.; Standaert, A.; Dons, E.; Holmstock, L.; Nawrot, T.; Boever, P. Short-term air pollution exposure decreases lung function: A repeated measures study in healthy adults. *Environ. Health* **2017**, *16*, 1–7. [CrossRef]
30. Makwana, A.H.; Solanki, J.D.; Gokhale, P.A.; Mehta, H.B.; Shah, C.J.; Gadhavi, B.P. Study of computerized spirometric parameters of traffic police personnel of Saurashtra region, Gujarat, India. *Lung India* **2015**, *32*, 457–461. [CrossRef]
31. Börekçi, S.; Demir, T.; Dilektaşlı, A.G.; Uygun, A.G.; Yıldırım, N. A Simple Measure to Assess Hyperinflation and Air Trapping: 1-Forced Expiratory Volume in Three Second/Forced Vital Capacity. *Balk. Med. J.* **2017**, *34*, 113–118. [CrossRef] [PubMed]
32. Maxwell, L.J.; Ellis, E.R. The effect on expiratory flow rate of maintaining bag compression during manual hyperinflation. *Aust. J. Physiother.* **2004**, *50*, 47–49. [CrossRef]
33. Schubauer-Berigan, M.K.; Dahm, M.M.; Erdely, A.; Beard, J.D.; Birch, M.E.; Evans, D.E.; Fernback, J.E.; Mercer, R.R.; Bertke, S.J.; Eye, T.; et al. Association of pulmonary, cardiovascular, and hematologic metrics with carbon nanotube and nanofiber exposure among U. S. workers: A cross-sectional study. *Part. Fibre Toxicol.* **2018**, *15*, 1–14. [CrossRef] [PubMed]
34. Oh, A.; Morris, T.A.; Yoshii, I.T.; Morris, T.A. Flow Decay: A Novel Spirometric Index to Quantify Dynamic Airway Resistance. *Respir. Care* **2017**, *62*, 928–935. [CrossRef]
35. Madsen, F.; Frizilund, L.; Ulrik, C.S.; Desken, A. Office spirometry: Temperature conversion of volumes measured by the Vitalograph-R bellows spirometer is not necessary. *Respir. Med.* **1999**, *93*, 685–688. [CrossRef]
36. Fan, D.; Yang, J.; Zhang, J.; Lv, Z.; Huang, H.; Qi, J.; Yang, P. Effectively Measuring Respiratory Flow With Portable Pressure Data Using Back Propagation Neural Network. *IEEE J. Transl. Eng. Heal. Med.* **2018**, *6*, 1–12. [CrossRef]
37. Plessis, E.; Swart, F.; Maree, D.; Heydenreich, J.; Heerden, J.V.; Esterhuizen, T.M.; Irusen, E.M.; Koegelenberg, C.N.F. The utility of hand-held mobile spirometer technology in a resource-constrained setting. *S. Afr. Med. J.* **2019**, *109*, 219–222. [CrossRef]
38. Vautz, W.; Baumbach, J.I.; Westhoff, M.; Züchner, K.; Carstens, E.T.H.; Perl, T. Breath sampling control for medical application. *Int. J. Ion Mobil. Spec.* **2010**, *13*, 41–46. [CrossRef]
39. Kecorius, S.; Jakob, L.; Wiedensohler, A.; Pfeifer, S.; Haudek, A.; Mardo, V. A new method to measure real-world respiratory tract deposition of inhaled ambient black carbon. *Environ. Pollut.* **2019**, *248*, 295–303. [CrossRef]
40. Kobler, A.; Hartnack, S.; Sacks, M. Assessing the accuracy of Tafonius® anesthesia machine in vitro and in vivo respiratory volume measurements. *Pferdeheilkunde* **2016**, *32*, 449–455. [CrossRef]
41. Wang, T.; Baker, R. Coriolis flowmeters: A review of developments over the past 20 years, and an assessment of the state of the art and likely future directions. *Flow Meas. Instrum.* **2014**, *40*, 99–123. [CrossRef]
42. Patrick, H.; Eisenberg, L. An Electronic Resuscitation Evaluation System. *IEEE Trans. Biomed. Eng.* **1972**, *BME-19*, 317–320. [CrossRef]
43. Li, H.N.; Wang, Z.T.; Li, X.B. Hardware design of CPR Simulation Control System based on SCM. In Proceedings of the 2011 International Conference on Electrical and Control Engineering, Yichang, China, 16–18 September 2011; pp. 4802–4805.

44. Leocádio, R.R.V.; Segundo, A.K.R.; Louzada, C.F. Sensor for Measuring the Volume of Air Supplied to the Lungs of Adult Mannequins in Ventilation Maneuvers during Cardiopulmonary Resuscitation. *Proceedings* **2018**, *4*, 39. [CrossRef]
45. Munson, B.R.; Yong, D.F.; Okiishi, T.H. *Fundamentals of Fluid Mechanics*, 4th ed.; Edgard Bluncher: Ames, IA, USA, 2004.
46. Mahan, B.M.; Myers, R.J. *University Chemistry*, 4th ed.; Addison-Wesley: Menlo Park, CA, USA, 1987.
47. Halliday, D.; Resnick, R.; Krane, K.S. *Fundamentals of Physics*, 8th ed.; Jearl Walker: Cleveland, OH, USA, 2009; Volumes 1–3.
48. Perry, R.H.; Green, D.W.; Maloney, J.O. *Perry's Chemical Engineers' Handbook*, 7th ed.; McGraw-Hill: Lawrence, KS, USA, 1997; Volume 27.
49. Fatehnia, M.; Paran, S.; Kish, S.; Tawfiq, K. Automating double ring infiltrometer with an Arduino microcontroller. *Geoderma* **2016**, *262*, 133–139. [CrossRef]
50. Fahmi, F.; Hizriadi, A.; Khairani, F.; Andayani, U.; Siregar, B. Clean water billing monitoring system using flow liquid meter sensor and SMS gateway. *J. Phys. Conf. Ser.* **2018**, *978*. [CrossRef]
51. Gosavi, G.; Gawde, G.; Gosavi, G. Smart water flow monitoring and forecasting system. In Proceedings of the RTEICT 2017 2nd IEEE International Conference on Recent Trends in Electronics, Information & Communication Technology, Bangalore, India, 19–20 May 2017; pp. 1218–1222.
52. Leocádio, R.R.V.; Segundo, A.K.R.; Louzada, C.F. Sistema de tempo real aplicado a simuladores de Ressuscitação Cardiopulmonar. In *14° Simpósio Brasileiro de Automação Inteligente*; Centro de Convenções da UFOP: Ouro Preto, Brazil, 2019; pp. 1–6.
53. Jamaluddin, A.; Harjunowibowo, D.; Rahardjo, D.T.; Adhitama, E.; Hadi, S. Wireless water flow monitoring based on Android smartphone. In Proceedings of the 2016 2nd International Conference of Industrial, Mechanical, Electrical, and Chemical Engineering (ICIMECE), Yogyakarta, Indonesia, 6–7 October 2016; pp. 243–247.
54. Jamaluddin, A.; Harjunowibowo, D.; Rahardjo, D.T.; Adhitama, E.; Hadi, S. Wireless water flow monitoring based on Android smartphone. *Proc. ImechE. Part I J. Syst. Control Eng.* **2017**, 220–224.
55. Mandal, N.; Rajita, G. An accurate technique of measurement of flow rate using rotameter as a primary sensor and an improved op-amp based network. *Flow Meas. Instrum.* **2017**, *58*, 38–45. [CrossRef]
56. Garmabdari, R.; Shafie, S.; Isa, M.M. Sensory system for the electronic water meter. In Proceedings of the ICCAS 2012 IEEE International Conference on Circuits and Systems, Kuala Lumpur, Malaysia, 3–4 October 2012; pp. 223–226.
57. Garmabdari, R.; Shafie, S.; Wan Hassan, W.Z.; Garmabdari, A. Study on the effectiveness of dual complementary Hall-effect sensors in water flow measurement for reducing magnetic disturbance. *Flow Meas. Instrum.* **2015**, *45*, 280–287. [CrossRef]
58. Sinha, S.; Banerjee, D.; Mandal, N.; Sarkar, R.; Bera, S.C. Design and Implementation of Real-Time Flow Measurement System Using Hall Probe Sensor and PC-Based SCADA. *IEEE Sens. J.* **2015**, *15*, 5592–5600. [CrossRef]
59. Gamage, S.K.; Henderson, H.T. A study on a silicon Hall effect device with an integrated electroplated planar coil for magnetic sensing applications. *J. Micromech. Microeng.* **2006**, *16*, 487–492. [CrossRef]
60. Di Lieto, A.; Giuliano, A.; MacCarrone, F.; Paffuti, G. Hall effect in a moving liquid. *Eur. J. Phys.* **2012**, *33*, 115–127. [CrossRef]
61. Urbański, M.; Nowicki, M.; Szewczyk, R.; Winiarski, W. Flowmeter Converter Based on Hall Effect Sensor. In *Automation, Robotics and Measuring Techniques*; Springer: Cham, Switzerland, 2015; pp. 265–276. [CrossRef]
62. Leonov, A.V.; Malykh, A.A.; Mordkovich, V.N.; Pavlyuk, M.I. An autogenerator induction-to-frequency converter circuit based on a field-effect Hall sensor with a regulated frequency. *Instruments Exp. Tech.* **2015**, *58*, 637–639. [CrossRef]
63. GUM, I. *Evaluation of Measurement Data: Guide to the Expression of Uncertainty in Measurement—GUM 2008*; JCGM: Sevres, France; BIPM: Sevres, France; IEC: Geneva, Switzerland; IFCC: Milano, Italy; ILAC: Toronto ON, Canada; ISO: Geneva, Switzerland; IUPAC: Zurich, Switzerland; IUPAP: Singapore; OIML: Paris, France, 2008.
64. De Bièvre, P. *International Vocabulary of Metrology—Basic and General Concepts and Associated Terms—VIM 2012*; JCGM: Sevres, France; BIPM: Sevres, France; IEC: Geneva, Switzerland; IFCC: Milano, Italy; ILAC: Toronto ON, Canada; ISO: Geneva, Switzerland; IUPAC: Zurich, Switzerland; IUPAP: Singapore; OIML: Paris, France, 2012.

65. Association, A.H. *Updating the CPR and ACE Guidelines: Highlights of the American Heart Association 2015*; American Heart Association: Dallas, TX, USA, 2015; pp. 4–22.
66. Jorge, A.d.S.; Ribeiro, A.; Gomes, C.; Frederico, C.; Aragão, D.; Gonçalves, F.; Júnior, F.F.; Dantas, I.; Vale, L.; Serafim, M. *Fundamentos de Física e Biofísica*, 1st ed.; FTC EaD: Salvador, BA, Brazil, 2008.
67. Kulish, V. *Human Respiration*; WIT Press: Billerica, MA, USA, 2006.
68. Sánchez, R.D.; López-Quintela, M.A.; Rivas, J.; González-Penedo, A.; García-Bastida, A.J.; Ramos, C.A.; Zysler, R.D.; Guevara, S.R. Magnetization and electron paramagnetic resonance of Co clusters embedded in Ag nanoparticles. *J. Phys. Condens. Matter* **1999**, *11*, 5643–5654. [CrossRef]
69. Mitra, S.; Mandal, K.; Kumar, P.A. Temperature dependence of magnetic properties of $NiFe_2O_4$ nanoparticles embeded in SiO_2 matrix. *J. Magn. Magn. Mater.* **2006**, *306*, 254–259. [CrossRef]
70. Chaudhuri, A.; Mandal, M.; Mandal, K. Preparation and study of $NiFe_2O_4/SiO_2$ core—Shell nanocomposites. *J. Alloy. Compd.* **2009**, *487*, 698–702. [CrossRef]
71. Ye, S.; Ney, V.; Kammermeier, T.; Ollefs, K.; Zhou, S.; Schmidt, H.; Wilhelm, F.; Rogalev, R.; Ney, A. Absence of ferromagnetic-transport signatures in epitaxial paramagnetic and superparamagnetic $Zn_{0.95}Co_{0.05}O$ films. *Phys. Rev. B* **2009**, *80*, 1–7. [CrossRef]
72. Sakamoto, S.; Anh, L.D.; Hai, P.N.; Shibata, G.; Takeda, Y.; Kobayashi, M.; Takahashi, Y.; Koide, T.; Tanaka, M. Magnetization process of the n-type ferromagnetic semiconductor (In, Fe)As: Be studied by x-ray magnetic circular dichroism. *Phys. Rev. B* **2016**, *93*, 1–6. [CrossRef]
73. Garcı, L.M.; Bartolome, F.; Bartolome, J. Strong Paramagnetism of Gold Nanoparticles Deposited on a Sulfolobus acidocaldarius S Layer. *Phys. Rev. Lett.* **2012**, *247203*, 1–5. [CrossRef]
74. Danielewicz-Ferchmin, I.; Ferchmin, A.R. Static permittivity of water revisited: E in the electric field above 10^8 Vm^{-1} and in the temperature range $273 \leq T \leq 373$ K. *Phys. Chem. Chem. Phys.* **2004**, *6*, 1332–1339. [CrossRef]
75. Capsal, J.; Galineau, J.; Guyomar, D. Evaluation of macroscopic polarization and actuation abilities of electrostrictive dipolar polymers using the microscopic Debye/Langevin formalism. *J. Phys. Appl. Phys.* **2012**, *45*, 9. [CrossRef]
76. Bichurin, M.; Petrov, V.; Zakharov, A.; Kovalenko, D.; Yang, S.C.; Maurya, D.; Bedekar, V.; Shashank, P. Magnetoelectric Interactions in Lead-Based and Lead-Free Composites. *Materials* **2011**, *4*, 651–702. [CrossRef] [PubMed]
77. Boggs, P.T.; Rogers, J.E. *Orthogonal Distance Regression Orthogonal Distance Regression*; Center for Computing and Applied Mathematics: Fullerton, CA, USA, 1990; NISTIR 89–4197.
78. Nocedal, J.; Wright, S.J. *Numerical Optimizatio*, 2nd ed.; Springer: Evanston, IL, USA, 1996; Volume 17.
79. Zwolak, J.W.; Boggs, P.T.; Watson, L.T. *ODRPACK95: A Weighted Orthogonal Distance Regression Code with Bound Constraints*; Virginia Tech: Blacksburg, VA, USA, 2004.
80. Zwick, D.S. Applications of orthogonal distance regression in metrology. *Double Star Res.* **2016**. Available online: https://www.researchgate.net/publication/262275537 (accessed on 12 June 2019).

MDPI
St. Alban-Anlage 66
4052 Basel
Switzerland
Tel. +41 61 683 77 34
Fax +41 61 302 89 18
www.mdpi.com

Sensors Editorial Office
E-mail: sensors@mdpi.com
www.mdpi.com/journal/sensors

www.ingramcontent.com/pod-product-compliance
Lightning Source LLC
LaVergne TN
LVHW071608170726
843515LV00008B/2346
9783039288496